4th International Conference on Aeronautical Materials and Aerospace Engineering (AMAE 2020)

AA007054

IOP Conference Series: Materials Science and Engineering
Volume 887

Chengdu, China
14 - 17 May 2020

ISBN: 978-1-7138-3623-0
ISSN: 1757-8981

Printed from e-media with permission by:

Curran Associates, Inc.
57 Morehouse Lane
Red Hook, NY 12571

Some format issues inherent in the e-media version may also appear in this print version.

This work is licensed under a Creative Commons Attribution 3.0 International Licence.
Licence details: http://creativecommons.org/licenses/by/3.0/.

No changes have been made to the content of these proceedings. There may be changes to pagination and minor adjustments for aesthetics.

Printed with permission by Curran Associates, Inc. (2021)

For permission requests, please contact the Institute of Physics at the address below.

Institute of Physics
Dirac House, Temple Back
Bristol BS1 6BE UK

Phone: 44 1 17 929 7481
Fax: 44 1 17 920 0979

techtracking@iop.org

Additional copies of this publication are available from:

Curran Associates, Inc.
57 Morehouse Lane
Red Hook, NY 12571 USA
Phone: 845-758-0400
Fax: 845-758-2633
Email: curran@proceedings.com
Web: www.proceedings.com

TABLE OF CONTENTS

AVIATION

DESIGN

SYSTEMS

IOP Conf. Series: Materials Science and Engineering **887** (2020) 011001 doi:10.1088/1757-899X/887/1/011001

PREFACE

Currently the entire world is struggling against the virulent pandemic COVID-19. Unfortunately, each of us is affected, either directly or indirectly. Our conference, 2020 4th International Conference on Aeronautical Materials and Aerospace Engineering (AMAE 2020) was not an exception.

In May, mass gatherings are not permitted by the government in China to protect people. It is uncertain when the COVID-19 would end, so it remains unclear for postponement date, while many scholars and researchers want to attend this long waited conference and have academic exchanges with their peers. Therefore, in order to actively respond the call of the government, and meet author's request, the AMAE 2020, which was planned to be held in Chengdu, China from May 14 to 17, 2020, was changed to be held on May 15, 2020 online through Tencent VooV software. This approach not only guarantee participants' safety and health but also meets their communication needs.

Each keynote speech lasted 40 minutes, invited speech 30 minutes and author presentation 15 minutes. Each presentation was packed with questions and answers. There were lively discussion at the meeting, which promotes the academic exchange. The success and prosperity of the conference is reflected high level of the papers received. AMAE 2020 became an effective communication platform for all the participants over the world.

AMAE 2020 was organized by Hong Kong Society of Mechanical Engineers. This conference aims to provide a platform for researchers and engineers to share their ideas, recent developments and successful practices in Aeronautical Materials and Aerospace Engineering. More than 70 participants attended the meeting, they were from USA, Italy, Germany, Canada, UK, Malaysia, India, Poland, Kuwait, Saudi Arabia, China and more. Five renowned speakers given speeches about their latest research and reports. They are: Prof. Ian McAndrew, Capitol Technology University, USA; Prof. Kretov Anatolii, Nanjing University of Aeronautics and Astronautics, China; Prof. Graham J. F. Hunt, Embry-Riddle Aeronautical University, USA; Assoc. Prof. Bing Li, Northwestern Polytechnical University, China and Assoc. Prof. Dmytro V. Tiniakov, Nanjing University of Aeronautics and Astronautics, China. The conference also had 2 technical sessions and 2 poster sessions.

The conference proceeding is a compilation of the accepted papers and represent an interesting outcome of the conference. This book covers 3 chapters: 1. Design; 2. Aviation; 3. Systems.

Content from this work may be used under the terms of the Creative Commons Attribution 3.0 licence. Any further distribution of this work must maintain attribution to the author(s) and the title of the work, journal citation and DOI.
Published under licence by IOP Publishing Ltd

IOP Conf. Series: Materials Science and Engineering **887** (2020) 011001 doi:10.1088/1757-899X/887/1/011001

We would like to acknowledge all of those who supported AMAE 2020. Each individual and institutional help were very important for the success of this conference. Especially we would like to thank the committee chairs, committee members and reviewers, for their tremendous contribution in conference organizing and peer review of the papers.

We sincerely hope that AMAE 2020 will be a forum for excellent discussions that will put forward new ideas and promote collaborative researches. We are sure that the proceedings will serve as an important research source of references and the knowledge, which will lead to not only scientific and engineering progress but also other new products and processes. Finally, we would like to thank the organization and committee for all their hard work and support and hope to meet you all in person at our next conference.

Prof. Ian McAndrew & Prof. Kretov Anatolii

Conference Co-Chairs

IOP Conf. Series: Materials Science and Engineering **887** (2020) 011002 doi:10.1088/1757-899X/887/1/011002

COMMITTEES

Conference Co-Chairs

Ian McAndrew, Capitol Technology University, USA

Kretov Anatolii, Nanjing University of Aeronautics and Astronautics, China

Program Committee Co-Chairs

Qiang Liu, Beihang University, China

Xiaohui Su, Dalian University of Technology, China

Local Committee

Chenglong Li, Civil Aviation Flight University of China, China

International Technical Committee

A Ram (Bella) Kim, Iowa State University, USA

Fairuz Izzuddin Romli, Universiti Putra Malaysia, Malaysia

Fidel Khouli, Carleton University, Canada

James Chen, University at Buffalo, USA

Jielong Wang, Beijing Aeronautical Science & Technology Research Institute, China

K. K. Viswanathan, Kuwait College of Science and Technology, Kuwait

Kin Hing Lo, University of Surrey, UK

M. Chandrasekaran, Vels University, India

Mohd Faisal Abdul Hamid, Universiti Putra Malaysia, Malaysia

Nickolay Zosimovych, Shantou University, China

Santanu Ghosh, Indian Institute of Technology Madras, India

Shinn-Liang Chang, National Formosa University, Taiwan, China

Content from this work may be used under the terms of the Creative Commons Attribution 3.0 licence. Any further distribution of this work must maintain attribution to the author(s) and the title of the work, journal citation and DOI.
Published under licence by IOP Publishing Ltd

IOP Conf. Series: Materials Science and Engineering **887** (2020) 011002 doi:10.1088/1757-899X/887/1/011002

Stephane Redonnet, The Hong Kong University of Science and Technology, HKSAR, China

Wael G. Abdelrahman, King Fahd University of Petroleum & Minerals, Saudi Arabia

Zhangming Wu, Cardiff University, UK

Peer review statement

All papers published in this volume of *IOP Conference Series: Materials Science and Engineering* have been peer reviewed through processes administered by the proceedings Editors. Reviews were conducted by expert referees to the professional and scientific standards expected of a proceedings journal published by IOP Publishing.

Content from this work may be used under the terms of the Creative Commons Attribution 3.0 licence. Any further distribution of this work must maintain attribution to the author(s) and the title of the work, journal citation and DOI.
Published under licence by IOP Publishing Ltd

Numerical Investigation of the Aerodynamic Characteristics of a Missile

Yu-shan Meng[1], Li Yan[1], Wei Huang[1†] and Xin-yu Tong[2]

[1] College of Aerospace Science and Engineering, National University of Defense Technology, Changsha, Hunan 410073, China
[2] Northwestern Polytechnical University, Xian, Shanxi 710072, China

[†]E-mail: gladrain2001@163.com

Abstract: During supersonic flight of supersonic missile, the missile configuration has great influence on the aerodynamic characteristics of the missile. This article evaluates the variation of the flow field results from the change of missile configuration by changing the shape of the head and the slenderness ratio of the body. The study is carried out by the CFD numerical method, and the structured grids are generated by ANSYS CFD software. We evaluate the aerodynamic characteristics of two-dimensional model and three-dimensional model by the flow solver FLUENT, and the standard Spalart-Allmaras one-equation is selected to simulate the flow around the missile, the results indicate that the drag coefficient of the oval head is less than parabolic head case and conical head case, and drag coefficient decreases with the increase of Mach number and slenderness ratio. With the increase of Mach number, the stagnation temperature increases correspondingly.

Keywords: Supersonic missile; Head shape; Drag coefficient; Aerodynamic characteristics

1. Introduction

The aerodynamic characteristic of missile is of great significance for the design and usage of missile. During the supersonic flight of missile, it encounters strong shock wave such as the bow shock which enables the rise of wave drag[1], especially for the supersonic aircraft with blunt body[2]. And there exists serious aerodynamic heating and large entropy gradient in the zoon close to the leading edge of the aircraft, the air around the warhead is strongly compressed and is subjected to severe friction. And the degree of air compression increases as the Mach number increases, meanwhile, the temperature around the missile increases correspondingly.

Numerical simulation is a branch of CFD technology, with the improvement of the computer performance, a series of solution algorithm based on Euler method were proposed , such as explicit Runge-Kutta method[3]. In recent years, numerical investigations of the vehicles flying at supersonic and hypersonic speeds have been widely carried out[4-8]. Qin et al. [9] adopted a fluid-thermal coupled strategy to study the aerodynamic characteristics of different flow fields and various body configurations. The results show that the shock intensity and the drag coefficient increase correspondingly as the free stream Mach number increases. Yue et al.[10] in virtue of the CFD software to carry out the numerical computations of aerodynamics of a missile with the Mach numbers and attack angle being different. The results show that the numerical simulation have special reference value for the design of a missile. Hao et al. [11] studied the transonic-supersonic wind tunnel

Content from this work may be used under the terms of the Creative Commons Attribution 3.0 licence. Any further distribution of this work must maintain attribution to the author(s) and the title of the work, journal citation and DOI.
Published under licence by IOP Publishing Ltd

experiment on the aerodynamics of the rockets and missiles with different fins. The results indicate that adopting more than six fins is beneficial for increasing the stability for low speed missiles while it is adverse for high speed missiles. Besides, Zhang et al. [12] has summarized parametric approaches specific to aerodynamic design process , and Huang et al. [13-16] numerically studied the performance of different thermal protection in details.

In this paper, we analyze the effect of the missile configuration on the aerodynamic characteristic, compare the impact of Mach number and discuss the flow characters around the missile. In Section 1, we briefly introduce the researches on the blunt body. Section 2 is an introduction to the governing equations, turbulence model and solving settings. The discussions and results are introduced in Section 3. And some remarks and conclusions are drawn in the last section.

2. Physical and numerical models

2.1 Governing equations and Turbulence model

The basic governing equations are continuity equation, momentum equation and energy equation, here we start from the three-dimensional steady NS equations to carry out the numerical simulation[17-18].

$$\frac{\partial U}{\partial t}+\frac{\partial F}{\partial x}+\frac{\partial G}{\partial y}+\frac{\partial H}{\partial z}=\frac{\partial F_v}{\partial x}+\frac{\partial G_v}{\partial y}+\frac{\partial H_v}{\partial z} \quad (1)$$

Where U is a conserved quantity, F,G,H are inviscid convective fluxes , and F_v, G_v, H_v are sticky fluxes.

$$\mathrm{U}=\begin{bmatrix}\rho\\ \rho u\\ \rho v\\ \rho w\\ E\end{bmatrix} \quad \mathrm{F}=\begin{bmatrix}\rho u\\ \rho u^2+p\\ \rho uv\\ \rho uw\\ (E+p)u\end{bmatrix} \quad \mathrm{G}=\begin{bmatrix}\rho v\\ \rho uv\\ \rho u^2+p\\ \rho vw\\ (E+p)v\end{bmatrix} \quad \mathrm{H}=\begin{bmatrix}\rho w\\ \rho uw\\ \rho vw\\ \rho w^2+p\\ (E+p)w\end{bmatrix}$$

Where ρ is the density, u ,v,w are used to measure the X-component, Y-component and Z-component of the velocity. Also, p is the static pressure, E is the total energy. The state equation is

$$\mathrm{P}=(\gamma-1)[\mathrm{e}-\frac{1}{2}\rho(u^2+v^2+w^2)] \quad (2)$$

The governing equations are discretized and solved by finite volume method. We adopt the standard Spalart-Allmaras one-equation to carry out the numerical simulation based on $\tilde{v}$, the kinematic coefficient of viscosity, and the follow is the transport equation

$$\rho\frac{D\tilde{v}}{Dt}=G_v+\frac{1}{\sigma_{\tilde{v}}}\left[\frac{\partial}{\partial x_j}\left\{(\mu+\rho\tilde{v})\frac{\partial\tilde{v}}{\partial x_j}\right\}+C_{b_2}\rho\left(\frac{\partial\tilde{v}}{\partial x_j}\right)\right]-Y_v \quad (3)$$

Where G_v denotes the production of the turbulent viscosity, $\sigma_{\tilde{v}}, C_{b_2}$ are the constant term, and Y_v denotes the decrease of the turbulent viscosity.

2.2 Geometry

In this section we mainly study the impacts that the head shape has on the aerodynamic characteristic of missile. We select a long-range air defense missile without main wings as the basic model, it has a cylindrical body with an oval head, and there exists a spherical transition from head to body. The total length of the body is 5.3 m, the length of the head is 0.98794 m, of which the radius of curvature is 10mm, and the diameter of the cylindrical body is 0.41 m, see Fig.1. In this paper, the aerodynamic interference among the structure would not be considered.

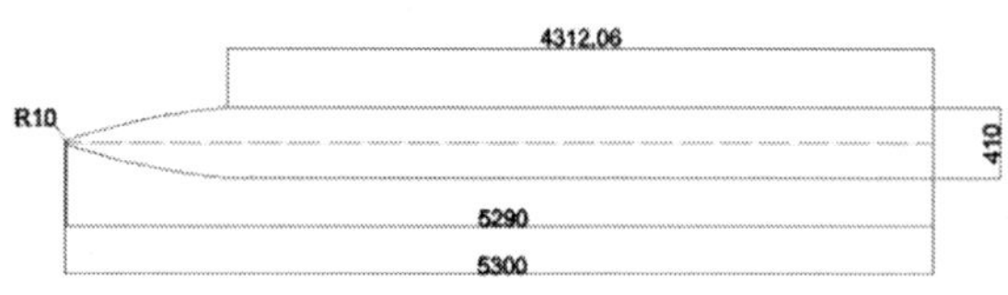

Figure1. The geometry model of the missile.

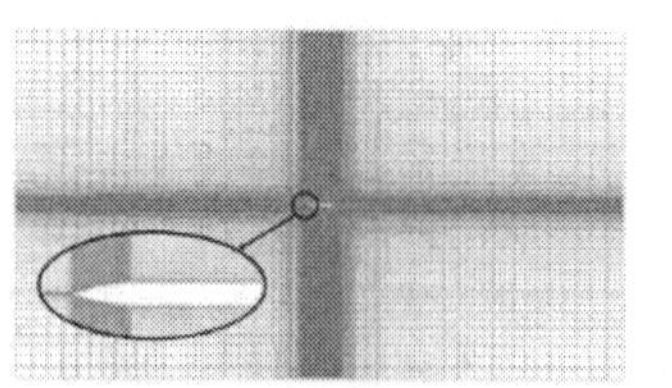

Figure2. Structured grids of the model.

2.3 Solver setting

We establish a rectangular far field with the dimension of twenty times as large as the basic model in ICEM-CFD. This paper uses the local grid refinement approach to refine the structured grids of the head and due to the assumption that the model is symmetrical, only half of the mesh needs to be generated. The Reynolds number 3.31472×106 is based on the diameterof the body[19]. We set the desired Y + value as 30, which is suitable to calculate the first cell thickness, the first layer grid height of the boundary is obtained as 4.59703×10^{-5}m, and the numerical simulation is carried out using the medium size of grids(1346860 cells), the accuracy of which is sufficient, see Fig. 2. After importing the mesh into FLUENT, we choose two-dimensional accuracy solver to carry out the calculation. Table 1 presents the parameters of the free stream. The analysis assumes that the fluid is idea-gas, there is no slip for velocity and adopts S-A turbulent model and second-order accuracy AUSM + spatial discretization scheme[20]. And the Courant number changes as the calculation goes on to accelerate the first layer grid height convergence. The monitors of drag coefficient and lift coefficient are defined and we assume the calculation is finished when the residuals decrease to five orders of magnitude.

3. Results and Discussions

3.1 Effect of Mach number

We obtain the aerodynamic contours of the missile with the angle of attack being 0° and the Mach number being 4 Ma. See Figs 3-4. Fig.5 depicts that the drag coefficient decreases as the free stream Mach number increases, and the slope decreases gradually as well.

After carrying out the analysis, it is of great significance to learn the flow characteristic around the missile. It can be observed that when the free stream flows around the missile, it encounters a strong compression which lead to the shock wave. The basic model has an oval head, and the radius of curvature becomes larger as the stream keep moving backward, thus a series of weak shock waves are formed. The velocity of the airflow encounters a sharply decline while the pressure of the warhead encounters a sharply increase. At the transition from head to body, a series of continuous sectorial expansion waves occur, so it is concluded that the distribution of the flow field is affected by a series of shock waves and expansion waves.

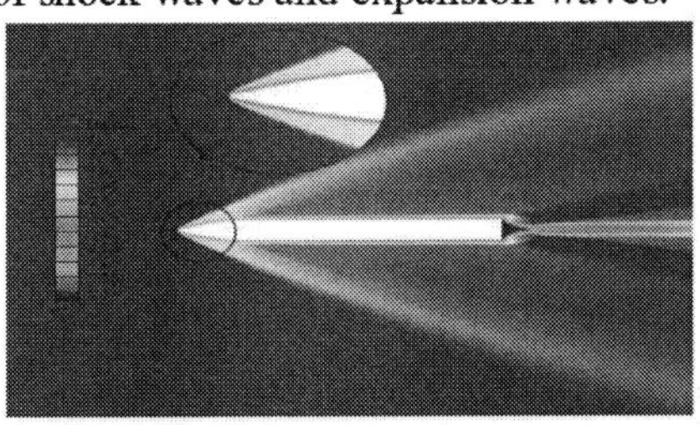

Figure 3. Mach cloud of reference model.

Table 1. Parameters of free stream.

Mach number	M_∞	4
Static pressure	P_∞ P_a	12111.8
Static Temperature	T_∞ K	216.65
Reynolds No.	Re_∞ $milliom/m$	3.31472
Density	ρ_∞ kg/m^3	0.194755
Velocity	u_∞ m/s	1180.28

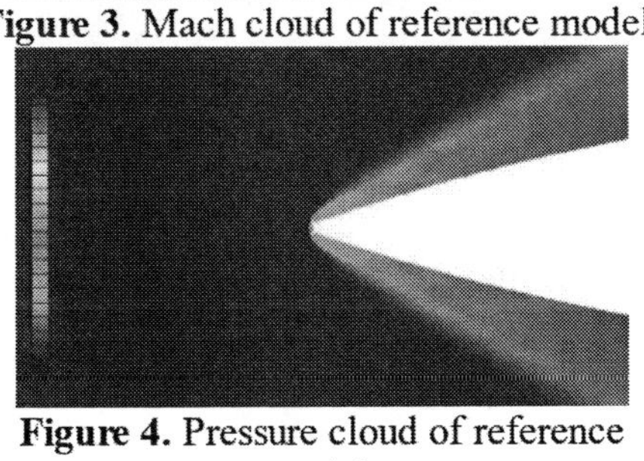

Figure 4. Pressure cloud of reference model.

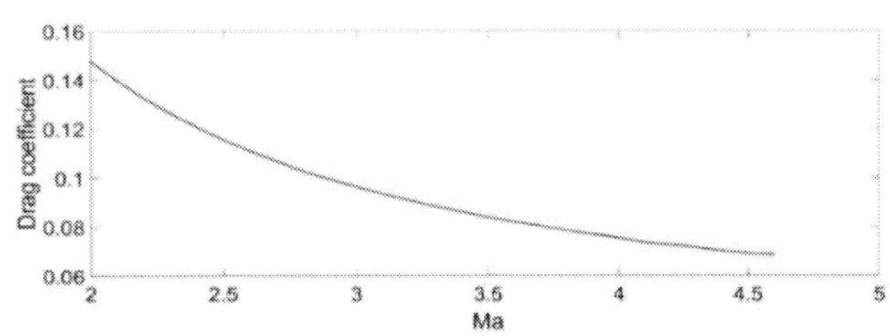

Figure 5. Effect of Mach number on Drag coefficient.

3.2 Effects of the head shape

In this section, we discuss the aerodynamic characteristics of missiles with different head shapes, see Fig.6. During the numerical simulation, Mach number is changed as 2, 2.5, 3, 3.5, 4, 4.5 respectively.

Fig. 7 presents that the variation trend of drag coefficient is similar with the basic model. The drag coefficient of the missile with Parabolic head is the largest while it of the missile with oval head (basic model) is the smallest. The strength of the shock wave increases as the radius of curvature of head decreases, it means that the blunter the head is, the stronger the shock waves are. The radius of curvature of parabolic head is the largest among these configurations, and the intensity increases correspondingly.

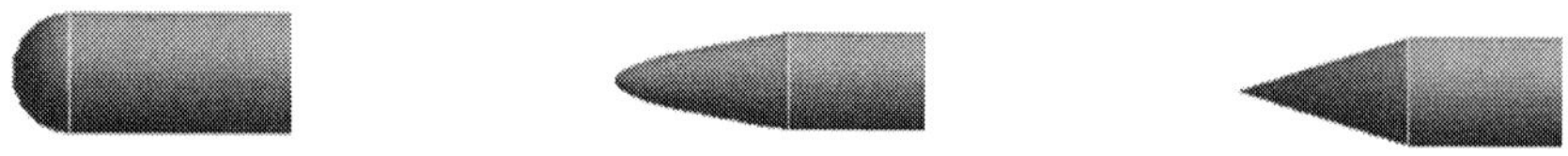

(a) Hemispherical head (b) Parabolic head (c) Conical head

Figure 6. Geometry of the models with different head shape.

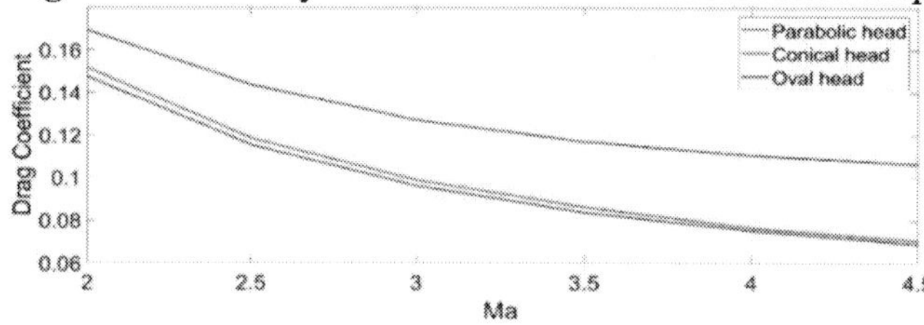

Figure 7. Comparison of Drag coefficient with respect to head shape

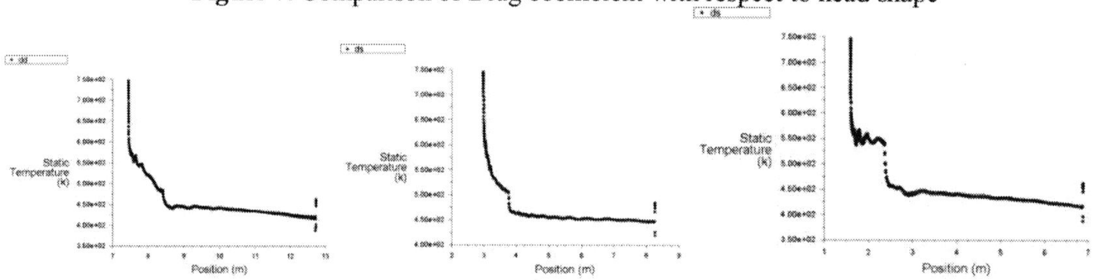

(a)Case 1: Oval head (b) Case 2: Parabolic head (c) Case 3: Conical head

Figure 8. Temperature distribution on the surface of the missile with the Mach number being 3.5 Ma.

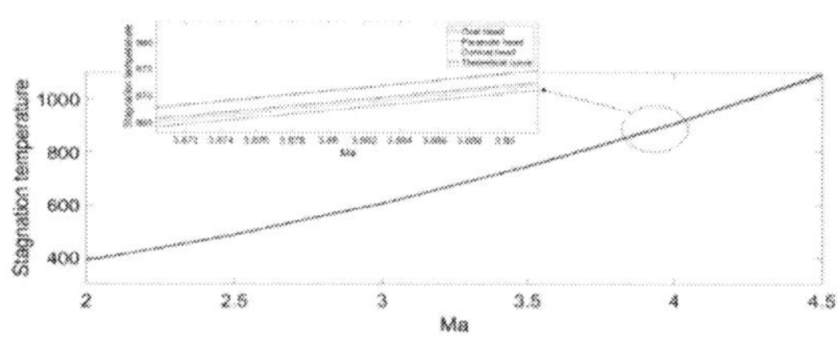

Figure 9. Comparison of stagnation temperature with respect to head shape.

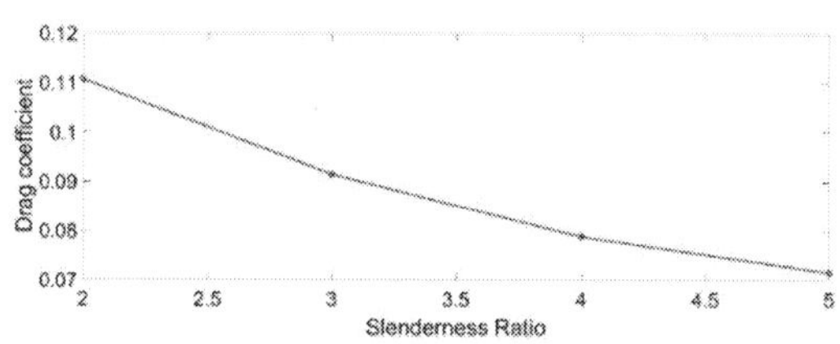

Figure 10. Effect of Slenderness ratio on Drag coefficient.

Fig. 8 presents the aerodynamic heating characteristics along the missile, and the variation trend from stagnation point to the junction of head and body is worthy of attention, We can observe that the variation trend of oval head tends to be linear, which, results in the linear variation of the thermal stress. Fig.9 presents that with the increase of Mach number, the stagnation temperature increases correspondingly. And the curves of different head shapes basically agree with the theoretical curve. It is concluded that the radius of curvature of missile head has little effect on the stagnation temperature. Fig.10 presents the effect of slenderness ratio on drag coefficient in MATLAB.

3.3 Effects of the slenderness ratio

IOP Conf. Series: Materials Science and Engineering **887** (2020) 012001 doi:10.1088/1757-899X/887/1/012001

Take the missile with parabolic head as the object of study, remain the Mach number being 4 Ma unchanged, and change the slenderness ratio as 2,3,4,5 respectively. The comparison of Mach number contour is presented in Fig.11. The results show that there exists strong detached bow shock wave in front of the head and a zone with high temperature and high-pressure area is formed, which, is the reason for the inevitable rise of drag coefficient, and the drag coefficient decreases as the slenderness ratio increases. In addition, the pressure and temperature increase significantly while the velocity decreases when the stream pass through the shock wave.

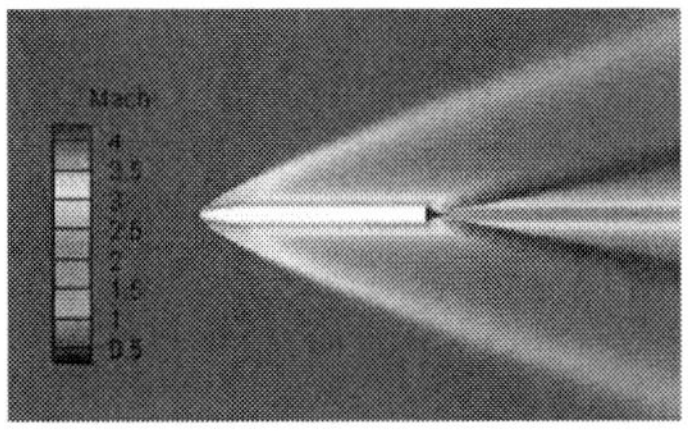

Case 1: Slenderness ratio=2

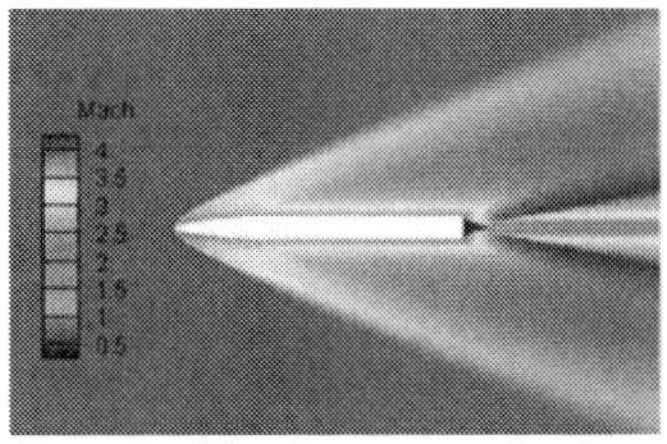

Case 2: Slenderness ratio=3

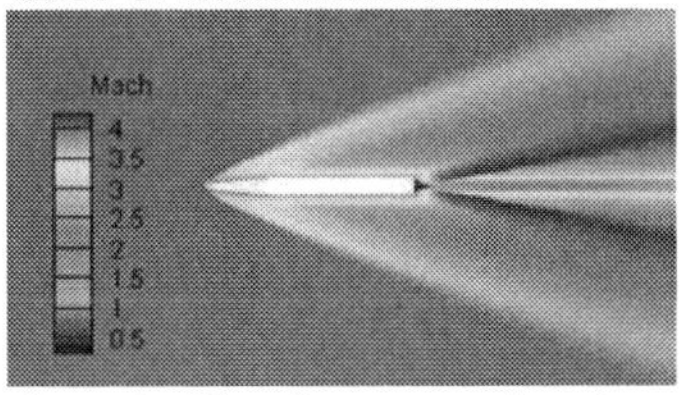

Case 1: Slenderness ratio=4

Case 1: Slenderness ratio=5

Figure 11. Comparison of Mach number contour with different Slenderness ratio.

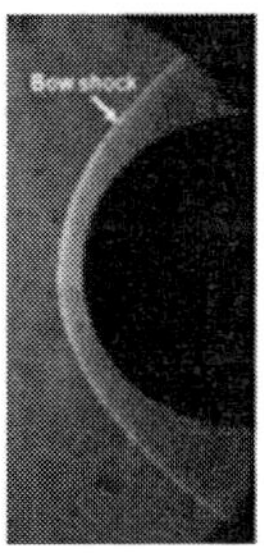

(a)Hemispherical head.[1].

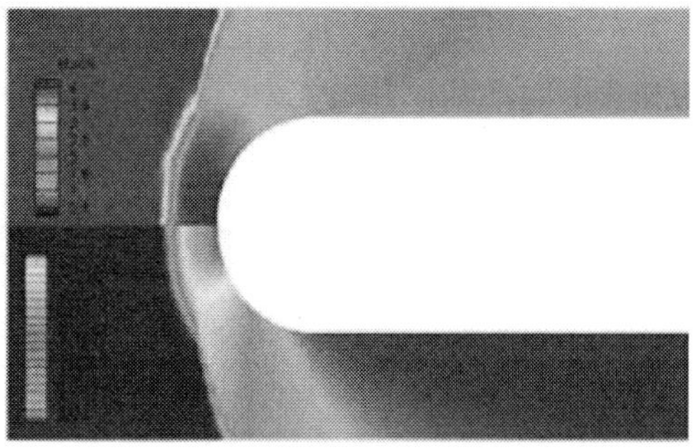

(b) Mach number and Pressure contour.

Figure 12. Aerodynamic features of the hemispherical head.

When there exists a hemispherical head, the aerodynamic feature is more remarkable, see Fig.12. The radius of curvature of the blunt head is large and it has big angle with the air flow. The head hinders the free stream from moving forward and there is an obvious detached bow shock wave, which, creates stronger energy consumption than the attached shock wave.

3.4 Y + Self-adaptive grid technique

FLUENT software has the self-adaptive grid technique, such as boundary self-adaptive technique, isoline self-adaptive technique and Y + self-adaptive technique, which, can improve the computational accuracy by adjusting the degree of grid density and refining the grid in the area with large gradient. In this paper, we use Y + self-adaptive grid technique to prove the accuracy of our calculation.

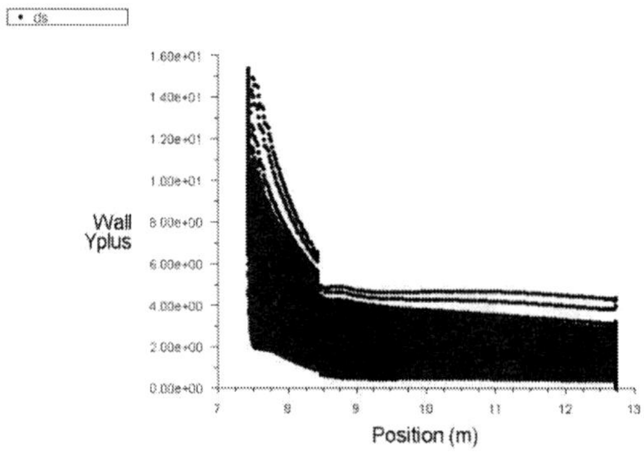

(a) Before self-adaption

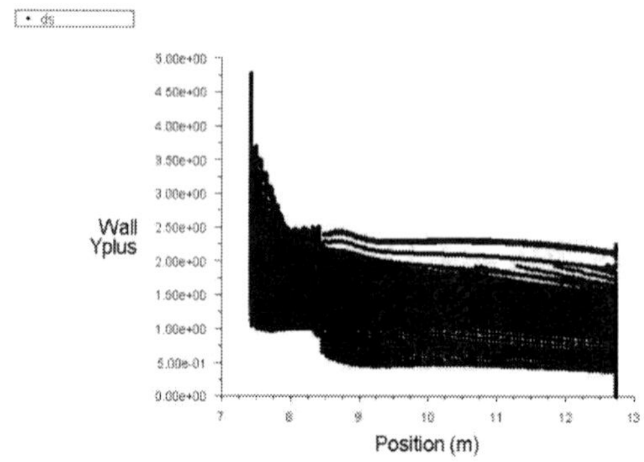

(b) First self-adaption

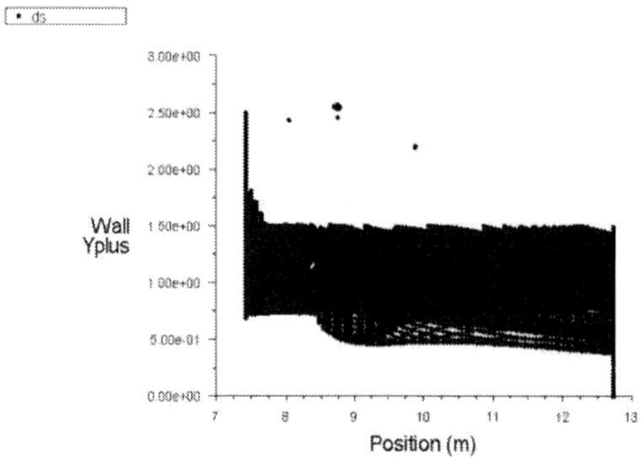

(c) Second self-adaption

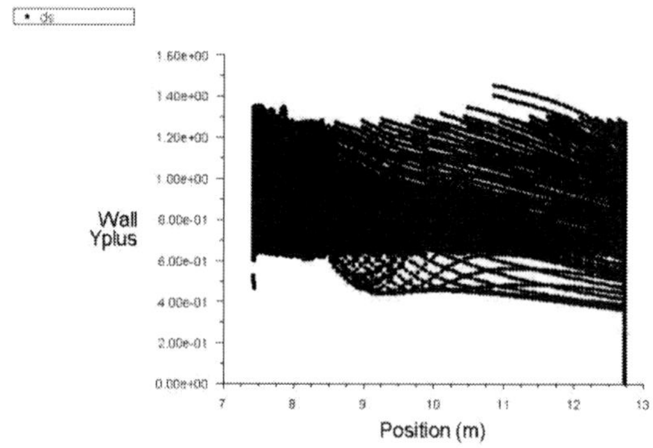

(d) Third self-adaption

Figure 13. Y + distribution before and after the self-adaption.

Fig.13 presents the Y + distribution before and after the self-adaption, we choose the basic oval head model to carry out the self-adaptive calculation with the Mach number being 4 Ma and the angle of attack being 15°. It is obtained that Y + value before self-adaption is higher than Y + value after self-adaption, which is mainly because that we set the first cell thickness as a fixed value and the head of missile suffers large pressure. After self-adaption, the Y + value is distributed around 1. Table 2 shows that before and after the self-adaption, there is little difference among the results, which, further verifies the accuracy of the analysis above.

Table 2. Aerodynamic data before and after the self-adaption.

Self-adaption times	Lift coefficient	Drag coefficient	Lift-drag ratio
0	0.14744	0.059625	2.473
1	0.14793	0.059763	2.475
2	0.14832	0.059804	2.480
3	0.14869	0.059853	2.484

3.5 Simulations of the three-dimensional missile

In order to further study the aerodynamics of the missile, based on the above analysis, in this section, simulations of the three-dimensional missile are done. The flow field is investigated with the angle of attack being 0°and 30°. Mach number contour with different angles of attack are presented in Fig.14, and the position of the stagnation point gradually moves down. When there is an angle of attack being 30°, a series of oblique shocks are generated, which are related to the deflection angle of the free stream. In addition, the lower surface is the windward side, which is the reason for the separation of airflow occurs on the upper surface of the missile.

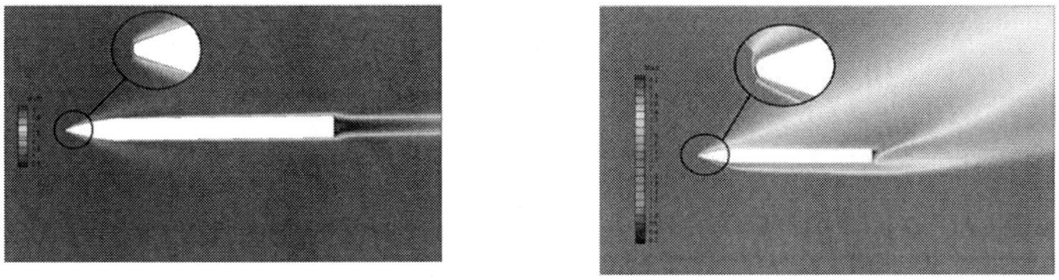

(a) Case 1: angle of attack being 0° (b) Case 2: angle of attack being 30°

Figure 14. Mach number cloud with different angles of attack.

Fig.15 presents the streamline in the rear of missile with angle of attack being 30°, a clockwise vortex can be seen clearly, which, may be the reason for the temperature decrease. There exists a large velocity gradient within the boundary layer region, and the temperature of boundary layer is high, which, is related to the viscous friction. We can observe the temperature distribution of the rear of missile with different angle of attack clearly in Fig.16. Temperature curves of upper and lower surfaces with different angle of attack are obtained in Fig.17. We can obtain that upper and lower surfaces are at the same temperature when the angle of attack is 0°, and the temperature of lower surface is higher than it of upper surface. The temperature of lower surface increases as the angle of attack increases, which, is related to the increase of windward area, and there is a downward trend of the curve as the lengthwise position changes and finally comes closer to balance since the distance from the stagnation point increases.

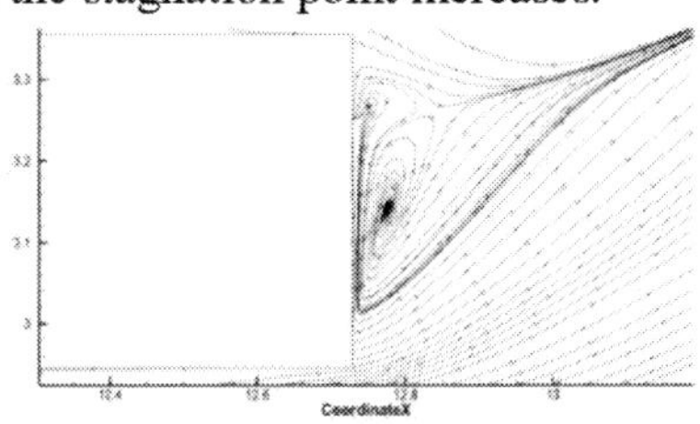

Figure 15. The streamline in the rear of missile with angle of attack being 30°.

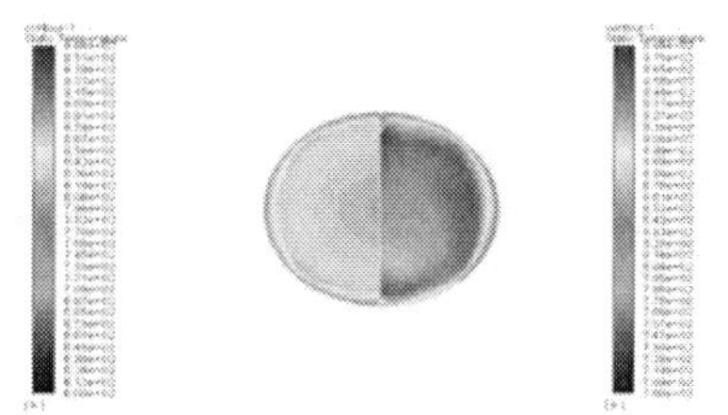

Figure 16. Temperature distribution of the rear with different cases.(Left: 0°, Right: 30°)

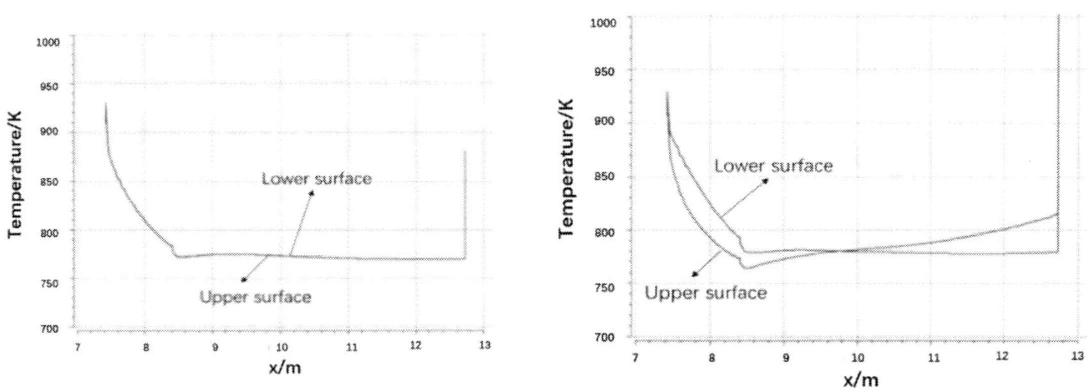

(a) Case 1: 0° (b) Case 2: 5°

IOP Conf. Series: Materials Science and Engineering **887** (2020) 012001 doi:10.1088/1757-899X/887/1/012001

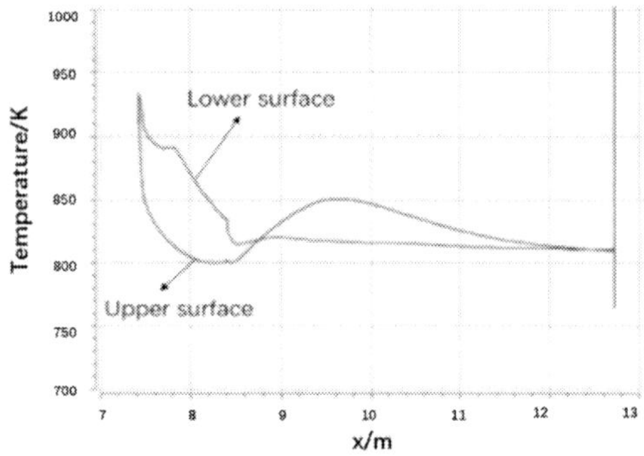

(c) Case 3: 10°

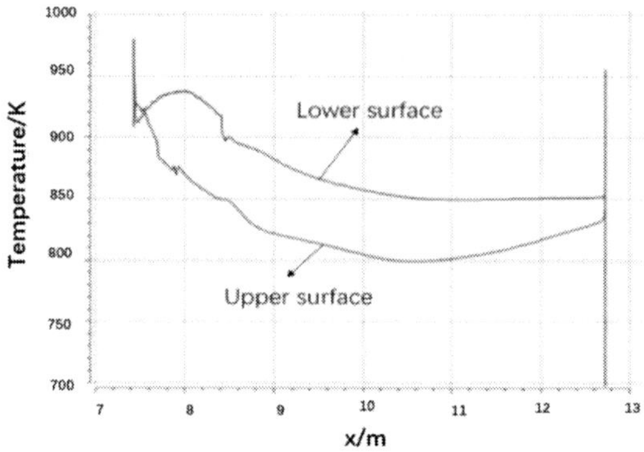

(d) Case 4: 15°

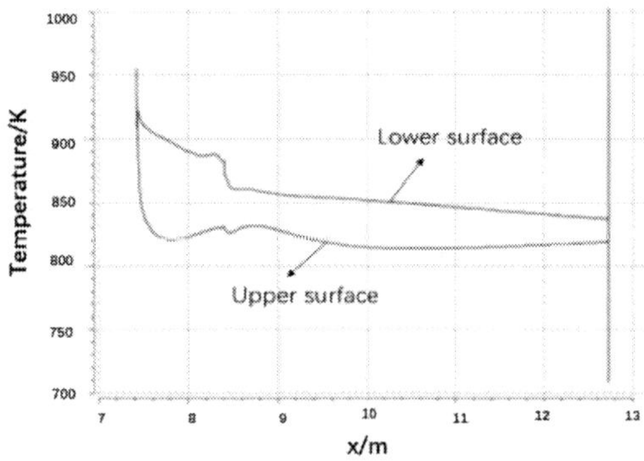

(e) Case 5: 20°

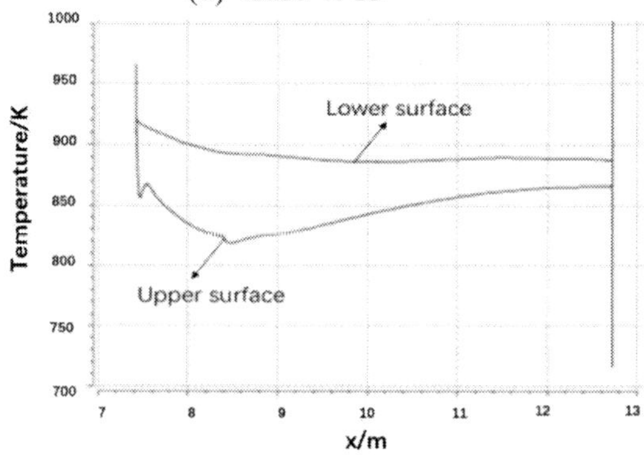

(f) Case 6:25°

Figure 17. Temperature curves of upper and lower surfaces with different angle of attack.

In order to study the process of aerodynamic heating more clearly, we use Fluent transient calculation to solve the external flow field of the basic model when the Mach number is 4 Ma and the angle of attack is 15°. The time step is set to 0.0001s, and the maximum iteration time is 50 times in each time step. We get temperature curves every 0.12 seconds as shown in Fig.18. It can be found that the temperature distribution on the upper and lower surface changes rapidly before 0.3s, but gradually slows down after 0.3s, and the whole flow field tends to be stable.

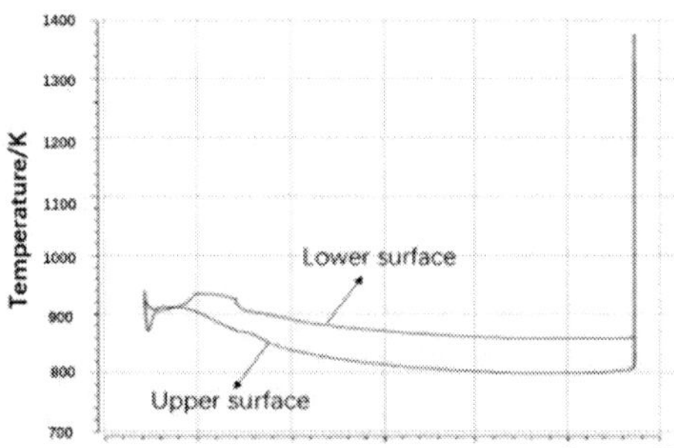

(a) 0.06s

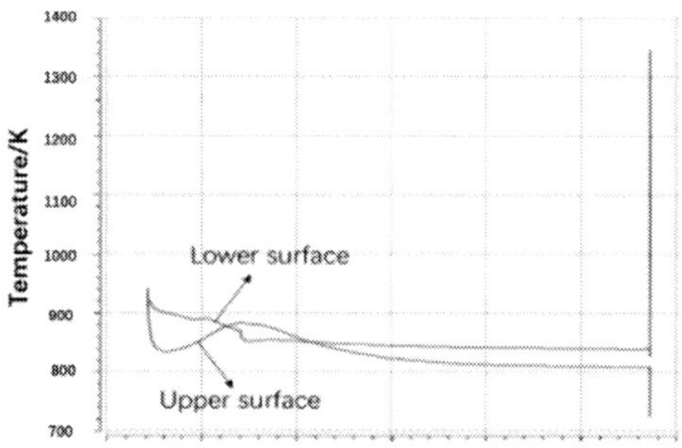

(b) 0.18s

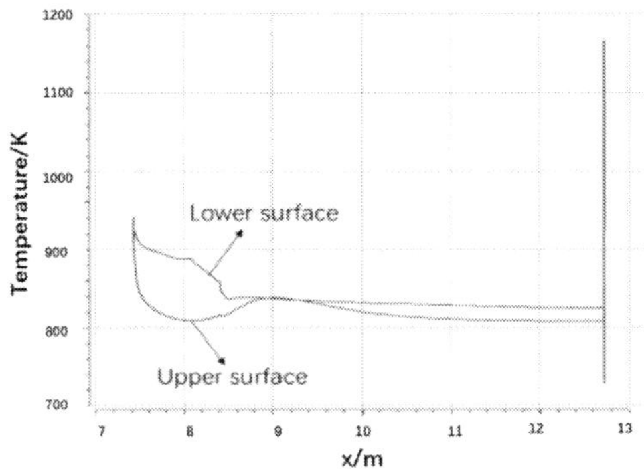

(c) 0.30s

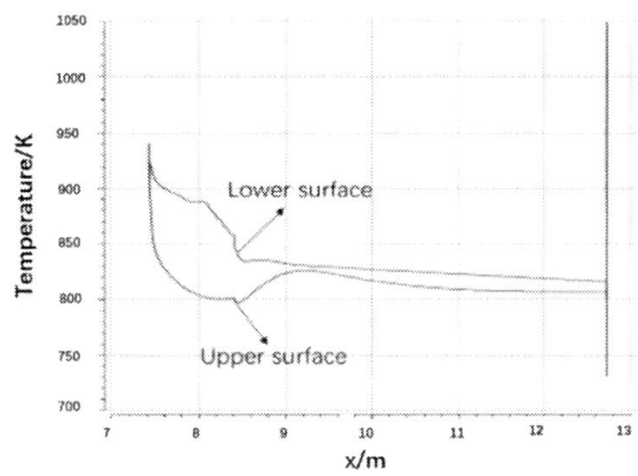

(d) 0.42s

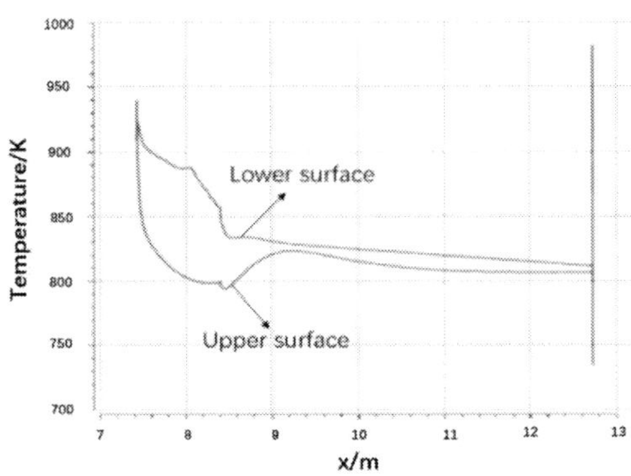

(e) 0.54s

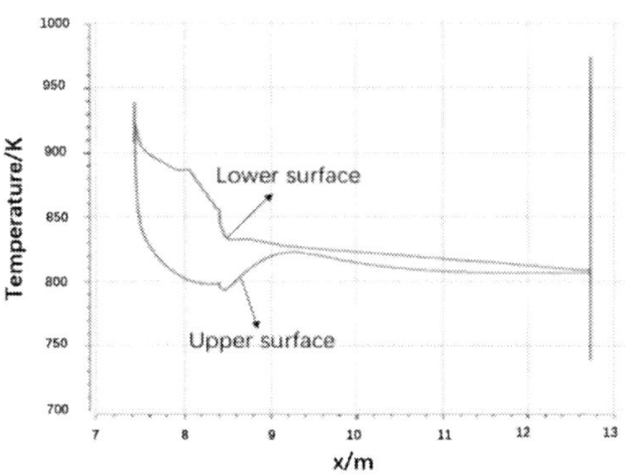

(f) 0.66s

Figure 18. Transient temperature curves of upper and lower surfaces.

4. Conclusions

In this paper, an investigation on the aerodynamic characteristic of the missile has been discussed in detail. The flow distribution is studied and the drag coefficient for different configuration is compared by the CFD numerical method. Based on the above discussion the conclusions are obtained as follows.

(1) During the supersonic flight, the missile encounters strong shock waves, and the velocity of the airflow has a sharp decline while the pressure of the warhead has a sharp increase. The drag coefficient decreases as the freestream Mach number increases from 2 to 4.5, and the shock wave strength increases correspondingly.

(2) The missile configuration has significant impact on the aerodynamic characteristics. The strength of the shock wave increases as the radius of curvature of head decreases, and as the slenderness ratio increases, the drag coefficient presents a gradually trend of decrease.

(3) The position of the stagnation point changes as the angle of attack changes. The air around the warhead is strongly compressed and is subjected to severe friction. Therefore, the temperature of the missile is influenced by shock waves as well as the viscous friction, and a vortex can be seen clearly in the rear of the missile.

(4) The temperature of lower surface is higher than the temperature of upper surface when there exists an angle of attack and it increases as the angle of attack increases. And curves obtained by transient calculation presents that temperature distribution on the upper and lower surface changes rapidly before 0.3s, but gradually slows down after 0.3s, and the whole flow field tends to be stable. Further investigation will mainly focus on the aerodynamic heating of the hypersonic missile.

Conflict of interest statement
The authors declare there is no conflict of interest in relation to this manuscript.

References

[1] Narayana G, Selvaraj S. Experimental investigation of heat transfer over double disk spike-blunt body at Mach 5.7. Experimental Thermal and Fluid Science, 2019, 102: 452-466
[2] Pezzella G. Aerodynamic and aerothermodynamic design of future launchers preparatory program concepts. Aerospace Science and Technology, 2012, 23(1): 233–249
[3] Jameson A. Numerical Simulation of the Euler Equations by Finite Volume Methods Using Methods Using Runge-Kutta Time-Marching Schemes. AIAA paper, 1981, 81-1295
[4] Persova M G, Soloveichik Y G, Belov V K, et al. Modeling of aerodynamic heat flux and thermo elastic behavior of nose caps of hypersonic vehicles. Acta Astronaut, 2017, 136: 312–331
[5] Huang W, Zhao Z T, Yan L, et al. Parametric study on the drag and heat flux reduction mechanism of forward-facing cavity on a blunt body in supersonic flows. Aerospace Science and Technology, 2017, 71: 619–626
[6] Suzuki T, Aoki T, Ogasawara T, et al. Nonablative lightweight thermal protection system for mars aeroflyby sample collection mission. Acta Astronaut, 2017, 136: 407–420
[7] Shoev G, Oblapenko G, Kunova O, et al. Validation of vibration-dissociation coupling models in hypersonic non-equilibrium separated flows. Acta Astronaut, 2018, 144: 147–159
[8] Belov I A, Isaev, Ju Mitin A. Calculation of separated flows within the framework of ideal fluid with allowance for a turbulent shear layer on the boundary of the separation region. Journal of Engineering Physics and Thermophysics, 1986, 51 (4): 1159–1164
[9] Qin Q H, Xu J L, Guo S. Fluid-thermal analysis of aerodynamic heating over spiked blunt body configurations. Acta Astronautica, 2017, 132: 230-242
[10] Yue C G, Chang X L, Zhang Y H, Yang S J. Numerical Calculation of a Missile's Aerodynamic Characteristic. Advanced Materials Research, 2011, 1157
[11] Hao L, Wu J S. Wind Tunnel Experimental Investigation on the Aerodynamic Characteristics of the Multifin Rockets and Missiles. Journal of Beijing Institute of Technology (English Edition), 2005(03):293-296
[12] Zhang T T, Wang Z G, Huang W, Yan L. A review of parametric approaches specific to aerodynamic design process. Acta Astronautica, 2018, 145: 319-331
[13] Huang W. A survey of drag and heat reduction in supersonic flows by a counterflowing jet and its combinations. Journal of Zhejiang University-Science A, 2015, 16(7): 551–561
[14] Huang W, Jiang Y, Yan L, Liu J. Heat flux reduction mechanism induced by a combinational opposing jet and cavity concept in supersonic flows, Acta Astronaut, 2016, 121: 164-171
[15] Sun X W, Guo Z Y, Huang W, Li S B, Yan L. A study of performance parameters on drag and heat flux reduction efficiency of combinational novel cavity and opposing jet concept in hypersonic flows. Acta Astronaut, 2017, 131: 204-225
[16] Huang W, Chen Z, Yan L, Yan B B, Du Z B. Drag and heat flux reduction mechanism induced by the spike and its combinations in supersonic flows: A review. Progress in Aerospace Sciences, 2019, 105: 31-39
[17] Qu F, Sun D, Bai J Q, Zuo G, Yan C. Numerical investigation of blunt body's heating load reduction with combination of spike and opposing jet. International Journal of Heat and Mass Transfer, 2018, 127: 7-15
[18] Zhao R, Wen C Y, Tian X D, et al. Numerical simulation pf local Wall heating and cooling effect on the stability of a hypersonic boundary layer. International Journal of Heat and Mass Transfer, 2018, 121: 986-998
[19] Crawford D H, Investigation of the flow over spiked-nose hemisphere-cylinder at Mach Number of 6.8. NACA, 1959, 6 (1): 112-118
[20] Liou M S. A sequel to AUSM: AUSM +. Journal of Computational Physics, 1996, 129(2): 364-382

Study on the Transverse Elastic Vibration Equation and the Generalized Aerodynamic Force of Space Vehicle

Guo Peng-fei, Li Zheng-xue, Wu Jian-hui, Dong Wei and Fei Wang-hua

China Academy of Launch Vehicle Technology, Beijing 100076, China

Corresponding author's e-mail address: gpfmail@126.com

Abstract. In order to avoid divergence of space vehicle attitude, structural elastic movements should be taken into consideration in the control system design. Space vehicle structure generally has a large length-diameter ratio, the corresponding rolling and longitudinal modal with higher natural frequency has a slight impact on the design of the control system, while the transverse elastic modal with lower natural frequency including pitching and yawing should be paid more attention in the design of space vehicle control system. In this paper, the influence of elastic vibration on the design of attitude control system of space vehicle is analyzed firstly. Then the transverse elastic vibration equation of spacecraft and the calculation method of generalized aerodynamic force are studied. Furthermore, the equation of elastic vibration considering the coupling effect of aerodynamic force between pitching and yawing control channels is proposed and the engineering calculation method of generalized aerodynamic force for typical stations in spacecraft is improved. The research results can provide significant reference for the design of the new generation of spacecraft control system.

1. Introduction

The attitude control system, which is one of the most important parts in spacecraft flight control system, is mainly responsible for stabilizing and controlling the pitching, yawing and rolling motion of spacecraft so as to guarantee the spacecraft to flight according to the predetermined trajectory. In the flight process, the combination effects of the aerodynamic force, the jet engine thrust and the servo driving force will lead to the structural deformation and elastic vibration, which will affect the motion of spacecraft around the center of mass and feedback to the control loop through inertial and gyro perception, and eventually have impact on the movements of aerial rudder, swing jet engines and gas rudder. In order to avoid divergence of space vehicle attitude, structural elastic movements should be taken into consideration in the control system design. However, the space vehicle structure generally has a large length-diameter ratio, the corresponding rolling and longitudinal modal with higher natural frequency has a slight impact on the design of the control system. Therefore, the transverse elastic modal with lower natural frequency including pitching and yawing should be paid more attention in the design of space vehicle control system.

The bending deformation of space vehicle caused by elastic vibration will introduce additional thrust components and additional attitude angle, which affect the space vehicle control system significantly. Since the late 1960s, extensive researches have been conducted by scholars from all over the world to study the influence of spacecraft structural flexibility on control system design and stability. Reis and Sundberg [1] studied the elastic bending of the rocket during launch and free flight by using two-rigid-body model and continuum model respectively. Cochran et al [2] analyzed the

Content from this work may be used under the terms of the Creative Commons Attribution 3.0 licence. Any further distribution of this work must maintain attribution to the author(s) and the title of the work, journal citation and DOI.
Published under licence by IOP Publishing Ltd

influence of elastic vibration on the attitude of free flight rocket, and concluded that the lateral bending would cause large transverse angular velocity. Choi and Bang [3] investigated the control system of KSR(Korea Sounding Rocket)-II sonic velocity rocket and found that the elastic vibration would be superimposed on the output of the controller through the feedback channel, which would affect the stable operation of the control system and even cause the system instability. So the adaptive notch technology was used to filter out this influence in the feedback channel so as to ensure the system stability. Wu and He [4] proposed the elastic vibration equation and the design method of attitude control system of large and medium-sized bundled rocket based on space modal, which took the coupling effects of rigid movements, fluid sloshing and elastic vibration into account. Wei et al[5] developed a full-component dynamic model, designed the self-adaptive attitude controller and variable gain scheduling controller based on frequency identification and realized the high-precision control of the elastic launch vehicle.

The traditional spacecraft, represented by launch vehicles and missile weapons, has symmetric aerodynamic shapes and structures and the corresponding pitching, yawing and rolling channels are independent. Thus, tri-channel control system with decoupled strategy is always employed [6]. However, with the development of new types of aircrafts and extensive applications of new types of flight trajectory, the coupling effect of the aerodynamic forces on the structure surface in the transverse pitching and yawing channels becomes serious, so it is necessary to fully consider the coupling effect and establish the corresponding transverse elastic vibration equations.

In this paper, in order to consider the coupling effect of spacecraft aerodynamic forces in pitching and yawing channels, the influence of spacecraft elastic vibration on the design of attitude control system of space vehicle is analyzed firstly. Then the lateral elastic vibration equation of spacecraft and the calculation method of generalized aerodynamic force are studied. Furthermore, the equation of elastic vibration considering the influence of aerodynamic coupling between pitching and yawing control channels is proposed and the engineering calculation method of generalized aerodynamic force for typical stations in spacecraft is improved. The research results can provide significant reference for the design of the new generation of spacecraft control system.

2. Transverse elastic vibration equation

The main structure of space vehicle is always axial symmetric and has a relatively large length-diameter ratio. Its elastic vibration modal shape mainly includes motions in longitudinal, pitching, yawing and rolling directions with pitching axis perpendicular to trajectory plane and yawing axis in the trajectory plane perpendicular to structural longitudinal axis. When transverse elastic vibrations in pitching and yawing directions are analyzed as shown in Figure 1, both computational fluid method and finite element method should be employed to obtain characteristics of the aerodynamic distribution and modal shapes respectively.

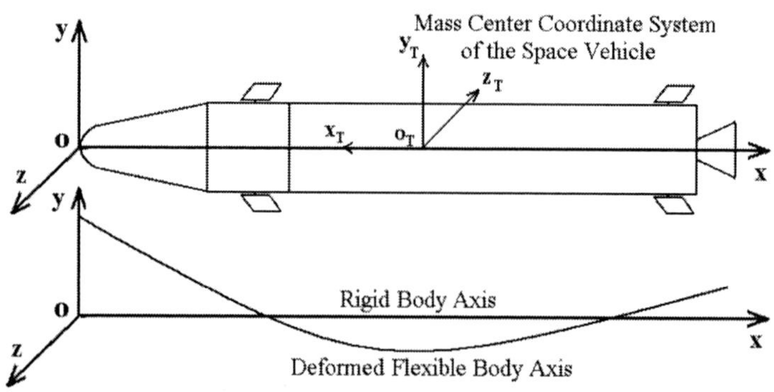

Figure 1. Elastic deformation of space vehicle structure

In modal analysis, the finite element method is always adopted and the elastic vibration equation of the spacecraft can be expressed by [7]:

$$M\ddot{X}(\mathrm{t}) + C\dot{X}(t) + KX(t) = F(t) \tag{1}$$

where M, C and K are mass matrix, damping matrix and stiffness matrix of the structural system respectively; $\mathrm{X(t)}$, $\dot{X}(t)$ and $\ddot{X}(t)$ are the finite element node displacement, velocity and acceleration vector respectively; $F(t)$ is the finite element node loading vector. Mode-superposition method is exploited to solve the equation, and it is assumed that

$$X(t) = \sum_{\mathrm{i}=1}^{n} \Phi_i q_i(t) \tag{2}$$

where Φ_i is the ith-order modal shape of the structural system, then Equation (1) can be decoupled as:

$$\ddot{q}_i(t) + 2\zeta_i\omega_i\dot{q}_i(t) + \omega_i^2 q_i(t) = F_i(t)/M_i \quad i = 1,2,\dots,n \tag{3}$$

where $\mathrm{q_i(t)}$, $\dot{q}_i(t)$ and $\ddot{q}_i(t)$ are the ith-order modal displacement, velocity and acceleration in generalized coordinates respectively; ω_i , ζ_i , $F_i(t)$ and M_i are the ith-order modal natural frequency, damping coefficient, generalized force and generalized mass of the structural system. M_i and $F_i(t)$ can be expressed as:

$$M_i = \Phi_i^{\mathrm{T}} M \Phi_i \tag{4}$$

$$F_i(t) = \Phi_i^{\mathrm{T}} F(t) \tag{5}$$

$$\omega_i = \sqrt{\Phi_i^{\mathrm{T}} K \Phi_i / \Phi_i^{\mathrm{T}} M \Phi_i} \tag{6}$$

$$2\zeta_i\omega_i = \sqrt{\Phi_i^{\mathrm{T}} C \Phi_i / \Phi_i^{\mathrm{T}} M \Phi_i} \tag{7}$$

In the control system of traditional launch vehicles and missile weapons, the influences of attack angle on lateral aerodynamic force and sideslip angle on normal aerodynamic force are usually ignored, and the elastic vibration equation in the direction of pitching and yawing can be obtained as [7]:

$$\ddot{q}_i(t) + 2\zeta_i\omega_i\dot{q}_i(t) + \omega_i^2 q_i(t) = D_{1i}\Delta\dot{\varphi} + D_{2i}\Delta\alpha + D_{3i}\Delta\delta_\varphi + D_{3i}''\Delta\ddot{\delta}_\varphi + Q_{iy} \tag{8}$$

$$D_{1i} = -\frac{180 \cdot qS_M}{\pi M_i V} \sum_{s=1}^{K} C_{nn}^{\alpha} \Phi_i(X)(X_{\mathrm{T}} - X_s)dX \tag{9}$$

$$D_{2i} = \frac{180 \cdot qS_M}{\pi M_i} \sum_{s=1}^{K} C_{nn}^{\alpha} \Phi_i(X)dX \tag{10}$$

$$D_{3i} = \frac{N_K H_K}{M_i}[F_{nn}^{K}\Phi_i(X_K) + m_K a_X l_K \Phi_i'(X_K)] \tag{11}$$

$$D_{3i}'' = \frac{N_K H_K}{M_i}[m_K l_K \Phi_i(X_K) + J_K \Phi_i'(X_K)] \tag{12}$$

where q is the dynamic pressure; S_M is the reference area; V is the flight velocity; M_i is general mass of the ith-order modal; Φ_i and Φ_i' are the modal shape and corresponding slope respectively; $\Delta\dot{\varphi}$ is the pitching (or yawing) rotational angular velocity; $\Delta\alpha$ is the attack angle (or sideslip angle); $\Delta\delta_\varphi$ is the angle of aerial rudder; $\Delta\ddot{\delta}_\varphi$ is the angular acceleration of aerial rudder; Q_{iy} is the general force corresponding to the disturbing force in the direction of main shaft vibration; C_{nn}^{α} is the coefficient derivative of transverse(in the direction of pitching and yawing) distribution of aerodynamic force; F_{nn}^{K} is the coefficient of the transverse(in the direction of the pitching and the yawing) aerodynamic force of aerial rudder; X_T is the longitudinal coordinate of space vehicle's mass center; $\mathrm{X_s}$, X_R are the longitudinal coordinates of all rocket(missile) body stations and connection location of the aerial rudder; a_X is the longitudinal acceleration; m_K is the mass of aerial rudder; l_K is the distance between

IOP Conf. Series: Materials Science and Engineering **887** (2020) 012002 doi:10.1088/1757-899X/887/1/012002

mass center of the aerial rudder and corresponding spinning shaft; J_K is the moment of inertia of aerial rudder; N_K is the number of aerial rudders; H_K is the layout coefficient of aerial rudders.

For the new generation of space vehicles, especially for those with lifting body shape, the change of sideslip angle will inevitably cause the significant change of lifting force, which means that the influence of the derivative of normal aerodynamic force to sideslip angle on the control system cannot be ignored. In addition, in the large sideslip condition, the change of attack angle, to some extent, will also cause the change of lateral force. Namely, the influence of the derivative of the lateral aerodynamic force to the attack angle on the control system should also be considered in the control system design.

Aerodynamic computational fluid simulation is conducted on some cabin parts of the American x-51 aircraft shown in Figure 2 when rudder angle are 0°and 25°, and the results of derivative of normal aerodynamic force to attack angle (alf) and sideslip angle (beta) are shown in Figures 3 and 4. It can be indicated from Figures 3(b) and 3(e) that when the aircraft flies in the state of zero sideslip and zero rudder deflection, the change of sideslip angle has slight impact on the normal aerodynamic force$\mathrm{Cy-beta}$, while the change of pitching angle has a greater impact on the normal aerodynamic force$\mathrm{Cy-alf}$, as shown in Equation (8). However, it can be seen from Figures 3 and 4 that when the aircraft flies with a large sideslip angle or rudder deflection, there are many states where$|\mathrm{Cy-beta}| \geq |\mathrm{Cy-alf}|$. Therefore, the change of sideslip angle is more important than the change of pitching angle and should not be ignored in these cases.

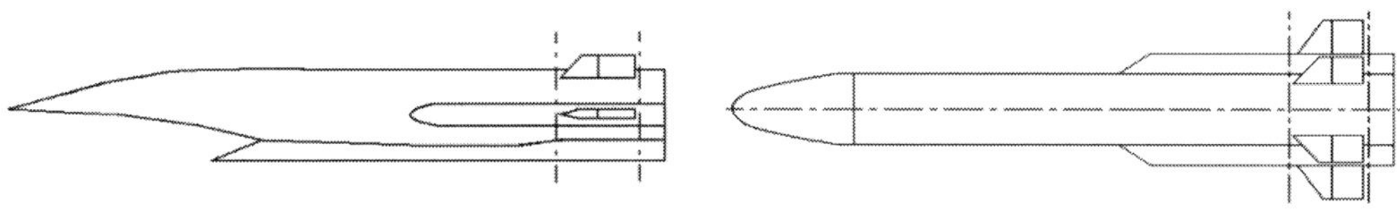

Figure 2. American X-51 Aircraft

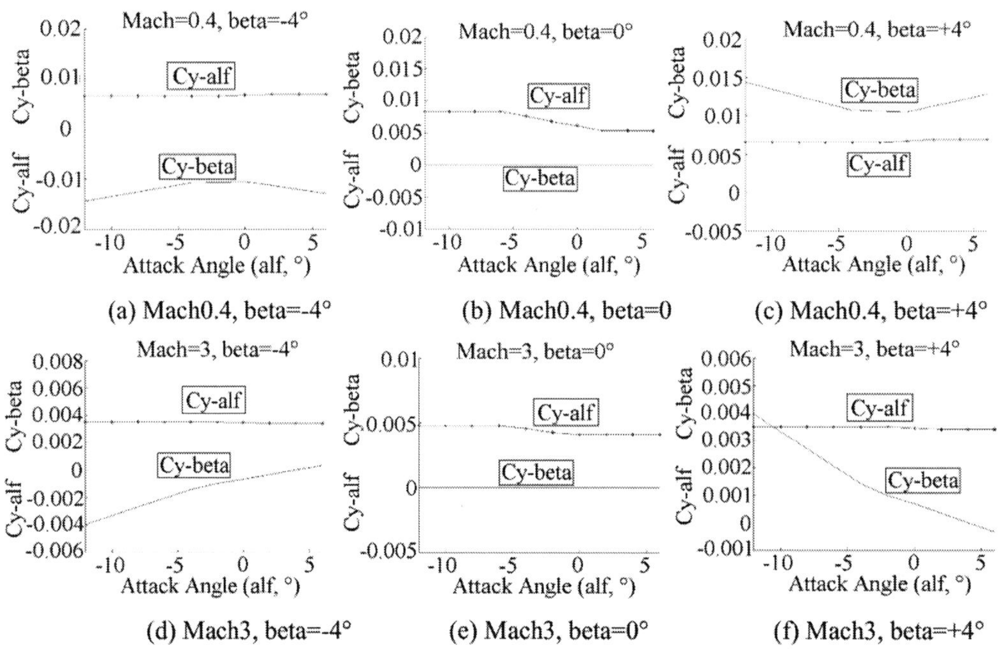

(a) Mach0.4, beta=-4° (b) Mach0.4, beta=0 (c) Mach0.4, beta=+4°

(d) Mach3, beta=-4° (e) Mach3, beta=0° (f) Mach3, beta=+4°

Figure 3. Normal aerodynamic force derivatives to attack angle alf and sideslip angle beta when the rudder angle is 0°

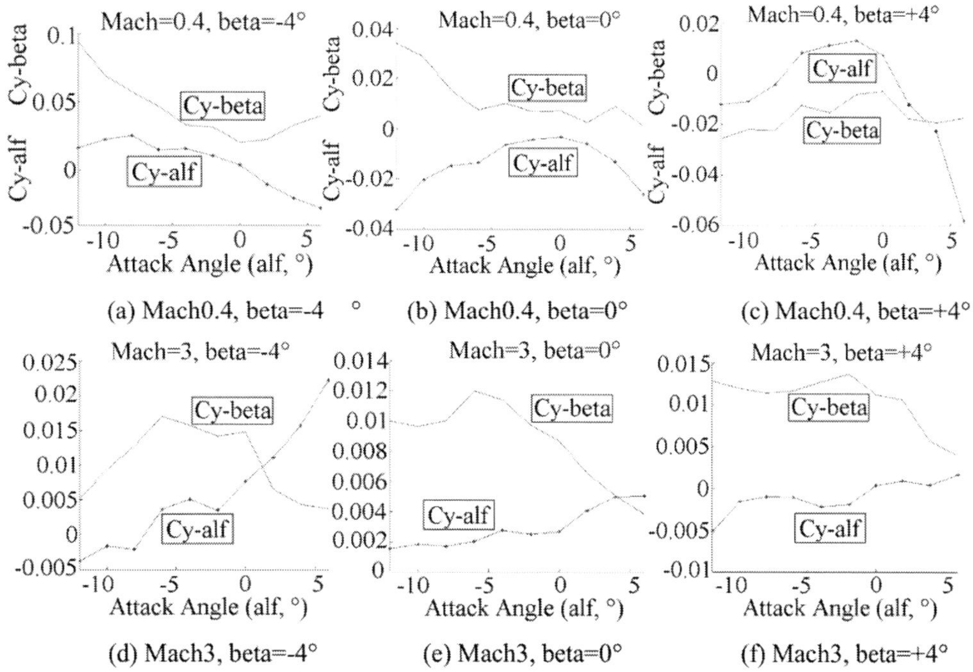

(a) Mach0.4, beta=-4 ° (b) Mach0.4, beta=0° (c) Mach0.4, beta=+4°

(d) Mach3, beta=-4° (e) Mach3, beta=0° (f) Mach3, beta=+4°

Figure 4. Normal aerodynamic force derivatives to attack angle alf and sideslip angle beta when the rudder angle is 25°

Consequently, both the influence of attack angle and sideslip angel should be taken into consideration in the design of the pitching and yawing channel control system of space vehicles. Thus, elastic vibration equation in the direction of the pitching and the yawing should be modified as follows.

$$\ddot{q}_i(t) + 2\zeta_i\omega_i\dot{q}_i(t) + \omega_i^2 q_i(t) = D_{1i}\Delta\dot{\varphi} + D_{2i}\Delta\alpha + D_{2i}^{*}\Delta\beta + D_{3i}\Delta\delta_{\varphi} + D_{3i}''\Delta\ddot{\delta}_{\varphi} + Q_{iy} \quad (13)$$

$$D_{2i}^{*} = \frac{180\cdot qS_M}{\pi M_i}\sum_{s=1}^{K} C_{nn}^{\beta}\Phi_i(X)dX \quad (14)$$

Where $D_{2i}^{*}\Delta\beta$ is the additional influential component caused by sideslip (or attack) angle to directional general force; C_{nn}^{β} is the coefficient of transverse (in the direction of the pitching and the yawing) aerodynamic force distribution.

3. Computation method of general aerodynamic derivative in traditional station

For the generalized force of aerodynamic distribution force derivative of traditional launch vehicle and missile weapon, the rocket or missile structure is always divided into several stations in engineering, and the generalized force is obtained by calculating the aerodynamic force derivative and the vibration mode on all stations.

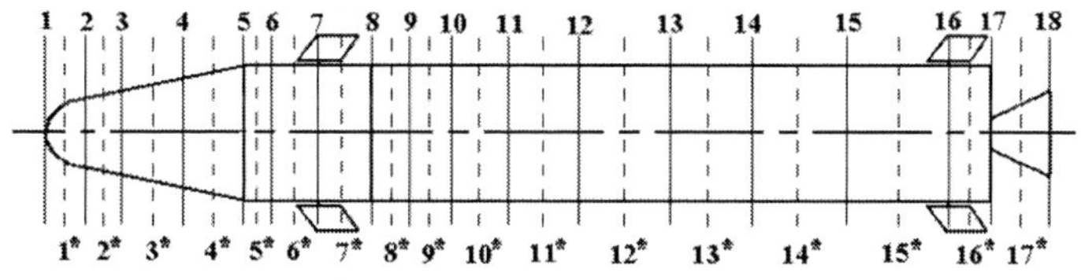

Figure 5. Station divisions of traditional launch vehicle and missile weapon

As shown in Figure 5, the whole body of traditional launch vehicle and missile can be divided into 18 stations. The cross sections between stations are marked 1* to 17* respectively. Besides, and the following conditions are required for station division:

(1) The variation of aerodynamic force derivative between stations is slight and can be linear;

(2) The variation of mode between stations is also slight and can be linear.

Based on the above station division conditions, if the aerodynamic distributed force distribution derivative between two adjacent stations with distance L is denoted by $F_X(x) = Ax + B$ in the direction of X and the vibration modal shape between stations is denoted by $\Phi_{i,X}(x) = Cx + D$, then the theoretical generalized force $F_{i,X}$ is

$$F_{i,X} = \int_0^L (Ax + B)(Cx + D)\mathrm{d}x = \frac{1}{3}ACL^3 + \frac{1}{2}ADL^2 + \frac{1}{2}BCL^2 + BDL \quad (15)$$

In the traditional design, the aerodynamic distribution characteristics of the rocket (missile) body surface are usually obtained by engineering calculation or computational fluid simulation, and the aerodynamic force derivatives $F^{(s)}$at each station are calculated by equally divided distance to obtain the generalized force [7]. In details, the aerodynamic force derivative on station 1 is the resultant force from station 1 to 1*. Similarly, the one on station 18 is the resultant force from station 17* to 18, and the aerodynamic force derivative on the middle station is the resultant force between its left and right equal sections. Thus, the generalized force obtained by this method is

$$F_{i,X}^{*} = D\int_0^{L/2}(Ax + B)\mathrm{d}x + (CL + D)\int_{L/2}^{L}(Ax + B)\mathrm{d}x = \frac{3}{8}ACL^3 + \frac{1}{2}ADL^2 + \frac{1}{2}BCL^2 + BDL \quad (16)$$

$$F_{i,X} - F_{i,X}^{*} = \frac{1}{24}ACL^3 \quad (17)$$

Therefore, the generalized force of typical stations based on the equal distance division should divide stations as much as possible so as to weaken the influence of station distance L, change slope of aerodynamic derivative A and change slope of vibration modal shape C, and eventually to improve the calculation accuracy.

In order to eliminate the influence of station division, aerodynamic characteristics and vibration modal shapes on generalized force, the generalized aerodynamic force derivatives of typical stations can be calculated by the following method of distribution of pressure center.

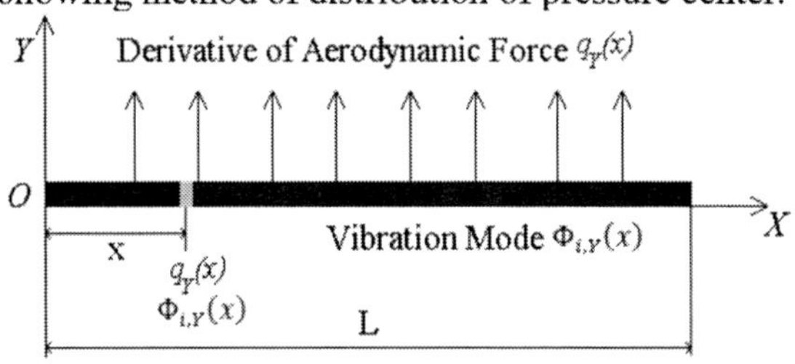

Figure 6. A distribution of station aerodynamic force and vibration modal shape

If $q_Y(x)$ is the distribution of the transverse aerodynamic force derivative in the direction of X and $\Phi_{i,Y}(x)$ is the distribution of the vibration modal shape in the direction of X as shown in Figure 6, aerodynamic force derivatives of right and left stations are derived as follows:

$$F_Y^{(s),left} = \int_0^L q_Y(x)\frac{x}{L}\mathrm{d}x \quad (18)$$

$$F_Y^{(s),right} = \int_0^L q_Y(x)\frac{(L-x)}{L}\mathrm{d}x \quad (19)$$

Because both changes of aerodynamic force and vibration modal shape between stations are linear, it can be assumed that

$$q_Y(x) = Ax + B \tag{20}$$

$$\Phi_{i,Y}(x) = Cx + D \tag{21}$$

Then, according to vibration modal shapes and aerodynamic forces on the left and right sides of the station, the following can be obtained:

$$F_{i,Y}^{**} = F_Y^{left} \cdot \Phi_{i,Y}^{left} + F_Y^{right} \cdot \Phi_{i,Y}^{right} = D\int_0^L (Ax+B)\frac{(L-x)}{L}\mathrm{d}x + (CL+D)\int_0^L (Ax+B)\frac{x}{L}\mathrm{d}x$$
$$= \frac{1}{3}ACL^3 + \frac{1}{2}ADL^2 + \frac{1}{2}BCL^2 + BDL \tag{22}$$

$$F_{i,X} = F_{i,X}^{**} \tag{23}$$

Therefore, the generalized aerodynamic force of typical station calculated by the method of distribution of pressure center is more accurate than that obtained by traditional engineering method as shown in Equation (16).

4. Conclusions

In this paper, the influence of the transverse elastic vibration equation on the attitude control system of spacecraft is analyzed, and the elastic vibration equation of the spacecraft cabin considering the coupling effect of the aerodynamic forces between pitching and yawing channels is developed according to the characteristics of the aerodynamic forces derivatives to the attack angle and sideslip angle. The improved calculation method of the generalized aerodynamic derivatives of typical stations based on the pressure center distribution is proposed. This computation method can reduce the requirement on the station density and improve the accuracy of calculation results. The research results in this paper can be widely used in the overall design of the new generation of missile weapons, launch vehicles and space vehicles, and can provide important reference for the design of control system and stability analysis.

References

[1] Reis G, Sundberg W. Calculated aeroelastic bending of a sounding rocket based on flight data [J]. *Journal of Space vehicle and Rockets*, 1967, **4**(11): 1489-1494.

[2] Cochran Jr J E, Castleberry G A, Rew S D. An analysis of the effects transverse vibrations on the attitude motion of free-flight rockets [R]. *Technical Report RL-CR-80-2, U.S. Army Missile Command*, Redstone Arsenal, 1979.

[3] Choi H D, Bang H. An adaptive control approach to the attitude control of a flexible rocket [J]. *Control Engineering Practice*, 2000, **18**(9): 1003-1010.

[4] Wu Y S, He L S. Attitude control technology of new generation launch vehicles [J]. *Journal of Beijing University of Aeronautics and Astronautics*, 2009, **35**(11): 1294-1297.

[5] Wei C Z, Ju X Z, Zhang L, Cui N G. The modeling and controller design for aeroelastic rocket launcher [J]. *Astronautical Systems Engineering Technology*, 2017, **1**(1): 21-26.

[6] Zhang W D, Liu Y X, Liu H B, Ding X F, Zhang K B. Development trend and prospect of attitude control technologies of launch vehicle. *Aerospace Control*, 2017, **3**(35): 85-89.

[7] Yin Y Y. Loads design foundation for solid rockets [B]. *China Aerospace Publishing House*, 2007.

Aerodynamic Analysis of a Gull-inspired Flapping Wing Glider

Dawei Bie, Shiyu Zuo, Huadong Li, Haoyuan Shao and Daochun Li*

School of Aeronautic Science and Engineering, Beihang University, 100191, Beijing, China

*Corresponding author: lidc@buaa.edu.cn

Abstract. This paper provided a gull-inspired flapping wing micro aerial vehicle (FMAV) concept followed with aerodynamic analysis. The FMAV supposed to implement both flapping and gliding mode during the flight to enlarge the endurance. A cambered airfoil is employed for wing cross section different with the traditional flat plate. Aerodynamic analysis is proposed through 3D computational fluid dynamics (CFD) method to get the lift characteristic of two different modes. Result shows that the FMAV is able to generate the lift force up to 5 N with an angle of attack 6 degrees and flight speed 6 m/s without flapping, which proves that the FMAV is able to achieve gliding in specific circumstance. Flapping motion is able to produce a higher lift during the flight. Comparison of the lift performance under a specific condition shows that the average lift generated with flapping is 65.3% higher than gliding mode. Research also found that the effect of airfoil camber on lift generation with flapping is not always positive, further analysis is needed when employs a cambered wing in design of FMAVs.

1. Introduction

Bionic flapping wing micro aerial vehicle (FMAV) has been widely concerned in recent years due to its high aerodynamic efficiency in low Reynolds number flight. Substantial design concepts of FMAV have been published. Researchers in Delft University of technology proposed a series of FMAVs concept named Delfly [1, 2] to process works such as stereo vision and flight control methodologies. Guo et al. [3] provided a design concept of flapping wing rotor MAV. The wing is activated through piezoelectric actuator. Prototype is produced and tested to verify the feasible of the design concept. Structure finite element analysis and aerodynamic simulation based on computational fluid dynamics (CFD) method is then proceeded. Ramezani et al. [4] designed a bat-inspired FMAV named Bat Bot (B2). The wing has a 5-DOFs mechanism to mimic the arm wing of the real bat which able to perform the asynchronously and mediolaterally motions. The employment of multi-DOF mechanism gives B2 ability to process more complex manoeuvre during flight.

To obtain the aerodynamic performance of the design concepts, Tay et al. [2] processed the analysis based on 3D CFD simulation with immersed boundary method, and then made a comparison of the lift and thrust of two types of FMAVs. Detailed vortex structure is revealed to illustrate the principle of lift and thrust generation of the FMAV. Xiang et al. [5] analysed the aerodynamic performance of the gliding mode of locust through 2D CFD simulations. The cross section of the forewing and hindwing is collected from biological data. The effect of wing corrugation is concluded with the simulation results. Mazaheri and Ebrahimi [6] conducted the wind tunnel test to analysis the lift and thrust properties under three design variables including angle of attack, wind speed and flapping frequency, respectively.

Content from this work may be used under the terms of the Creative Commons Attribution 3.0 licence. Any further distribution of this work must maintain attribution to the author(s) and the title of the work, journal citation and DOI.
Published under licence by IOP Publishing Ltd

Low range and endurance are always the problem which restricted the application of FMAVs. To solve the problem in a view of biomimicry, among the series of birds, gulls are specialized on fast and long-distance migration, and are able to employ different flying strategy respect to different environment conditions during their journey. The gull is able to accomplish a long sea crossing ranges up to 700 km in an average speed 10.7 m/s. [7] The imitation of the wing shape and flight strategy of gulls is a possible way to solve this problem. In this paper, aerodynamic analysis of a gull-inspired flapping wing glider concept is provided. Analysis of flapping mode and gliding mode are proposed based on CFD simulations. Conclusions have been made based on simulation results.

2. Simulation setup

2.1 Model description

The concept of the gull-inspired FMAV employs a traditional layout which consist of wing, tail and fuselage. The CAD model for simulation is shown in Figure 1. Only lift characteristic is discussed in this paper thus the tail structure is ignored within the simulation model. The wing span of the model is 1.6 m and the aspect ratio is 8.

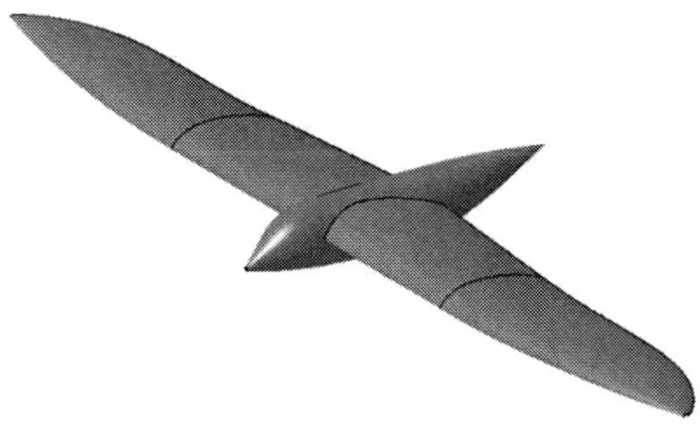

Figure 1. Simplified simulation model

The aerodynamic analysis of the flapping mode and gliding mode is provided. Analysis is processed based on 3D CFD simulation. Three dimensional Reynolds-averaged Navier-Stokes (RANS) equation is solved through finite volume method.

2.2 Gliding mode simulation

Structured grid is employed as shown in Figure 2 for the simulation of gliding mode in order to obtain more accurate results. The grid generated is shown in Figure 2. The size of the far field is 100c*25c*50c. The grid thickness of boundary first layer is 10^{-4} and the growth rate is 1.1. Symmetry boundary is employed at x=0. The total grid number is 7.3 million.

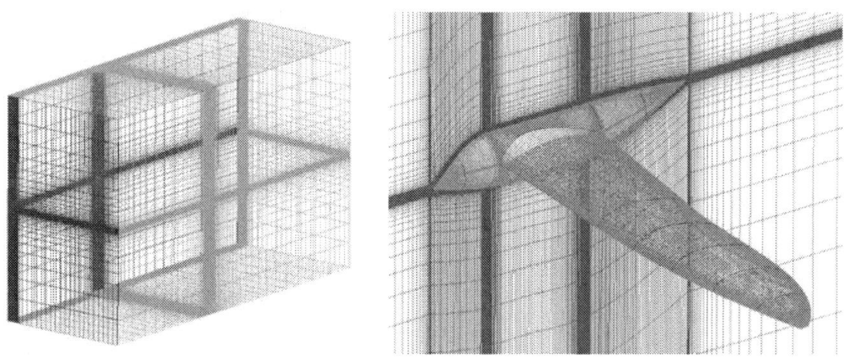

Figure 2. Structured grid for simulation of gliding mode

The simulation is processed based on pressure based steady solver. SST k-ω model is selected as the viscos model. The boundary conditions are velocity inlet, pressure outlet and no-slip wall. SIMPLE scheme is employed for pressure-velocity coupling scheme. Second order upwind scheme is employed for spatial discretization. The initial simulation is processed with five different angles of attack with three free stream velocities including 3 m/s, 6 m/s and 10 m/s. Static lift and thrust at different angle of attack is obtained with the simulations.

2.3 Flapping mode simulation
Unstructured grid is employed for analysis of flapping mode because the unstructured grid is more stable in simulation with dynamic mesh. The generated grid is shown in Figure 3.

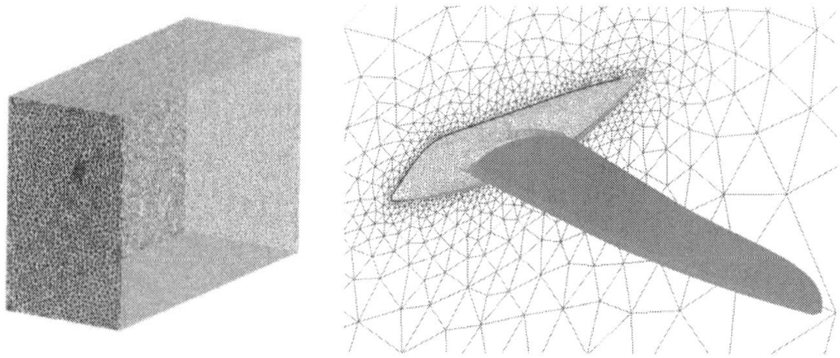

Figure 3. Unstructured grid for simulation of flapping mode

The size of the far field is same as gliding mode. The minimum grid size is 0.05c on the fuselage. Density increment is applied to the leading edge of the wing. The minimum grid size is 0.007c for the leading edge and 0.025c for rest of the wing. The total grid number of the model is 2.54 million. Dynamic mesh is implemented to fit the flapping motion of the wing. The wing motion is defined with a sinusoidal function as follows,

$$\theta = \theta_{max} \cdot sin\,(2\pi f \cdot t) \tag{1}$$

where θ is the flapping angle at each timestep, θ_{max} is the flapping amplitude, and f represent the flapping frequency.

The flight speed is selected to be 6 m/s for simulation. The variables selected during the simulation are flapping amplitude, angle of attack and flapping frequency.

3. Result discussion

3.1 Gliding mode results
The lift respect to angle of attack in different flight speed are shown in Figure 4. Result shows that the selected airfoil has a good aerodynamic performance. Results indicate that the designed FMAV has a property of glide in variance of flight speed. The lift slope at 10 m/s is 89% higher than the slope of 3 m/s, which shows the wing has a better lift performance at relatively higher speed.

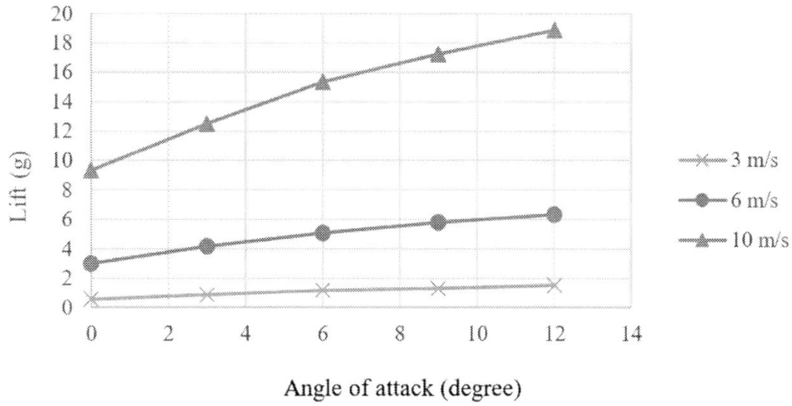

Figure 4. Lift respect to angle of attack in gliding mode

CFD result shows that the FMAV is able to generate lift force up to 5 N with a flight speed 6 m/s and angle of attack 6 degrees. Pressure distribution with streamline around the wing cross section at half wing span is take for analysis. Data along the flow direction is provided in Figure 5. It can be seen that the air flow around the wing is adhered well at the angle of attack 3 degrees, only a small area of separation at trilling edge. While a large separation appears on the upper surface at the angle of attack 12 degrees, resulted a decrease of the slope of the lift curve.

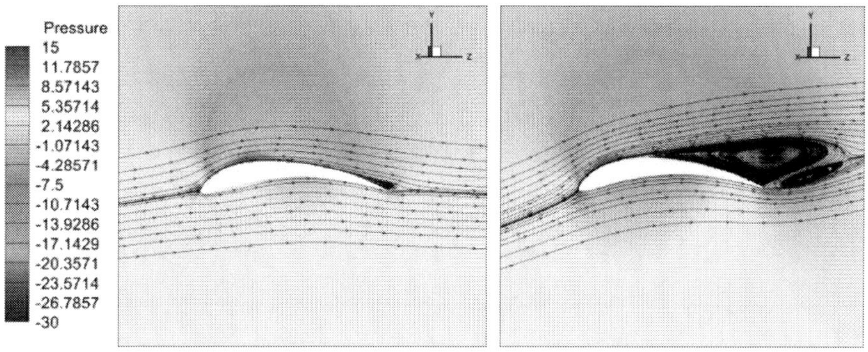

(a) AoA = 3 degrees (b) AoA = 12 degrees

Figure 5. Streamlines around the wing

3.2 Flapping mode results

3.2.1 Variation of flapping amplitude. The lift characteristic with the flapping amplitude 15 degrees, 30 degrees and 45 degrees are proceeded. Other variables are fixed including angle of attack 3 degrees and flapping frequency 6 Hz. Data of eight points are picked within one flapping cycle for analysis which are described in Figure 6. Lift results in one flapping cycle are shown in Figure 7. Results of average lift with three different flapping amplitude are 3.86 N, 4.90N and 7.29 N, respectively. It is able to obtain from the results that the average lift is growing with the increasing of the flapping amplitude. Serious care must be taken that the average lift with the flapping amplitude 15 degrees is lower than the steady lift without flapping. Pressure distribution with streamline along the flow direction with flapping amplitude 15 degrees are provided in Figure 8. Results show that the leading edge vortex during downstroke is not fully developed, while a strong vortex appears under the lower surface during upstroke, which produced a larger pressure difference compared with the downstroke at the same position.

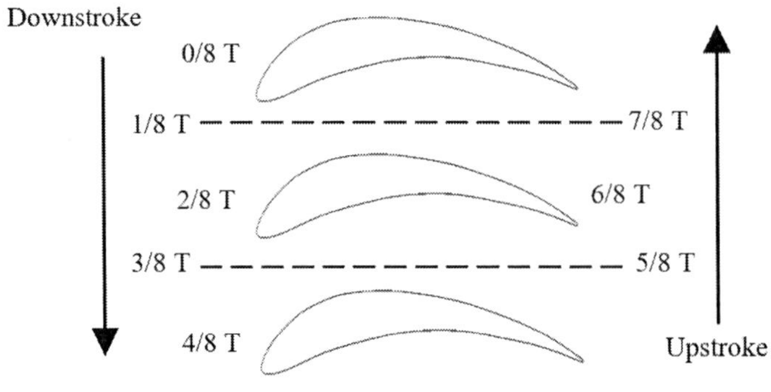

Figure 6. Schematic drawing of the time points in one flapping period

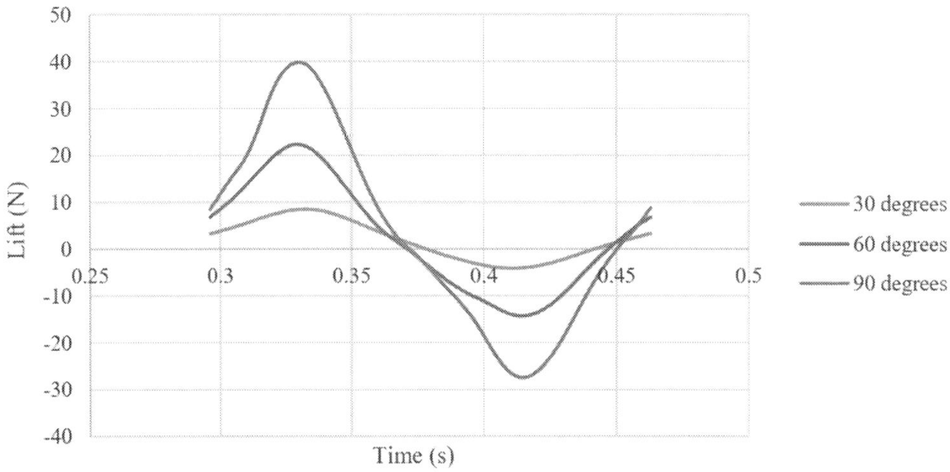

Figure 7. Effect of flapping amplitude to aerodynamic lift

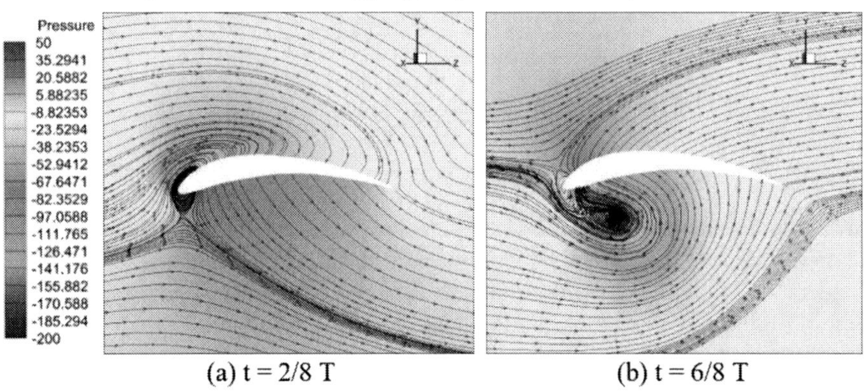

Figure 8. Pressure distribution with streamlines with flapping amplitude of 15 degrees

Pressure distribution with streamline along the flow direction with flapping amplitude 45 degrees are provided in Figure 9. Results show that the structure and position of the leading edge vortex during downstroke and upstroke has no obvious difference, but the pressure difference at 2/8 T during downstroke is dramatically larger than the result at 6/8 T. The appearance of this circumstance is probably due to the effect of the airfoil camber. Detailed reason may need further analysis in the future.

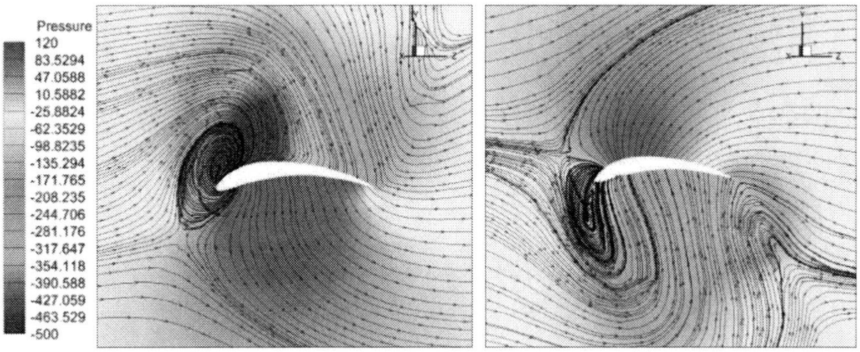

(a) t = 2/8 T (b) t = 6/8 T

Figure 9. Pressure distribution with streamlines with flapping amplitude of 45 degrees

3.2.2 Variation of flapping frequency and angle of attack. The simulation of lift characteristic respect to different flapping frequency in three angle of attack conditions from 3 to 9 degrees are processed. The flapping amplitude in simulations are fixed to 45 degrees. Results are shown in Figure 10. It can be obtained that the increase of angle of attack and flapping frequency can all resulted a growth of lift at a large flapping amplitude. The high frequency flapping motion is able to increase the lift compared with the steady lift of gliding mode. Take the case with angle of attack 3 degrees as example, the lift increment with 4 Hz and 6 Hz flapping are 8.8% and 65.3% compared with the steady lift in gliding mode. But the lift of the case with flapping frequency 2 Hz resulted a lift decrement of 24%. Results show that a low flapping frequency could also induce a lift reduction as same as the effect of low flapping amplitude.

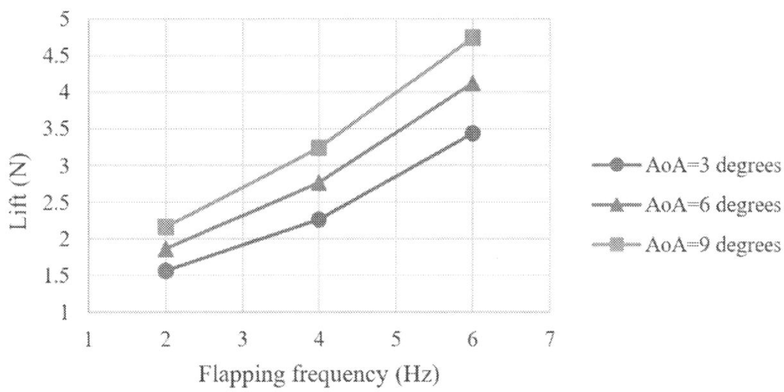

Figure 10. Lift respect to flapping frequency

4. Conclusions

The aerodynamic analysis of a gull-inspired FMAV concept is provided in this paper. The FMAV is designed to flight with two different flying modes, flapping and gliding. 3D CFD simulations has been proceeded for both flapping and gliding modes. Analysis concluded that firstly, the FMAV is able to generate lift in traditional fixed wing mode with no flapping. Secondly, flapping mode is able to support a higher lift compared with the gliding mode at same flight conditions. The implement of flapping motion provides the FMAV a higher manoeuvrability compared with traditional fixed wing glider. Thirdly, the effect of airfoil camber on lift generation with flapping motion is not always positive. Simulation results in this paper show that the lift characteristics performs a decrement with the flapping amplitude 15 degrees. Cases employs a flapping frequency 2 Hz with the flapping amplitude 45 degrees has obtained the same effects. Further analysis is needed when employs a cambered wing in future design works of FMAV.

Acknowledgment
This work was supported by Beijing Advanced Discipline Center for Unmanned Aircraft System.

Reference

[1] Karásekl M, Muijres F, Wagter C, Remes B and Croon G 2019. A tailless aerial robotic flapper reveals that files use torque coupling in rapid banked turns. *Science.* **361** 1089
[2] Tay W, Oudheusden B, Bijl H 2015. Numerical simulation of a flapping four-wing micro-aerial vehicle. *J. Fluids Struct.* **55** 237
[3] Guo S, Li D and Wu J 2012. Theoretical and experimental study of a piezoelectric flapping wing rotor for micro aerial vehicle. *Aerosp. Sci. Technol.* **23** 429
[4] Ramezani A, Chung S and Hutchinson S 2017. A biomimetic robotic platform to study flight specializations of bats. *Sci. Robot.* **2** eaal2505
[5] Xiang J, Du J, Li D and Liu K 2016. Aerodynamic Performance of the Locust Wing in Gliding Mode at Low Reynolds Number. *J. Bionic Eng.* **13** 249
[6] Mazaheri K and Ebrahimi A 2011. Experimental investigation on aerodynamic performance of a flapping wing vehicle in forward flight. *J. Fluids Struct.* **27** 586
[7] Klaassen, R, Ens B, Shamoun J, Exo K and Bairlein F 2012. Migration strategy of a flight generalist, the Lesser Black-backed Gull Larus fuscus. *Behav. Ecol.* **23** 58

Research on TCH and RDH of Glide Path Equipment

Ye Jiaquan[1], Liang fei[*], Yang Xiaojia[1] and Li Yuankai[1]

[1]The Second Research Institution of CAAC, 610041, Cheng Du, China

[*]liangfei@caacsri.com

Abstract. There is a certain difference between the RDH obtained from the flight inspection and the TCH calculated from geometry terrain. The problem is always neglected and someone attribute the cause of poor RDH to equipment issues, actually equipment is a problem, but the electromagnetic environment is also a very important factor overlooked by owner. When the electromagnetic environment around glide path antenna is worse, the difference of TCH and RDH is even larger, and even cannot meet the requirements for safe operation. This article study the calculation method of the actual measured RDH, revealing fundamental reason of the difference, and further explaining the importance of protecting the electromagnetic environment of the instrument landing system, .providing analysis ideas and solutions for equipment construction and maintenance engineers.

1. Introduction

In recent years, with the increase of airports, there are some cases where the threshold crossing height (TCH) [1-2] cannot support the operational requirements of Instrument Landing System (ILS). TCH is a key parameter during approach and landing stage. The parameter is beyond 15m ~ 18m cannot pass the flight inspection, the consequence is ILS category I approach cannot be approved. That means the pilot can only fly visually during the final approach phase. The effect is bad.

TCH is the height of the glide path above the threshold of the runway. Internationally, Annex 10 "Aeronautical Telecommunications" of ICAO and Order 6750-16E "Siting Criteria for Instrument Landing System" of Federal Aviation Administration (FAA) introduce TCH is calculated by longitudinal slope in front of the glide path reflection plane, glide path angle and setback distance of glide path equipment. Civil Aviation Administration of China cited the methods of ICAO and FAA to calculate TCH.

Actually, the TCH released by AIP is not the calculated value, it is the measured TCH got through flight inspection maybe has more accurate professional name reference datum height (RDH) [1-2]. So in some airports RDH is not equal to the calculated TCH value, some even exceed the standard requirements, and the engineer can just adjust the antenna offset and shift of glide path equipment to optimizing the RDH, but extreme cases often have little effect. The underlying reason is TCH is calculated by geometric formula, but RDH is derived by computing the glide path in flight inspection between points “A” and “B” and by projecting an extension of this glide path through the threshold. Point "A" on the ILS glide path measured along the extended runway centre line in the approach direction a distance of 7500 m from the threshold. Point “B”. A point on the ILS glide path measured along the extended runway centre line in the approach direction a distance of 1050 m from the threshold, show as Figure 1.

Content from this work may be used under the terms of the Creative Commons Attribution 3.0 licence. Any further distribution of this work must maintain attribution to the author(s) and the title of the work, journal citation and DOI.
Published under licence by IOP Publishing Ltd

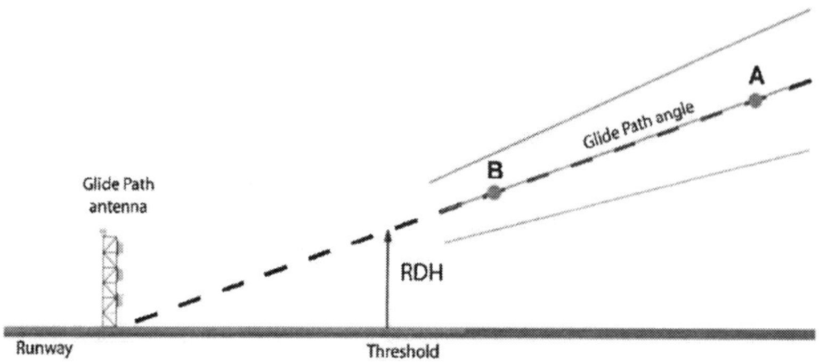

Figure 1. relationship between RDH and points "A" and "B"

2. Calculation of TCH

According to "Electromagnetic environment requirements for aeronautical radio navigation stations"(GB6364-2013) of China, the glide path antenna setback distance can be decided by these parameter include glide path angle, TCH and longitudinal slope in front of the glide path reflection plane.

$$D = \frac{TCH+Y}{\tan(\theta+\alpha)} \tag{1}$$

Glide path antenna setback distance is D, glide path angle is θ, longitudinal slope in front of the glide path reflection plane is α, the vertical height of the runway threshold above P' is Y, when the sideway slope is less than 1%, it is not necessary to introduce the Y. Glide path antenna siting figure for sloping runway show as Figure 2[3].

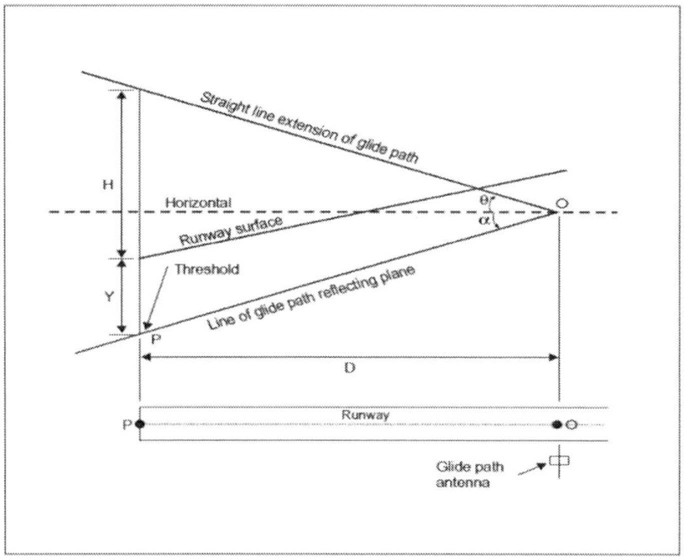

Figure 2. Glide path antenna siting figure for sloping runway

One airport is designed α is 4‰ downslope, glide path angle θ is 3°, D is 292, according to formula 1, the calculated TCH is 16.5 m.

3. Calculation of Reference datum height (RDH)

In order to calculate the RDH, the specified portion of the descent path, called ILS Zone 2, is used between ILS Point A (4 NM from threshold) and ILS Point B (3500 feet from threshold). The Best Fit Straight Line (BFSL) [2] must be fitted by measured glide path, RDH is the line extends down to the threshold. The measured glide path and BFSL show as Figure 3 is got by flight inspection.

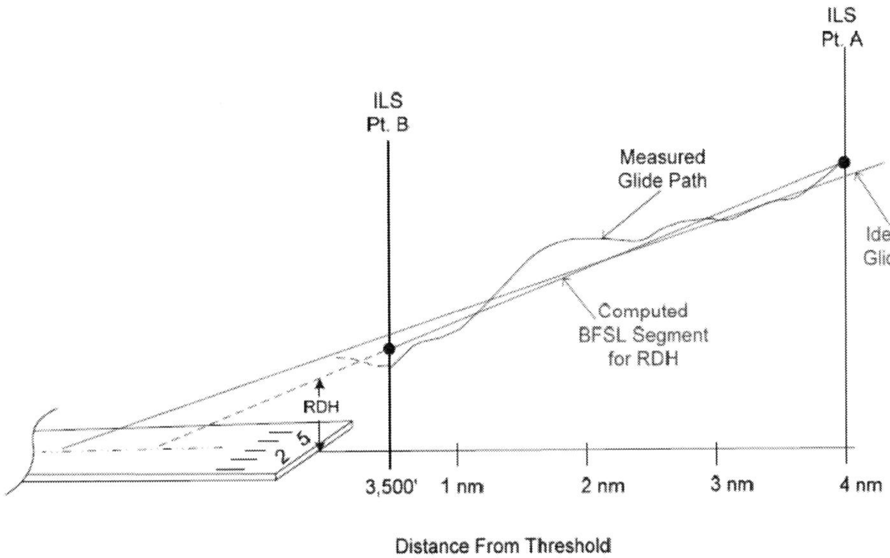

Figure 3. Best-Fit-Straight-Line and RDH

How to get the BFSL? Every glide path have different DDM value, obviously, the best choice is least square method.

3.1 least square method

The method of least squares assumes that the best-fit curve of a given type is the curve that has the minimal sum of the deviations squared (least square error) from a given set of data. Suppose that the data points are (x_1, y_1), (x_2, y_2), ... , (x_n, y_n) where x is the independent variable and y is the dependent variable. The fitting curve f(x) has the deviation (error) d from each data point, i.e., $d_1=y_1-f(x_1)$, $d2=y_2-f(x_2)$ … $d_n=y_n-f(x_n)$, show as Figure 4.

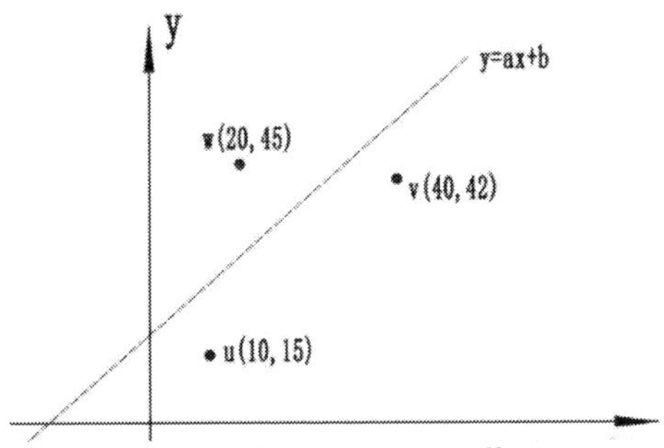

Figure 4. Least squares diagram

According to the method of least squares, the best fitting Line has the property show as formula 2 , formula 3 and formula 4.

$$b = \frac{\Sigma(X_i-\bar{X})(Y_i-\bar{Y})}{\Sigma(X_i-\bar{X})^2} \tag{2}$$

$$a = \bar{Y} - b\bar{X} \tag{3}$$

$$\bar{Y} = bX + a \tag{4}$$

3.2 parameter of flight inspection

There are many parameters during flight inspection. For glide path antenna, setback distance, sideway distance, glide path elevation, threshold elevation can be got, For aircraft, the flight distance and height of aircraft relative to glide path antenna can be got, show as Figure 5 [4].

IOP Conf. Series: Materials Science and Engineering **887** (2020) 012004 doi:10.1088/1757-899X/887/1/012004

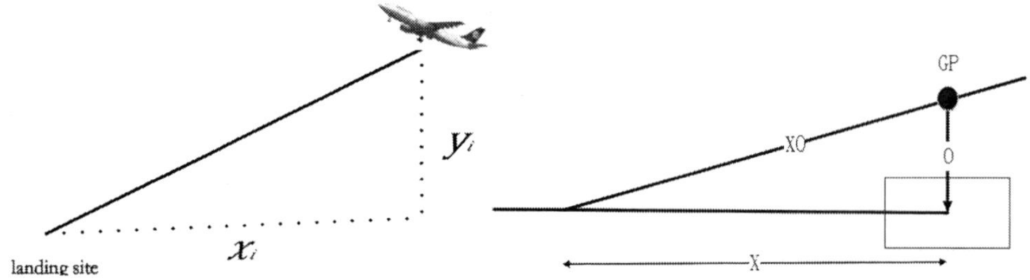

Figure 5 location and height of aircraft and glide path antenna

According to formula 2, x_i is distance between of landing point to aircraft, y_i is flight height of aircraft. Because glide path antenna is sited on side of runway, actually X in formula 2, 3 and 4 can be replaced by X_{oi} that could be calculated from formula 3

$$X_{Oi} = \sqrt{X_i^2 + O^2} \tag{5}$$

$$Y_i = \tan\theta \times X_{oi} \tag{6}$$

$$\overline{X} = \frac{\sum_1^n X_{oi}}{n} \tag{7}$$

$$\overline{Y} = \frac{\sum_1^n Y_i}{n} \tag{8}$$

$$x_i = X_{oi} - \overline{X} \tag{9}$$

3.3 RDH Calculation

One airport the glide path antenna site is ideal, down slope in front of glide path antenna is 4‰, lateral slope from runway to glide path antenna is less than 1%. Setback distance is 293 m from threshold, lateral distance is 120 m from runway centre line and glide path angle is 3°. The calculated TCH is 16.5 m according to formula 1, but the measured RDH by flight inspection and simulation is 16.79 m, shown as Figure 6.

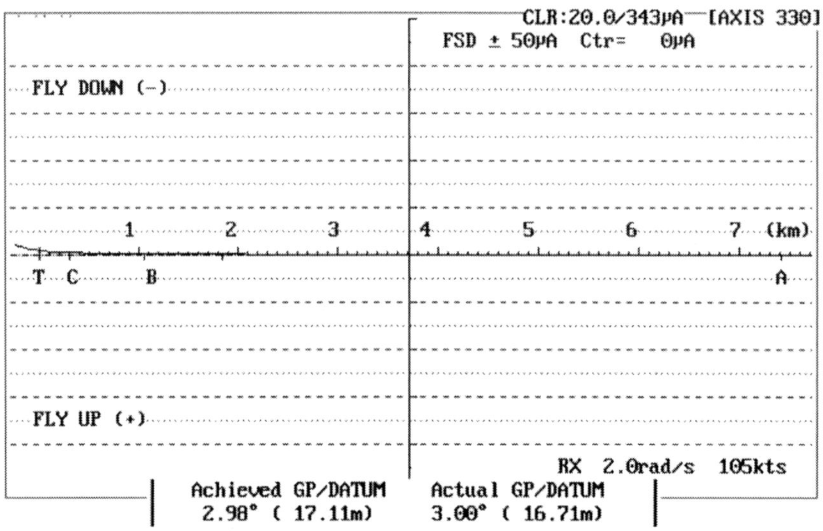

Figure 6 RDH by simulation

In actual flight inspection, the distance and height of aircraft, location of glide path antenna is definite, so all parameter can be got ease show as Table 1.

Table 1. parameter for calculation RDH(ft)

Sample	θ_m(°)	X	X_0	Y	x
Pt "A"	3.00048	24600	24603	1290	10496
1	3.00048	23616	23619	1238	9512
2	3.00048	22632	22635	1186	8528
3	3.00048	21648	21652	1135	7544
4	3.00048	20500	20504	1075	6396
5	3.00048	19516	19520	1023	5412
6	3.00048	18532	18536	972	4428
7	3.00048	17548	17552	920	3444
8	3.00048	16564	16569	868	2460
9	3.00048	15580	15585	817	1476
10	3.00048	14596	14601	765	492
11	3.00048	13612	13618	714	-492
12	3.00048	12628	12634	662	-1476
13	3.00048	11644	11651	611	-2460
14	3.00048	10660	10667	559	-3444
15	3.00048	9512	9520	499	-4592
16	3.00048	8528	8537	447	-5576
17	3.00096	7544	7554	396	-6560
18	3.00096	6560	6572	345	-7544
19	3.00096	5576	5590	293	-8528
20	3.00144	4592	4609	242	-9512
Pt "B"	3.00144	3608	3629	190	-10496

The slope of the BFSL is 0.0524 according to formula 3, corresponding to 2.998° glide path angle(θ_b). The correspondingly RDH is 16.79 m that very close to flight inspection and simulation value.

4. RDH and site environment

Throughout the calculation process, the RDH is related to the aircraft flight height, the aircraft flight height is related to the glide path structure, which is closely related to the site environment [2, 5]. So, electronic magnetic environment around glide path antenna is very important, same longitudinal and transverse slopes for same glide path antenna. A cabin is 260 meters in front of the glide path antenna, 22 meters to the side, length is 4 meters and height is 4 meters, the flight inspection and simulation RDH is 16.82 m which is big different, show as Figure 7

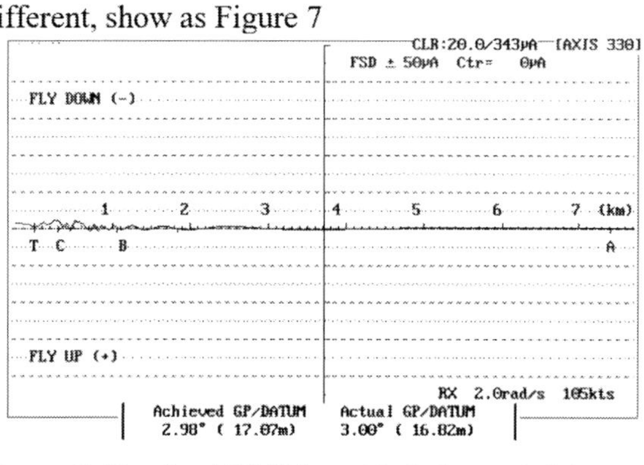

Figure 7. Simulated RDH for cabin in front of glide antenna

5. Conclusion

The RDH verified by actual flight inspection has always been inconsistent with the TCH calculated by geometry terrain, and even the measured RDH does not meet the requirements for airport safe operation.

The main reason is because the electromagnetic environmental protection area around the glide antenna is not done well. At present, there are too many airports built at high speed in China, the protection area in front of glide antenna are not handled well. At commissioned flight inspection, all parameter (including RDH) are good, as the time goes, rainwater washed away and the protection area in front of settlement severely, which resulted in the deterioration of the glide structure, which affected the RDH. As a result, the equipment engineer cannot meet the equipment operation requirements no matter how to debug equipment.

During constructed the airport, especially in the construction and levelling of the site in front of glide antenna, the site should be designed and constructed in strictly accordance with the requirements of "Electromagnetic environment requirements for aeronautical radio navigation stations"(GB6364-2013) and "Specification for aeronautical communication navigation and surveillance station siting criteria—Part 1:Navigation" (MH / T4003.1-2014) to compact and secured to ensure the equipment continued safe operation.

Acknowledgments
First of all, I would like to extend my sincere gratitude to my director Yang Xiaojia, for his instructive advice and useful suggestions on my thesis. I am deeply grateful of their help in the completion of this thesis. I am also deeply indebted to all the other tutor in my ILS studies for their direct and indirect help to me. This work was supported by National Key Research and Development Project (Grant no. 2018YFC0809505).

References

[1] ICAO Annex 10, Aeronautical telecommunications Volume 1 Radio Navigation Aids (2018).
[2] FAA, Siting Criteria for Instrument Landing System(2014)
[3] GB6364, Electromagnetic environment requirements for aeronautical radio navigation stations (2013)
[4] FAA, Assessment of the effectiveness of the RDH/ARDH evaluation methodology for the ILS glide slope (2008)
[5] CAAC, Specification for aeronautical communication navigation and surveillance station siting criteria—Part 1: Navigation" (2014)

Research on Evolution Process and Control Method of External Shapes in the Life-cycle of Modern Civil Jets

Ding Jin and Yue Wang

Shanghai Aircraft Design and Research Institute, Shanghai, China

E-mail: jinding@comac.cc

Abstract. External shape is a key property of aircraft. With the wide application of digital technology, various external shapes are produced in different states and different uses in the life-cycle of modern civil jets. If these external shapes are not clearly distinguished, they might be used improperly. Therefore, in this paper, various shapes in the full life-cycle of modern civil jets are defined, and the evolution process of these external shapes is researched. Based on the control of these external shapes, some advanced management methods are put forward, which have been successfully applied in civil jet development in China, and effectively improve the efficiency of civil jet product development.

1. Introduction

Modern civil jet is an extremely complex industrial product. As an important product property of aircraft, external shape is used by all parties concerned. In the life-cycle of modern civil jets, various states and various uses of aircraft external shapes will be produced.

There is a complex evolution relationship between various external shapes. If these external shapes is not clearly defined and the relationship between them is not clarified, these external shapes may be used improperly, which will not only affect the safety and economy of aircraft, but also sometimes even affect the success or failure of the whole project. For instance, in the early sorties of Boeing's B787 aircraft, due to the different definitions of the theoretical shape and tolerance boundary at abutting station of the fuselage between the manufacturers in the United States and Japan, there was a 7mm step difference at the abutting station, which made it difficult for the fuselage to be assembled, so it was necessary to readjust the design, which affected the project progress.

Therefore, it is of great necessity to have a clear definition of various external shapes, clarify the evolution relationship between these shapes, and develop effective control means to ensure that the product shape is always the same as expected.

2. Classification of External Shapes

The full life-cycle of modern civil jets can be generally divided into market phase, concept phase, definition phase, manufacture & test phase, delivery & service phase and recycle phase. From the concept phase to the end of the service phase, the external shape of aircraft will be used all the time.

The external shape used in different phases is different, because the external shape is constantly evolving. The different shapes in different phases includes 1G flight shape, jig shape, engineering shape, manufacturing shape, parking shape, actual flight shape and operation shape, see Figure 1. Some of them are theoretical, the others are practical.

Content from this work may be used under the terms of the Creative Commons Attribution 3.0 licence. Any further distribution of this work must maintain attribution to the author(s) and the title of the work, journal citation and DOI.
Published under licence by IOP Publishing Ltd

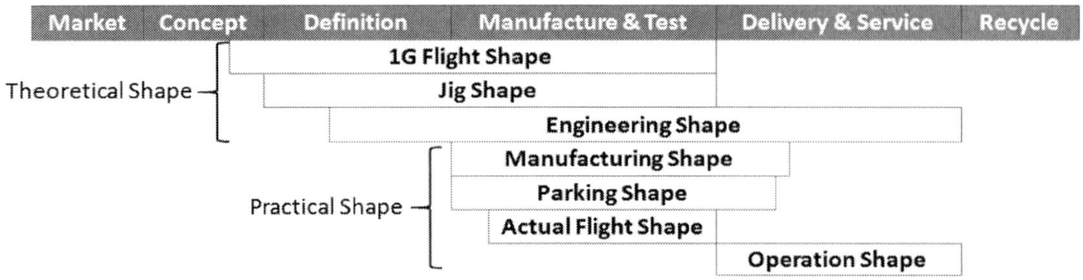

Figure1. Different external shapes in their corresponding application phases

3. Evolution of External Shape

3.1. 1G Flight Shape

1G flight shape refers to the shape of an aircraft during cruise under specific 1G load condition. The specific 1G load condition refers to specific cruise speed, altitude and weight. The 1G flight shape is mainly used for the aerodynamic design and analysis, and its input includes 2D drawing, performance target and other general technical constraints. The 1G flight shape starts from the conceptual design phase, and is finally frozen in the definition phase. During this period, the 1G flight shape is constantly optimized, so it needs to go through several stages of freezing to facilitate the evaluation work of other parties. After 1G flight shape is finally frozen, it will continue to be used until the end of the test phase.

3.2. Jig Shape

Jig shape is the shape for the aircraft theoretical definition without the effects of weight, lift and other aerodynamics. Jig shape plays a critical role in the aircraft definition phase, and it is the basis of the downstream design work. Jig shape is converted from 1g flight shape. Some sections of jig shape and 1G flight shape are the same, such as fuselage and tail. Some parts of the jig shape are obtained from the 1G flight shape transformation, such as wing, high lift devices, flap fairing, pylon and nacelle. The jig shape may include pre-deformation for aerodynamic effect.

Figure 2 is the basic process of wing transition from 1G flight shape to jig shape[1]. For each periodic freeze of 1G flight shape, jig shape needs to be updated, and the transition from 1G flight shape to jig shape is becoming increasingly accurate. For the wing, generally, only torsion is considered in the transformation. With the application of composite materials, the aspect ratio of the wing increases gradually, the elastic deformation increases accordingly, therefore the bending deformation of the wing must be considered.

For the high lift devices, the elastic deformation should be considered when they are opened, and pre-deformation for aerodynamic effect should be added to the jig shape. For jumble jet, due to the fact that the chord length of the wing is relatively longer, the bending deformation outside the front baem and rear beam of the wing should also be considered properly, see Figure 3. For flap fairing, pylon and nacelle, they can be regarded as rigid body connected to the elastic wing, and their shapes in the jig condition are only affected by the wing deformation.

Jig shape should not be changed greatly after preliminary design review, and should be basically frozen after detailed design review. Jig shape will be in use continuously until the end of the manufacturing & test phase, during which little adjustments can still be made to jig shape.

IOP Conf. Series: Materials Science and Engineering **887** (2020) 012005 doi:10.1088/1757-899X/887/1/012005

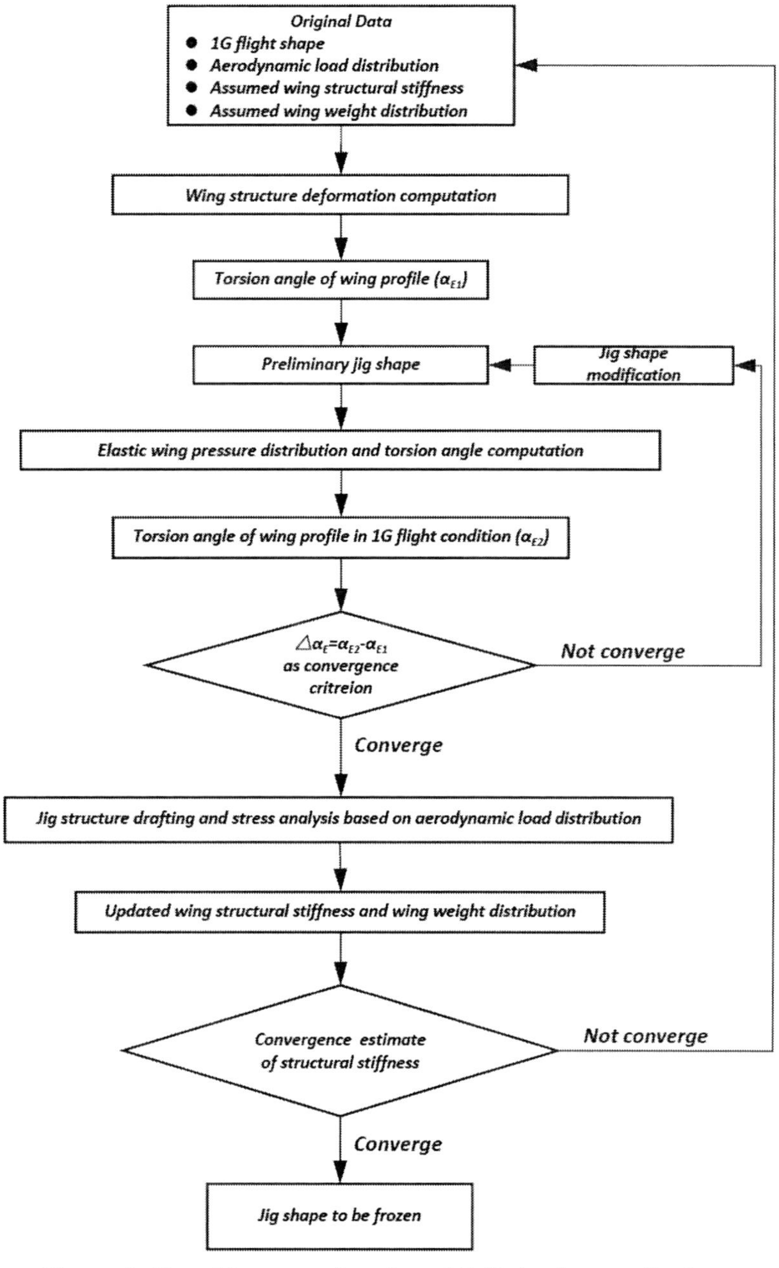

Figure 2. Transition procedure from 1G flight shape to jig shape

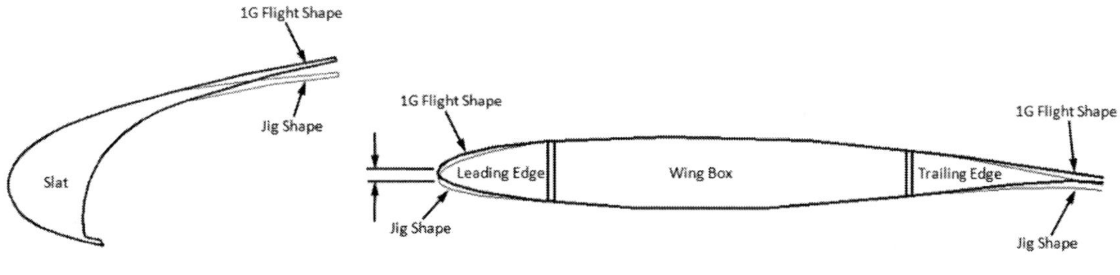

Figure 3. Example for pre-deformation of aerodynamic effect

3.3. Engineering Shape

Jig shape is the ideal smooth shape of aircraft, based on which, the aerodynamic, structural and system teams carry out detailed design, and the ideal smooth shape will become less smooth. The reasons include skin lapping, antenna installation, reinforced structure, etc. And finally the less smooth shape forms the theoretical product shape of the aircraft, which we also call engineering shape. The engineering shape is still theoretical and needs to reflect the detailed shape of product. Compared with jig shape, the added elements of the engineering shape are shown in Table 1.

Table 1. Added elements of engineering shape

Items	Elements
Skin	thickened skin, skin lapping, external reinforcement
Protuberance	antenna, sensor, lampshade, wiper, discharge brush, flow deflector, local bump
Opening/Hole/gap	inlet and outlet of air, liquid outlet, gap of moving parts, grille

From jig shape to engineering shape, the drag will increase, as the external shape will become less smooth. According to engineering calculation and experience, this part of drag accounts for about 10% of the aircraft zero lift drag coefficient[2]. If added elements are not under effective control, the fuel consumption of the aircraft will increase and the competitiveness will become worse. Therefore, in the early stage of the definition phase, the design requirements of the engineering shape should be formulated, including the design requirements of skin thickening, gap, sealing, protuberance and opening. The downstream departments should carry out the detailed design of the aircraft according to the requirements so as to restrict the transformation process from jig shape to engineering shape.

In the definition phase, in order to reduce the drag and optimize the function, the engineering shape is constantly optimized. The optimization means include CFD and wind tunnel test. The engineering shape will be frozen at the end of the test phase. In the manufacturing and delivery phase, engineering shape is the basis of manufacturing. In addition, engineering shape provides support for customer service, such as manual writing and maintenance engineering evaluation.

3.4. Manufacturing shape

After the definition phase, the aircraft is manufactured by the manufacturing department according to the drawings. Due to the influence of manufacturing error and self weight deformation, the actual shape will change further, which is called manufacturing shape. Manufacturing shape is the real shape of an aircraft, which contains various errors. For these errors, the corresponding tolerance control requirements should be provided to control manufacturing deviation. See Table 2 for basic elements of tolerance requirements. For various tolerance requirements, the aircraft shape with tolerance range can be created on the basis of engineering shape, which is used for tolerance analysis and decomposition, and is also convenient to evaluate the out of tolerance.

Table 2. Basic elements of tolerance requirements

Items	Elements
Partial tolerance	profile , waviness, roughness, step, gap, fastener flushness
Overall tolerance	twist, incidence, dihedral, sweep, symmetry

The manufacturing shape is obtained by digital measurement technology in the jig state or jacking state, which is used to evaluate whether the manufacturing shape meets the tolerance requirements. Limited by measuring instruments, the manufacturing profile data are almost in the form of measuring points. For each aircraft, a corresponding manufacturing shape will be generated for deviation evaluation. The manufacturing shape is generated in the manufacturing stage, and some data of the manufacturing shape will also be used as a datum for deformation comparison in the customer service

IOP Conf. Series: Materials Science and Engineering **887** (2020) 012005 doi:10.1088/1757-899X/887/1/012005

phase, such as leveling and symmetry datum data, RVSM area surface data. The manufacturing shape will be permanently kept as files of each aircraft.

3.5. Parking shape

After an aircraft is manufactured, it will deform under the influence of gravity in the landing gear support state. At this time, the aircraft shape is called parking shape. The parking shape is also obtained by measurement. Generally, only some areas that need to be concerned are measured, so parking shape is not a complete aircraft external shape. The parking shape is mainly used to optimize structural parameters and prepare customer service manual. For instance, the ground clearance of some components under different aircraft weight needs to be provided in the structural repair manual or recovery manual. The parking shape is generated after the first aircraft is manufactured and lasts until the early delivery phase.

3.6. Actual Flight Shape

The actual flight shape is the external shape obtained by measuring in flight state, similar to the parking shape. The actual flight shape is not a complete aircraft shape, only some areas that need to be concerned are measured, such as the wing deformation, steps of opening and door, gaps around moving parts, etc. The purpose of measuring actual flight shape is to check the design results of jig shape and 1g flight shape for further optimization. The actual flight shape is generated after the first flight and lasts until the end of the test phase.

3.7. Operation Shape

The external shape of the aircraft put into airline operation is called operation shape. The operation shape starts from the delivery to the customer and continues to be used until it is scrapped and recycled. The operation shape is not a fixed shape, as the aircraft changes from new to old, the operation shape will gradually become not ideal, and this process is called deterioration. The main causes of deterioration are natural aging, wrong operation, improper maintenance and changes due to repairs. Deterioration mainly affects the drag, resulting in increased fuel consumption. For example, the fuel consumption increment caused by the resistance of a high-density paint peeling over a $1m^2$ area of leading edge of wing exceeds 2500 US gallon oil per year for single channel aircraft. Based on economic considerations, the operation shape shall be continuously concerned, items to be concerned include poor gap sealing, paint peeling, external patches, mismatched doors, etc. Whether to rectify these items immediately requires a trade-off between maintenance cost and fuel consumption increment. The trade-off is based on a large number of theoretical research data from CFD method or wind tunnel test and numerical performance audits data from airlines[3].

In most cases, only visual inspection or rough measurement shall be conducted for the operation shape, and necessary maintenance and records shall be made. In some special cases, it is necessary to obtain local data of operation shape through accurate measurement, such as waviness of static port area, misrigging of moving surface, deformation of main parts after overloading, which will be used for safety and airworthiness evaluation and recorded permanently.

3.8. Other Shape

In the life-cycle of modern civil jets, there are also some special shapes, such as CFD & wind tunnel test shape, flight test shape, sales & display shape, which can be obtained on the basis of the shape in Figure 1. These shapes are not in the same dimension as the shape in Figure 1, so this paper will not discuss them.

4. Control Methods

4.1. Requirement Management Based on System Engineering

With more and more advanced technologies applied in the advanced civil jet development, the system integration plays more and more important role in the aircraft development industry. Due to this, a new design method, called Systems Engineering, is formed and studied, and finally achieves great evolvement [4].

Research on the Systems Engineering method starts in the second half of 20th century. Under the guidance and promotion of Mr. Tsien Huseshen, this method has been applied extensively, and great achievements had been obtained[5].One of the core elements of systems engineering method is requirements definition, which makes it different from the conventional design principle. The requirements are divided into three levels, they are aircraft level, system level and items level.

Each kind of external shapes has corresponding requirement levels, as shown in the Figure 4. The requirements of different external shapes at different levels may include: airworthiness regulations, aircraft design requirements and objectives, aerodynamic smoothness requirements, structure and strength requirements, space requirements, manufacturing requirements, service requirements, etc. These requirements shall be decomposed and transmitted to the downstream according to the corresponding requirements levels of different external shapes.

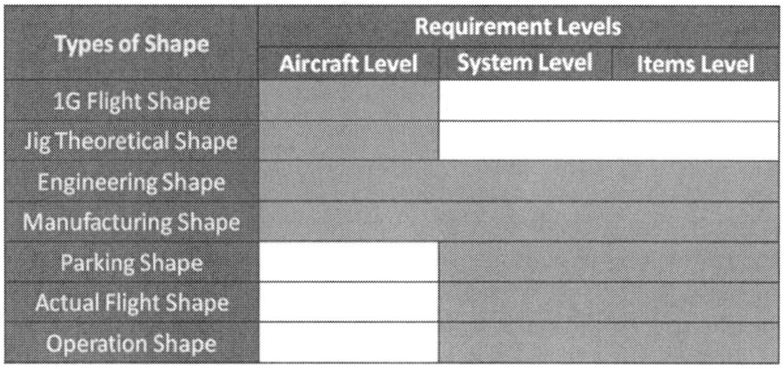

Figure 4. Requirement levels of different external shapes

Whether the requirements are met or not needs to be verified by some means, the requirement verification method of theoretical external shape is inspection and review, meanwhile, the requirement verification method of practical external shape is mainly measurement and comparison. In some cases, the practical external shape can provide reference for theoretical external shape requirement verification.

A series of requirement management activities including requirements capture, requirements definition, requirements validation, requirements decomposition, requirements transfer and requirements verification ensure the aircraft shape meets expectation.

4.2. Digital Mockup of Full Life-cycle

In 1990, Boeing first fully adopted 3D digital design technology in the Boeing 777 aircraft project. The world's first fully digital defined mockup has been completed through paperless design using computer and software[6]. Using digital mockup can greatly improve manufacturing quality, reduce the cost and shorten the project period. Airbus A380 aircraft project extends greatly the application range of digital mockup from design to the full life-cycle of aircraft, such as sales, manufacturing and customer service.

The application of digital mockup in the full life-cycle of modern civil jets has become the mainstream trend. Through the full life-cycle and quantitative maturity of digital mockup management, the scope and depth of digital mockup management can be expanded, and the coordination and continuity of model data can be improved[7]. At the same time, it can ensure that the activities of engineering design, manufacturing and assembly, flight test and customer service are organized and planned, and the process and results are effective, correct and traceable.

Based on the same purpose, the concept of external shape mockup is proposed. External shape mockup is a series of models generated in the process of aircraft shape transforming from theoretical shape to practical shape. It is a part of the full life-cycle digital mockup, as shown in Figure 5.

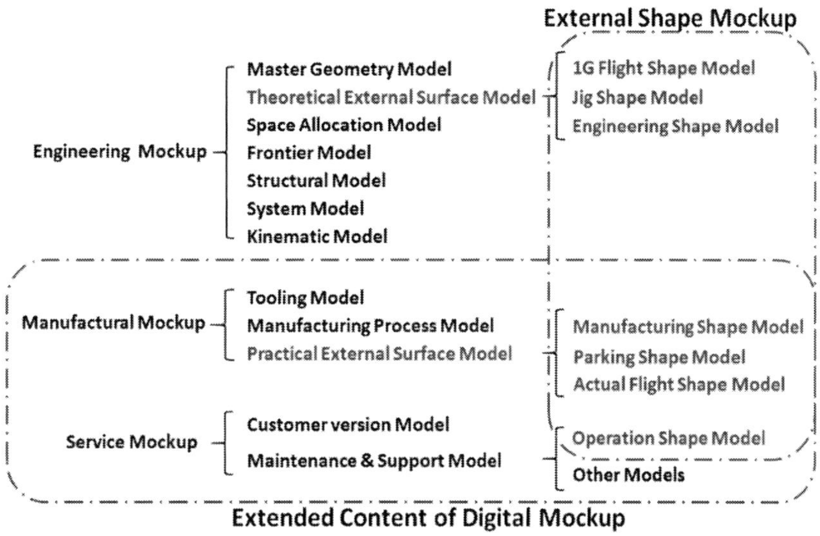

Figure 5. Full Life-cycle Mockup

The functions of external shape mockup mainly includes:
a) the evolution process control and record of the full life-cycle external shape;
b) access management of the external shape data;
c) the configuration difference comparison of different external shapes;
d) historical data trace of the external shapes in different phases.

4.3. Detailed Database

For manufacturing shape and operation shape, corresponding files shall be established for each aircraft to form databases, so as to grasp the shape status of each aircraft. The databases are part of the external shape mockup.

In the manufacturing phase, a manufacturing shape database system should be established to record all information of manufacturing shape, including sorties, date, measurement data, out of tolerance information, repair results or concession details, which are used for historical traceability, airworthiness safety review and documents of delivery configuration. Furthermore, the manufacturing process that needs to be improved can be easily found through the statistics of manufacturing shape data.

In the service phase, an operation shape database system shall be established. Through this system, users can report external shape related problems. The system can record the reported information and feedback maintenance suggestions. The information recorded in the database includes registration number, date, problem description, measurement data and maintenance suggestion.

5. Conclusion

To meet the needs of modern civil jet development, the evolution process of external shape in the life-cycle of modern civil jets has been researched. Based on the control of external shape, advanced management methods have been proposed. The conclusions are as follow:

a) The use of external shape is throughout the life-cycle of modern civil jet, meanwhile, external shape is evolving continuously;

b) Various external shapes are generated during the evolution, they should be effectively controlled to prevent using improperly;

c) Each of different external shapes has been clearly defined, and the evolutionary relationship between these external shapes has been researched;

d) Three external shape management and control methods have been put forward. The methods are requirement management based on system engineering, digital mockup of full life-cycle and detailed databases, which have been successfully applied in civil jet development in China.

References

[1] Fang B 1997 *Aircraft Aerodynamic Layout Design* Aviation Industry Press Beijing 1142–44
[2] Francois D, Miriam L and Alexis R 2009 Performance and Design of the Airbus A320 Analysis of a Subsonic Aircraft *Fundamental of flight* **SD2601**
[3] Robert S, Edward B, David C, Raymond C, Christopher M and C.P.van Dam 2013 Realistic Leading-Edge Roughness Effects on Airfoil Performance 31st *AIAA Applied Aerodynamics Conference* **2013-2800**
[4] Scott J writing, Qian Z and Zhao Y translating 2016 *Systems Engineering for Commercial Aircraft–A Domain-Specific Adaptation* Shanghai Jiao Tong University Press Shanghai 65–93
[5] SAE. ARP4754A 2010 *Guidelines for Development of Civil Aircraft and Systems* 22–28
[6] Fan Y, Mei Z and Tao J 2011 *Aircraft Digital Manufacturing Engineering* Aviation Industry Press Beijing 173–74
[7] Xu J 2018 Development of Digital Mockup in the Aircraft Digital Cooperation *Civil Aircraft Design & Research* **130** 87–91

IOP Conf. Series: Materials Science and Engineering **887** (2020) 012006 doi:10.1088/1757-899X/887/1/012006

Study on Outer Rotor FSCW Permanent Magnet Machines for Unmanned Aerial Vehicle Electric Green Taxiing Systems

Hua Xu, Yaohua Hu, and Jinpeng Yang

Aviation Key Laboratory of Science and Technology on Aero Electromechanical System Integration, Nanjing Engineering Institute of Aircraft Systems, Nanjing 211106, China

Email: huyaohua413@126.com

Abstract. This paper considers the possibility of adopting an outer rotor fractional slot concentrated winding (FSCW) permanent magnet synchronous machine (PMSM) for Unmanned Aerial Vehicle Electric Green Taxiing (EGT) application. The electromagnet performance such as no-load air-gap flux density, flux linkage distribution, electromotive force (EMF), average torque, torque ripple, and iron losses are studied by finite-element analysis (FEA). The results show the iron loss is mainly effected by the machine frequency. The unmanned aerial vehicle EGT structure are introduced, which includes the outer rotor FSCW PM machine, the reducer, the clutch, and the tire.

1. Introduction

The electric machines have more and more attractive for the aircraft traction during taxi phase [1]-[3]. The Unmanned Aerial Vehicle Electric Green Taxiing (EGT) is try to increase the service life of the engine. There are several types of electric machines that can be used for EGT applications such as the interior permanent magnet machines and surface-mounted permanent magnet machines[4].

Fraction-Slot Concentrated-Winding (FSCW) Permanent Magnet (PM) synchronous machines have been gaining interest over the last few years. This is mainly due to the several advantages, such as high efficiency, short end turns, and low cogging torque [5]-[6]. The outer rotor PM machines are widely used for the traction applications due to its high torque density and power density.

This paper presents a possibility of adopting an outer rotor fractional slot concentrated winding (FSCW) permanent magnet synchronous machine (PMSM) for Unmanned Aerial Vehicle Electric Green Taxiing (EGT) application. The electromagnet performances are studied by finite-element analysis (FEA), the FEA results are analysed, and the unmanned aerial vehicle EGT structure are introduced.

2. Machine Specification

Table 1 shows the machine specification The power rating of the outer rotor PM machine is 1.6 kW continuous, the over load torque is 6.5 Nm, the rated speed is 2400 r/min, the maximum speed is 7800 r/min, and the reduction ratio of the gears is 70. The outer rotor PM machine and gears are installed on main land wheel. The structural diagram of the outer rotor FSCW PM machine is shown in Figure 1. The stator has the iron core and three phase windings, and the rotor has the magnets, and iron.

Table 1. Machine specification

Content from this work may be used under the terms of the Creative Commons Attribution 3.0 licence. Any further distribution of this work must maintain attribution to the author(s) and the title of the work, journal citation and DOI.
Published under licence by IOP Publishing Ltd

Design Parameters	Value	Design Parameters	Value
Phase	3 phases	Stack length	25 mm
Maximum rotational speed	7800 r/min	Thickness of steel sheet	0.2 mm
Rated rotational speed	2400 r/min	Inverter	f_c= 20 kHz, V_{DC}= 270 V
Phase current (RMS)	21 A	Number of stator teeth	24
Torque	> 6.5 Nm	Number of poles	20
Air gap length	1 mm	Magnets (Samco)	H_c = -830000 A/m
Diameter of stator and rotor	128 mm, 146 mm		B_r = 1.13 T

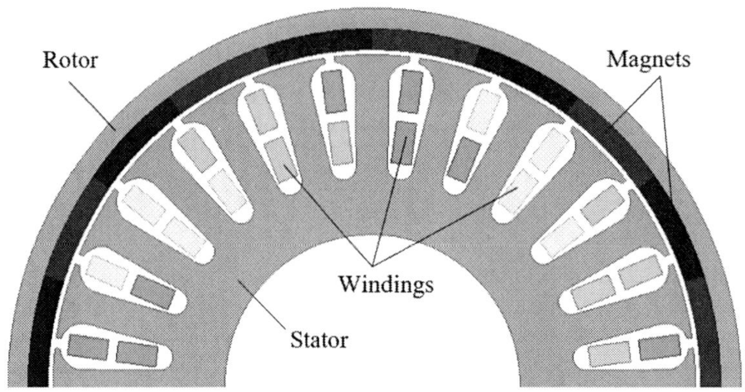

Figure 1. Proposed outer rotor PM machine for electric green taxiing systems.

3. Study on Electromagnet Performance

3.1. No-load air-gap flux density

The no-load air-gap flux density and flux linkage distribution are shown in Figure 2. The flux density of stator teeth is 1.6 T.

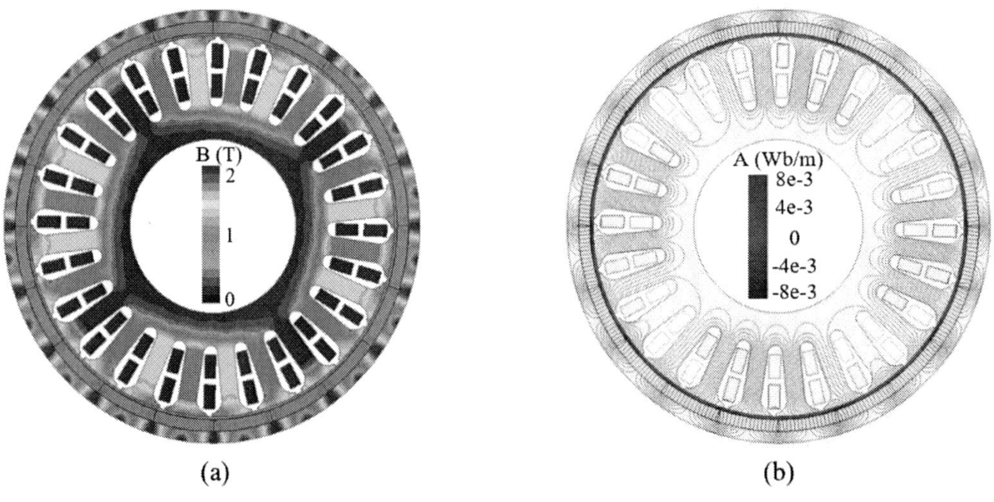

Figure 2. No-load air-gap flux density of the PM machine. (a) Flux density. (b) Flux linkage.

The FEA predicted no-load air-gap flux density distribution of the outer rotor FSCW PM machine is shown in Figure 3. The total harmonic distortion (THD) of air-gap flux density is 27.5 %.

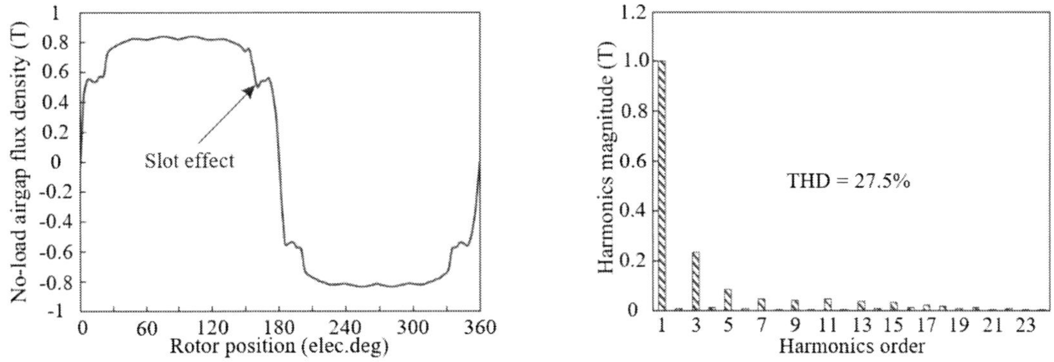

(a) (b)

Figure 3. Air-gap flux density distribution of the outer rotor FSCW PM machine. (a) Waveform of air-gap flux density. (b) Spectral components.

Figure 4 shows FEA predicted the electromotive force (EMF) of the outer rotor FSCW PM machine. The sinusoidal of the EMF is good, the THD of EMF is only 3.15 %.

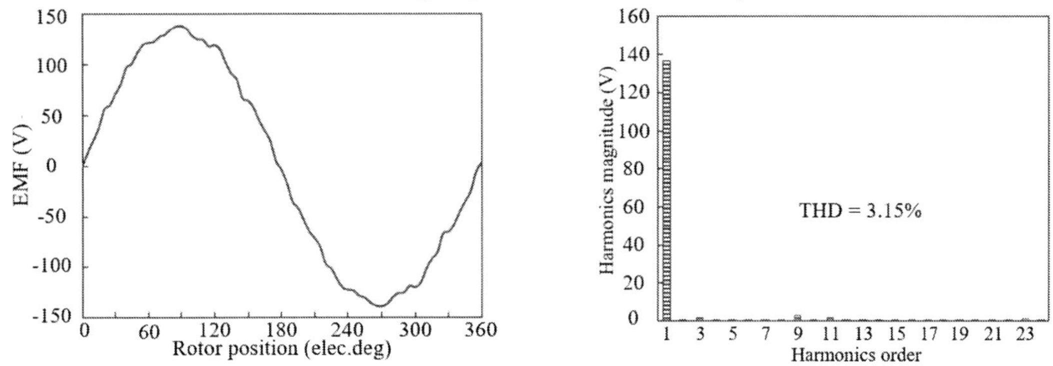

(a) (b)

Figure 4. EMF of the outer rotor FSCW PM machine. (a) Waveform of EMF. (b) Spectral components.

3.2. Average torque and torque ripple

Figure 5 shows the torque waveform of the outer rotor FSCW PM machine when the speed is 2400 r/min and the current is 21 A.

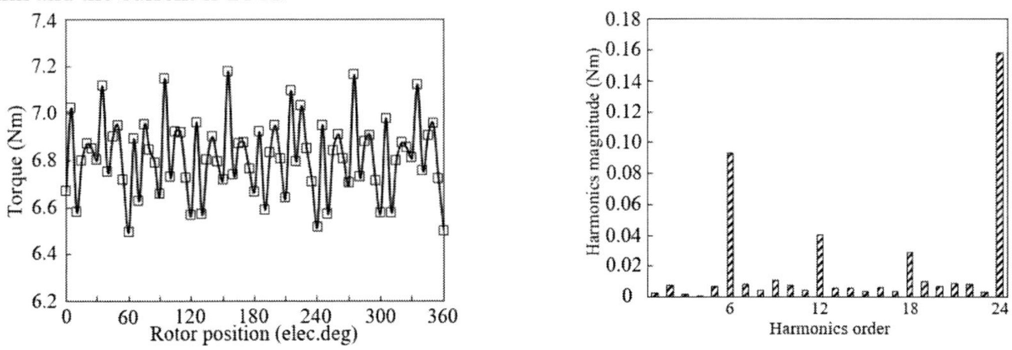

(a) (b)

Figure 5. Torque waveform of the outer rotor FSCW PM machine when the speed is 2400 r/min and the current is 21A. (a) Torque waveform. (b) Spectral components.

Figure 6 shows the torque waveform of the outer rotor FSCW PM machine when the speed is 7800 r/min and the current is 7 A.

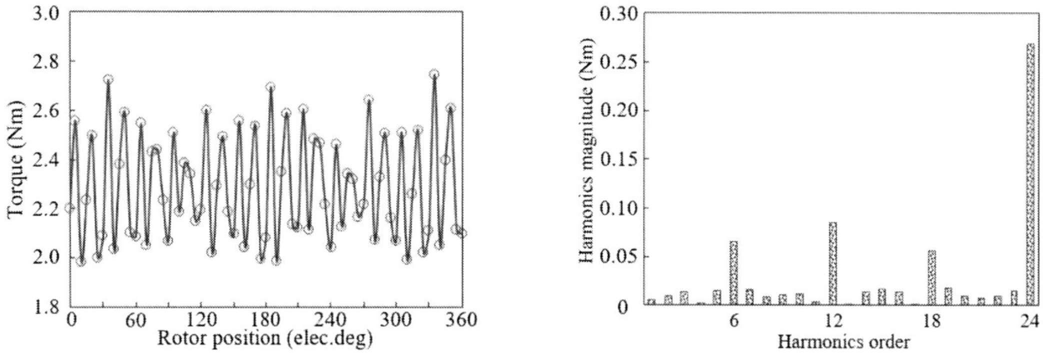

(a) (b)

Figure 6. Torque waveform of the outer rotor FSCW PM machine when the speed is 7800 r/min and the current is 7 A. (a) Torque waveform. (b) Spectral components.

The torque ripple can be expressed as follows

$$T_{\text{ripple}} = \frac{T_{\max} - T_{\min}}{T_{ave}} \times 100 = \frac{\Delta T_{peak-peak}}{T_{ave}} \times 100 \quad (1)$$

Table 2 shows the average torque, peak to peak torque, and torque ripple of two machines. Where $T_{\max}$ is maximum torque, $T_{\min}$ is minimum torque, T_{avg} is average torque, $\Delta T_{\text{peak-peak}}$ is peak to peak torque, T_{ripple} is torque ripple.

Table 2. Torque Components

Parameter	Speed = 2400r/min I = 21 A	Speed = 7800r/min I = 7 A
6*th* (Nm)	0.093	0.065
12*th* (Nm)	0.04	0.085
18*th* (Nm)	0.029	0.056
24*th* (Nm)	0.158	0.268
T_{max} (Nm)	7.17	2.76
T_{min} (Nm)	6.49	1.98
T_{ave} (Nm)	6.82	2.29
$\Delta T_{peak\text{-}peak}$ (Nm)	0.68	0.76
T_{ripple} (%)	10.05	33.29

3.3. Iron losses

Iron losses are the main component of losses in the outer rotor FSCW PM machine under high-speed operation. Iron losses include the core loss and the eddy current loss. The iron loss with the Fourier transformation of the flux-density waveform at each point can be expressed as

$$W_c = W_e + W_h = \sum_{n=1}^{n_{\max}} k_{\text{en}}(nf_1, B_n) \cdot (nf_1)^2 + \sum_{n=1}^{n_{\max}} k_{\text{hn}}(nf_1) \cdot (nf_1) \cdot B_n^{\,a(nf_1, B_n)} \quad (2)$$

where W_e represents the eddy-current losses, W_h represents the hysteresis losses, f_1 is the fundamental frequency, k_{en} is the eddy-current component coefficient, k_{hn} and a are the hysteresis component coefficient respectively, and n is the nth harmonic component.

Figure 7 shows the iron loss density distribution of the outer rotor FSCW PM machine at different speeds and currents. When the speed is 2400 r/min, I is 21 A, the iron loss density is low. When the speed is 7800 r/min, I is 7 A, the iron loss density is relatively high. Therefore, the iron loss is mainly effected by the machine frequency.

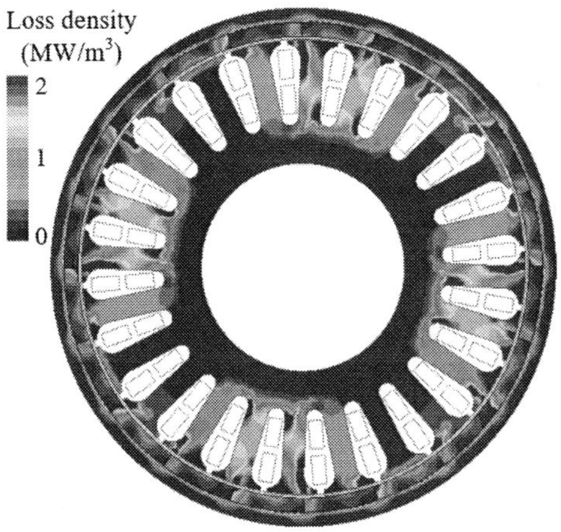

(a)

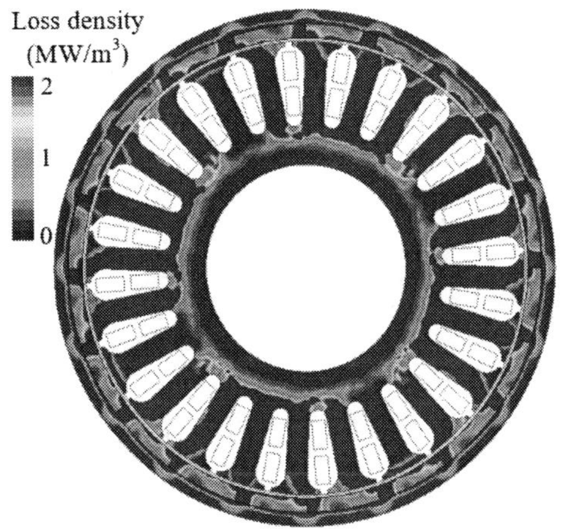

(b)

Figure 7. Iron loss density distribution of the outer rotor FSCW PM machine at different speeds and currents. (a) *Speed* = 2400r/min, I = 21 A. (b) *Speed* = 7800r/min, I = 7 A.

Figure 8(a) shows the iron loss of the outer rotor FSCW PM machine at different speeds and currents. When the speed is 2400 r/min, I is 21 A, the core loss is 87.18 W, the rotor eddy current loss is 25.11 W. When the speed is 7800 r/min, I is 7 A, the core loss is 659.67 W, the rotor eddy current loss is 100.63 W. Figure 8(b) shows the unmanned aerial vehicle electric green taxiing structure, which includes the outer rotor FSCW PM machine, the reducer, the clutch, and the tire. The electric green taxiing structure is compact. The maximum and the minimum speed of the unmanned aerial vehicle at the electric green taxiing phase is 10 km/h and 3 km/h, which corresponding the 7800 r/min and 2400 r/min of the outer rotor FSCW PM machine.

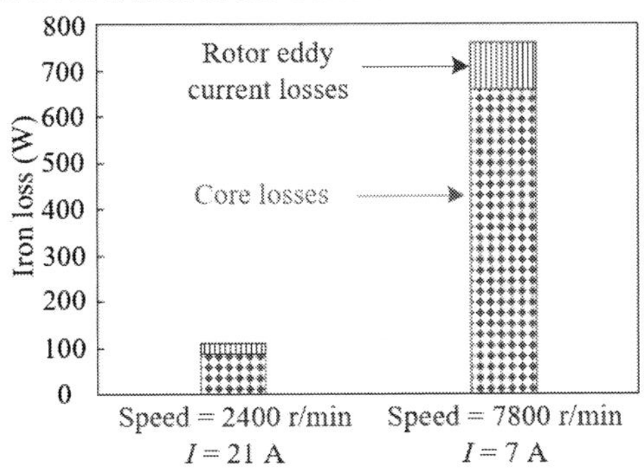

(a)

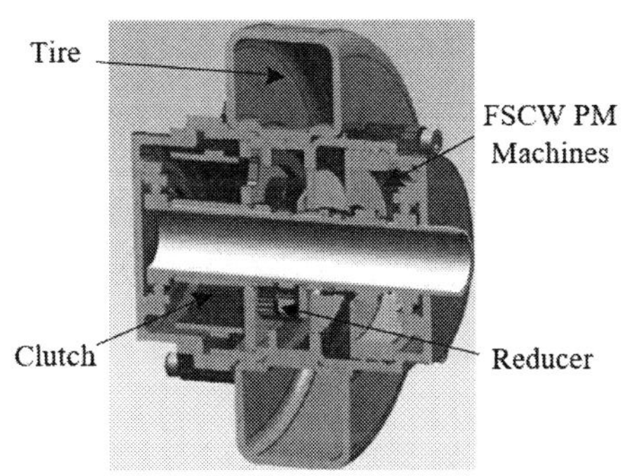

(b)

Figure 8. Iron loss and unmanned aerial vehicle electric green taxiing structure. (a) Iron loss. (b) Unmanned aerial vehicle electric green taxiing structure.

4. Conclusion

This paper presents a feasible scheme with outer rotor fractional slot concentrated winding (FSCW)

PMSM for Unmanned Aerial Vehicle EGT applications. The electromagnet performance such as no-load air-gap flux density, flux linkage distribution, EMF, average torque, torque ripple, and iron losses are studied by FEA. The results show the iron loss is mainly effected by the machine frequency. The unmanned aerial vehicle EGT structure are introduced, which includes the outer rotor FSCW PM machine, the reducer, the clutch, and the tire.

Acknowledgments

This work was supported by the Aeronautical Science Foundation of China under Project 2018ZC09002.

References

[1] A. Teo, K. Rajashekara, J. Hill, and B. Simmers, “Examination of aircraft electric wheel drive taxiing concept,” *The Soc. Automotive Engineers Power Syst. Conf.*, Seattle, WA, 2008.
[2] F. Re, “Assessing environmental benefits of electric aircraft taxiing through object-oriented simulation,” *Soc. Automotive Engineers Int. J. Aerosp.*, vol. 5, no. 2, pp. 503-512, 2012.
[3] H. Oyori and N. Morioka, “Power management system for the electric taxiing system incorporating the more electric architecture,” *The SAE Aero Tech Congress. Exhibition*, Montréal, Canada, 2013.
[4] E. D. Ganev, "Electric Drives for Electric Green Taxiing Systems: Examining and Evaluating the Electric Drive System," IEEE Electrification Magazine, vol. 5, no. 4, pp. 10-24, Dec. 2017.
[5] Shi-Uk C, Jong-Moo K, Dae-Hyun K, et al., “Fractional Slot Concentrated Winding Permanent Magnet Synchronous Machine With Consequent Pole Rotor for Low Speed Direct Drive,” *IEEE Trans. Magn.*, vol. 48, no. 11, pp. 2965-2968, June. 2012.
[6] Tsampouris EM, Beniakar ME, Kladas AG, “Geometry Optimization of PMSMs Comparing Full and Fractional Pitch Winding Configurations for Aerospace Actuation Applications,” *IEEE Trans. Magn.*, vol. 48, no. 2, pp. 943-946, June. 2012.

IOP Conf. Series: Materials Science and Engineering **887** (2020) 012007 doi:10.1088/1757-899X/887/1/012007

Method for Determining Takeoff Weight and Thrust-To-Weight Ratio of Aircraft Variants by Decision Speed with Engine Failed at Takeoff Run

L Kapitanova[1], D Tiniakov[2], L Makarova[3]

[1]Department of Airplane and Helicopter Design, National Aerospace University, Kharkiv, Ukraine
[2]Department of Civil Aviation Engineering, Nanjing University of Aeronautics and Astronautics, Nanjing, China
[3]Department of Helicopter Engineering, Nanjing University of Aeronautics and Astronautics, Nanjing, China

E-mail: Tiniakov_d@nuaa.edu.cn

Abstract. In modern time the aircraft variants creation is main trend in the aviation industry. One the aspect of the aircraft safety is its safety take-off [1]. The aircraft parameters like relative take-off weight and thrust-to-weight ratio can determine many its performance include take-off. Their influence on decision speed during take-off of modified transport category airplanes are explored. The analytic model is simple and high efficiency way for the first approach of airplane`s safety analyses. The problem is solved by analytics modelling airplane motion on runway during take-off. Quantitative estimation of influence of the airplane weight and thrust-to-weight ratio variation on decision speed of continued or aborted take-off with critical engine failure is given on the base of example of heavy cargo airplane. The application of this way for the airplanes safety analysis is important for each types of aircraft. Estimation model is given in dimensionless parameters and so is applicable for any type of transport category airplanes.

1. Introduction

The development process of modifications is widely used in aircraft engineering [2-12]. In the process of designing modifications are changed to obtain required competitive performance, such as: chart "cargo-distance", fuel efficiency, cost of a ton-kilometer of air transportation [2,4,13].

The required competitive performance is achieved by changing the basic parameters of modifications such as their relative takeoff weight and takeoff thrust-to-weight ratio relative value [4,11,14].

Changes to these determining parameters affect virtually on all assemblies and systems of modifications, including takeoff and landing characteristics (TOLC), i. e., on takeoff and climb distance and on one of the most important parameters — the decision speed to stop and continue takeoff (Figure 1). The importance of this parameter is determined not only by keeping the modifications' TOLC close to ones of the basic version of the aircraft, but also by the possibility of aircraft modifications at declared airbases.

Content from this work may be used under the terms of the Creative Commons Attribution 3.0 licence. Any further distribution of this work must maintain attribution to the author(s) and the title of the work, journal citation and DOI.
Published under licence by IOP Publishing Ltd

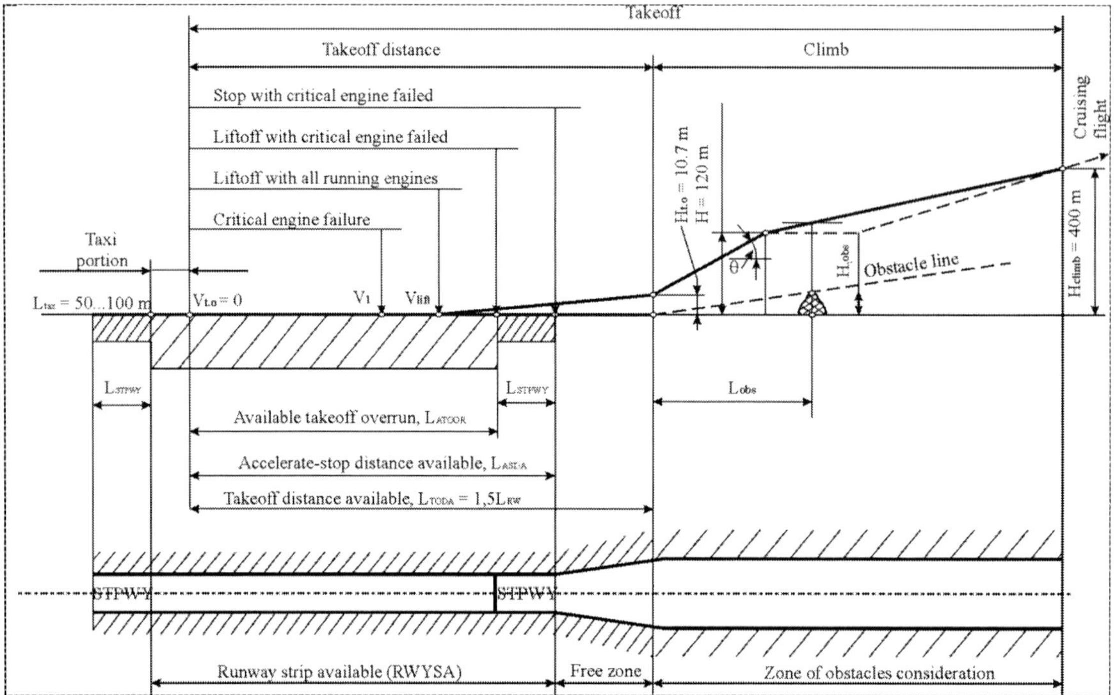

Figure 1. Main Parameters of Take-off Distance and Climb of Transport Category Airplane.

2. Objective

To quantify the influence of the changed parameters $\bar{G}_{t.o}$ and $\bar{t}_{t.o}$ on the value of decision speed V_1 during take-off by modelling the movement of the aircraft on the ground.

3. Solving the Problem

At the takeoff run portion the plane moves along the runway (RWY) at a rate from $V_{t.o} = 0$ to liftoff speed V_{lift}.

When calculating the takeoff run [12,15] we assume that this procedure is performed by using main landing gear. Angle of attack ($\alpha_{t.o}$) and thrust turning angle (φ_T) in the process of takeoff are considered constant that allows to accept $\alpha_{t.o} - \varphi_T \cong 0$, and $\cos(\alpha_{t.o} + \varphi_T) \cong 0$, and $\sin(\alpha_{t.o} + \varphi_T) \cong 0$.

Integral equation of takeoff run length with the assumptions accepted is as follows:

$$L_{t.o} = \frac{1}{2g}\int_0^{V_{lift}} \frac{dV^2}{K_1 t_{t.o} - f_{wheel} - \frac{\rho_0 S_w V^2}{2G_{t.o}}\left(C_{Dt.o} - f_{wheel} C_{Lt.o}\right)} \quad (1)$$

where V_{lift} is liftoff speed; K_1 is factor considering the engine thrust decay by speed and loss in air intakes; $t_{t.o}$ is thrust-to-weight ratio; f_{wheel} is coefficient of rolling friction; ρ_0 is air density at 0 km altitude; S_w is full wing area; $G_{t.o}$ is takeoff weight of airplane; $C_{Dt.o}$ is drag ratio at takeoff; $C_{Lt.o}$ is lift ratio at takeoff. The integration is carried out graphically from $V_{t.o} = 0$ to

$$V_{lift} = \sqrt{\frac{2G_{t.o}}{\rho_0 C_{Llift} S_w}}\,. \quad (2)$$

To evaluate the effect of growth of takeoff weight of the modification compared to the reference plane let us use fairly accurate analytical expression of takeoff run length [11,12,16,17]

$$L_{t.o} \cong \frac{V_{lift}^2}{2g\left(K_1 t_{t.o} - f_{wheel} - \frac{\rho_0 C_{Dt.o} S_w}{6G_{t.o}} V_{lift}^2\right)} \quad (3)$$

where K_1 is a factor considering the engine thrust decay by speed and loss in air intakes.

For turbojet engines with the standard atmosphere $K_1 \cong 0.9$ for $t_H^0 = +30^o C$ and 730 Hg mm $K_1 \cong 0.813$; f_{wheel} is coefficient of rolling friction $0.02 \le f_{weel} \le 0.08$, 0.02 – on dry concrete, 0.04 – on hard solid; $C_{Dt.o}$ is drag coefficient during takeoff run.

If reference and modified airplane takeoff lengths are equal $\bar{L}_{t.o} = L_{t.o} / L_{t.o.b} = 1$ and K_1, f_{weel}, ρ_0, $C_{Dt.o}$, C_{Llift}, S_w are constants, we have an equation

$$\frac{1}{K_1 t_{t.o.b.} - f_{weel} - \frac{\rho_0 C_{Dt.o} S_w}{6G_{t.o.b}} V_{lift.b}^2} = \frac{\bar{V}_{lift}^2}{K_1 t_{t.o.b} \bar{t}_{t.o} - f_{weel} - \frac{\rho_0 C_{Dt.o} S_w}{6G_{t.o}} V_{lift}^2}. \quad (4)$$

Substituting the expression $V_{lift.b}^2$, V_{lift}^2 from (2), we get an equation to assess mass variation of potential modifications

$$\bar{G}_{t.o} = \frac{\frac{K_1}{f_{weel} + C_{Dt.o} / 3C_{Llift}} .t_{t.o.b} .\bar{t}_{t.o} - 1}{\frac{K_1}{f_{weel} + C_{Lt.o} / 3C_{Llift}} .t_{t.o.b} - 1}. \quad (5)$$

Decision speed (V_1) at takeoff run should ensure the safe termination or continuation of the takeoff with a critical engine failed [9,18-20].

Rejected takeoff distance ($L_{RTOD} \le L_{ASDA}$) and required distance of complete takeoff run $\left(L_{RDCTO} \le L_{ATOOR}\right)$ are respectively equal:

$$L_{ASDA} \ge L_{RTOD} = L_{0 \to Vfail} + L_{Vfail \to V} + L_{V \to 0}, \quad (6)$$

$$L_{ATOOR} \ge L_{RDCTO} = L_{0 \to Vfail} + L_{Vfail \to V} + L_{V \to Vlift}, \quad (7)$$

where $L_{0 \to Vfail}$ is takeoff run length with all operative engines from start to the moment of critical engine failure at speed of V_{fail}; $L_{Vfail \to V}$ is takeoff run length with one non-operative engine and with standard operation of the rest up to getting the decision speed; $L_{V \to 0}$ is deceleration segment with non-operative engine from speed V_1, up to full stop; $L_{Vfail \to V}$ is takeoff run length with one non-operative engine and with standard operation of the rest from speed V_1, up to liftoff speed V_{fail}.

To determine V_1 it is required to equal the available rejected takeoff distance (6), reduced by runway overrun length $\left(L_{STPWY}\right)$ to the required complete takeoff run distance (7) we get

$$L_{V \to Vlift} = L_{V \to 0} - L_{STPWY}. \quad (8)$$

With the accepted assumptions we get fairly accurate analytical expression by using integral equation (1).

$$\frac{V_{lift}^2 - V_1^2}{2q\left[K_1\left(1-\frac{1}{n_{eng}}\right)t_{t.o} - f_{weel} - \frac{\rho_0 C_D S_w}{6G_{t.o}}\left(\frac{V_{lift}^3 - V_1^3}{V_{lift} - V_1}\right)\right]} = \frac{K_2 V_1^2}{2q\left[K_1\left(1-\frac{1}{n_{eng}}\right)r_T t_{t.o} + f_{red.weel} + \frac{\rho_0 C_{Dt.o} S_w}{6G_{t.o}} V_1^2\right]} - L_{STPWY}, \tag{9}$$

where n_{eng} is quantity of engines; $C_{Dt.o}$ is drag coefficient during takeoff run; K_2 is coefficient that takes into account the decision time required for a pilot and the time of engagement of braking devices of aircraft; r_T is thrust reverser to engine forward thrust ratio; $f_{red.weel}$ is reduced friction ratio of wheels during takeoff run (mean value).

$$f_{red.weel} = \vartheta\left[\frac{\bar{l}_{NLG} f_{fr} + \bar{l}_{MLG} f_s}{1+\bar{h}_{LG}(f_{fr} - f_s)}\right], \tag{10}$$

where ϑ =0.75—0.95 is coefficient depending on the quality (in particular, the inertia) of anti-skid device of main landing gear wheels. With nose brake wheels $f_{red.weel} = \vartheta \cdot f_{fr}$; f_{fr} is friction coefficient of the main landing gear braked wheels, f_{fr} = (V, RWY coating); $\bar{l}_{NLG} = l_{NLG} / l_{LG}$, $\bar{l}_{MLG} = l_{MLG} / l_{LG}$, $\bar{h}_{LG} = h_{LG} / l_{LG}$ are relative linear dimensions (to wheel base (l_{LG})) of nose (l_{NLG}) and main (l_{MLG}) landing gear offset from aircraft CG, h_{LG} is linear dimension of aircraft CG from RWY plane.

Let us transform the equation (9) in relative values and get the equation (11).

The equation (11) helps to estimate the relative decision speed dependence on takeoff weight relative increase and relative weight-to-thrust of aircraft potential modifications.

Rejected takeoff is performed as standard one up to the moment of failure of critical engine or aircraft systems influencing the takeoff performance. On taking the decision by the pilot the takeoff is rejected and the aircraft is braked up to full stop [5,9,19,21,22]

$\bar{t}_{t.o} = t_{t.o} / t_{t.o.b}$, $\bar{V}_{lift}^2 = V_{lift}^2 / V_{lift.b}^2 = \bar{G}_{t.o} = G_{t.o} / G_{t.o.b}$, $\bar{V}_1 = V_1 / V_{1b}$.

$$\frac{V_{lift.b}^2 \bar{G}_{t.o} - V_{1b}^2 \bar{V}_1^2}{K_1\left(1-\frac{1}{n_{eng}}\right)t_{t.o.b}\bar{t}_{t.o} - f_{weel} - \frac{\rho_0 C_{Dt.o} S_w}{6G_{t.o.b}\bar{G}_{t.o}}\left(V_{lift.b}^2 \bar{G}_{t.o} + V_{1b} V_{lift.b} \bar{V}_1 \sqrt{\bar{G}_{t.o}} + V_{1b}^2 \bar{V}_1^2\right)} =$$
$$= \frac{K_2 V_{1b}^2 \bar{V}_1^2}{K_1\left(1-\frac{1}{n_{eng}}\right)r_T t_{t.o.b}\bar{t}_{t.o} + f_{red.weel} + \frac{\rho_0 C_{Dred} S_w}{6G_{t.o.b}\bar{G}_{t.o}} V_{1b}^2 \bar{V}_1^2} - 2gL_{STPWY}. \tag{11}$$

Known, quite accurate analytical equation for determination of rejected takeoff distance is as follows [9,18,19]:

$$L = \frac{V_1^2}{2g}\left[\frac{1}{K_1 t_{t.o} - f_{weel} - \frac{\rho_0 C_{Dt.o} S_w}{6G_{t.o}} V_1} + \frac{K_2}{K_1\left(1-\frac{1}{n_{eng}}\right)r_T t_{t.o} + f_{red.weel} + \frac{\rho_0 C_{Dred} Sw}{6G_{t.o}} V_1^2}\right]. \tag{12}$$

We transform the equation (12) in relative values $\bar{t}_0, \bar{G}_{0,} \bar{V}_1$ and under condition when

$\bar{L}_{rej.b} = L_{rej} / L_{rej.b} = 1$.

$$L_{rej.b}=\frac{V_{1b}^2\bar{V}_1^2}{2g}\times\left[\frac{1}{K_1 t_{t.o.b}\bar{t}_{t.o}-f_{weel}-\frac{\rho_0 C_{Dt.o}S_w}{6G_{t.o.b}\bar{G}_b}V_{1b}^2\bar{V}_1^2}+\frac{K_2}{K_1\left(1-\frac{1}{n_{eng}}\right)r_T t_{t.o.b}\bar{t}_{t.o}+f_{red.weel}+\frac{\rho_0 C_{Dred}S_w}{6G_{t.o}\bar{G}_{t.o.b}}V_{1b}^2\bar{V}_1^2}\right]. \quad (13)$$

We substitute the second additive component in equation (13) from equation (11), and get relationship of three relative values

$$L_{rej.b}=\frac{1}{2g}x\left[\frac{V_{lift.b}^2\bar{G}_{t.o}-V_{1b}^2\bar{V}_1^2}{K_1\left(1-\frac{1}{n_{eng}}\right)t_{t.o.b}\bar{t}_{t.o}-f_{weel}-\frac{\rho_0 C_{Dt.o}S_w}{6G_{t.o.b}}\bar{G}_{t.o}\left(V_{lift.b}^2\bar{G}_{t.o}+V_{1b}V_{lift.b}\bar{V}_1\sqrt{\bar{G}_{t.o}}+V_{1b}^2\bar{V}_1^2\right)}+\right.$$

$$\left.+\frac{V_{1b}^2\bar{V}_1^2}{K_1 t_{t.o.b}\bar{t}_{t.o}-f_{weel}-\frac{\rho_0 C_{Dt.o}S_w}{6G_{t.o.b}\bar{G}_{t.o}}V_{1b}^2\bar{V}_1^2}+2qL_{STRWY}\right]. \quad (14)$$

From relationship (14) it is simple to express $\bar{t}_{t.o}=f\left(\bar{G}_{t.o},\bar{V}_1\right)$.

We introduce the following designations:

$$\alpha_1=f_{weel}+\frac{1}{3}\frac{C_{Dt.o}V_{1b}^2}{C_{Llift}V_{lift.b}^2}\frac{\bar{V}_1}{\bar{G}_{t.o}}, \quad (15)$$

$$\alpha_2=\alpha_1+\frac{1}{3}\frac{C_{Dt.o}}{C_{Llift}}\left(1+\frac{V_{1b}}{V_{lift.b}}\frac{\bar{V}_1}{\sqrt{\bar{G}_{t.o}}}\right). \quad (16)$$

We substitute α_1 and α_2 into equation (14), and get relationship relative $\bar{t}_{t.o}$.

$$2g\left(L_{rej.b}-L_{STPWY}\right)=\frac{V_{1b}\bar{V}_1}{K_1 t_{t.o.b}\bar{t}_{t.o}-\alpha_1}+\frac{V_{lift.b}^2\bar{G}_{t.o}-V_{1b}^2\bar{V}_1^2}{K_1\left(1-\frac{1}{n_{eng}}\right)t_{t.o.b}\bar{t}_{t.o}-\alpha_2}, \quad (17)$$

$$\bar{t}_{t.o}^2-\frac{2g\left(L_{rej.b}-L_{STPWY}\right)\left(\alpha_1-\frac{\alpha_1}{n_{eng}}+\alpha_2\right)-\frac{V_{1b}^2\bar{V}_1^2}{n_{eng}}+V_{2lift.b}^2\bar{G}_{t.o}}{2g\left(L_{rej.b}-L_{STPWY}\right)K_1\left(1-\frac{1}{n_{eng}}\right)t_{t.o.b}}\bar{t}_{t.o}+$$

$$+\frac{2g\left(L_{rej.b}-L_{STPWY}\right)\alpha_1\alpha_2+V_{1b}^2\left(\alpha_2-\alpha_1\right)\bar{V}_1^2+V_{lift.b}^2\alpha_1\bar{G}_{t.o}}{2g\left(L_{rej.b}-L_{STPWY}\right)K_1^2\left(1-\frac{1}{n_{eng}}\right)t_{t.o.b}^2}=0. \quad (18)$$

To analyze the relationship (18) let us consider as an example the basic version of the aircraft having the following parameters: $G_{t.o.b}=300\ 000$ kg; $S_w=600$ m²; $n_{eng}=4$; $t_{t.o.b}=0.30$; $K_1\cong 0,813$; $r_T=0.4$;

$g = 9.81$ m/s^2; $f_{weel} = 0.02$; $f_{red.weel} = 0.25$; $L_{lift.b} = 1263$ m; $C_{Dt.o} = 0.08$; $C_{Dred} = 0.30$; $C_{Dlift} = 1.70$; $L_{rej.b} = 1716$ m; $L_{STRWY} = 300$ m; $V_{1b} = 65$ m/s (234 km/h); $V_{lift.b} = 71.828$ m/s ($\cong$ 258.581 km/h). Let us note that parameters of the aircraft basic version correspond to equations (2), (9), (10) and (12).

We substitute the values of parameters of the basic version into equation (18):

$$\alpha_1 = 0.02 + 0.0128457\,\frac{\bar{V}_1^2}{\bar{G}_{t.o}};$$

$$\alpha_2 = 0.0356862 + 0.0128457\frac{\bar{V}_1^2}{\bar{G}_{t.o}} + 0.0141951\,\frac{\bar{V}_1}{\sqrt{\bar{G}_{t.o}}};$$

$$\alpha_2 - \alpha_1 = 0.0156862 + 0.0141951\,\frac{\bar{V}_1}{\sqrt{\bar{G}_{t.o}}};$$

$$\alpha_1 - \frac{\alpha_1}{n_{eng}} + \alpha_2 = 0.0506862 + 0.0141957\,\frac{\bar{V}_1}{\sqrt{\bar{G}_{t.o}}} + 0.0224799\,\frac{\bar{V}_1^2}{\bar{G}_{t.o}};$$

$$\alpha_1\alpha_2 = 10\text{-}7(7137.24 + 2839.14\,\frac{\bar{V}_1}{\sqrt{\bar{G}_{t.o}}} + 7153.2821\times\frac{\bar{V}_1^2}{\bar{G}_{t.o}} + 1823.537\,\frac{\bar{V}_1^3}{\bar{G}_{t.o}\sqrt{\bar{G}_{t.o}}} + 1650.12\,\frac{\bar{V}_1^4}{\bar{G}_{t.o}^2};$$

$$2g\left(L_{rej.b} - L_{STPWY}\right) = 277.92 \text{ m}^2/\text{s}^2;$$

$$2g\left(L_{rej.b} - L_{STPWY}\right)K_1\left(1 - \frac{1}{n_{eng}}\right)t_{t.o.b} = 5082.0075 \text{ m}^2/\text{s}^2;$$

$$2g\left(L_{rej.b} - L_{STPWY}\right)K_1^2\left(1 - \frac{1}{n_{eng}}\right)t_{t.o.b}^2 = =1239.5014 \text{ m}^2/\text{s}^2;$$

$$\frac{V_{1b}^2}{n_{eng}} = 1056.25 \text{ m2/s2};\ V_{1b}^2 = 4225 \text{ m2/s2};\ V_{lift.b}^2 = 5159.2615 \text{ m}^2/\text{s}^2.$$

After such substitution of parameters of the basic version and their approximation to the third sign we get the following values:

$$\bar{f}_{t.o}^2 - (0.277 + 0.077\,\frac{\bar{V}_1}{\sqrt{\bar{G}_{t.o}}} + 0.123\,\frac{\bar{V}_1^2}{\bar{G}_{t.o}} - 0.208\,\bar{V}_1^2 + 1.015\,\bar{G}_{t.o})\bar{t}_{t.o} + 0.016 + 0.006\,\frac{\bar{V}_1}{\sqrt{\bar{G}_{t.o}}} +$$
$$0.016\,\frac{\bar{V}_1^2}{\bar{G}_{t.o}} + 0.004\,\frac{\bar{V}_1^3}{\bar{G}_{t.o}\sqrt{\bar{G}_{t.o}}} + 0.004\,\frac{\bar{V}_1^4}{\bar{G}_{t.o}^2} + 0.107\,\bar{V}_1^2 + 0.048\,\frac{\bar{V}_1^3}{\bar{G}_{t.o}} + 0.083\,\bar{G}_{t.o} = 0. \qquad (19)$$

When substituting the values $\bar{G}_{t.o} = 1$ and $\bar{V}_1 = 1$ into (19), we get: $\bar{t}_{t.o}^2 - 1.284\bar{t}_{t.o} + 0.284 = 0$, that confirms the correctness of the equation (19).

Let us represent the relationship (19) in graphic form for values $\bar{V}_1$ = 0; 0.5; 0.8; 0.9; 1.0; 1.1; 1.2 and 1.3 if $\bar{G}_{t.o}$ = 0.4; 0.5; 0.6; 0.7; 0.8; 0.9; 1.0; 1.1; 1.2; 1.3 and 1.4. (see Figure 2).

Also, the relationship (5) is shown in Figure 2 if f_{weel} = 0.02 and K_1 = 0.813 (dot-dash line).

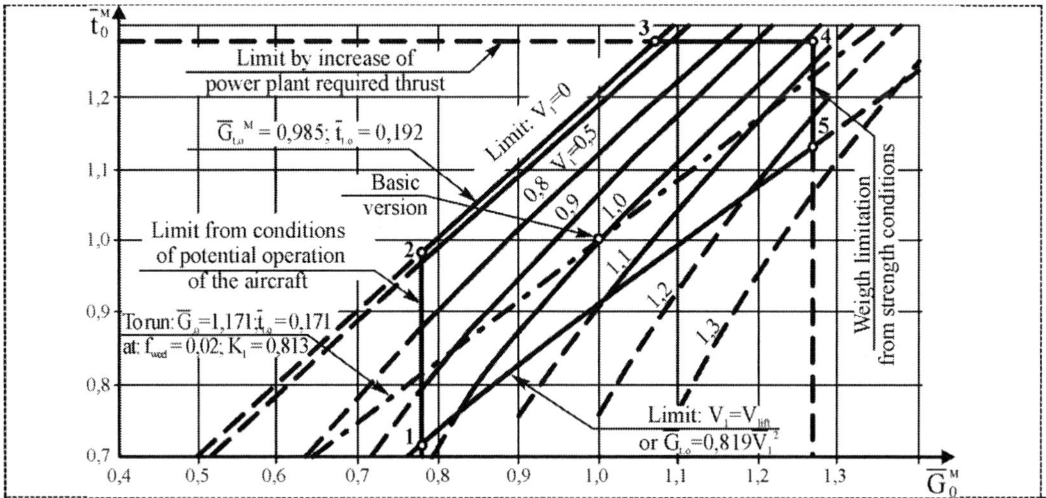

Figure 2. Influence of Modification Weight and Decision Speed Variation on the Required Value of Thrust/Weight Value.

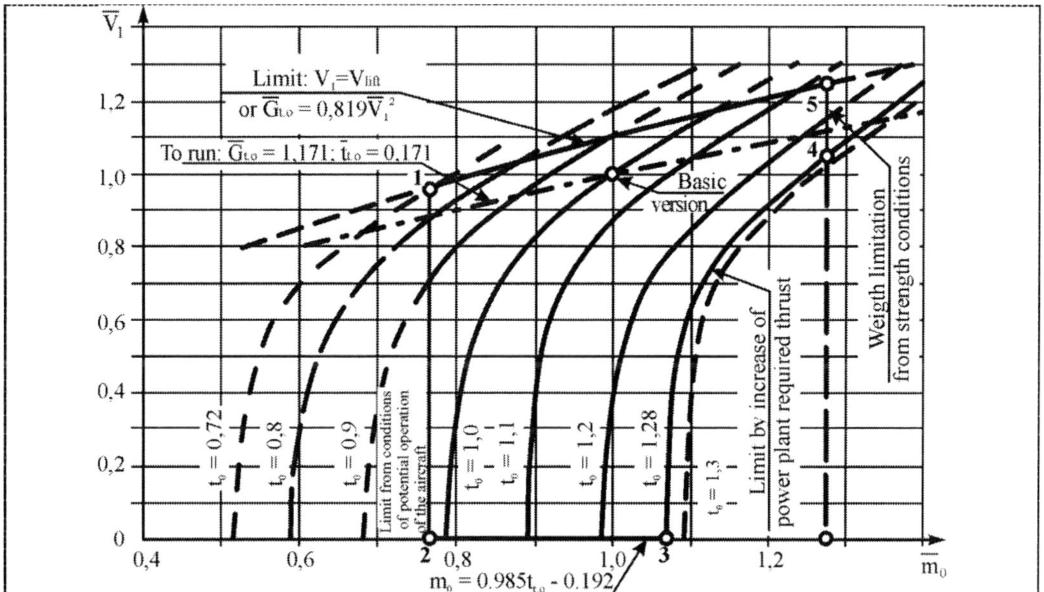

Figure 3. Influence of Take-off Weight and Thrust/Weight of Potential Modifications on Decision Speed at Take-off of Transport Category Aircraft.

Selected zone (1-2-3-4-5-1) of the relationship $\bar{t}_{t.o} = f\left(\bar{G}_{t.o}, \bar{V}\right)$ at $\bar{L}_{rej} \cong 1.0$ is determined by the following restrictions (boundary conditions):

1. Line 1-2 — by restriction from conditions of probable aircraft operation. For example, decrease of the takeoff weight of the aircraft basic version due to partial fueling or payload. In that case we accept restriction $\bar{G}_{t.o\min} = 0.78$, i.e. $G_{t.o\min} = 234$ t instead of $G_{t.o} = 300$ t.

2. Line 2-3 — by restriction $\bar{V}_1 = 0$.

Equation (19) is transformed into relationship

$$\bar{G}_{t.o} = 0{,}985\bar{t}_{t.o} - 0.192, \tag{20}$$

which can be considered as completed takeoff run with one failed critical engine at takeoff.

In considered example the required distance of complete takeoff run is equal to $L = L_{rej} - L_{STPWY} = 1716 - 300 = 1416\,m$, both for basic version and for other modifications under consideration.

3. Line 3-4 — by restriction of available possibility to increase the required thrust of power plant, and consequently power-weight ratio of the aircraft $\bar{t}_{t.o\max} = 1.28$.

Increase of the required thrust of aircraft power plant in considered ratios $1.0 \le \bar{t}_{t.o} \le 1.28$ can be provided both by availability of afterburner mode engaged automatically (in response to engine failure signal) and by possibility of installation of more high-power engines.

4. Line 4-5 — by restriction of aircraft weight from strength conditions determined by the worst cases of loading of different aircraft components (wing, tail unit, landing gear, etc.) by modes of operation. In this case it is accepted that $\bar{G}_{t.o\max} = 1.26$.

5. Line 5-1 – by restriction $V_1 = V_{lift}$. This restriction gives the relationship

$$V_{1b}\bar{V}_1 = V_{lift.b}\bar{V}_{lift} = V_{lift.b}\sqrt{\bar{G}_{t.o}},$$

$$\bar{G}_{t.o} = \frac{V_{1b}^2}{V_{lift.b}^2}\bar{V}_1^2 \quad (21)$$

$$\text{or } \bar{G}_{t.o} = 0.819\bar{V}_1^2, \quad (22)$$

which can be considered as the complete takeoff run with one failed critical engine at liftoff (see para. 2 above).

Using the data given in Figure 2 we construct the dependence $\bar{V}_1 = f\left(\bar{G}_{t.o}, \bar{t}_{t.o}\right)$ shown in Figure 3 on coordinates $\bar{G}_{t.o}$ and $\bar{V}_1$ at different values of $\bar{t}_{t.o}$.

4. Conclusions

Decision speed V_1 is specified in the Airplane Flight Manual and should be greater or equal to minimum maneuvering speed at takeoff run $\left(V_{\min.maneuv}\right)$, at which after critical engine failure aircraft control is provided by means of aerodynamic controls to maintain straight motion $V_1 \ge V_{\min.maneuv}$, and less or equal to nose landing gear lift speed which is also specified in the Airplane Flight Manual $V_1 \le V_{NLG}$ [19,22]. The method of quantitative estimation of influence of modifications takeoff mass and thrust-to-weight ratio variation on decision speed relative value with maintaining one of basic parameters – rejected takeoff distance, i.e. of assigned base airfields has been proposed.

References

[1] Medvedev A 2013. Airplane catastrophe as a result of operational errors and violations. *Aviation*. 17(2), 70-75. DOI:https://doi.org/10.3846/16487788.2013.805866
[2] Sheynin V M 1983 *Purpose of Modification in Development of Aircraft*. Moscow: Nauka p. 226
[3] Alonso J J, LeGresley P, Pereyra V 2009 Aircraft design optimization *Mathematics and Computers in Simulation*. Vol. 79, iss. 6. P. 1948–1958.
[4] Kapitanova L V 2016 Conceptual aspects for providing the take-off and landing performance of the transport category aircraft variants *Aerospace equipment and technology*. 7 (134). p. 11–16.
[5] McCullers L A 1984 Aircraft configuration optimization including optimized flight profiles // *Recent Experiences in Multidisciplinary Analysis and Optimization*: proc. of symp. (Apr. 24–26, 1984) NASA Langley Research Center. Hampton, Part 1. p. 395–412.

[6] Sobieszczanski-Sobieski J, Haftka R T 1997 Multidisciplinary aerospace design optimization: survey of recent developments *Structural optimization*. Vol. 14, iss. 1. p. 1–23.
[7] Shirley C M, Schetz J A, Kapania R K and Haftka R T 2014 Tradeoffs of wing weight and lift/drag in design of medium-range transport aircraft *Journal of Aircraft* Vol. 51, no. 3. p. 904–912.
[8] Malone B, Mason W H 1995 Multidisciplinary optimization in aircraft design using analytic technology models *Journal of Aircraft* Vol. 32, no. 2. P. 431–438.
[9] Kapitanova L V, Kuznetsova N V and Dzhurinsky A N 2012 Modeling of transients at the aircraft landing stage *Visnik dvigunobuduvannya* № 2. p. 44–49
[10] Torenbeek E 2013 *Advanced Aircraft Design*. Netherlands: Delft University of Technology p. 410
[11] Tiniakov D V 2016 The Analysis of the Choice Methods of the Airplane Design Parameters. *Problems of Aircraft Creation and Service Life Ensuring*. National Aerospace University "KhAI", Kharkiv, Ukraine p. 39
[12] Pierluigi D V, Fabrizio N 2014 Aerodynamic guidelines in the design and optimization of new regional turboprop aircraft. *Aerospace Science and Technology*, 38 p. 88-104
[13] Balabuyev P V, Bichkov S A, Grebenikov A G, Zjeldochenko V N, Kobilyanskiy A A, Myalitsa A K, Ryabkov V I and Tseplyaeva T P 2013 *Principles of designing of airplanes with gas turbine engines* Kharkov: National Aerospace University "KhAI" p. 731
[14] Kapitanova L V 2007 The main conditions of keeping the basic takeoff performances at the stage of the airplane modifications developing. *The Designing and Manufacturing* of Aircraft National aerospace university "KhAI" Kharkiv, Ukraine. 2(49). p. 11-19
[15] Bochakarev A F 1985 *Aerodynamic of airplane*. Moscow: Mashinostroenie p. 360
[16] Process for verifying the coherence between aircraft take-off parameters and an available runway length *Electronic resource*: Pat. 8380372B2 USA: Int. Cl. G06F19/00. Appl. No. 12/537478; fil. 07.08.2009; publ. 19.02.2013. 13 p. URL: http://www.freepatentsonline.com/8380372.pdf. (last accessed: 11.09.2015).
[17] Wei X., Nie H 2005 Dynamic analysis of aircraft landing impact using landing-region-based model *Journal of Aircraft* Vol. 42, 6. p. 1631–1637.
[18] Prelshteyn V N 1972 *To derivation of the airplane movement equations at takeoff run and rollout with taking to account the elasticity of its structure and the shock-absorption characteristic of the landing gear*. Moscow: Tr. Gos. NIIGA 17. p. 34 – 42
[19] Valdés R M A, Comendador F G, Gordún L M and Nieto F J S 2011 The development of probabilistic models to estimate accident risk (due to runway overrun and landing undershoot) applicable to the design and construction of runway safety areas *Safety science* Vol. 49, 5. p. 633–650.
[20] Zammit-Mangion D., Eshelby M 2008 Simplified algorithm to model aircraft acceleration during takeoff *Journal of Aircraft* Vol. 45, 4. p. 1090–1097.
[21] Kapitanova L V 2007 Determination of admissible changes of weight and required thrust-weight ratio of modifications of heavy aircraft of transport category under the terms of their interrupted take-off. *Visti akademii inzhenernih nauk Ukraini*. Kiev: NTUU KPU. 3(33). p 99-106.
[22] *Aviation Rules. part 25. Airworthiness for aircraft of transport category*. 1994 Moscow: MAK

Design and Simulation Analysis of Deployable Structure of a Single-unit Radar

JinWei Wang, XiFang Zhao and BaoFu Tang

Nanjing Research Institute of Electronics Technology, Nanjing 210039, China

zxfang203@sohu.com

Abstract. The deployable structure has the characteristics of polymorphism, and it appears as the function of the mechanism during the deployment process, and as the function of the structure during the deployment and locking stage. Since the 21st century, it has been increasingly adopted by the military field. Based on the tactical requirements of individual radars for rapid withdrawal, off-road marching, rapid erection, and light weight in the war, a soldier radar with an expandable structure consisting of several active linear array antennas arranged uniformly on the circumference to form a cylindrical antenna array is designed. In order to ensure a smooth and reliable deployment of the deployable structure, a link mechanism design with hinge gap is used. According to the corresponding technical requirements,the rigidity of the single-armor radar's cylindrical box and antenna unit panel was simulated and calculated. The results show that the single-armor radar can meet the technical requirements under the corresponding working conditions without damage.

1. Introduction

At present, the collection of circular array radar array carried by individual soldiers at home and abroad mainly adopts separate installation and folding, which has a lot of process and slow speed. There is still a large gap between the requirements of rapid collection and cross-country March from the combat readiness state of one position to another position for quick installation and transition to combat readiness state [1]. For example, the TPQ-48 single soldier radar in the United States is installed separately in pieces, with a weight of 55kg. It takes about 10min for two to three people to set up and spread the radar, and the speed of frame collection is slow. In order to meet the needs of rapid response in modern war, a new type of single soldier radar array structure is urgently needed

In order to solve the problem of slow collection speed of single soldier radar array, it is necessary to design a developable structure to realize the rapid installation and precise positioning of active linear array antenna with the cooperation of two people, so as to greatly improve the efficiency of collection; at the same time, aiming at the problem of heavy weight and poor portability, a new lightweight material is adopted to control the weight of the product within 50kg on the premise of meeting the rigid strength of the product. In this paper, the rapid development and retraction of the single radar array is realized by using the shear hinge structure, the lightweight materials and weight reduction design are adopted, and the finite element analysis is combined to ensure that the whole radar meets the requirements of lightweight quality and stiffness under the corresponding working conditions, which can provide certain reference for the design of similar radars.

Content from this work may be used under the terms of the Creative Commons Attribution 3.0 licence. Any further distribution of this work must maintain attribution to the author(s) and the title of the work, journal citation and DOI.
Published under licence by IOP Publishing Ltd

2. The design of deployable structure of individual radar

In order to meet the technical requirements of rapid unwheeling and simple unwheeling process of single round array radar, the whole radar array is designed with shear hinged structure [2] [3]. As shown in Figure 1, the radar is mainly composed of a cylindrical box, a bottom plate, a scissor hinged deployable array surface, a hanging plate, etc. The shear hinged deployable array is composed of the upper hinged deployable structure, the lower deployable structure and the pin shaft of the connecting plate. The antenna unit is installed on the connecting plate in a circular distribution, forming a cylindrical antenna array. In addition, the corresponding reflective cloth is installed on the outside of the two connecting plates, which can expand and collapse with the array surface. As shown in figure 2 and figure 3, the bottom plate with reserved mounting holes is buckled to the corresponding predetermined position at the bottom of the cylindrical box. After the deployable structure is expanded to the corresponding position, the hanging plate is installed on the bottom plate through the fixed scissor hinge, so that the bottom of the deployable array is fixed. Then, the top cover plate and the hanging plate on the upper part of the cylindrical box are hinged to form a whole. When the individual radar is in working state, the upper and lower parts of the deployable structure are in developing state. The stability of the array is ensured by fixing the hanging plate and the scissor hinge. In addition, in order to ensure the technical requirements of the whole radar, such as light weight, good portability, etc., the material used for the radar is mainly aluminum, and the weight reduction design is adopted for the parts with small force, so as to ensure the radar's quality is below 50kg.

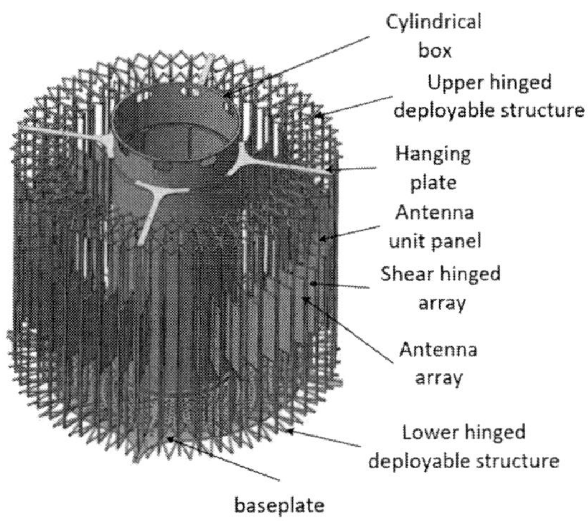

Figure 1. Antenna structure model

Figure 2. Unfolding state of bottom deployable structure

Figure 3. Unfolding state of top deployable structure

3. Establishment of finite element model

On the premise of satisfying the lightweight of the whole radar, it is also necessary to verify the mechanical performance of the radar under the corresponding working conditions. According to the design requirements, it is necessary to ensure that the individual radar can work normally under the wind speed of 15m / s, that is to say, when the individual radar is deployed, the cylindricity of the box and the evenly distributed antenna unit panel in the circular state shall be less than 1.5mm. Under the wind speed of 30 m / s, the whole array will not be damaged. In order to ensure the accuracy and reliability of the finite element analysis results, this paper mainly analyzes the stress characteristics, material properties and external load conditions of the whole cylinder box and the array panel when the individual radar is working, and determines the reasonable grid elements and load conditions to reduce the error between the calculation results and the actual results [4].

3.1. Model simplification

There are many components and complicated structures in the single soldier radar. If the real antenna structure model is used, it will lead to a great amount of calculation. Therefore, it is necessary to simplify the model of the whole single soldier radar and reduce the calculation amount [5]. After analysis, the wind load on the antenna mainly acts on the circular array facing the wind direction cylinder surface, and finally acts on the cylinder box through the conduction of force, which has no effect on the structure and position of components inside the cylinder. Therefore, the cylinder body with equal volume and mass is used to replace the actual model of the cylinder box inside the antenna for simulation calculation. In addition, because the reflection cloth between the two adjacent antenna elements is in the expanded state when the whole single soldier radar is in the expanded state, and the external wind load directly acts on the reflection cloth, the reflection cloth outside the cylinder is simplified into a film cloth structure with equal volume and mass when the model is simplified. The simplified single soldier radar model is shown in figure 4, which is mainly composed of outer membrane cloth, antenna unit panel, cylinder box, etc.

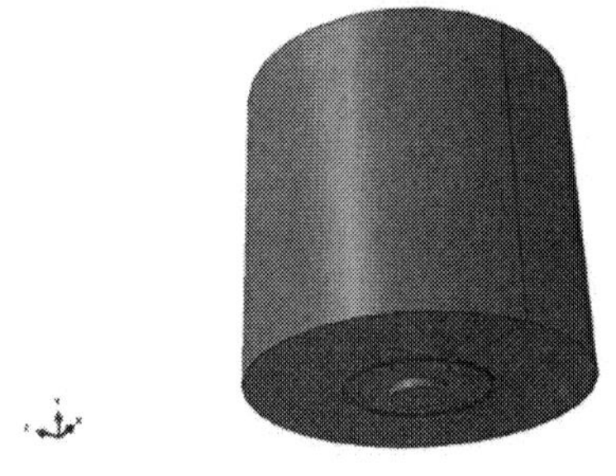

Figure 4. Simplified model

3.2. Unit type and attribute

After simplification, the whole radar consists of three parts: outer membrane, circular uniform antenna unit panel and cylinder box. The outer membrane is shell unit, and the antenna unit panel and cylinder box are solid unit. According to the structural design requirements, stress characteristics and material requirements of the individual radar, the material of antenna unit panel and cylinder box is aluminum, the material of outer membrane is tin membrane, and the material of antenna unit blade is aluminum. The elastic modulus of aluminum is 72000Mpa, Poisson's ratio is 0.33, and density is 2.7e-9 (t / mm^3). The elastic modulus of Tin film is 106000Mpa, Poisson's ratio is 0.25, and density is 3.7e-9 (t / mm^3).

3.3. Grid division

The small holes in the model do not affect the overall calculation results of the structure to be omitted, the rigid element equivalent treatment is used for hinge connection, and the mesh division of the cylindrical box and the antenna unit panel is carried out automatically by software. According to the

IOP Conf. Series: Materials Science and Engineering **887** (2020) 012008 doi:10.1088/1757-899X/887/1/012008

different structures, the mesh is automatically divided into tetrahedral and hexahedral mesh as shown in figure 5 and figure 6, and the maximum size of the mesh is 3 mm. The outer membrane structure is represented by shell element, and the middle surface is used for mesh processing. The results show that the grid accuracy can meet the analysis requirements

Figure 5. Cylindrical box grid

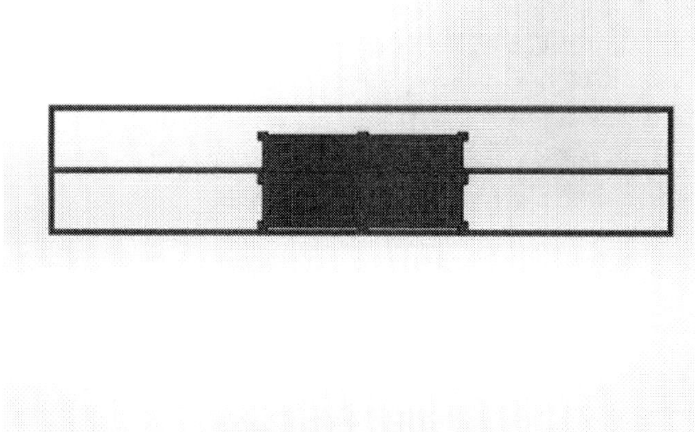
Figure 6. Antenna unit panel grid

4. Loads and constraints

The load of the single soldier radar in the working state mainly comes from the wind load. The load directions of the individual radar and the antenna unit array of the single soldier radar in the working state are shown in figure 7 and figure 8 respectively. In the actual situation, the wind load moves irregularly with time [6]. According to the design requirements, the individual radar needs to work normally under the wind speed of 15m / s and not be damaged under the wind speed of 30m/s. therefore, in the simulation calculation, the wind load is calculated and checked according to the static load, taking the wind speed of 15m/s and 30m/s respectively. The calculation formula of wind load is shown in (1)

$$F = C_{Fq} A \tag{1}$$

Where, C_{Fq} is the wind pressure value and A is the characteristic area of the object.

After calculation, when the wind speed is 15m / s, the surface load on the outer membrane material is 0.302 KN/m^2. When the wind speed is 30m / s, the surface load on the outer membrane material is 1.208 KN/m^2. After loading the corresponding load, according to the actual test situation, the lower surface of the cylindrical box is set as a fixed constraint, and finally the solution is calculated.

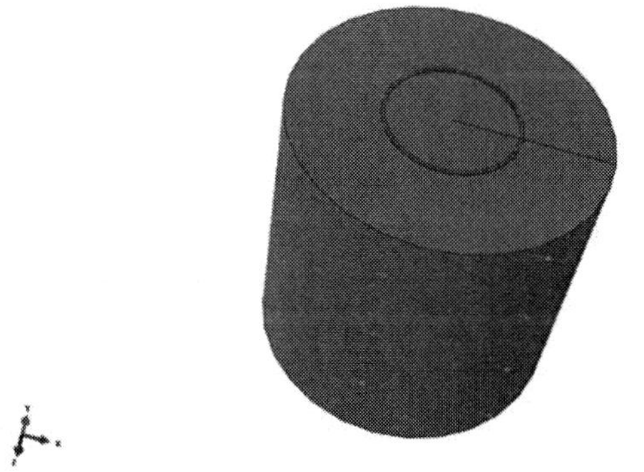
Figure 7. Wind load direction of individual radar

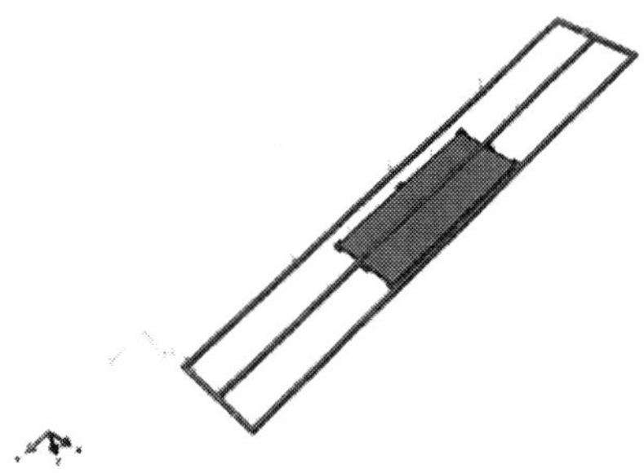
Figure 8. Wind load direction of antenna unit panel

IOP Conf. Series: Materials Science and Engineering **887** (2020) 012008 doi:10.1088/1757-899X/887/1/012008

5. Calculation results and analysis

Through calculation and analysis, the deformation results of the cylindrical box and the antenna array panel of the single soldier radar under the wind speed of 15m/s and 30m/s are obtained, and the displacement nephogram of the cylindrical box and the antenna array panel under the corresponding working conditions are extracted respectively, as shown in figure 9, figure 10, figure 11 and figure 12. The deformation of cylinder box and antenna unit panel under different wind load conditions are shown in table 1.

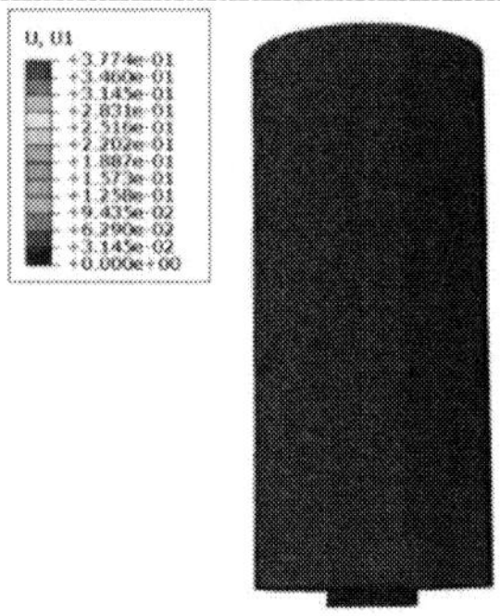

Figure 9. Cloud chart of cylindrical box displacement under the wind speed of 15m/s

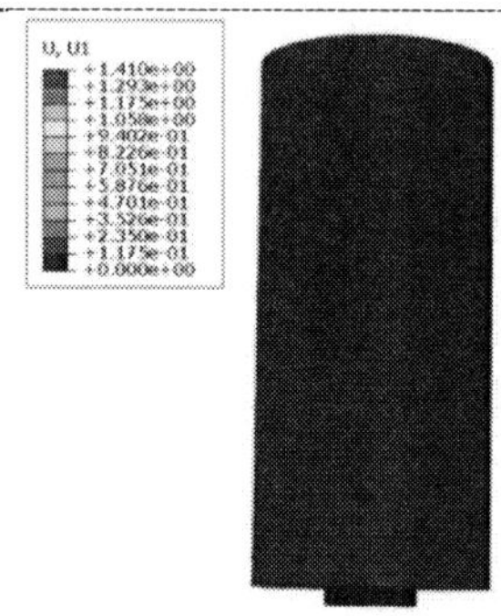

Figure 10. Cloud chart of cylindrical box displacement under the wind speed of 30m/s

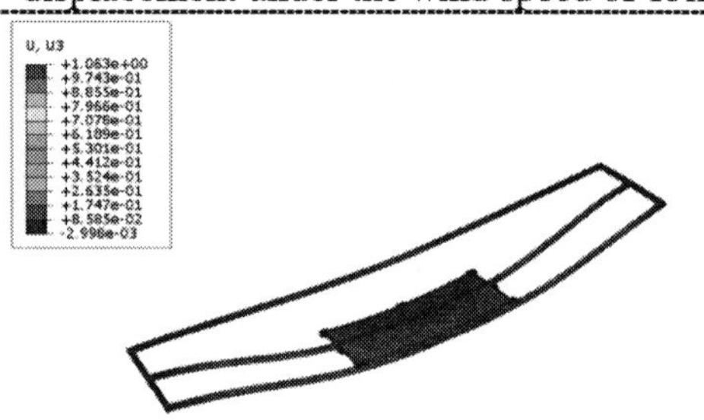

Figure 11. Cloud chart of antenna unit panel displacement under wind speed of 15m/s

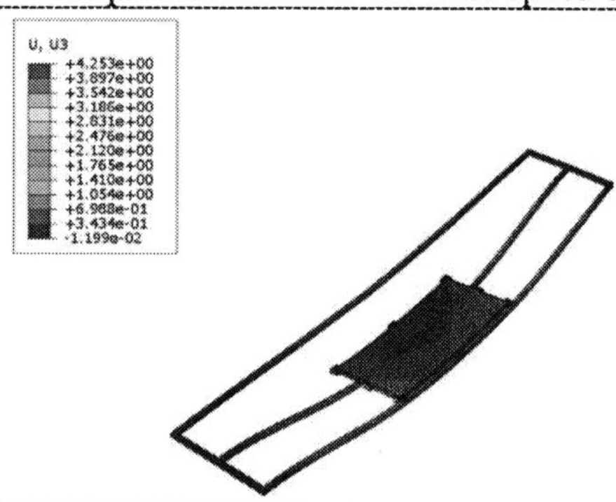

Figure 12. Cloud chart of antenna unit panel displacement under wind speed of 30m/s

Table 1. The deformation of cylinder box and antenna unit panel under different wind load conditions

	Maximum displacement (mm)	
wind speed	X-direction displacement of cylindrical box	Z-direction displacement of antenna unit panel
15m/s	0.37	1.06
30m/s	1.41	4.25

Combined with the calculation results, it can be seen that under the wind load of 15m/s, the maximum displacement of the cylindrical box in X direction is 0.37mm, and the maximum displacement of the antenna array in Z direction is 1.06mm. The cylindricity of the box and the evenly distributed antenna unit panel in the circular state can be ensured to be less than 1.5mm, which can meet the requirements of normal work Requirements in working state. When the wind speed is 30 m/s, the maximum displacement of the cylindrical box in X direction is 1.41 mm, and the maximum displacement of the antenna array in Z direction is 4.25 mm. The whole single radar will not be damaged. To sum up, the design of individual radar can meet the corresponding design requirements.

6. Conclusion

Combined with the calculation results, it can be seen that under the wind load of 15m/s, the maximum displacement of the cylindrical box in X direction is 0.37mm, and the maximum displacement of the antenna array in Z direction is 1.06mm. The cylindricity of the box and the evenly distributed antenna unit panel in the circular state can be ensured to be less than 1.5mm, which can meet the requirements of normal work Requirements in working state. When the wind speed is 30 m/s, the maximum displacement of the cylindrical box in X direction is 1.41 mm, and the maximum displacement of the antenna array in Z direction is 4.25 mm. The whole single radar will not be damaged. To sum up, the design of individual radar can meet the corresponding design requirements.

References

[1] Wang Jinwei, Tang Baofu, Zhao Xifang, et al. Error analysis of deployable structure of a single radar [J]. Modern radar, 2014, 36 (4)
[2] Zhang Shujie. Design of deployment mechanism and thermal structure coupling analysis of semi-rigid solar array in space laboratory [D]. 2004
[3] Cui Qifeng, Ni Bo, Shanghai Institute of aerospace systems engineering, Shanghai. Topological characteristics analysis of space articulated extension mechanism [J]. 2009 joint academic exchange meeting of space electromechanical and space optics Professional Committee and space materials professional committee of Chinese society of space science, 2011
[4] Yan Rongjun, Li Weitian. Stress analysis of a simple radar structure [J]. Naval Electronic Engineering (8)
[5] Structural design of light antenna for man on back radar [J]. Fiber composite, 2011 (3): 36-39
[6] Gao Yanlong. Research on numerical simulation method of wind load characteristics of radar antenna [D]. University of Electronic Science and technology, 2009

IOP Conf. Series: Materials Science and Engineering **887** (2020) 012009 doi:10.1088/1757-899X/887/1/012009

The Initial Weight Estimation of New Aircraft Design Solutions

Anatolii Kretov

Nanjing University of Astronautics and Aeronautics, College of Aerospace Engineering, 29 Yudao St., Nanjing 210016, P.R.China

kretov-ac@nuaa.edu.cn

Abstract. The work is devoted to the rapid assessment of the weight of the aircraft and its design. This problem occurs when creating new aircraft or in the process of constructive improvement of existing ones. To determine the mass of the entire aircraft, the method of "mass change coefficients" (growth coefficients) is used, which is based on the equation of the existence of the aircraft. To analyze the mass of initial design changes, the Komarov weight formula based on the concept of "load-carrying factor" is used. The article considers a numerical example of estimating changes in the mass of an aircraft due to the use of composite materials in the fuselage structure.

1. Introduction

Flight of any flight vehicle (FV) is the process of countering the gravity force with aerostatic, aerodynamic or rocket-dynamic ones. Therefore, for any FV its weight (mass) is a paramount parameter. The mass evaluation of FV from the very beginning of the project, during its consistent refinement in the design process and constant monitoring throughout the life cycle of the FV is the most important task. In the struggle to reduce the weight of the FV, their creators are looking for new technologies that are associated with both design and production.

For the design evaluation of the mass of FV and its aggregates, semi-empirical weight formulas are usually used [1-3].

The proposed work considers the problem of rapid assessment of the total change in the mass of an existing aircraft a prepared project due to a possible local change in the mass of its individual components, including its design due to new design solutions, for example, as a result of the use of a new material. In this case, there are generally three scenarios for the implementation of such changes:

– the target return remains unchanged, for which the weight of the airframe structure, engines and fuel are changed accordingly;

– the target load, power plant, fuel mass and geometric dimensions of the aircraft do not change, but due to changes in the mass of the structure, the take-off weight and flight properties of the aircraft change, which means that the target return changes;

– the change in the mass of the structure is compensated only by the change in the mass of the target load.

In this paper we are talking about the first scenario.

In the struggle to reduce the weight of FV, their creators are always striving for new designs and technologies. The last major achievements in this direction are related to the use of composite materials

Content from this work may be used under the terms of the Creative Commons Attribution 3.0 licence. Any further distribution of this work must maintain attribution to the author(s) and the title of the work, journal citation and DOI.
Published under licence by IOP Publishing Ltd

(CM). In the design of the most modern mainline aircraft (Boing 787, Airbus A350), about 50% is accounted for by CM. This allows you to reduce the cost of a mile-flight by up to 15% and consequently reduce operating costs. However, the proportion of CM in the fuselages of these aircraft is significantly less than in the wing. In this regard, the study of the use of CM in the fuselage is certainly an urgent and important topic.

2. Initial estimation of the mass of FV

The aircraft is considered as a system having a certain hierarchical structure and its mass is represented as the sum of the masses of the following five parts: target load m_g, airframe structure m_{str}, power plant m_e, fuel system m_f, equipment and control system m_{eq}

$$m = m_g + m_{str} + m_{en} + m_f + m_{eq} = \sum m_c. \quad (1)$$

Goal load is a subsystem whose elements are determined by the purpose of the aircraft: commercial (paid) or combat load; crew, equipment; equipment of passenger and baggage-cargo facilities, weapons.

The airframe design is a subsystem, the elements of which are determined by the implementation of the corresponding principle of flight in terms of reliable perception of operating loads. For aircraft, this wing, tail, fuselage and landing gear.

Power plant is a subsystem whose elements provide the creation of the required thrust: engines and devices for their installation.

Fuel system is a subsystem whose elements ensure the operation of engines during a given flight time on the provided flight modes: fuel and devices for its placement and pumping.

The equipment and control system ensures reliable operation of the aircraft its control during the flight.

Representation of the mass of the aircraft in the form of (1) is in accordance with the functional isolation of the five subsystems listed, each of which has its own specifics of mass formation. The target load is usually specified or determined independently of the total mass. The other subsystems depend on m, and the dependencies $m_{str}(m)$, $m_e(m)$ are quite close to directly proportional. Dividing equation (1) by the mass of the aircraft yields an equation in the form of relative masses $\overline{m}_c = m_c/m$

$$1 = \sum_c \overline{m}_c = \overline{m}_g + \overline{m}_{str} + \overline{m}_{en} + \overline{m}_f + \overline{m}_{eq}. \quad (2)$$

Equation (2) determines the balance of relative masses and the conditions of existence FV in terms of weight. With the help of equation (2) and a given mass of the target load m_g we can find the mass of the whole FV

$$m = m_g / (1 - \sum_c \overline{m}_c) \cdot \quad (3)$$

To determine the mass in the first approximation, statistical data are usually used with a refined estimate of the relative mass of the fuel, based on a given range and speed, aerodynamic quality and specific fuel consumption (Breguet formula, adjusted for non-cruising modes). According to statistics, the relative masses of subsonic passenger aircraft are: $\overline{m}_{str}$ =0.25-0.32; $\overline{m}_{en}$=0.08-0.14; $\overline{m}_f$=0.18-0.4; $\overline{m}_{eq}$ =0.09-0.12.

3. Estimation of initial mass changes by the total mass of the aircraft

Let as a result of some changes there was a small initial change of mass $\Delta m_{i\,0}$, then from (1) follows

$$\Delta m_{i\,0} = m - (m_g + m_{str} + m_{en} + m_f + m_{eq}) \,. \quad (4)$$

According to the chosen scenario, we must preserve the mass and layout of the target load, which means that the geometry of the fuselage of the passenger aircraft, which houses the target load, remains unchanged. To fulfill this condition, the mass components related to m_{en} and m_f are written in the form suggested in [4]

$$m_{en} + m_f = (m_{en} + m_f) C_{D\,fus}/C_D + (m_{en} + m_f)\,(1 - C_{D\,fus}/C_D) \quad (5)$$

where $C_{D\,fus}$, C_D are the drag coefficients of the fuselage and the entire aircraft.

The first term of the right part (5) determines the outlay of thrust and fuel for the transportation of the fuselage of the original dimensions in cruising flight and when we have $\Delta m_{i\,0}$, they will not change. Taking into account this remark $\Delta m_{i\,0}$ can be written as follows (index old means the unchanged value)

$$\Delta m_{i\,0}=m-[m_g+\bar{m}_{str}m+\overline{\Delta m}_{i\,0}m+\bar{m}_{eq}m_{old}+(\bar{m}_e+\bar{m}_f)(C_{Dfus}/C_D\, m)_{old}+(\bar{m}_e+\bar{m}_f)(1-C_{Dfus}/C_D)_{old}m] \quad (6)$$

We take the partial derivative of (6) by mass

$$\frac{\partial(\Delta m_{i\,0})}{\partial m}=1-\bar{m}_{str}-\overline{\Delta m}_{i\,0}-(\bar{m}_{en}+\bar{m}_f)\,(1-C_{D\,fus}/C_D)_{old}\,m). \quad (7)$$

From the expression (7) we can find the inverse derivative, which is called the mass change factor (MCF) of the aircraft

$$\mu_{m_i}=\frac{\partial(m)}{\partial\Delta m_{i0}}=\frac{1}{1-[\bar{m}_{str}+\overline{\Delta m}_{i0}+(\bar{m}_{en}+\bar{m}_f)(1-C_{Dfus}/C_D)]} \quad (8)$$

MCF shows how many times the total mass change of the aircraft greater than the partial change $\Delta m_{i\,0}$. If to pass to final changes (Figure 1), then at initial small change of weight on $\Delta m_{i\,0}$

$$\Delta m=\mu_{mi}\,\Delta m_{i\,0}\,. \quad (9)$$

Taking into account (2), the MCF can be presented in the form

$$\mu_{mi}=1/[\bar{m}_g-\overline{\Delta m}_{i\,0}+\bar{m}_{eq}+(\bar{m}_{en}+\bar{m}_f)C_{D\,fus}/C_D\,]. \quad (10)$$

4. Assessment of initial weight changes as a result of constructive measures

4.1. Basics of a new weight estimation method

To calculate the initial mass change, we use the approach proposed by V. A. Komarov [5], which is based on the so-called integral criterion "load-carryig factor" (G-factor), which is generally written as

$$G=\int_V \sigma^{\mathrm{Eqv}}dV \quad (11)$$

where σ^{Eqv} is equivalent stress acting in the structure, V is the volume of the material of the load-carrying structure. The discretization of the design, in particular in the FEM variant, gives us

$$G=\sum_i \sigma_i^{\mathrm{Eqv}}V_i \quad (12)$$

where σ_i^{Eqv} , V_i° are the equivalent stresses and the volume of the ith element in which °σ_i^{Eqv} is assumed constant.

When using the same type of material with specific strength $\bar{\sigma}=\sigma_P/\rho$, and in the case of reaching the maximum permissible stresses in each element σ_P, the mass of the structure can be estimated through the G-factor

$$m=\varphi\, G/\,\bar{\sigma} \quad (13)$$

The coefficient φ in this expression takes into account various really existing additional mass gains (joints and non-force elements in the structure, deviations from the optimal distribution of material in favor of simplicity and manufacturability of structures, and much more), which are usually not taken into account in the design evaluation or which is difficult to take into account in initial design steps. So if the coefficient φ=1, we have a so- called theoretical mass m^{Theor}.

As a result, all factors affecting the load-carrying mass are divided into three fairly independent groups [6]:

– level of constructive and technological perfection and target filling of a design (φ);
– design and power circuit and operating loads (G);
– physical and mechanical properties of the material ($\bar{\sigma}$).

For convenience, the parameter G can be represented as three cofactors: the dimensionless coefficient C_K, the characteristic load P and the characteristic dimensional parameter L

$$G= C_K P L \qquad (14)$$

As a result the final form of the weight formula of V. A. Komarov will have the form

$$m=\varphi\ C_K P L/\ \bar{\sigma} \qquad (15)$$

One of the main advantages of this approach is that the C_K coefficient in the design process can serve as a dimensionless criterion of the weight perfection of structures, and its reduction will contribute to the reduction of mass.

Figure 2 shows the simplest example to explain this property: a symmetric two-rod truss loaded in the first case (solid line) by a horizontal force P_1, in the second – by a vertical force P_2 (dashed line); and the values of the coefficients C_K for these cases as a function of the angle α. The minimum values of C_K, and accordingly the minimum mass, will be achieved in the first case of loading at α=0° (C_K=1) and α=45° in the second (C_K =2).

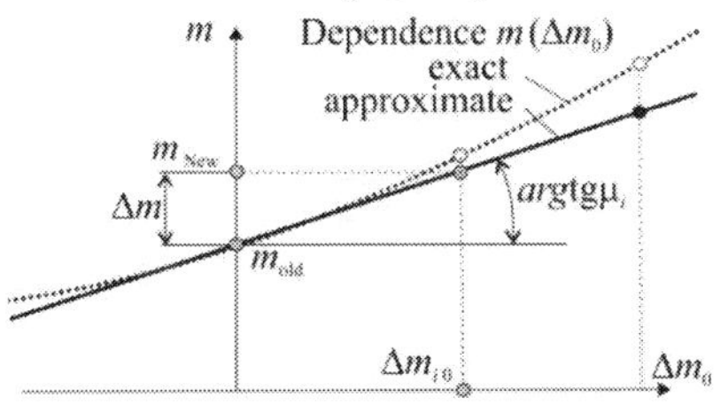

Figure 1. Dependence of the final mass on changes in the initial mass

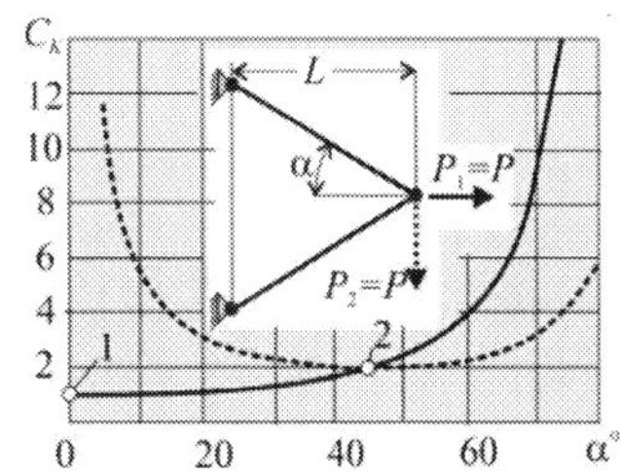

Figure 2. Dependence coefficient C_K from the angle α for the symmetric truss

5. Example on the weight estimation of fuselage

To combine approaches using MCF and G-factor, consider the problem of approximate design evaluation of the change in weight of the aircraft as a result of the replacement of the skin and longitudinal fuselage set from aluminum alloy to CM. The fuselage structure is represented as a cylindrical shell with length l_{fus} width b_{fus} and height h_{fus}; with a longitudinal-transverse set and a skin. The fuselage is reduced to a structurally axisymmetric with a diameter D=(b_{fus}+h_{fus})/2=2R. Figure 3 shows the plots of typical loads acting on the fuselage of a passenger aircraft: generalized axial N_x and shear Q_y forces, bending M_z and torque M_x moments, as well as excessive internal pressure p. By this the fuselage is considered fixed by the cross section of the location of the main wing attachment points and is divided into two cantilever compartments: the front with a length of $l_{fus.f}$ and back – $l_{fus.b}$. In the design evaluation we will compare the mass ratio of the metal variant with the composite one only by two main load factors which need the main part of the load-carrying mass: excessive internal pressure and bending moment (they are highlighted by contour lines).

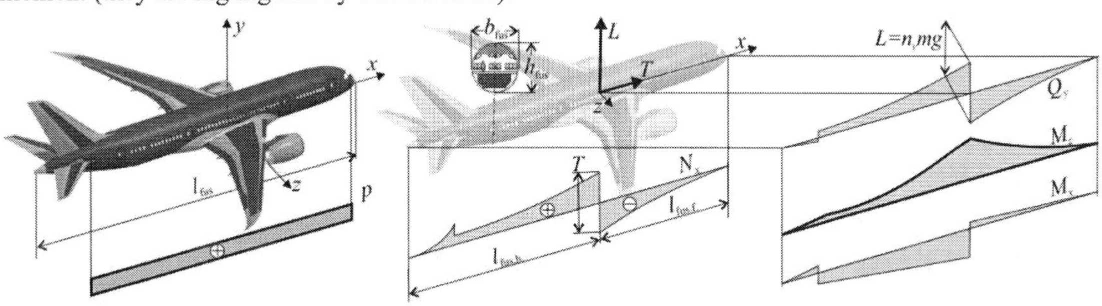

Figure 3. Loads acting on the fuselage structure

5.1. Metal fuselage.

For design analysis, we use the thin-walled shell with the generalized thickness δ, includes thickness of really skin and distributed area of the longitudinal set. The excessive internal pressure generates the circumferential $\sigma_\beta = pR/\delta$ and meridional linear stresses $\sigma_\alpha = pR/(2\delta)$. By Mises the expression for equivalent stresses by the momentless theoryby these stresses will be

$$\sigma^{Eqv} = (\sigma_\alpha^2 + \sigma_\beta^2 - \sigma_\alpha\sigma_\beta) = 3^{1/2}/2\, pR/\delta \leq \sigma_P. \quad (16)$$

The theoretical mass (φ=1) of such shell by its nominal thickness will be

$$m_p = \pi R^2 l\, \rho\, p 3^{-1/2} / \sigma^{Eqv} = 3^{-1/2}\, p\, V/\bar{\sigma} \quad (17)$$

where V is the shell volume.

Comparing the expressions (17) and (15) it is obvious that if we take as a characteristic load overpressure $P=p$, and the characteristic size of the volume of the fuselage pressurized compartment $L=V$, and assuming φ=1, than the coefficient of force factor $C_K = \sqrt{3}$.

The second main load is the bending moment. Consider the front part of the fuselage. From its action in the cross section of the fuselage there are normal stresses $\sigma = M_z R \cos\beta / I$, where β – the angle between the design point in cross-section and the z axis, I is moment of inertia, for a thin-walled circular ring – $I=\pi R^3\delta$. From these relations, the reduced thickness of the shell is determined for the perception of bending stresses $\delta(\beta) = M_z R \cos\beta/(\pi R^2 \sigma_P)$. Bend moment taking into account its variability, is represented as $M_z = M_{z\ max} F(\bar{x})$, where $M_{z\ max}$ is the maximum moment value, $F(\bar{x})$ is the function that determines a moment distribution on the fuselage. $\bar{x}$ is relative coordinate, for nose part it is $\bar{x} = x/l_{fus.f}$. Then the mass of the construction for perception of M_z is

$$m_{M_z} = 2\int_{\bar{x}=0}^{1}\int_{\beta=0}^{\pi} M_z R \cos\beta/(\pi R^2) d\beta d\bar{x}/\bar{\sigma}_{met} = 8/\pi \int_{\bar{x}=0}^{1} F_{M_z} d\bar{x}\, M_{z\ max} \lambda_{fus}/\bar{\sigma}_{met} = C_K P L/\bar{\sigma}_{met} \quad (18)$$

Here $\int_{\bar{x}=0}^{1} F_{M_z} d\bar{x} = 1/3$ since the bending moment is characterized by a parabolic dependence of the change in the length of the cantilever half of the fuselage; $P=M_{z\ max}$; $L=\lambda_{fus,f} = l_{fus.f}/D$; $C_K = 8/(3\pi)$.

We can do the similar reasonings for the tail part of the fuselage.

5.2. Fuselage from CM.

The design of the fuselage from CM has a lot of specifics [7-10]. The composite fuselage consists from unidirectional structures (the longitudinal ribs: stringers, shelves of longitudinal beams, and shelves of frames), and a skin and walls which are given a laminate structure with the optimization of the directions of reinforcing elements (fibers, threads, tapes).

For a CM structure with annular layers of total thickness δ_R and spiral layers of thickness δ_S with angles $\pm\varphi$, we calculate the mass of the equal-strength structure. In this case, the layered structure is considered an equivalent homogeneous shell. Internal longitudinal and annular forces flows are related to the stresses in the spiral and annular layers by the ratios $N_\alpha = \sigma_{1S}\, \delta_S \cos^2\varphi$, $N_\beta = \sigma_{1S}\, \delta_S \sin^2\varphi + \sigma_{1R}\, \delta_R$, hence

$$\sigma_{1S} = pR/(2\delta_S \cos^2\varphi), \quad \sigma_{1R} = pR/(2\,\delta_R)(2 - \mathrm{tg}^2\varphi). \quad (19)$$

If the allowable stresses reach in all layers $\sigma_{1S} = \sigma_{1R} = \sigma_{f\,P}$, then the thickness of the fiber package at which the mass will be minimal

$$\delta_f = \delta_S + \delta_R = p\, R/(2\, \sigma_{f\,P})/(\cos^2\varphi + 2 - \mathrm{tg}^2\varphi) = 3/2\, p\, R/\sigma_{f\,P}) \quad (20)$$

At the same time, in the general package of CM of thickness δ_{CMp}, it is necessary to take into account the presence of a matrix, which according to the accepted model does not participate in the work. If f is the volume content of fibers in CM $f = \delta_f / \delta_{CM}$, then the thickness of the matrix will be $\delta_m = \delta_f(1-f)/f$.

The mass of the cylindrical shell from CM, which works only on the internal overpressure will be

$$m_p = 2\pi R l (\rho_f \delta_f + \rho_m \delta_m) = 3\, p\, V/(\bar{\sigma}_f)\ [1 + \rho_m (1-f)/(\rho_f f)] = 3\, pV/(\bar{\sigma}_{CM}) \quad (21)$$

where ρ_m, ρ_f are density of matrix and fibers; $\bar{\sigma}_{CM}$ – is thee specific strength of CM $\bar{\sigma}_{CM} = \bar{\sigma}_f / [1 + \rho_m / \rho_f\ (1-f)/f]$, taking into account the accepted model of its work.

From (21) it follows that regardless of the angles of laying layers coefficient of force factor C_K=3.

Analyzing the work of the cylindrical shell from CM from the bend and assuming that the reinforcing fibers are directed along the longitudinal x axis of the fuselage, we obtain a similar formula like (18), but with the corresponding specific strength $\bar{\sigma}_{CM}$.

6. Numerical example

Let 's make an approximate estimate of the weight of the fuselage structure using a traditional metal structure (for testing the technique) and compare it with what will give the transition to a composite one. We do this on the example of a conventional medium-haul aircraft similar in characteristics to the Airbus A-320 aircraft and the new Russian MC-21 being developed , for which we will take the following values of the parameters necessary for calculating: m=80 t; l_{fus} =40 m; D_{fus} =4 m; T=240 kN; $\bar{m}_g$=0.25; $\bar{m}_{str}$=0.3; $\bar{m}_{eq}$=0.1; $\bar{m}_{en}$=0.1; $\bar{m}_f$=0.25; the ratio of the aerodynamic drag coefficients is $C_{D\ fus}/C_D$ =0.3. From $\bar{m}_{str}$=0.3 the third part is accounted for the design of the fuselage – m_{fus}≈80 t·0.1 =8 t. The maximum operational load factor is assumed n= 2.5. Permissible stresses must be set taking into account the resource (100,000 flight hours), flutter, survivability, lack of loss of stability, residual strength after corrosion: specific strength for aluminum alloy elements $\bar{\sigma}_{met}$ =0.3 GPa/2700 kg/m^3; for carbon composites with σ_f =2GPa, with f=0.5 and a polymer matrix $\bar{\sigma}_{CM}$ =0.7 GPa /1500 kg/m^3. The minimum safety factor is 1.5. Additional values of the safety coefficient: for compartments working under pressure – 3.5; for structures made of CM –1.5. To simplify the problem, we also assume that the plane of attachment of the front and rear fuselage is located in the middle ($l_{fus.f}$=$l_{fus.b}$=l_{fus}/2=20m).

We obtained theoretical load-carrying mass from two types of loads m_{met}=3.5 t and m_{CM}=2 t. Performing the analysis for the remaining loads, as we did in sections 5.1 and 5.2, gives a small clarification of the theoretical mass of the load-carrying elements: m_{met}≈4.2 t and m_{CM}≈2.4 t.

We will assume that not bearing mass of the fuselage, which for the metal version was m_{fus}– m_{fus}^{Theor} =8 t–4.2 t=3.8 t, and that it has not changed in the case of the composite version. Then we have $\Delta m_{str\ 0}$ =$m_{fus\ met}$ – $m_{fus\ CM}$≈1.8 t. The paper [11] contained an assessment of such a transition from aluminium load-carrying elements to carbon-composite ones. A reduction in the weight of the fuselage was by 2.5 t, but this value significantly depends on the amount of permissible stresses, which were not resulted.

Let's analyze the decrease in the total weight of the aircraft with these parameters due to the transition from a metal fuselage to a composite one. According (10) we have $\mu_{m\ str}$ =2.2, then

$$\Delta m = \mu_{m\ str}\, \Delta m_{str\ 0} = 3.5\ \text{t}. \qquad (22)$$

7. Conclusion

The considered method of combining two approaches, which are based on the application of the mass change factor and the load-carrying factor, allowed to develop a method of operative[design evaluation of the aircraft mass due to the initial change in the mass of any elements while maintaining the flight performance properties of the aircraft.

References

[1] Edited by M A Poghosyan 2018 *Aircraft Design* (Moscow Innovative engineering) 864 p [In Russian]
[2] Torenbeek E 2013 *Advanced Aircraft Ddesign: Conceptual Design, Analysis, and Optimization of Subsonic Civil Airplanes* (John Wiley and Sons Ltd.) 410 p
[3] Raymer D P 2006 *Aircraft Design. A Conceptual Approach* (4th Edition, AIAA, USA) 838 p
[4] Gogolin V P 1973 *Determination of the growth coefficient of take-off weight changes in the implementation of initial changes in the weight of the structure* (Kazan, Works of KAI) vol 160 pp 11-14 (In Russian)
[5] Komarov V A and Weisshaar T 2002 *A New Approach to Improving the Aircraft Structural Design process* (Journal of Aircraft) March–April vol 39 Num 2 pp 227-233

[6] Komarov V A 2018 *Dimensionless criterion of load-carrying perfection of constructions* (Mechanics of solid body) No 4 pp 33-46 [In Russian]

[7] Vasiliev V V, Vsehvalnov V K, Nikitiuk V A and Puhov A A 2001 *Composite Lattice Geodesic Structure for a Commercial Airplane Fuselage* (Russian scientific and sechnical journal Polyot) Num 11 p 22–25

[8] Boldyrev A V, Kozlov D M, Pavelchuk M V 2017 *Evaluation of Anisogric Composite Lattices Structures Weight Effectiveness Using the Load-carrying Factor* (Proc. Engineering) vol 185 p 153-159

[9] Schwinn D B, Kohlgrüber D, Scherer J, and Siemann M H 2016 *A Parametric Aircraft Fuselage Model for Ppreliminary Sizing and Crashworthiness Applications* (CEAS Aeronaut J) 7 p 357

[10] Quatmann M and Reimerdes H 2011 *Preliminary Design of Composite Fuselage Structures Using Analytical Rapid Sizing Methods* (CEAS Aeronaut J) 2 pp 231–241

[11] Achternbosch, Matthias, Bräutigam and others 2003 *Material Flow Analysis – a Comparison of Manufacturing, Use And Fate of CFRP-fuselage Components Versus Aluminum-components for Commercial Airliners* (Fresenius Environmental Bulletin January).

Estimation Method of Pollutant Emissions in Certain Aircraft Flight

H Song[1], L Yue[1], X Han[2] and Z Wei[2]

[1] Beijing Aircraft Technology Research Institute, Commercial Aircraft Corporation of China, Beijing 102211, China
[2] College of Air Traffic Management, Civil Aviation University of China, Tianjin 300300, China

songhanbing@comac.cc

Abstract: This paper intends to calculate the emissions of pollutants during the route flight quantitatively. In this paper, the flight dynamics model of the climb, cruise and descent phases are established respectively. Then, the pollutant emission estimation model of the aircraft is established based on the ICAO LTO model and the benchmark emissions data of engine manufacturers. Take certain civil aviation aircraft into account, a typical route is simulated, then the emissions of various pollutants at different flight stages are calculated and analyzed. Finally, several typical cruise strategies are selected for further comparative analysis. The results show that CO2 and NOX are the main pollutants during route flight. In addition, the pollutant emissions during the cruise phase are the highest, and the pollutant emissions during route flight can be effectively reduced by choosing the appropriate cruise strategy. The research results can be used to quantify and analyze the impact of flight performance optimization on civil aviation energy conservation and emission reduction, so as to better satisfy the requirements of green operations while ensuring flight safety.
Keywords: environmental engineering; route flight; fuel consumption; pollutant emissions

1. Introduction

With the rapid development of the civil aviation industry, pollutants such as CO_2, NO_x, CO, HC and SO_2 generated by civil aircraft in flight have also attracted more and more attention from the industry and the public. According to the distribution area of emissions, the environmental impact of aircraft engines can be divided into two major areas: the global greenhouse effect and air quality problems within the airport area [1]. In order to control and reduce the impact of the aviation industry on the greenhouse effect, the EU has incorporated the airline's aircraft carbon emissions into the carbon emissions trading system [2]. Although there are still objections to who should be the charging entity, the air industry has agreed that the aircraft emissions should be controlled by market means. The International Civil Aviation Organization (ICAO) has also begun to organize experts to discuss and study global aviation emissions trading mechanisms and control issues based on market models. At the same time, Committee on Aviation Environmental Protection (CAEP) has also established a carbon emission assessment indicator system for civil aviation transportation and design in 2013 [3].

Accurate and rapid estimation of aircraft pollutant emissions is fundamental to emission control and environmental impact assessment, and is also the key to energy efficiency assessment of air traffic control. Based on the engine manufacturer's certification data, the International Civil Aviation

Content from this work may be used under the terms of the Creative Commons Attribution 3.0 licence. Any further distribution of this work must maintain attribution to the author(s) and the title of the work, journal citation and DOI.
Published under licence by IOP Publishing Ltd

Organization (ICAO) has developed an emission database for various aircraft types, which uses the standard take-off and landing cycle (LTO) to calculate the CO, HC and NOx emissions for taxi, take-off, climb, approach and landing phases of aircraft below 1 000 m within the airport area. However, the route emission which is account for nearly 80% time of the flight is not considered [4]. Based on the ICAO emission model, Sherry [5] used discrete data to improve the estimation accuracy by study the increase of airport pollutant emissions when reducing thrust. In order to estimate the impact of actual flight and meteorological conditions on emissions, Boeing proposed the BM2 method to implement the correction of the ICAO model [6], and made a basic study for accurate estimation of pollutant emissions in flight.

Based on the BM2 method, Turgut et al. [7] proposed a method for evaluating the NOx distribution in different directions during the take-off and climb phases of flight. Based on the Australian air traffic control record in 2008, Pham et al. [8] established a 4D emission database list with precision unit of 1 × 1 × 1 000 ft. Mitchell [9] assessed the environmental impact of the departure operation under the speed limit. According to the ICAO database, the BM2 model and the Basic Aircraft Data (BADA) model, the European Air Traffic Control Administration developed the Advanced Emission Model (AEM) to estimate actual emissions data based on 4D track data [10]. Wei Zhiqiang et al. [11], estimated the different pollutant emissions in each flight phase based on the established emission index BM2 correction model. in 2008.

In order to consider the impact of non-carbon pollutant emissions, Cook et al. [12] studied the weight relationship between different pollutants in 2009, proposed the concept of dynamic cost index with emission costs, and used it for flight delay analysis. Schumann [13] established route optimization methods based on trail cloud and fuel consumption, in order to minimize the environmental impact of flight operations. Cruise is the major flight phase of a civil aircraft, which contributes the vast majority of fuel consumption, flight time and pollutant emissions. Wei Zhiqiang et al. [14] established a model of pollutant emission and optimization during the cruise phase, trying to reduce the impact of emissions by optimizing flight parameters.

This paper studies the estimation method of aircraft energy consumption and emissions based on flight dynamics model. Based on ICAO's LTO model, engine manufacturer's baseline emissions data, and Boeing's BM2 model, an aircraft fuel consumption and pollutant emissions estimation model covering all phases of flight was constructed. Visual Studio programming was used to calculate aircraft performance parameters and pollutant emissions at different phases. Based on the above method, a typical route is taken into account for calculation and analysis, and the pollutant emissions under different cruise modes are compared and analyzed in order to consider the effect of aircraft performance optimization on energy saving and emission reduction while ensuring the safe operation of civil aviation.

2. Flight dynamics model

The flight route mainly includes climb, cruise and descent phases. Based on the established flight dynamics model at each stage, the flight parameters of each flight state and the corresponding pollutant emissions can be calculated.

2.1 Climb Phase

When the aircraft is during en-route climbing, the external forces received are engine thrust FN, Drag D, lift L, and gravity W. Its dynamic equation is as follows:

$$\left.\begin{aligned} \frac{W}{g}\frac{dV}{dt} &= F_N \cos\alpha - D - W\sin\theta \\ \frac{W}{g}V\frac{d\theta}{dt} &= F_N T\sin\alpha + L - W\cos\theta \end{aligned}\right\} \tag{1}$$

where α is the angle of attack; θ is the climb track angle; *V* is the flight speed; *t* is the flight time. During the en-route climb, the angle of attack is relatively small, then $\sin\alpha \approx 0$, $\cos\alpha \approx 1$.

Climb rate formula is as follow:

$$r/c = \frac{dh}{dt} = V \sin\theta \tag{2}$$

Substituting formula (1) into formula (2), we get:

$$r/c = \frac{\frac{F_N - D}{W} \cdot V}{1 + \frac{VdV}{gdh}} \tag{3}$$

where $1 + \frac{VdV}{gdh}$ is the acceleration factor.

Climb time:

$$\Delta t = t_2 - t_1 = \frac{h_2 - h_1}{(r/c)_{avg}} = \frac{\Delta h}{(r/c)_{avg}} \tag{4}$$

where Δt is the climb time used by the aircraft to climb the altitudeΔh; $(r/c)_{avg}$ is the average climb rate from h_1 to h_2 during the time Δt.

Under non-standard atmospheric conditions, the true height (geometric height) is not equal to the air pressure height. Δt can be obtained from the relationship between geometric height and air pressure height. The calculation formula is as follows:

$$\Delta t = \frac{h_{p2} - h_{p1}}{(r/c)_{avg}} \left(\frac{T_{ns}}{T_{std}}\right) \tag{5}$$

Then the total climb time is:

$$t = \sum \Delta t \tag{6}$$

Where T_{ns} and $T_{std}$are the absolute temperatures at the non-standard atmosphere and the standard atmosphere, respectively.

The climb track angle $\theta$of a civil airliner is very small, i.e $cos\theta \approx 1$, then the horizontal distance to climb a short altitude is:

$$\Delta S = V_{avg} \Delta t \tag{7}$$

where V_{avg} is the average true airspeed of climb from h_1 to h_2 during Δt.

The total horizontal distance of the climb:

$$S = \sum \Delta S \tag{8}$$

Fuel consumption:

$$\Delta F = W_{F_{avg}} \times \frac{\Delta t}{60} \tag{9}$$

where W_{Favg} is the average fuel flow for the selected altitude intervals.

The total fuel consumption:

$$F = \sum \Delta F \tag{10}$$

2.2 Cruise Phase

Assume the aircraft can meet the conditions of straight and horizontal flight at a constant speed at each moment, and the following dynamic equations are obtained:

$$\left.\begin{aligned} F_N \cos\alpha - D = 0 \\ W - D - F_N \sin\alpha = 0 \end{aligned}\right\} \tag{11}$$

Assume the aircraft maintains a level flight when cruising in static atmosphere, then the lift and the gravity of the aircraft could be balanced, i.e.:

$$L = W = \frac{1}{2}\rho V^2 C_L S_W = \frac{1}{2}\rho_0 M^2 {a_0}^2 \delta C_L S_W \tag{12}$$

where ρ is the air density at the local altitude, C_L is the lift coefficient, S_W is the wing area, ρ_0 is the air density at standard sea level, a_0 is the speed of sound at standard sea level, and δ is the pressure ratio at the local altitude.

2.3 Descent Phase

Descent rate equation:

$$r/d = -\frac{dh}{dt} = V \sin\theta \tag{13}$$

Flight dynamics equation:

$$\left.\begin{aligned} F_N - D + W\sin\theta - \frac{W}{g}\left(\frac{dV}{dt}\right) = 0 \\ L + \frac{W}{g}V\left(\frac{d\theta}{dt}\right) - W\cos\theta = 0 \end{aligned}\right\} \quad (14)$$

Considering that when the aircraft descends, the rate of change of the descent angle is small, hence $\frac{d\theta}{dt} \approx 0$, and the normal descent angle is small, hence $cos\theta \approx 1$, then L≈W.

The descent time is:

$$\Delta t = t_2 - t_1 = \frac{\Delta h}{(r/d)_{avg}} = \frac{(h_{p2} - h_{p1})\frac{T_{ns}}{T_{std}}}{(r/d)_{avg}} \quad (15)$$

$$t = \sum \Delta t \quad (16)$$

The horizontal distance during the descent is:

$$\Delta S = V_{avg}\Delta t \quad (17)$$

$$S = \sum \Delta S \quad (18)$$

3. Pollutant emission calculation model

3.1 ICAO emission index and calculation model

Through statistical analysis of the take-off and landing data for each aircraft type, the International Civil Aviation Organization (ICAO) defines a standard take-off and landing cycle (LTO), including four phases of approach landing, taxiing, take-off and climb. On this basis, engine manufacturers provide emission index data for International Standard Atmosphere Model (ISAM) on sea level based on flight test data. The baseline emissions for certain aircraft with 17GE180-Genx-1B76A/P2 engine is shown in Table 1 (Sample) [15].

Table 1. Basic emission indexes of 17GE180-Genx-1B76A/P2 engine

Flight Phase	Time /s	Thrust Level /%	Fuel Flow /(kg·s^{-1})	Emission Index/(kg·kg^{-1})				
				NO_x	HC	CO	SO_2	CO_2
Take-Off	0.7	100	2.790	0.04690	0.0003	0.0005	0.001	3.15
Take-Off Climb	2.2	85	2.262	0.02678	0.0002	0.0009	0.001	3.15
Approach Landing	4.0	30	0.709	0.01178	0.0004	0.00153	0.001	3.15
Taxiing	26.0	7	0.223	0.05040	0.0036	0.01344	0.001	3.15

According to the definition of the above emission index, an emission calculation model can be obtained.

$$E_{m,i} = nt_iI_{m,i}\mathrm{FF}_i \quad (19)$$

where m is the type of pollutant; i is the flight time; $E_{m,i}$ is emission of m pollutants at time i, kg; n is the number of engines; t_i is the flight time step, s; $I_{m,i}$ is the emission index; FF_i is the fuel flow.

3.2 Fuel flow correction model

The baseline flight state is the flight state assumed to meet the airworthiness certification, which is different from the actual flight state of the aircraft. Therefore, it is necessary to amend the emission index according to actual flight conditions to obtain more accurate pollutant discharge costs. When the types of aviation fuel are specified, the CO_2 and SO_2 emission index is independent of other factors and free of corrections. Since the actual cruise state of the aircraft is different from the baseline flight state, the other pollutant emission index needs to be corrected according to ICAO's BM2 method, in order to consider the impact of cruise parameters such as fuel flow, flight altitude, temperature deviation etc.

1) Correct the fuel flow. Firstly, the fuel flow of the aircraft needs to be converted to the modified fuel flow at ISA and sea level.

2) Calculate the emission index. ICAO provides thrust ratings and emission index for four typical flight conditions, but in actual flight, the emission index would be different from the Table 1 due to the difference between the aircraft's flight status and the ICAO data validation basis. So, the baseline emission index for each pollutant is calculated using a piecewise linear fit with the modified fuel flow based on Table 1.

3) Amend the emission index. The aircraft's emission index is also related to flight altitude, flight speed, atmospheric temperature and humidity.

The specific correction process is shown in Figure 1.

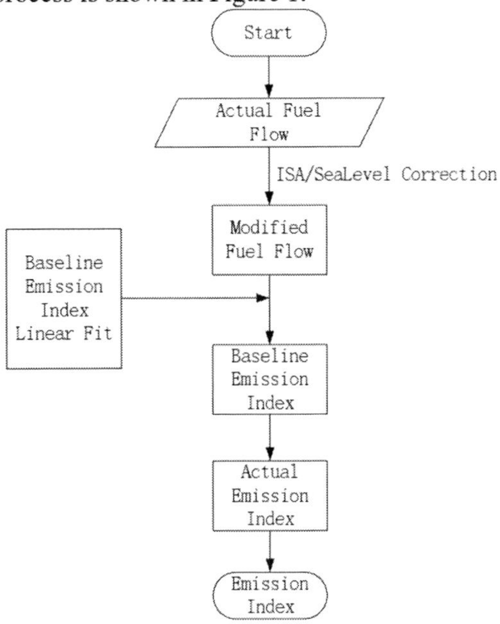

Figure 1. Flow chart of fuel flow correction model

4. Example Analysis

4.1 Calculation and analysis of pollutants on a typical route

Taking a typical route as an example, the flight parameters of the entire route are calculated using the flight dynamics and the pollutant emission model, in order to get flight distance, flight time, fuel quantity and emissions of each pollutant in each flight phase. The calculation conditions are as follows: the flight distance is 9000km, the initial weight of the aircraft is 230,000kg, the temperature deviation is 0°C, the climb strategy is constant speed / equal Mach number (250kt / 300kt / 0.8 / 0.85) climbing, and the cruise strategy is step cruise with fixed Mach number (FL370- FL410 / M0.85), the descent strategy is symmetric to the climb. The detailed calculation results are shown in Table 2.

Table 2. Aircraft performance parameters and emissions calculation results on a typical route

Flight phase	Flight distance /km	Time/s	Fuel/kg	CO_2/kg	NO_x/kg	CO/kg	HC/kg	SO_2/kg
Climb	336.60	1560.77	4737.98	14948.33	96.13	2.03	0.18	4.74
Cruise	8399.40	33489.38	51586.81	162756.40	525.77	125.94	3.80	51.59
Descent	264.00	1367.17	468.35	1477.64	1.98	15.51	0.42	0.47
Sum	9000.00	36417.32	56793.14	179128.40	623.88	143.49	4.40	56.79

From the aspect of pollutant emissions, CO_2 emissions are the most in the entire route, followed by NO_X, and the emission of CO, SO_2 is relatively small and HC emission is the least. From the aspect of flight phase, the emissions of pollutants during the cruise phase are the highest. For CO_2, NO_X, and SO_2, the emissions during the climbing phase are greater than those during the declining phase, while CO and HC are the opposite. The calculation results show that the pollutant emissions are different in different flight stages. For the cruise stage with the largest emissions, the cruise parameters can be optimized to reduce the pollutant emissions of the entire route.

4.2 Impact of cruise strategy on pollutant emissions

The aircraft performance parameters and emissions at different cruise altitudes are calculated and analyzed based on the calculation model established. Table 3 shows the detailed calculation results at different cruise altitudes. The rest of the calculation conditions are the same as in Section 4.1.

Table 3. Aircraft performance parameters and emissions at different cruise altitudes

Cruise Strategy	Phase	Distance /km	Time/s	Fuel/kg	Pollutant Emissions/kg				
					CO_2	NO_X	CO	HC	SO_2
Step Cruise (FL350-FL410/M0.85)	Climb	260.63	1255.57	4066.18	12828.80	86.94	1.39	0.14	4.07
	Cruise	8475.19	33745.77	52003.79	164071.96	538.45	125.04	3.77	52.00
	Descent	264.12	1367.89	468.60	1478.43	1.99	15.62	0.42	0.47
	Overall	8999.95	36369.23	56538.56	178379.16	627.39	142.05	4.33	56.54
Fixed Altitude Cruise (FL350/M0.85)	Climb	260.63	1255.57	4066.18	12828.80	86.94	1.39	0.14	4.07
	Cruise	8512.34	33771.91	52697.02	166259.10	564.08	113.09	3.51	52.70
	Descent	227.06	1219.63	428.48	1351.85	1.86	12.73	0.34	0.43
	Overall	9000.03	36247.11	57191.68	180439.75	652.89	127.21	4.00	57.19
Fixed Altitude Cruise (FL370/M0.85)	Climb	336.60	1560.77	4737.98	14948.33	96.13	2.03	0.18	4.74
	Cruise	8424.70	33590.26	51670.16	163019.35	530.62	122.87	3.70	51.67
	Descent	238.67	1266.15	441.00	1391.36	1.90	13.58	0.37	0.44
	Overall	8999.97	36417.18	56849.13	179359.01	628.65	138.48	4.25	56.85

Table 3 shows that the cruise strategy does have obvious impact on the time and pollutant emissions. Fixed altitude cruise (FL350 / M0.85) offers the least flight time, yet the step cruise (FL350-FL410 / M0.85) gives the least fuel consumption and CO_2/NOX emissions. Therefore, given the same flight Mach number, the pollutant emissions under step cruise can be reduced up to 1.14% of CO_2/SO_2, and 3.91% of NO_X compared with the cruise at a fixed altitude.

Further analysis of the impact of cruise Mach number on pollutant emissions during route flight is carried out. Aircraft performance parameters and emissions are calculated at different cruise Mach numbers. The other calculation conditions are the same as in Section 4.1. Table 4 shows the detailed calculation results under different cruise Mach numbers.

Table 4. Aircraft performance parameters and emissions at different cruise Mach numbers

Cruise Strategy	Phase	Distance/km	Time/s	Fuel/kg	Pollutant Emissions/kg				
					CO_2	NO_X	CO	HC	SO_2
Fixed Altitude Cruise (FL350/M0.8)	Climb	260.63	1255.57	4066.18	12828.80	86.94	1.39	0.14	4.07
	Cruise	8518.91	35910.35	52026.38	164143.23	528.54	132.67	3.70	52.03
	Descent	220.46	1312.59	451.41	1424.20	1.85	12.17	0.33	0.45
	Overall	9000.00	38478.50	56543.97	178396.23	617.33	146.23	4.17	56.54
Fixed Altitude Cruise	Climb	260.63	1255.57	4066.18	12828.80	86.94	1.39	0.14	4.07
	Cruise	8519.09	34822.90	52341.15	165136.33	545.85	122.95	3.58	52.34

(FL350/M0.825)	Descent	220.27	1311.61	451.07	1423.13	1.81	12.10	0.33	0.45
	Overall	8999.99	37390.08	56858.39	179388.22	634.61	136.44	4.05	56.86
Fixed Altitude Cruise (FL350/M0.85)	Climb	260.63	1255.57	4066.18	12828.80	86.94	1.39	0.14	4.07
	Cruise	8512.34	33771.91	52697.02	166259.10	564.08	113.09	3.51	52.70
	Descent	227.06	1339.63	458.48	1446.50	1.86	12.73	0.34	0.46
	Overall	9000.03	36367.11	57221.68	180534.40	652.89	127.21	4.00	57.22

As can be seen from Table 4, FL350 / M0.85 offers the least flight time, yet FL350 / M0.8 obtains the least fuel consumption and CO_2, NO_X, SO_2 emissions, and FL350 / M0.85 gets the least CO and HC emissions. Because CO_2 and NO_X are the main pollutants during flight, at the same cruising altitude, the pollutant emissions can be reduced up to 1.18% of CO_2, SO_2 and 5.45% NO_X under low flight Mach numbers compared with high flight Mach numbers.

5. Conclusion

Based on the ICAO LTO model and Boeing's BM2 method, this paper constructs an aircraft engine emissions estimation method. Taking certain aircraft as example, the performance simulation of typical route is carried out to calculate the flight performance parameters and pollutant emissions during different flight phases. In addition, pollutant emissions under different cruise strategies are compared and analyzed.

The conclusion is summarized below:

(1) CO_2 emissions account for an absolute proportion of various pollutants, exceeding 99%, followed by NO_X, CO, SO_2, and HC.

(2) Pollutants are emitted mostly during the cruise phase during route flight.

(3) Under the same flight Mach number, the step cruise can effectively reduce the pollutant emissions during route flight compared with the fixed cruise strategy.

(4) Flying at the same altitude, the low Mach number emits less pollutants compared with the high Mach number.

Reference

[1] Kofei B, Szopa S, Cozic A 2010 Present and future impact of aircraft, road traffic and shipping emissions on global tropospheric ozone *Atmospheric Chemistry&Physics Discussions*: 11681–11705

[2] Robert M, Dominic M, Niven W 2012 The impact of the European union emissions trading scheme on US aviation *Journal of Air Transport Management*, pp 36-41

[3] Michelle K, Dongwook L, Taewoo N 2013 Investigation of Notional CO2 Standard Applicability to Fleet-wide Fuel Efficiency *Proc. Int. Conf. on Aviation Technology, Integration, and Operations Conference, Los Angeles*, CA: American Institute of Aeronautics and Astronautics

[4] ICAO 2009 Engine Exhaust Emissions Data Bank Doc 9646-AN/943, Issue 16A

[5] Sherry L 2015 Improving the accuracy of airport emissions inventories using disparate datasets *IIE Transactions*, pp 577-585

[6] Martin RL 1995 Boeing Method 2 Fuel Flow Methodology Description *CAEP Working Group III Certification Subgroup*

[7] Turgut E T, Usanmaz O, Rosen M A 2013 Estimation of vertical and horizontal distribution of takeoff and climb nox emission for commercial aircraft *Energy Conversion and Management*, pp 121-127

[8] Pham V, Tang J, Alam S, Lokan C 2010 Aviation emission inventory development and analysis *Environmental Modelling & Software*, pp 1738–1753

[9] Mitchell D, Ekstrand H, Prats X, et al. 2012 An Environmental assessment of air traffic speed constraints in the departure phase of flight: a case study at Gothenburg Landvetter Airport, Sweden *Transportation Research Part D: Transport and Environment*, pp 610-618

[10] Eurocontrol. AEM[EB/OL]. [2019-7-24] http://www.eurocontrol.int/environment/public/standard_page/AEM.html
[11] Wei Zhiqiang, Wang Chao 2010 Estimating method of pollution emissions for scheduled flight in different phases *Journal of Traffic and Transportation Engineering*, pp 48-52
[12] Cook A, Tanner G, Williams V, Meise G. 2009 Dynamic cost indexing Managing airline delay costs *Journal of Air Transport Management*, pp 26-35.
[13] Schumann U, Graf K 2011 Mannstein H, Potential to reduce the climate impact of aviation by flight level changes *Proc. Int. Conf. on 3rd. AIAA Atmosphere Space Environments*, Honolulu, Hawaii
[14] Wei Zhiqiang, Diao Huazhi, Han Bo. 2014 Research on calculating of aircraft pollution emissions in cruise phase *Science Technology and Engineering*, pp 122-127
[15] European Union Aviation Safety Agency. ICAO aircraft engine emissions databank [EB/OL] 2019 https://www.easa.europa.eu/easa-and-you/environment/icao-aircraft-engine-emissions-databank#group-easa-downloads

Boundary Layer Transition Measurement Flight Test Based on Infrared Image Technique

Hai Zhong, Shenjun Qi and Yalong Wang

Chinese Flight Test Establishment, Xi'an, China

E-mail address: yalong2015@163.com

Abstract. According to the requirements of boundary layer transition detection measurement flight test, the study on the detection technique of infrared thermal image in real flight conditions was carried out. In this paper, the basic principle of infrared measurement technique was introduced firstly. Through the analysis of the main factors, such as emissivity, radiation incidence angle and temperature difference between the measured object and environment, the test means were determined, which includes surface treatment, infrared light path adjustment, solar radiation and internal heating of tested parts. Then the whole flight test method of boundary layer transition measurement was designed, including selecting test object and test instrument, determining test procedure and image processing procedure, and giving transition location criterion. Finally, the validity of this flight test method was demonstrated through testing results, the research of which would be helpful and meaningful in boundary layer transition measurement.

1. Introduction

Drag reduction has always been one of most concerned problems in aviation, which is more significant for increasing civil aircraft efficiency. Recent years, the technology of drag reduction in laminar flow has attracted more and more attention, especially natural laminar flow wing design and hybrid laminar flow control method, which would be one of most important way of drag reduction in future civil aircraft design.

In order to demonstrate the effect of hybrid laminar flow control and laminar wing design in flight condition, flight test should be design and conducted. In early stage of flight test for laminar aerodynamic characteristics, boundary layer transition is frequently detected by oil-flow method[1] or sublimation techniques[2]. Measurement results of these methods may be affected by the visualization material, especially the operation of sublimation techniques is hard to ensure test duration and difficult to apply and control in flight test. Comparing with other test method, Infrared image technique has its advantages as one non-contact global measuring method, such as no interference to flow, high temperature resolution, high space resolution and easy operation. Infrared image technique has been conducted flight test in test aircraft[3-6], like LFU 205,VFW614/ATTAS, Do 228,Fokker 100, but boundary layer transition application was only implemented in wind tunnel in China[7-9]. As civil aircraft program like C919 is proceeding, it is urgent to demonstrate drag reduction technique in laminar flow and carry out boundary layer transition flight test.

In this paper, infrared image technique was applied in boundary layer transition flight test to explore transition location during real flight conditions. Firstly, the article introduces the principle of measurement for boundary layer transition using infrared image techniques. And the main factors

Content from this work may be used under the terms of the Creative Commons Attribution 3.0 licence. Any further distribution of this work must maintain attribution to the author(s) and the title of the work, journal citation and DOI.
Published under licence by IOP Publishing Ltd

affecting boundary layer transition are analyzed, such as infrared emissivity, radiation incidence angle and temperature difference between test object and background environment. Moreover, certain solutions are proposed, including matted painted on surface of test object, infrared ray direction adjusting, solar radiation and internal heating, which are useful for boundary layer transition location exploration in flight test demonstration. Furthermore, flight test method is introduced and results are analyzed respectively. Through research of boundary layer transition using infrared image technique, an integrated theory of transition flight test has been established, which is helpful and meaningful for aircraft design.

2. Principle of measurement

2.1 Skin friction and heat transfer

When aircraft fly in the atmosphere, boundary layer transition would be caused due to instability mechanism in wing surface and boundary layer was divided into laminar and turbulent flow, which are two totally different flow form. Smooth and continuous laminar flow move along main flow direction, and do regular and stratified movement, while the turbulent flow mix unordered and violent. Hence, under the same condition, the heat transfer of turbulent is greater than laminar. According to Reynolds analogy, the heat transfer coefficient α_h is determined by following parameter in a simplified form[10]:

$$\alpha_h = \frac{1}{2} C_f^{'} \frac{U\lambda}{\upsilon} \tag{1}$$

Where, $C_f^{'}$ is the skin friction coefficient, U is the local velocity, λ is the heat conductivity of the fluid and υ is the kinematic viscosity.

For the civil aircraft design, λ does not change with pressure. The formula (1) explains that heat transfer in a laminar boundary layer is small since skin friction coefficient is small. The surface temperature in laminar boundary layer could only be transferred to the temperature of free flow slowly. However, ignoring some transition range conditions, skin friction coefficient in turbulent boundary layer would increase dramatically compared with laminar boundary layer, and heat transfer is much increased. Hence, turbulent boundary layer would absorb heat quickly from outside free flow and reach the temperature of free flow.

In a certain amount of time, the temperature of laminar boundary layer is higher than turbulent boundary layer. Through measurement, transition location of boundary layer could be estimated from the temperature difference showed in temperature distribution of test zone.

2.2 Infrared image technique

Infrared image technique is one non-contact passive measuring method. The principle of Infrared image technique is using that the object in natural world which temperature is higher than absolute temperature will emit infrared radiation automatically, which can be used to estimate temperature distribution and temperature of test target.

Infrared image technique was applied in boundary layer transition measurement, and the principle of measurement mechanism is that heat transfer would cause surface temperature change of test object when air flow pass the surface of measured piece. When the flow past the wing surface is transferred from laminar to turbulent, heat transfer is different between flow and surface due to different status of flow and temperature of surface would be changed.

Through Infrared detection equipment Infrared radiation signal emitted by test object was converted to electric signal according to infrared photoelectric effect. After a series of transformation, like signal amplification, setting standard black-body, correction and calibration[11], the electrical signal was converted to visual signal so that infrared image system could stimulate spatial distribution of surface temperature of test object. After system processing, infrared image visual signal is shown in gray scale or pseudo-color, which is called infrared image. Hence, temperature field of test object could be obtained directly by infrared image which shows measured temperature of test object in real time and boundary layer location could be estimated by temperature difference.

3. Influence factors analysis

3.1 Radiation coefficient

Boundary layer transition based on infrared image technique needs to explore infrared radiation of test object and background environment. So the definition of radiation coefficient is that the absolute value of exitance difference between test object and background environment divides the maximum of radiation exitance of test object and background environment [12]. The radiation coefficient can be described as follows[12]:

$$C = \frac{|M_T - M_B|}{\max\{M_T, M_B\}} \tag{2}$$

Where C is radiation coefficient, M_T is radiation exitance of measured piece, M_B radiation exitance of background environment.

From Stefan-Boltzmann's law, effective radiation exitance of test object and background can be expressed in following form:

$$M_T = \xi_T \sigma T_T^4 + (1-\xi_T)\sigma T_B^4 \tag{3}$$

$$M_B = \xi_B \sigma T_B^4 + (1-\xi_B)\sigma T_T^4 \tag{4}$$

where ξ_T is emissivity of test object, ξ_B is emissivity of background environment, σ is Stefan-Boltzmann's constant, T_T is surface temperature of test object, T_B is temperature of background environment.

Assuming that temperature of background environment is constant, using equation (3) and (4), formulation (2) can be rewritten as follows:

$$C = \frac{4}{1+\frac{1-\varepsilon_T}{\varepsilon_T}\left(\frac{T_B}{T_T}\right)^4}\frac{\Delta T}{T_T} + \frac{1-\left(\frac{T_B}{T_T}\right)^4}{1+\frac{1-\varepsilon_T}{\varepsilon_T}\left(\frac{T_B}{T_T}\right)^4}\frac{\Delta\varepsilon}{\varepsilon_T} \tag{5}$$

Where ΔT is temperature difference between test object and background environment, $\Delta\varepsilon$ is emissivity difference of test object and background environment.

From equation (5), it can be seen that the main factors affecting radiation coefficient are temperature difference between test object and background environment and infrared emissivity difference between test object and external environment.

3.2 Emissivity and Radiation incidence angle

Emissivity is the ratio of radiation energy of object to that of blackbody in theory at a certain temperature, which falls between 0 and 1. During flight test of boundary layer transition exploration, emissivity of atmospheric environment could not be changed generally, and emissivity of test object is related to material of object, surface condition, reflectivity and opacity. Among these factors, material of object is the most important factor affecting emissivity and oxygenation efficiency, surface treatment and surface contamination affect emissivity of test object too. In general, emissivity of nonmetallic and material with low surface gloss and low opacity is higher than metallic and material with high gloss and high transparency.

Meanwhile, energy distribution of infrared radiation is totally different in different direction, so emissivity $\xi(\theta)$ was introduced for further expression, which is defined by

$$\xi(\theta) = \frac{L(\theta)}{L_b} \tag{6}$$

IOP Conf. Series: Materials Science and Engineering **887** (2020) 012011 doi:10.1088/1757-899X/887/1/012011

Where θ is radiation incidence angle, $L(\theta)$ is directional radiation intensity which is tilted θ relative to normal radiant surface, L_b is directional radiation intensity of black-body at the same temperature.

For nonmetallic material, directional emissivity changes a little when θ=0°~60° and directional emissivity decreases obviously when θ exceeds 60°. For Metallic material, directional emissivity has a little change when θ varies from 0° to one certain angle, which is relative with material, and then directional emissivity decreases rapidly when θ increases. But When θ=0°~90°, $\xi(\theta)$ is rather small and could not achieve the requirement of infrared emissivity ($\xi(\theta)$>0.5). Directional radiation properties of nonmetallic and metallic surface are shown in figure 1.

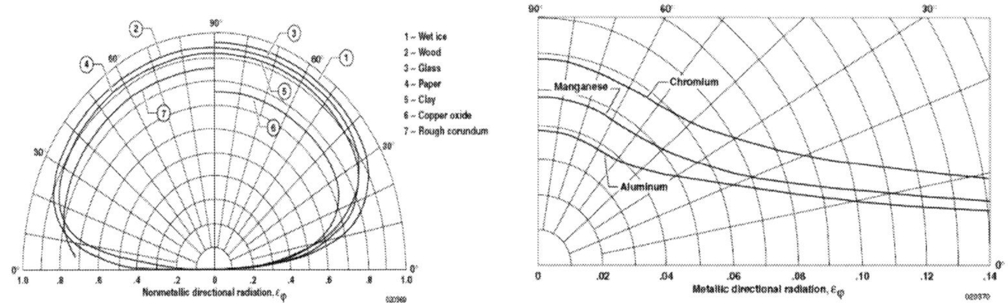

(a) Nonmetallic b) Metallic

Figure 1. Directional radiation properties of nonmetallic and metallic surface

Hence, considering flight test condition, there are certain actions that could be taken during flight test. Firstly, under the premise of structural strength during research process of test object, Nonmetallic with high infrared reflectivity like composite material could be used as surface structure material. If test object is metallic or the surface is painted smoothly, black paint with high infrared reflectivity should be chosen for spray. If infrared ray direction could be selected properly, small radiation angle of incidence could help get high radiation intensity.

3.3 Temperature difference

Temperature difference between test object surface and background environment, could be obtained through certain measures such as changing flight altitude, solar radiation and internal heating.

(1) changing altitude. flight height is changed based on the temperature gradient of different pressure heights. Therefore, temperature difference could be obtained through changing flight height where heat exchange could be made through temperature changing of external free flow. The advantage of this measure is requiring no additional equipment, but this measure demands that the range of altitude change greatly to acquire larger temperature difference. And the aircraft changes altitude during flight, heat of aircraft transfer to atmosphere through thermal transmission, and finally effect is not ideal by this kind of measure to obtain temperature difference.

(2) External solar radiation heating. External solar radiation heating through heating surface of test object by solar radiation makes the temperature of surface increase, and temperature difference is obtained. The advantage of external heating is same with that of changing altitude. But image could not be recognized due to overexposure, which caused by sunlight reflecting off the lens of infrared image camera. When solar radiation intensity is not enough, such as clouds, temperature difference could not be got. Meanwhile, it is necessary to paint surface dark in order to acquire better endothermic effect.

(3) Internal heating. Generally, internal heating is laying heating film or heater strip in interior surface of test object to achieve quick heating. Surface temperature of test object increases due to ohmic heating, so temperature difference is formed. The advantage of internal heating is that temperature can be controlled quantitatively by pilot or engineer, but the disadvantage is that cost and workload of design and modification are increasing dramatically. Meanwhile, it is worth noting that

transition location may be affected by convection heat exchange between external free flow and test object. According to [9], there is no influence on transition location when temperature difference is controlled within 20°C.

4. Conducting Flight test

4.1 Test object

Test aircraft is one low-winged and tail mounted engine business aircraft and the middle of left wing was used as the test area which was shown in figure 2.

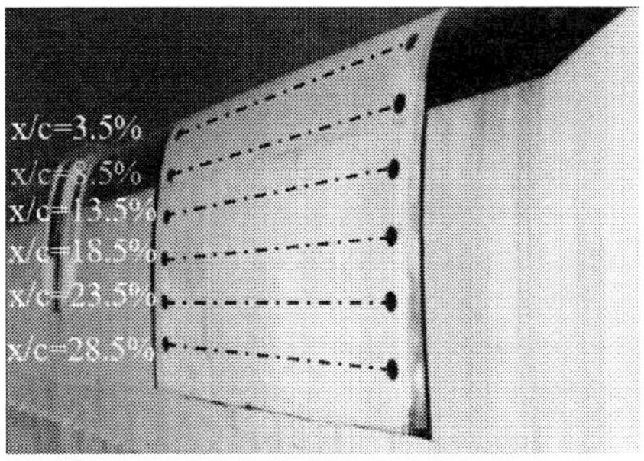

Figure 2. Boundary layer transition measurement region

Table 1. Flight test condition

Pressure altitude	Mach	Reynolds number	no heating measurement	changing altitude	solar radiation heated	Internal heated
7000m	0.65	1.595×10^7	-	-	-	√
7000m	0.60	1.472×10^7	-	-	-	√
7000m	0.55	1.347×10^7	√	-	-	-
7000m	0.55	1.347×10^7	-	√	-	-
7000m	0.55	1.347×10^7		-	√	-
7000m	0.55	1.347×10^7	-	-	-	√
6500m	0.65	1.682×10^7	-	-	-	√
6500m	0.60	1.553×10^7	-	-	-	√
6500m	0.55	1.424×10^7	-	-	-	√
6000m	0.65	1.766×10^7	-	-	-	√
6000m	0.60	1.637×10^7	-	-	-	√
6000m	0.55	1.491×10^7	-	-	-	√

The material of wing cover in test region is aluminum alloy and the surface of wing cover is white, polished and painted, the chord length of middle test zone is 2.1 meter. The span-wise range is 1.91 to 3.21 meters between test zone and wing-root and the leading edge sweep angel is 9°. In chord position x/c=0%~33% of test zone, the surface was painted black coating with high infrared emissivity. In boundary layer test zone besides wingtip and wing-root, aluminum foil was pasted as reference marks at intervals of 5 percent of chord length of the wing, which are x/c=3.5%, 8.5%, 13.5%, 23.5%, 28.5%. All of these marks are used for infrared image geometric coordinate transformation and transition location identification during data processing.

4.2 Testing instrument

One FLIR A655cs infrared image camera was mounted besides cabin window, shown in figure 3. The optical field of view is 45°× 45°,the measurement accuracy of absolute temperature is 0.1°C,the range of measurement is -40°C to 650°C, the range of response wavelength is 7.5μm to 14μm,Pixel resolution is 640 × 480.

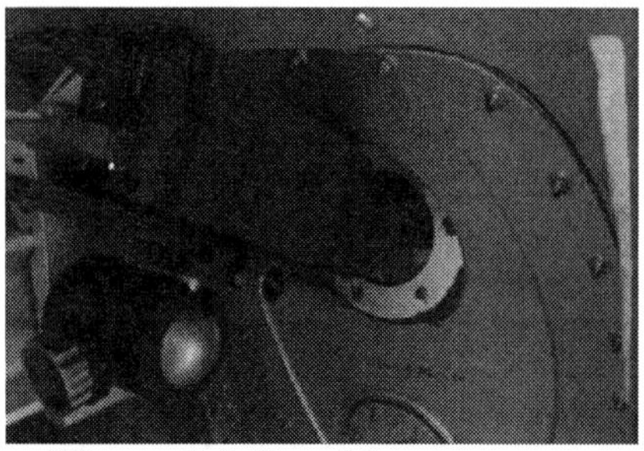

Figure 3. infrared image camera

4.3 Testing Method

Three flight levels were chosen between 6000 and 7000 meters, which is pressure altitude. Three velocities were selected between 0.5 and 0.65 Mach Number. The range of Reynolds number is 1.3×10^7 to 1.7×10^7. The measurement of boundary layer transition was conducted in steady level flight, and the whole flight conditions are shown in table 1.during flight test, aiming at temperature difference between test area and background environment, the following steps are taken respectively:

(1) No heating measure: test area of wing under shadow of fuselage and no sun radiation.

(2) Altering flight altitude: test aircraft climbs to the altitude which is over than scheduled test height, then descend rapidly to determined altitude. Test area under shadow of fuselage and no sun radiation.

(3) sun radiation: through changing heading of aircraft, let test area leave shadow of fuselage and adopt sun radiation

(4) internal heating: test area under shadow of fuselage and no sun radiation, using heating unit mounted inside leading edge to heat test area for 30 seconds, the heating location is about x/c=0%~8%.

4.4 Infrared image processing method

The whole process of infrared image is shown as follows. The original infrared images were obtained from flight test, which were given normalized treatment. Then infrared image gray field was calculated, and the range of gray field is 0 to 1. Furthermore, in order to eliminate lens distortion, geometrical deformation was given to infrared images so that sweep-back wing image could be transformed to rectangular wing image and redundance of image was cropped for next step. Furthermore, temperature gradient field was calculated based on infrared images and each transition location of span-wise section was estimated according to the extreme value of temperature. Statistics are made for all transition locations of span-wise section, and most precise location are indicated based on curve of maximum probability density. The procedure of infrared image processing is shown in figure 4.

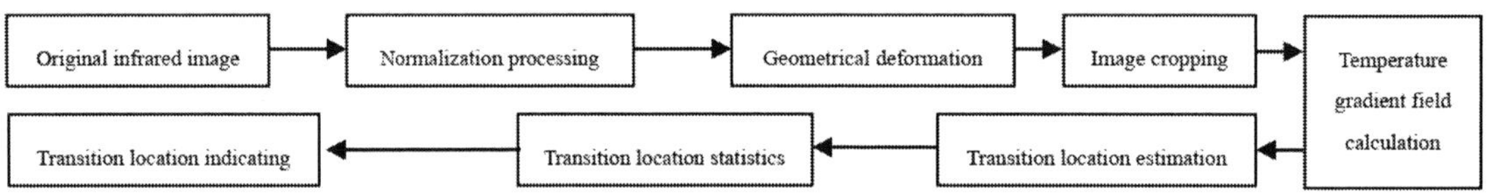

Figure 4. infrared image processing procedure

4.5 Transition location measurement accuracy

In order to verify the accuracy of boundary layer transition measurement results, One trip strip was applied on the surface used rough strips with 0.1mm thick, 1.2mm diameter and 2mm at interval at *x/c*=3.5% and one third span location near wing-let, which is based on the acquired natural transition location *x/c*=7.4%~7.8% in measured region. After this measure taken, boundary layer transition was happened in this region and the location is *x/c*=3.5%. In the subsequent processing of infrared image, transition location measurement accuracy could be given on the basis of difference between x/c=3.5% and indicated transition location estimated by measured transition location accuracy zone.

5. Results analysis

The pressure height of the flight test is 7000m and Mach number of the flight test is 0.55, the flight test is conducted under the condition of no heating measurement, changing flight height, sun radiation heating and using the leading edge deicing equipment to heat the measured region x/c=0%~8% respectively, the results under these four conditions are given by figure 5-figure 8. Figure 7 also shows the indicated transition position based on the accuracy of the transition position measurement.

From figure 7, it can be seen that the measured fixed transition position under accuracy region of the transition position measurement is x/c=3.12%. Meanwhile, the fixed transition position of other flight conditions is calculated, of which the difference value with the forced transition roughness strip is 0.2%~0.4%. The results show that it's credible to measure the transition position by infrared image technique.

According to figure5-figure8, it can be concluded that under the certain flight condition that pressure height is 7000m and the Mach number is 0.55, the extreme value of the temperature of the span-wise profile is scattered under non heating measure condition, and the transition location can't be given by statistics. When the flight height is changed, infrared image can be used to identify the temperature boundary between laminar boundary and turbulent boundary before and after transition happened. Although some extreme values of temperature of the span-wise profile are scattered, the transition location can still be given as x/c=7.44%. Under solar radiation condition, temperature difference can be identified by infrared image, the transition location can be given as x/c=6.30%. In the measured region, there are three extreme values of temperature of span-wise profile which are close to leading edge, and it's probably induced by the contamination of the measured object's surface. Under the condition of leading edge heating, the radiation contrast of infrared image is obvious and clear, where the extreme value of temperature of the span-wise profile is uniform, and the transition position is given as x/c=7.69%. Under the condition of heating the inside of the measured object, the heating power can be calculated by Newton cooling law which is given by equation (7); Temperature difference is set as 5°C and area power is set as 400W/m^2, the heating power is estimated by reference[10].

$$Q = \bar{h}A(T_0 - T_\infty) = 2\frac{k}{L}\sqrt{\mathrm{Re}_L}(0.332\,\mathrm{Pr}^{1/3})A\Delta T \tag{7}$$

Where Q is heating power ; $\bar{h}$ is convective hear transfer coefficient; A is the area of the measured region; T_0 is the temperature of background environment; k is thermal conductivity; L is characteristic length of the heat transfer surface; Re_L is Reynolds number corresponding to the characteristic length; Pr is Prandtl number; ΔT is the temperature difference between the surface of the measured object and background environment.

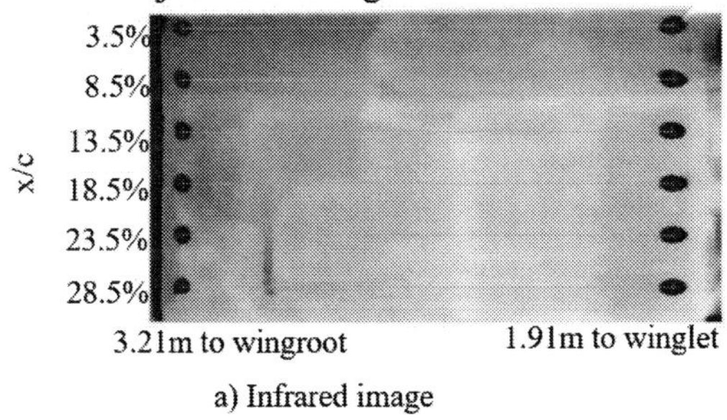

a) Infrared image

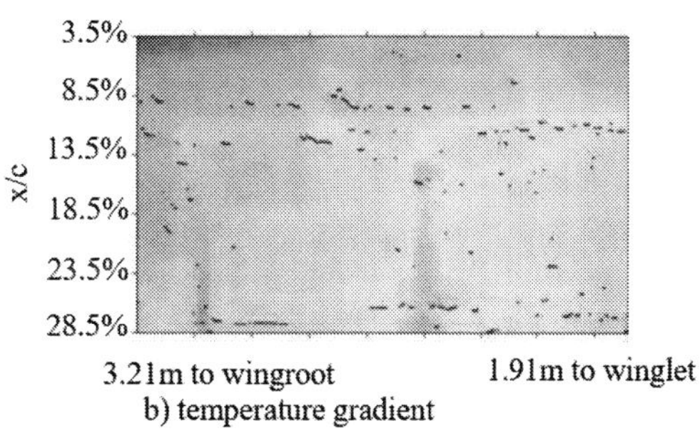

b) temperature gradient

Figure 5. Hp=7000m, M=0.55, no heating

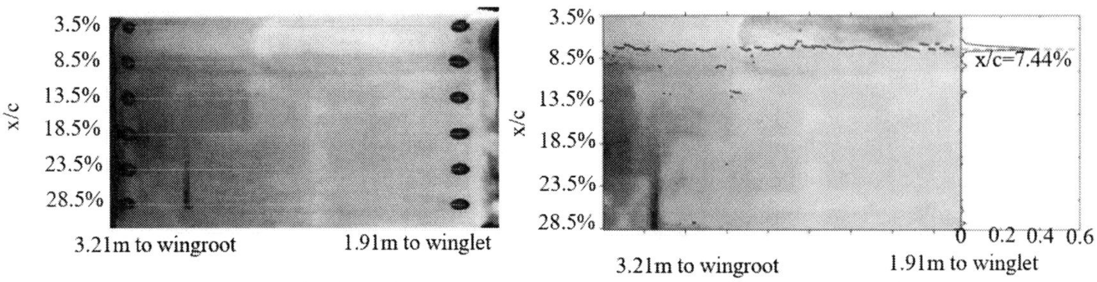

a) Infrared image b) temperature gradient

Figure 6. *Hp*=7000m, *M*=0.55, alter flight altitude

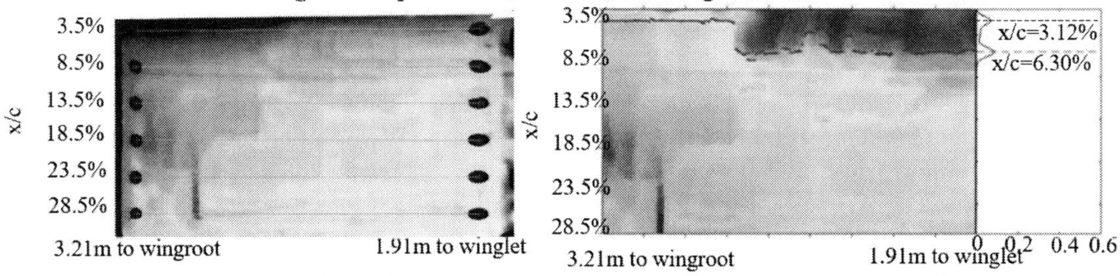

a) Infrared image b) temperature gradient

Figure 7. *Hp*=7000m, *M*=0.55, sun radiation

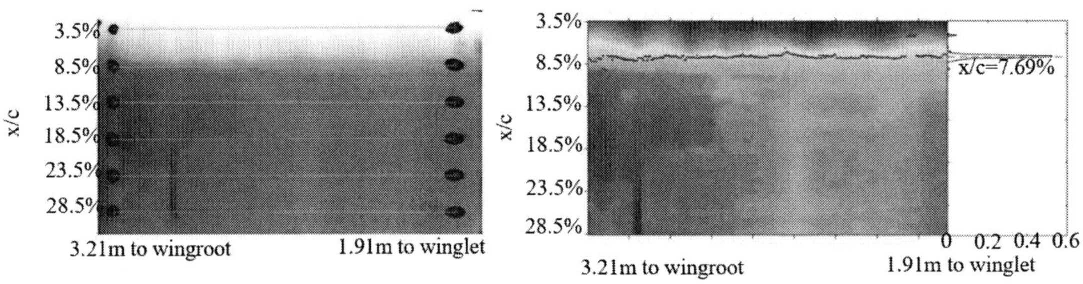

a) Infrared image b) temperature gradient

Figure 8. *Hp*=7000m, *M*=0.55, internal heating

6. Conclusions

Considering the problem of the application of infrared image technique on boundary layer transition measurement in real flight conditions, the main influence factors, such as radiation coefficient, radiation incidence angle, temperature distinction of measured object and background environment, are analyzed firstly in this paper. Then the flight test is designed and established, and according to flight test data analysis, the following conclusions can be given: in flight test conditions, by spraying the black paint on surface, adjusting reasonably infrared light path and heating the measurement object, infrared image technique can be used for flight test of boundary layer transition measurement, results of measured transition position are credible. Under the same flight condition, it is suggested that heating test object inside is the best method to get radiation contrast effect in order to get temperature difference of the measured object and background. In practical engineering application, calculation equation of heating power or area power can be used to calculation or estimate the needed heating power.

Reference

[1] Curry R E,Meyer R Jr. The use of oil for in-flight flow visualization[R]. NASA TM 84915 2001.
[2] Main-Smith J D. Chemical solids as diffusible coating films for visual indication of boundary layer transition in air and water[R]. ARC British R&M No.2755,1950.
[3] Horstmann K H,Quast A,Redeker G. Flight and wind-tunnel investigation on boundary layer transition at reynold numbers up to 10 to the 7th[R]. Proceedings of the 16th ICAS Congress Vol. 2,Jerusalem,Israel,1988:979-986.
[4] Horstmann K H,Redeker G,Quast A,Dressler U,Bieler H. Flight tests with a natural laminar flow glove on a transport aircracft[R]. AIAA 90-3044,1990.
[5] Horstmann K H,Komer H,Wagner B,Welte D. Natural laminar flight test investigaion for commuter aircraft application[R]. Proc. Second European Forum on Laminar Flow Technoloyg,1996.
[6] Voogt N. Flight testing a fokker 100 test aircraft with laminar flow glove[R]. Proc. Second European Forum on Laminar Flow Technoloyg,1996.
[7] Bian yuzhong, Xu Tiejun, Fan Jiechuan et. Investigation of synthetic diagnosis technique of boundary layer[J]. Experiments and Measurements in Fluid Mechanics,2000,14(4):7-13.
[8] Li Rongrong, Zhangjin, He Chunhua. Application of test measurement technique on Low-temperature High Reynolds Number wind tunnel [C]. Proc. Sixth conference on aerodynamic measurement techniques .Hu Bei:China aerodynamic society, 2013:24-31.
[9] Geng Zihai, He Xianzhong, Wang Xunnian et. Non-intrusive test technique investigation of transition measurement with infrared image in low speed wind tunnel [J]. Journal of experiments in fluid mechanics,2010,24(6):77-82.
[10] Fisher D,Horstmann K H,Riedel H. Flight test measurement techniques for laminar flow[M].RTO of NATO,2003:4.3-4.5.
[11] Tian Yupeng, Wang Tiecheng and Zhao Zhongfu. Infrared imaging technique for surface flow visualizaition[J]. Measurement in Fluid Mechanics, 1998,12(1): 88-91.
[12] Zhang Jianqi, Fang Xiaoping. Infrared Physics[M]. Second Edition. Xi'an:Xidian University Press, 2013:86-92.

IOP Conf. Series: Materials Science and Engineering **887** (2020) 012012 doi:10.1088/1757-899X/887/1/012012

Research on Reliability Evaluation Method of Multi - rotor UAV Airborne Controller

Gaowei Zhang [1,2] and Xiaoqun Dong [3]

[1] Sch. of Auto., Beijing Inst. of Technol., Beijing, China
[2] Unit 92785 of the Chinese PLA, Qinhuangdao, China
[3] The fifth research institute of MIIT, P.R.China, Guangzhou, China

zhanggaowei001@126.com

Abstract. The reliability of multi-rotor Unmanned Aerial Vehicle (UAV) and the airborne controller is one of the research focus in UAV filed. In order to evaluate the reliability life of multi - rotor UAV airborne controller, this paper studied the basic principle of reliability life qualification firstly. Secondly, this paper focused on the analysis of the task profile of the multi-rotor UAV airborne controller, as well as the temperature, electricity and vibration stress of each task stage. Thirdly, the test profiles and specific parameters of each stress in the reliability test were determined. Lastly, some tests were done to verify the method proposed in the article. The reliability evaluation method of the multi - rotor UAV airborne controller proposed in this paper can be used in the practical application of UAV enterprises. It has good guiding significance.

1. Introduction

Common aircraft are usually divided into fixed wing, helicopter and multi-rotor (four rotor or six rotor is the mainstream). At present, the multi-rotor Unmanned Aerial Vehicle (hereinafter referred to as UAV) is widely used in civil and commercial use. People can use UAV to carry out aerial photography, express delivery, road mapping, disaster monitoring, agricultural plant protection, power inspection and other work. UAV has gradually become a part of people's daily life. Although the application of UAV is very wide, but due to the relatively less technology accumulation, the relatively weak key materials and processes and other reasons, the current domestic UAV in the process of carrying out the task, there are often some failures, which will affect the execution of the task, and will lead to the damage of parts or the whole machine of UAV, endangering personal and property safety. How to evaluate and improve the reliability of components and the whole UAV has become one of the important research contents in the UAV industry.

The main modules of multi-rotor UAV include flight platform, power device, navigation and flight controller, electrical system, communication system, necessary mission equipment and ground controller, etc. As shown in Figure.1. The sensors mainly include gyroscopes, accelerometers, magnetometers, barometers, GPS and so on, which are placed separately or integrated into the flight controller.

Due to the vibration, high temperature and other factors in the flight process of UAV, it will cause UAV failure. UAV faults can be roughly divided into software faults and hardware faults. There are two kinds of software faults: sensor data error and communication interruption. The solution of this kind of fault can only be realized on the basis of improving the fault tolerance and self-test function of the control program. Hardware failure can be divided into two categories: control system hardware failure

Content from this work may be used under the terms of the Creative Commons Attribution 3.0 licence. Any further distribution of this work must maintain attribution to the author(s) and the title of the work, journal citation and DOI.
Published under licence by IOP Publishing Ltd

and power system hardware failure. The types of control system hardware failure include sensor failure, single-chip or similar processor crash, pin damage, etc. the types of power system failure include motor abnormality, blade unexpected flying out and damage, motor drive abnormality and power system circuit abnormality (power supply abnormality of power system), etc. Most of these hardware faults can be identified by simulating the vibration, temperature and other multi stress tests of UAV in normal flight, so as to realize the reliability life assessment of multi rotor UAV. According to the basic principle of reliability, after the product is finalized, its inherent life will be determined accordingly. But how can the user of the product, such as the manufacturer of UAV, determine the reliability life of the product when purchasing the product to meet the requirements? Therefore, it is necessary to use the reliability identification method to identify the reliability life of products.

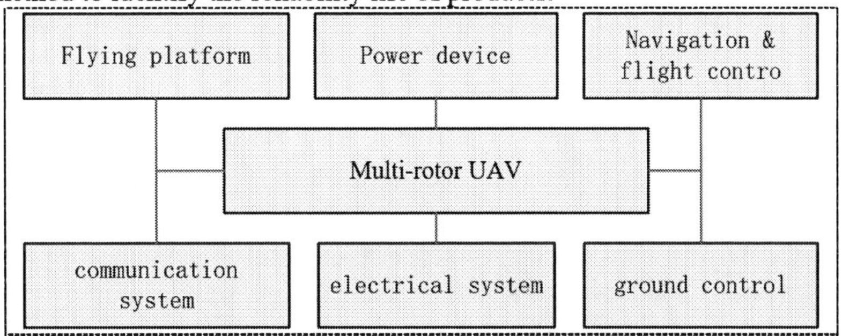

Figure 1. The main modules of multi rotor UAV.

Literature [1-3] studies the life of UAV body, literature [4-6] studies the reliability identification test scheme of UAV recording equipment, focusing on the analysis of the parameters selection in the reliability identification test scheme, such as test time t, sample number n, failure number r, manufacturer's risk α and user's risk β, as well as the determination method, and literature [7-8] Studies rotor UAV The fault safety system of UAV is studied, some common faults of UAV are given, and the fault detection and corresponding decision-making are studied, but the specific test methods and test sections for reliability evaluation of UAV airborne equipment are not involved. From the point of view of reliability life appraisal, this paper studies the whole process of reliability life appraisal scheme of airborne controller of rotorcraft, and focuses on the research and analysis of parameter selection method of test section, in order to provide enlightenment for reliability appraisal of airborne controller of rotorcraft.

2. Basic principle of reliability life assessment

It is assumed that the life distribution of UAV airborne equipment conforms to the exponential law, that is, the loss efficiency is constant. In reliability practice, this assumption is mostly satisfied for electronic products.

Because the reliability of the whole batch of products is judged by checking the reliability of the samples during the reliability appraisal test, there may be two types of errors: one is to criticize the qualified products as the unqualified products (the first type of errors); the other is to criticize the unqualified products as the qualified products (the second type of errors). The first type of error will cause loss to the producer, so the probability of the first type of error is the producer's risk (α); the probability of the second type of error is the user's risk (β). When designing test plans, α and β are usually between 5% and 30%.

When determining the reliability evaluation scheme, α and β are negotiated by the manufacturer and the user, while the expected life MTBF value of the product is specified in the production contract. By selecting the appropriate upper limit (θ_0) and lower limit (θ_1) of MTBF inspection, the parameters in the evaluation scheme can be determined.

Take the statistical scheme of sequential test type reliability appraisal test as an example. For exponential products with unknown MTBF value θ, the probability of r failures in the accumulated working time t is:

$$P_r(r) = \left(\frac{t}{\theta}\right)^r \left(\frac{e^{-t/\theta}}{r!}\right) \tag{1}$$

If the actual MTBF of the product is equal to the lower limit of test θ_1 of MTBF, the probability of r failures within the working time t is:

$$P_1(r) = \left(\frac{t}{\theta_1}\right)^r \left(\frac{e^{-t/\theta_1}}{r!}\right) \tag{2}$$

If the actual MTBF of the product is equal to the upper limit of the inspection θ_0, the probability of r failures within the working time t is:

$$P_0(r) = \left(\frac{t}{\theta_0}\right)^r \left(\frac{e^{-t/\theta_0}}{r!}\right) \tag{3}$$

Let $P(r)$ be the ratio of the probability $P_1(r)$ when the MTBF value of the product is equal to the lower limit of test θ_1 of MTBF to the probability $P_0(r)$ when the MTBF value of the product is equal to the upper limit of test θ_0 of MTBF:

$$P(r) = \frac{P_1(r)}{P_0(r)} = \left(\frac{\theta_0}{\theta_1}\right)^r e^{-[(1/\theta_1)-(1/\theta_0)]t} \tag{4}$$

During the test, the ratio was calculated continuously and compared with two predetermined constants A and B, using the following decision:

- If $P(r) < B$, receive the product and stop the test.
- If $P(r) > A$, reject the product and stop the test.
- If $B < P(r) < A$, continue the test.

Where the constants A and B are:

$$A = \frac{(1-\beta)(d+1)}{2\alpha d} \tag{5}$$

$$B = \frac{\beta}{1-\alpha} \tag{6}$$

Among them, α is the producer's risk, β is the user's risk, d is the identification ratio, that is, $d = \theta_0/\theta_1$.

During the test, if the failure failure number r is taken as the judgment basis, then

$$\frac{\ln B}{\ln(\theta_0/\theta_1)} + \frac{(1/\theta_1 - 1/\theta_0)}{\ln(\theta_0/\theta_1)} t < r < \frac{\ln A}{\ln(\theta_0/\theta_1)} + \frac{(1/\theta_1 - 1/\theta_0)}{\ln(\theta_0/\theta_1)} t \tag{7}$$

That is to say, when the value of failure number r is between the left and right sides of the inequality, the test will continue. If r is not greater than the left side, the acceptance judgment will be made and the test will be terminated; if r is not less than the right side, the rejection judgment will be made and the test will be terminated.

For the maximum test time T_0 in the qualification test scheme, it is derived from the estimated value of MTBF one-sided lower confidence limit:

$$T_0 = \frac{\theta_0 \chi^2_{(1-\alpha),2r_0}}{2} \tag{8}$$

r_0 is selected by the following formula:

$$\frac{\chi^2_{(1-\alpha),2r}}{\chi^2_{\beta,2r}} \geq \frac{\theta_1}{\theta_0} \tag{9}$$

Where, $\chi^2_{(1-\alpha),2r}$ and $\chi^2_{\beta,2r}$ are the upper quantiles of the χ^2 distribution with a degree of freedom of $2r$. Through the determined α, β and looking up the χ^2 distribution table, we can get the values of the two until the ratio of the two values is not less than θ_1/θ_0, then the r value at this time is r_0.

3. Research on reliability test profile

After determining the parameter value of reliability qualification test, it is to develop appropriate test conditions (test profile) and carry out the test according to the selected reliability qualification test parameters.

In the reliability evaluation of UAV flight controller, the applied environmental stress includes electrical stress, vibration stress, temperature stress, humidity stress, etc. In order to make a comprehensive investigation, take the typical mission profile of rotorcraft, i.e. high altitude demonstration flight as an example, without external equipment, only to ensure the operation of the UAV's own equipment. During the operation, the ground control station is not included in the assessment scope. Accordingly, the mission profile of rotorcraft is formulated as shown in Figure.2:

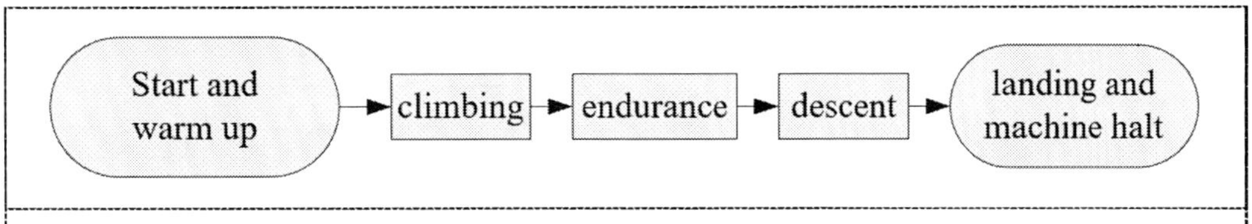

Figure 2. Mission profile of rotor UAV.

For the controller on the rotor UAV, its mission profile is consistent with the UAV body itself.

3.1. Electrical stress test profile

The electrical stress test profile of the UAV flight controller is generally applied according to the input voltage change requirements specified by the controller. If it is not specified in the equipment specification, the electrical stress of the flight controller can be considered to be cyclic variation with the amplitude of ± 10% of the nominal input voltage, i.e. the input voltage of the first test cycle is 110% of the nominal voltage; the input voltage of the second test cycle is the nominal voltage (UA); and the input voltage of the third test cycle is 90% of the nominal voltage. Among them, the third test cycle is used in start-up and warm-up phase, the first test cycle is used in climb phase, the second test cycle is used in cruise phase, and the third test cycle is used in descent, landing and shutdown phase. Corresponding to the task profile, the change of input voltage of multiple test cycles constitutes a complete electrical stress cycle. The test section is shown in Figure.3:

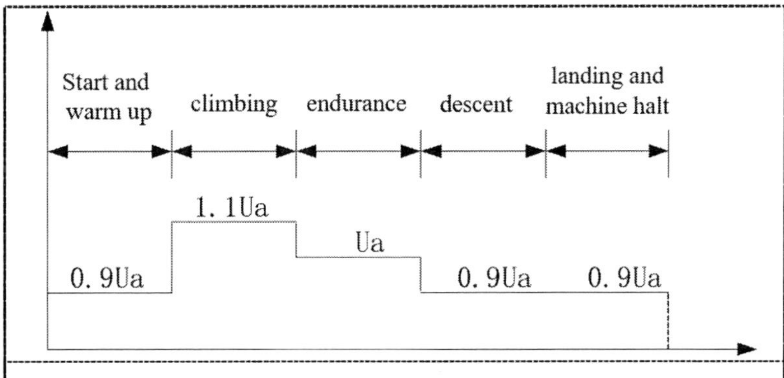

Figure 3. electrical stress test profile.

3.2. Vibration stress

The vibration characteristics of helicopter cargo transportation are the superposition of strong narrow-band peaks on the background of low-order continuous wide-band random vibration. This environment is a combination of many sinusoidal or nearly sinusoidal components caused by the main rotor, tail rotor and rotating machinery and low-order random components caused by the flow field. Figure 4 shows the vibration environment spectrum of the helicopter [9-10].

Because of the flight principle and structure principle of the current multi rotor UAV, the vibration generation is similar to that of the helicopter. Therefore, the vibration stress profile of the multi rotor UAV can be made according to the helicopter vibration environment. Considering the current use conditions of civil vibration equipment, it is determined that the continuous logarithmic scanning vibration of $5Hz \sim 2000Hz \sim 5Hz$ is used. One scanning cycle of each axis of three axes is $36\boldsymbol{min}$, the maximum duration is $3h$, and the maximum stress of the equipment is $50 m/s^2$. The rotorcraft is an equipment without shock absorber, and its test cycle is as follows:

The acceleration from $5Hz$ to 33 Hz is ± 20 m/s^2, and the acceleration from 33 Hz to $2000Hz$ is ± 50 m/s^2.

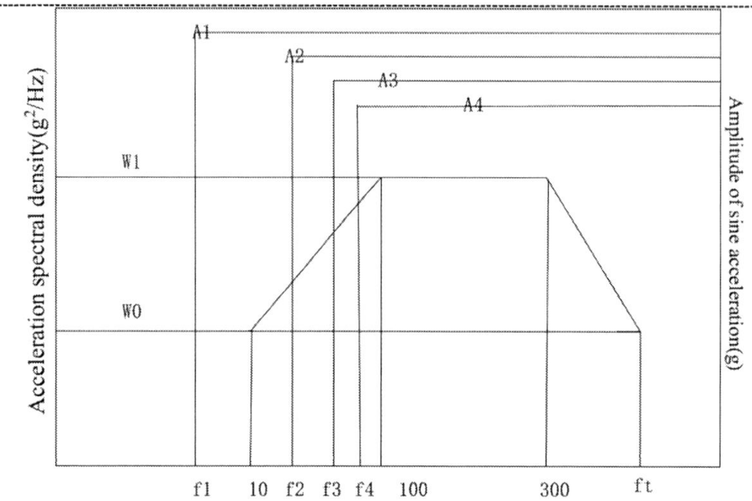

Figure 4. Spectrum of helicopter vibration environment.

3.3. *Thermal stress*

The temperature distribution data of a four rotor UAV (direct flight, low altitude $3m$) is shown in Table.1:

Table 1. Flight test temperature data of a certain four rotor UAV.

Measuring position	Maximum temperature (℃)	Maximum temperature rise(K)
Shell (upper)	29.8	0.7
IMU Shell	36.6	7.5
Main control panel shell	35.1	6.0
Power chip shell	32.3	3.2
PCB	32.1	3.0
Motor housing	34.3	5.2
Shell (side)	29.6	0.5

Among them, the temperature of IMU inertial navigation module shell, main control board module shell and drive motor shell changes greatly, reaching 7.5 ℃, 6.0 ℃ and 5.2 ℃ respectively, and actually reaching 36.6 ℃, 35.1 ℃ and 34.3 ℃ (the ambient temperature is 29.1 ℃, and the relative humidity is 51% ***RH***). Considering the difference of ambient temperature in different regions, the standard temperature is adopted for the test, which meets the actual requirements and is more strict, that is, high temperature + 55 ℃, low temperature - 50 ℃.

3.4. *Humidity stress*

Humidity should be able to simulate warm and humid atmospheric conditions, especially the hot and humid conditions prevailing in tropical climate. The dew point in the test chamber shall be maintained at 31℃ or above during the hot soak stage and the ground working stage of the climate. In other stages, the dew point is not controlled, but the air in the test chamber cannot be dried.

4. Test verification

In order to verify the reliability life appraisal method of the airborne controller of the multi rotor UAV proposed in this paper, a certain type of four rotor UAV is selected for test. According to the test section studied in this paper, the test is carried out. Temperature, electric stress and vibration stress are combined into one test cycle, and humidity is not controlled. During the test, the output signal of the controller shall be monitored, including direction signal, throttle signal, indicator light, etc. [11]. Once there is an unrecoverable output abnormality, the onboard controller shall be judged as faulty and the life test shall be terminated. When the test lasts for 156h, the direction signal output of the controller is abnormal and there is continuous noise. It is determined that the fault occurs and the test is terminated. The reliability life of the airborne controller can be obtained by combining the relevant formula.

5. Conclusion

This paper studied the reliability evaluation method of the airborne controller of the multi rotor UAV, analyzed the basic principle of the reliability evaluation, and focused on the comprehensive environmental stress in the reliability evaluation test of the airborne controller, and finally carries out the test verification. The reliability evaluation method of the multi rotor UAV airborne controller proposed in this paper can be applied in the UAV enterprises, and can be extended to other applicable UAV products.

References

[1] W. min. research on failure safety system of six rotor UAV. Nanjing University of information engineering, 2015.
[2] Q. juntong, W. Zhenwei, H. Jianda. Fault diagnosis and fault tolerant control technology overview of rotorcraft. 2008 UAV Summit Forum, pp.245-252.
[3] X. bin, H. Wei. Fault tolerant control method for partial actuator failure of four rotor UAV, cn104965414a. 2015.
[4] L. Gencheng, Z. Hu. Design and analysis of reliability evaluation test scheme for an airborne product. Aviation standardization and quality, 2004 (4): pp.41-44.
[5] W. Hao, X. Ming. Discussion on determination method of statistical scheme for reliability evaluation test of airborne equipment. Aviation standardization and quality, 2005 (2): pp.39-41.
[6] X. Jianglin. Reliability evaluation test of airborne electronic equipment. Beijing University of Aeronautics and Astronautics, 2000.
[7] W. Guchang, W. Yibo. Fault analysis and spare parts optimization of UAV control and navigation system. Automation of military engineering, 2015 (1): pp.79-81.
[8] Y. Bin, Y. Jianguo, X. Xiaojun, et al. Hardware fault diagnosis method of UAV flight control computer. Computer measurement and control, 2011, 19 (10): pp.2407-2409.
[9] GJB 899a-2009 reliability appraisal and acceptance test.
[10] GJB 150.16a-2009 environmental test methods for military equipment laboratory Part 16: vibration test.
[11] Z. Yanhong. Design of intelligent monitoring system for UAV's twin rotor failure in severe environment. Science and technology and engineering, 2016, 16 (17): pp. 225-229.

Aerodynamic Analysis and Drag-reduction Design for Ultra-low-orbit Satellite

Shao-ting Yu[1, a] and Cai-zhi Fan[2, b]

[1] Master Candidate, College of Aerospace and Engineering, National University of Defense Technology, Changsha 410073, China.
[2] Corresponding Author, College of Aerospace and Engineering, National University of Defense Technology, Changsha 410073, China.

E-mail: [a]1762520115@qq.com; [b] caizhifan@nudt.edn.cn

Abstract. When the satellite is operating at an orbital height of 120-300km, aerodynamic drag becomes the most important factor affecting ultra-low orbital attenuation and attitude disturbance. Based on the theory of free-molecular flow, the model of aerodynamic drag of a panel in free-molecular flow is established firstly, and then the influence of the angle of attack, adaptation coefficient, and the molecular velocity ratio on the drag coefficient of the panel is analysed. Finally, A reduced drag design method for a ultra-low-orbit satellite is proposed based on the analysis results. And it is validated by the simulation results. The approaches presented in this paper can be applied to aerodynamic design and calculation of ultra-low-orbit satellites.

Key words: Ultra-low-orbit satellite, Free-molecular flow, Aerodynamics analysis, Reduced drag design, Satellite configuration.

1. Introduction

Ultra-low orbit usually refers to satellite orbits outside the dense atmosphere but below the height of ordinary spacecraft orbits. In this paper, ultra-low orbits are defined as flight orbits with a spatial range of more than 120 kilometres and less than 300 kilometres from the earth's surface[1], Satellites operating in this orbit have the advantages of high resolution, high cost efficiency, and low launch cost[2]. In the space range of the ultra-low orbit, the atmosphere is very thin, but compared with the flying height of traditional satellites, there are a large number of gas molecules. When the satellite is operating in ultra-low orbit, these gas molecules collide with the surface of the satellite to generate aerodynamic drag. The aerodynamic drag perturbation has become the most important factor affecting the attenuation and attitude disturbance of ultra-low orbit. Currently, the ultra-low orbit satellite has attracted the attention of the United States, Japan, ESA and so on. The major Aerospace countries have successively formulated the ultra-low orbit satellite project plan, and actively carried out feasibility study, key technology research and flight test[3-5].

The aerodynamic problem of ultra-low orbit satellite belongs to the field of free-molecular flow. At present, there are four methods to simulate the movement and action process of free-molecular flow: Panel method, Ray-Tracing Panel method, Direct Simulation Monte Carlo method and Test-Particle Monte Carlo method[6]. The Panel method does not consider the problem of occlusion and multiple reflections, and considers that the aerodynamic forces generated by each part of the satellite do not interfere with each other, so It can be divided into several parts, and the aerodynamic force is calculated

Content from this work may be used under the terms of the Creative Commons Attribution 3.0 licence. Any further distribution of this work must maintain attribution to the author(s) and the title of the work, journal citation and DOI.
Published under licence by IOP Publishing Ltd

by distribution. This paper focuses on the configuration design and optimization of satellites. When calculating the aerodynamic forces of satellites, Ray-Tracing Panel method, Direct Simulation Monte Carlo method and Test-Particle Monte Carlo method need a lot of computing resources and take a long time, while Panel method has the advantages of small amount of computation and high efficiency, and can analyse and optimize the configuration parameters through analytical solutions. Therefore, this paper chooses Panel method.

In this paper, firstly, based on the theory of free-molecular flow, the aerodynamic drag model of the panel is established, and the aerodynamic drag coefficient of the panel is analysed systematically. Then, the cross-sectional shape, slenderness ratio, and head configuration of the ultra-low orbit satellite are calculated and analysed by the Panel method, and the design method of the drag reduction configuration of the ultra-low orbit satellite is proposed.

2. Aerodynamic drag model of panel in free-molecular flow

2.1. Judgment of free-molecular flow

According to the degree of leanness, the flow of lean gas can be divided into four zones[7], as shown in Table 1.

Table 1. Knudsen number corresponding to different mobile fields

Continuous medium zone:	$K_n < 0.01$
Slipstream zone:	$0.01 < K_n < 0.1$
Transition zone	$0.1 < K_n < 10$
Free-molecular flow zone	$10 < K_n$

Where $K_n = \lambda / L$ is the ratio of the mean free path (λ) of the molecule to the value of the flow characteristic length(L).

2.2. Density properties of ultra-low orbit atmosphere

The atmospheric density of the area where the ultra-low orbit is located changes drastically with the influence of solar activity, geomagnetic activity, season and other factor[8]. Based on the NRLMSISE atmospheric model, the changes in atmospheric density with orbital height in different years of solar activity are shown in Figure 1.

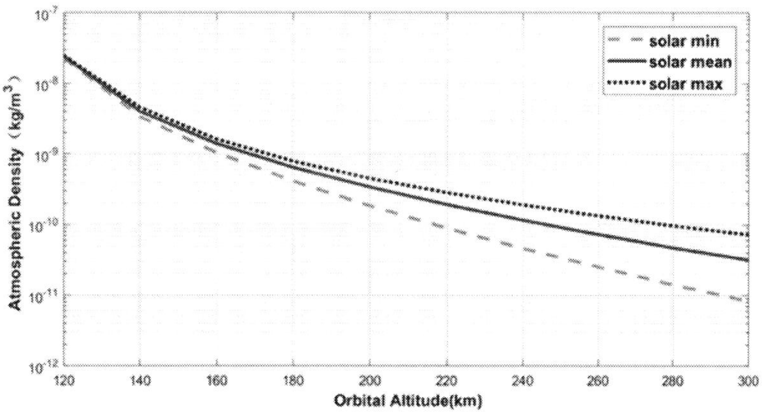

Figure 1. Variation of atmospheric density in different solar activity years atmospheric density.

It can be seen that when the orbit height is 120km, the magnitude of atmospheric density is 10^{-8}, when the orbit height is 300km, the magnitude of atmospheric density is 10^{-11} in the low year of solar activity and 10^{-10} in the high year of solar activity. When the orbit is low, the influence of solar activity on the atmospheric density is small. With the increase of orbit, atmospheric density ratio of high and low solar activity also increase, and the ratio is 8.76 at the height of 300km. In the calculation in this paper, the average density at a specific height is taken as the calculated value.

2.3. Gas-surface interaction model

The aerodynamics force beating for the panel in the free molecule flow is shown in Figure 2. Where p, τ respectively represent the normal and tangential momentum flow components; the subscript i indicates that the quantity belongs to the incoming flow, and the subscript r indicates that the quantity belongs to the reflected flow.

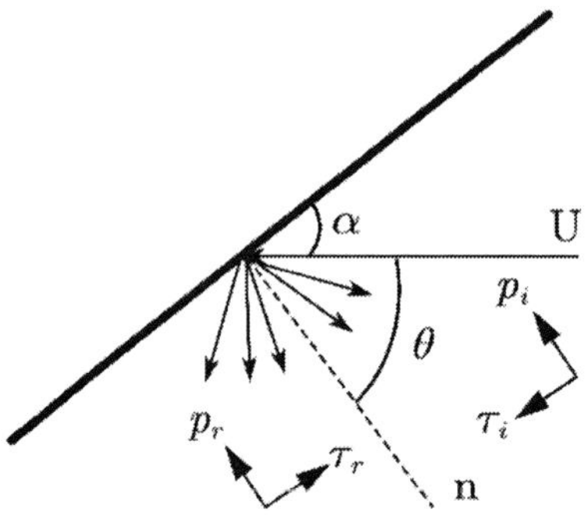

Figure 2. The aerodynamics force beating for the plate in the free molecule flow

According to the Maxwell model[9], the expressions of the positive pressure and shear force on the surface of the object are calculated as:

$$P_a = \frac{\rho_\infty U^2}{2S^2}\left\{\begin{array}{c}\left(\frac{2-\sigma}{\sqrt{\pi}}Ssin\theta + \frac{\sigma}{2}\sqrt{\frac{T_r}{T}}\right)e^{(-(Ssin\theta)^2)} + \\ \left\{(2-\sigma)\left[(Ssin\theta)^2+\frac{1}{2}\right]+\frac{\sigma}{2}\sqrt{\frac{\pi T_r}{T}}Ssin\theta\right\}[1+erf(Ssin\theta)]\end{array}\right\} \tag{1}$$

$$\tau_a = -\frac{\sigma\rho_\infty U^2 cos\theta}{2\sqrt{\pi}S}\left\{e^{(Ssin\theta)^2}+\sqrt{\pi}(Ssin\theta)[1+erf(Ssin\theta)]\right\} \tag{2}$$

Where ρ_∞ is the inflow density, U is the inflow velocity, $S=\frac{U}{\sqrt{2RT}}$ is the molecular velocity ratio, R is the gas constant, T is the inflow temperature, T_r is the reflected gas temperature, which is considered equal to the wall temperature, $erf(a)=\frac{2}{\sqrt{\pi}}\int_0^a \exp(-x^2)dx$ is the error function; θ is the angle between the inflow U and the normal direction of the surface.

Taking the area of one side of the panel as the characteristic area, the aerodynamic drag coefficient with the angle of attack α can be calculated from equations (1) and (2) as:

$$\begin{aligned}C_D &= \frac{2}{\rho_\infty U^2}\left[P_a|_{\theta=\frac{\pi}{2}-\alpha}\right]\cdot sin\alpha - \left[\tau_a|_{\theta=\frac{\pi}{2}-\alpha}\right]\cdot cos\alpha \\ &= \frac{1}{S^2}\left\{\begin{array}{c}\left(\frac{2S}{\sqrt{\pi}}sin^2\alpha+\frac{\sigma S}{\sqrt{\pi}}cos2\alpha+\frac{\sigma}{2}\sqrt{\frac{T_r}{T}}\sin\alpha\right)\cdot e^{-(Ssin\alpha)^2}+ \\ sin\alpha\cdot[1+erf(Ssin\alpha)]\cdot\left[\sigma S^2\cos2\alpha+2S^2sin^2\alpha+1-\frac{\sigma}{2}+\frac{\sigma}{2}\sqrt{\frac{\pi T_r}{T}}S\sin\alpha\right]\end{array}\right\}\end{aligned} \tag{3}$$

3. Analysis of influencing factors of panel aerodynamic drag coefficient

3.1. Change law of aerodynamic drag coefficient with angle of attack

From the formula (3), it can be seen that the magnitude of the panel aerodynamic drag coefficient C_D is closely related to the angle of attack (α), the adaptation coefficient (σ), and the molecular speed ratio (S). In order to analyze the relationship between them, C_D is divided into:

$$C_D = C_{D\sigma} + C_{D^*} \quad (4)$$

Where $C_{D\sigma}$ is related to the adaptation coefficient:

$$C_{D\sigma} = \frac{\sigma}{S^2}\left\{\begin{matrix}\left(\frac{S}{\sqrt{\pi}}cos2\alpha + \frac{1}{2}\sqrt{\frac{T_r}{T}}\sin\alpha\right)e^{-(Ssin\alpha)^2} + \\ sin\alpha[1+erf(Ssin\alpha)]\left[S^2\cos 2\alpha - \frac{1}{2} + \frac{1}{2}\sqrt{\frac{\pi T_r}{T}}S\sin\alpha\right]\end{matrix}\right\} \quad (5)$$

C_{D^*} is uncorrelated to the adaptation coefficient:

$$C_{D^*} = \frac{1}{S^2}\left\{\frac{2S}{\sqrt{\pi}}sin^2\alpha \cdot e^{-(Ssin\alpha)^2} + sin\alpha[1+erf(Ssin\alpha)]\,[2S^2 sin^2\alpha + 1]\right\} \quad (6)$$

Taking the adaptation coefficient $\sigma = 0.9$, the orbit height is 200 km, the incoming temperature is 841K, the wall temperature is equal to the incoming temperature, and the change law of the drag coefficient with the angle of attack is analysed, as shown in Figure 3.

It can be seen that as the angle of attack increases, C_D and C_{D^*} gradually increase, while $C_{D\sigma}$ increases firstly and then decreases, and a critical angle (β) appears, so that $C_{D\sigma}$ is 0, that is, the angle of attack is at β, the change of the adaptation coefficient has no effect on the aerodynamic drag. In addition, when the angle of attack is 0, the total aerodynamic drag coefficient is not 0. This is due to the thermal motion of the molecules, and the airflow produces a shearing force on the plate, and the shearing force is $\tau_0 = -\frac{\sigma\rho_\infty U^2}{2\sqrt{\pi}S}$.

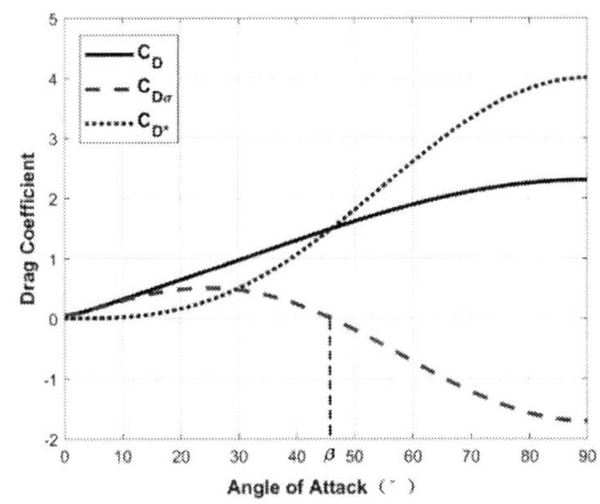

Figure 3. Variation of drag coefficient with angle of attack.

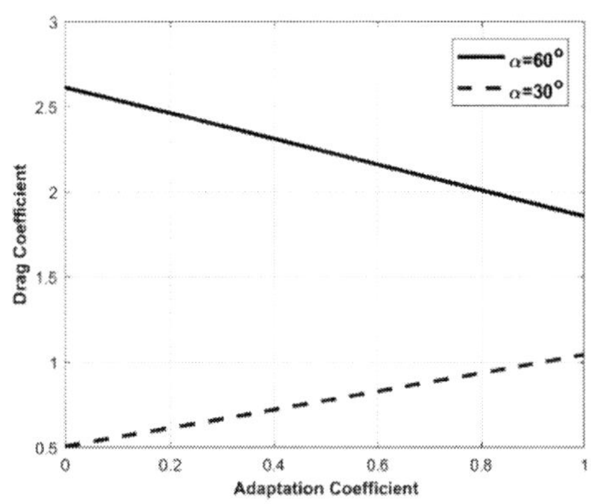

Figure 4. Variation of drag coefficient with adaptation coefficient.

3.2. Change law of aerodynamic drag coefficient with adaptation coefficient

When the angle of attack is 30° (less than β) and 60° (more than β), the change rule of aerodynamic drag coefficient with the adaptation coefficient is shown in Figure 4. It can be seen that when the angle of attack is 30° (less than β), the drag coefficient increases with the increase of adaptation coefficient; when the angle of attack is 60° (more than β), the drag coefficient decreases with the increase of adaptation coefficient. In the satellite configuration design, the angle of attack of the windward side is generally large. As the adaptation coefficient increases, the aerodynamic drag decreases. Therefore, the windward side can use ordinary industrial processing technology, and the general adaptation coefficient is about 0.9. But the side attack angle is small, the side smoothing technology should be used to reduce the side drag.

3.3. Change law of aerodynamic drag coefficient with molecular velocity ratio

From the definition of the molecular speed ratio, it can be seen that the molecular speed ratio is directly proportional to the satellite's operating speed and inversely proportional to the square of the incoming gas temperature. Taking the adaptation coefficient $\sigma = 0.9$,and the angle of attack of 90 °, the change law of the panel drag coefficient and critical angle with the molecular velocity ratio is analysed, as shown in Figure 5.

It can be seen that as the molecular velocity ratio increases, the drag coefficient and the critical angle both decrease, because the drag coefficient decreases with the increase of the molecular speed ratio, indicating that after the molecular speed ratio increases, The influence of the adaptation coefficient on the drag coefficient will be reduced.

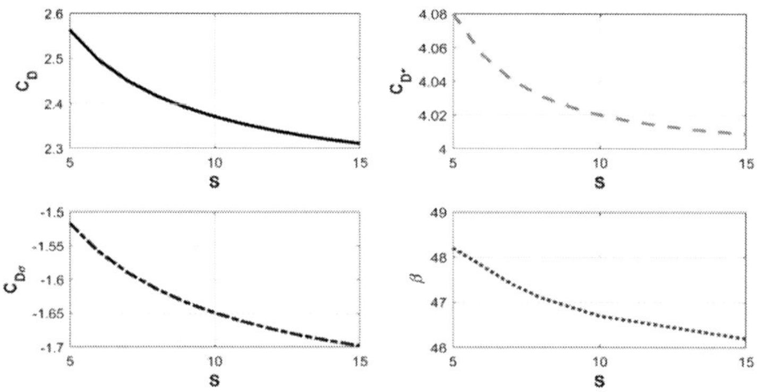

Figure 5. Variation of drag coefficient and critical angle of attack with molecular velocity ratio.

4. Ultra-low orbit satellite drag reduction configuration design

4.1. Satellite cross-section shape design

The slender body configuration satellite is selected to study the aerodynamic drag effect. Inscribed quadrilateral, Inscribed hexagon, Inscribed octagon, and circle with a diameter of 1 meter in cross section are taken as slender bodies with a length of 4 meters. Solar sails are not considered. When the solar array and other expansion devices are not considered, calculate the aerodynamic drag and volume drag ratio under the ultra-low orbit, as shown in Table 2.

Table 2. Aerodynamic drag and volume drag ratio under different cross-section.

Cross-section	Dynamic pressure (P_a)	Cross-sectional area(m^2)	Volume (m^3)	Aerodynamic drag(N)	Volume drag ratio(m^3/N)
Quadrilateral	0.007945	0.500	2.000	0.01742	114.8
Hexagon	0.007945	0.650	2.827	0.02158	125.5
Octagon	0.007945	0.707	2.828	0.02195	128.8
Circle	0.007945	0.785	3.142	0.02364	132.9

It can be seen that with the increase of the number of sides of the cross-sectional shape (the circle has an infinite number of sides), the aerodynamic drag of a satellite operating in an ultra-low orbit increases, the volume increases, and the volume drag ratio also increases. Considering only the aerodynamic drag and volume, the slender satellite with a circular cross section has the best aerodynamic performance.

4.2. Satellite slenderness ratio design

Select the shape of the satellite section as a circle, the circle diameter is D, the volume is V, and the total length is L. The total aerodynamic drag of the satellite is expressed as:

$$F_D = F_1 + F_2 = qC_{D1}S_1 + qC_{D2}S_2 = \frac{1}{2}\rho_\infty U^2 \cdot C_{D1}\left(\frac{\pi D^2}{4} + \frac{C_{D2}}{C_{D1}}\frac{4V}{D}\right) \tag{7}$$

Where F_1 is the windward drag, F_2 is the side drag, C_{D1} is the windward drag coefficient, C_{D2} is the side drag coefficient, S_1 is the windward area, S_2 is the side area.

It can be found that when the slenderness ratio is $L/D = C_{D1}/2C_{D2}$, the total aerodynamic drag of the spacecraft is the smallest.

Take a cylindrical satellite with a volume of 1 Cubic meter and the adaptation coefficient is 0.9. The variation rule of aerodynamic drag with the slenderness ratio of the satellite is calculated when the orbit height of the satellite is 200 km, as shown in Figure 6.

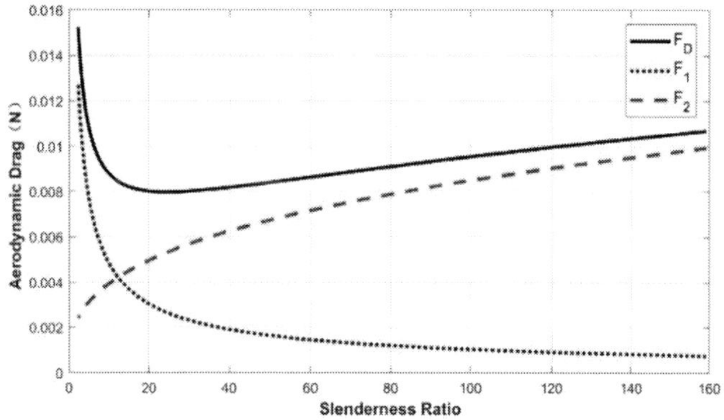

Figure 6. Variation of aerodynamic drag with satellite slenderness ratio.

It can be seen that: (1) with the increase of slenderness ratio, the aerodynamic drag decreases first and then increases; when the slenderness ratio of satellite is small, the side area has little influence on the total drag of satellite, and the total drag of spacecraft tends to be the drag only considering the windward area, so the aerodynamic drag decreases with the increase of slenderness ratio; when the slenderness ratio increases to a certain extent, the drag effect of side area is obvious, The total drag of the spacecraft increases with the increase of slenderness ratio; (2) when slenderness ratio of the cylindrical spacecraft is about 23, the aerodynamic drag is the smallest, and the aerodynamic drag on the side is about twice that on the windward side. (3) In the practical design of ultra-low orbit satellite, considering the limitation of launch vehicle and the maneuverability of satellite in orbit, the actual slenderness ratio of ultra-low orbit satellite is generally far less than 23, so it is only necessary to reduce the windward side of spacecraft to reduce its aerodynamic drag.

4.3. Satellite head optimization

Select the satellite with a slender cylinder configuration as the research object. The diameter of the cylindrical cross section is 1 meter and the length is 4 meter. The first 5th of the length of the satellite is set as the satellite head. In order to study the influence of the satellite head on the aerodynamics of the satellite, the head is modified and designed into different shapes, as shown in Figure 7.

IOP Conf. Series: Materials Science and Engineering **887** (2020) 012013 doi:10.1088/1757-899X/887/1/012013

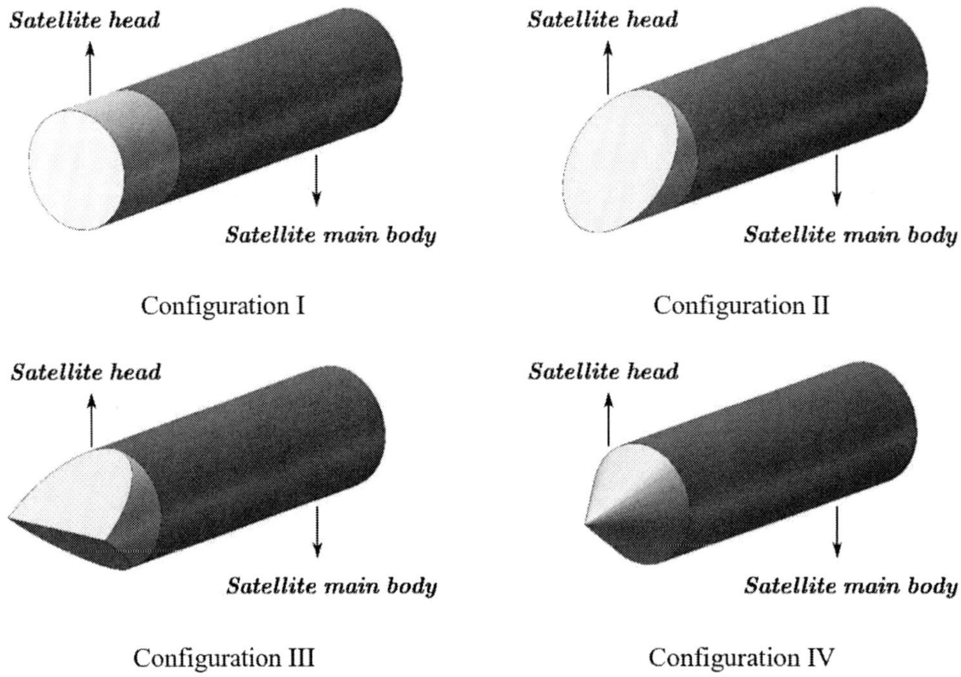

Figure 7. Schematic diagram of satellite head optimization.

Taking the adaptation coefficient $\sigma = 0.9$ and the orbital height of 200km, the aerodynamic drag and volume drag ratio of the four configurations under moderate solar activity conditions are calculated and analysed, as shown in Table 3.

Table 3. Aerodynamic drag and volume drag ratio for the four configurations

Configuration	Volume (m^3)	Head aerodynamic drag(N)	Total aerodynamic drag(N)	Volume drag ratio(m^3/N)
Configuration I	3.142	0.01646	0.02364	132.9
Configuration II	2.827	0.01439	0.02158	131.0
Configuration III	2.875	0.01311	0.02029	141.7
Configuration IV	2.723	0.01245	0.01964	136.7

It can be seen that the optimization of the satellite head can effectively reduce the aerodynamic drag and increase the volume drag ratio. Configuration IV has the minimum drag, which can be reduced by 16.92% compared with configuration I when the orbit height is 200km; configuration III has the optimal volume drag ratio, which can be increased by 6.64% compared with configuration I when the orbit height is 200km. Thus, the rationality of the configuration of NanoEye satellite is proved.

From the analysis in Section 3.1, it can be concluded that the size of the adaptation coefficient directly affects the aerodynamic drag. Through the special treatment of the satellite head (improving the surface smoothness, using special materials, etc.), the adaptation coefficient of the head can be changed to achieve the drag reduction effect.

Taking the orbital height of 200km and the adaptation coefficient of the satellite main body as 0.9, the aerodynamic drag and volume drag ratio of the four configurations with different adaptation coefficients at the head are calculated and analyzed, as shown in Figure 8.

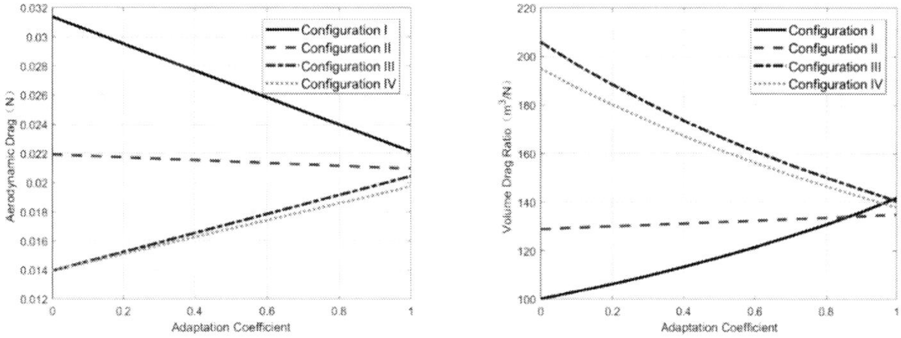

Figure 8. Aerodynamic drag and volume drag ratio of the four configurations under different adaptation coefficients.

It can be seen that with the increase of the adaptation coefficient, the configuration I and configuration II decrease the aerodynamic drag and increase the volume drag ratio; the configuration III and configuration IV increase the aerodynamic drag and decrease the volume drag ratio. This is because the head cone angles of configuration I and II are larger than the critical angle, and the head cone angles of configuration III and IV are smaller than the critical angle. Therefore, for the configuration I and the configuration II, the head can adopt ordinary processing technology, and the general adaptation coefficient is 0.9, and no special processing and processing technology is required. For Configuration III and Configuration IV the head should try to improve the surface smoothness and use special materials to reduce the adaptation coefficient.

5. Conclusion

Based on the theory of free-molecular flow, this paper uses the Panel method to calculate the aerodynamic drag of a simple-shaped satellite, the effects of the angle of attack, the adaptation coefficient, and the molecular velocity ratio on the panel drag coefficient are systematically analysed, and the configuration design method of ultra-low orbit satellite is proposed. The conclusion is as follows:

(1) The drag coefficient is closely related to the angle of attack, the molecular speed ratio, and the adaptation coefficient of the material. With the increase of angle of attack and molecular velocity ratio, the drag coefficient increases, and there is a critical angle, which makes the change of adaptation coefficient have no effect on aerodynamic drag.

(2) For a slender satellite, with the increase of the number of sides of the cross-section shape, the windward drag and side drag of the satellite increase, the volume of the satellite increases, and the volume drag ratio of the satellite also increases. Considering only drag and satellite volume, the circular section slender satellite has the best aerodynamic performance

(3) For the slender satellite with circular cross section, the aerodynamic drag decreases first and then increases with the increase of slenderness ratio; When the slenderness ratio of satellite is small, the influence of side area on the total drag of satellite is small, and the aerodynamic drag decreases with the increase of slenderness ratio; when slenderness ratio is larger than $C_{D1}/2C_{D2}$, the drag effect of side area is obvious, and the total drag of spacecraft increases with the increase of slenderness ratio;

(4) After the satellite head is optimized and improved, the drag decreases and the volume decreases, but the volume drag ratio increases. In the case of considering only drag and satellite volume, the configuration III proposed in this paper has the best aerodynamic performance, and its head should be as smooth as possible, which can effectively reduce aerodynamic drag.

References

[1] Hu Lingyun, Zhang Lihua, Cheng Xiaoli. *Method of Aerodynamic Design and Calculation for Ultra-LEO Spacecraft* [J]. Spacecraft Engineering,2016,25(1):10-18

[2] Wen Shenglin, Yan Ye, Yi Teng. *Analyzing perturbation characteristic and orbital maintenance strategy for super low altitude satellite*[J]. Journal of National University of Defense Technology, 2015, (2):128-134
[3] Strizzi J D. *An improved algorithm for satellite orbit decay and re-entry prediction*[D]. Massachusetts Institute of Technology, 1993
[4] Steiger C, Romanazzo M, Emanuelli P P, et al. *The deorbiting of ESA's gravity mission GOCE–spacecraft operations in extreme drag conditions*[C]. International Conference on Space Operations, Pasadena, USA, 2014
[5] Wertz, J R, Nicola S-A, Shao A, etc. *Moderately elliptical very low orbits (MEVLOs) as a long-term solution to orbital debris*[C]. 26th Annual. AIAA/USU Conference on SmallSats, Logan, UT, August 13-16, 2012: 1-12
[6] Benjamin P Graziano. *Computational modeling of aerodynamic disturbances on spacecraft within a concurrent engineering framework* [D]. London: Cranfield University.2007
[7] Tsien H S. *Superdynamics, mechanics of rarefied gases*. J. AERO. Sci. 13,12,653—664(1964)
[8] Du Heng, Ye Zonghai. *Space Environment Manual for Low Orbit Spacecraft* [M]. Beijing: National Defense Industry Press,1996:28~75
[9] Shen Qing. *Rarefied gas dynamics* [M]. Beijing: National Defense Industry Press, 2003: 141~150

IOP Conf. Series: Materials Science and Engineering **887** (2020) 012014 doi:10.1088/1757-899X/887/1/012014

Numerical Study on Flow Regulation Characteristics of High Temperature Jet Vortex Valve

L Zhang* and Z J Wei

School of Aerospace and Engineering, Beijing Institute of Technology, Beijing 100081, China

* 1143733228@qq.com

Abstract. The use of a high-temperature jet vortex valve without moving parts for the secondary jet of the vector nozzle can meet the needs of the secondary jet flow regulation of the vector nozzle in a high temperature and high pressure environment. The high-temperature jet vortex valve is a valve device with a certain amplification effect that uses a small cold flow to regulate a large flow of high-temperature gas flow. This paper uses numerical simulation to investigate the flow field distribution and flow adjustment characteristics of the vortex valve in different control pressure ratio when the main gas pressure of the vortex valve is 16 MPa and the temperature is 2500K. The results show that the high-temperature jet vortex valve relies on different radial pressure gradients generated by different control flow pressures could achieve the effect of adjusting the flow rate of high-temperature and high-pressure gas. The results proved the feasibility of the vortex valve in adjusting the secondary jet flow rate of the nozzle; further exploring indicated that the flow adjustment ratio TD will change when the ratio of the control flow port area to the outlet area changes.

1. Introduction

Due to the increasing mobility of weapons in modern warfare, in order to meet the requirements of aircraft higher attack speed and high maneuver, there is an urgent need for engine thrust vector adjustment technology that can adjust the thrust magnitude and direction in real time.

At present, there are mainly two types of thrust vector adjustment schemes: Mechanical Thrust Vector Control(MTVC) and Fluid Thrust Vector Control(FTVC). MTVC scheme is implemented by the mechanical thrust vector nozzle. This scheme adjust the nozzle by directly changing the nozzle mechanical structure to achieve Mainstream deflection. FTVC scheme adjusts the mainstream direction of the nozzle by injecting a secondary flow into the nozzle expansion section to achieve thrust vector adjustment [1]. Among them, there are many problems and deficiencies in the mechanical thrust vector nozzle scheme: there are many moving parts, the mechanical structure is complex, and the weight gain is obvious. In contrast, the fluid thrust vector nozzle has no large servo mechanism, and has a stable surface structure, with the advantages of good sealing performance and simple structure, the fluid thrust vector control technology has gradually become the research hotspot and the development frontier of aircraft thrust vector control. It also has a wider application prospect and engineering value.

The working principle of the FTVC is shown in Figure 1 and Figure 2. In the expansion section of nozzle, the main flow of the nozzle flows at supersonic speed, and a secondary flow is injected into the opening of the nozzle expansion wall to compress the main flow of the nozzle. The oblique shock wave is generated in the flow field, and the main flow of nozzle deviates from the original direction after the

Content from this work may be used under the terms of the Creative Commons Attribution 3.0 licence. Any further distribution of this work must maintain attribution to the author(s) and the title of the work, journal citation and DOI.
Published under licence by IOP Publishing Ltd

oblique shock wave, and forms a certain angle with the nozzle axis, so that the nozzle generates a certain thrust vector.

According to the source and nature of the secondary flow, the secondary flow supply system is divided into two types: Gas Split type and Cold Air type. As shown in Figure 1, the secondary flow used in the Gas Split-type solution takes the high-temperature gas in the combustion chamber of the engine. However, this solution has the difficulty of developing a flow regulating gas valve in a high-temperature and high-pressure environment and is not suitable for long-term work. The secondary flow used in Cold Air type comes from an independent gas cylinder, as shown in Figure2. The flow rate is adjusted by adjusting the pressure of the gas cylinder. In this scheme, but the secondary flow comes from the compressed gas in the gas cylinder, so the larger gas cylinder needs to carry more that provides a secondary flow, which undoubtedly increases the negative mass of the engine [2].

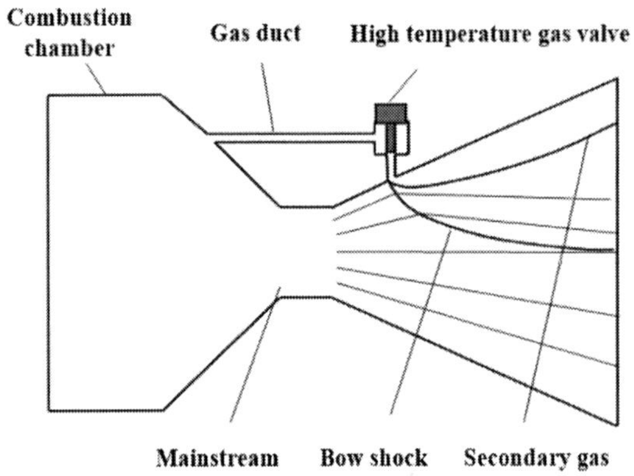

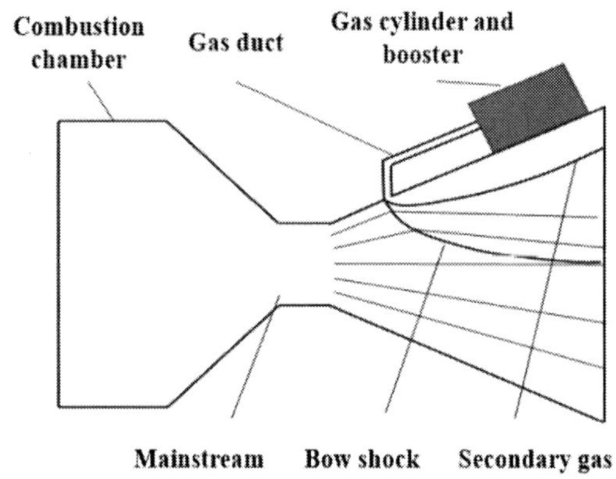

Figure 1. Schematic diagram of gas split scheme. **Figure 2.** Schematic diagram of cold air scheme.

Combining the gas split type and cold air type solutions, this paper proposes the application of a high-temperature jet vortex valve to a secondary jet supply system of thrust vectors. As shown in Figure 3, the high-temperature jet vortex valve uses a micro gas cylinder to provide a small amount of control flow for adjusting the high temperature and high pressure gas flow which from the combustion chamber. Compared with the mechanical valve in the gas split scheme, this structure does not require a gas valve under high temperature and high pressure working conditions. At the same time, it can achieve high temperature and high pressure down injection by carrying a relatively small amount of external gas. The secondary jet flow of the tube is adjusted to achieve 'big and small'. The high-temperature jet vortex valve scheme is completely adjusted by the gas, and has the advantages of fewer mechanical moving parts and compact structure.

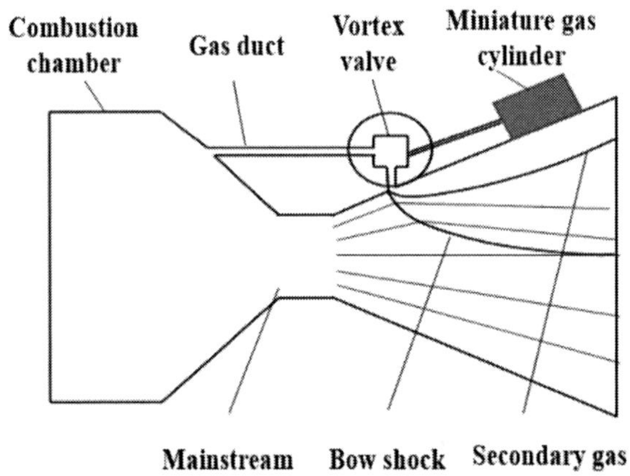

Figure 3. Schematic diagram of the vortex valve supply scheme.

The research shows that the flow regulation characteristics of the vortex valve have certain engineering application value. **R S Lin** [3] performed flow regulation ability analysis on 11 different structures of the vortex valve used in the aero-engine fuel supply system. Results shown that when the ratio of the control flow pressure to the mainstream pressure is 1.008, the TD ratio is 2.9, and when the nozzle is increased at the outlet, the TD ratio is increased to 4.2. **A Blatter** [4-5] applied the vortex valve concept to the variable thrust adjustment of the solid rocket motors, and carried out experimental research by building an axial jet vortex valve platform. The results show that adjusting the combustion chamber pressure through the vortex valve can achieve the thrust control of the engine, And put forward the conclusion that increasing the temperature of the control flow will increase the vortex valve flow adjustment range. **R J Woolhouse** [6] analyzed the eddy current field in the gas turbine by using numerical simulation and experimental comparison methods, and obtained a turbulence model and a physical model that can be used to calculate the vortex. The results show that the SST model is more suitable for high speed vortex than the model than $k-\omega$ model. **Jawarneh** [7] established a new method to study the radial pressure distribution of the angular velocity of the vortex chamber using the vortex model with a vortex valve with two supply ports, by using the continuous equation and the energy equation.

This article takes the high temperature vortex valve applied to the secondary jet of the nozzle as the research background, uses CFD numerical simulation to analyze the flow amplification mechanism of the vortex valve under high temperature and high pressure working conditions, and explores the main influence of the control flow pressure, control flow port area, which have the main influence on the flow adjustment characteristics. The purpose of this paper is to provide a reference for the subsequent design of the high-temperature jet vortex valve applied to the secondary jet of the nozzle.

2. Computational models and method

2.1. Vortex Valve Models

This paper uses a high temperature vortex valve structure consisting of a short cylindrical vortex chamber (usually the height of the vortex chamber is less than the radius of the vortex chamber) and three ports. They are the outlet end coaxial with the vortex chamber, the supply gas flow port of the radial jet and the control flow port of the tangential jet, which play a regulating role.

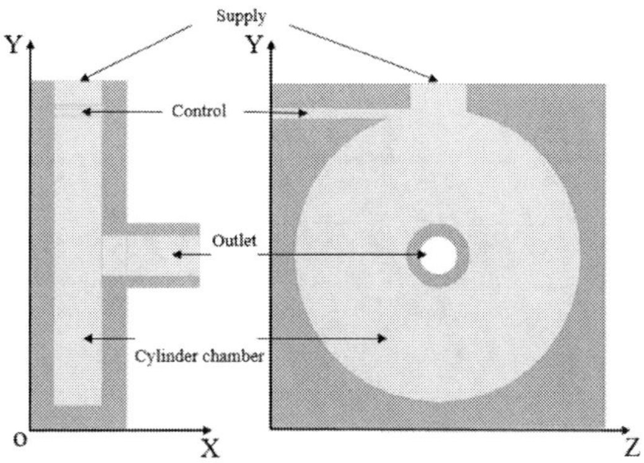

Figure 4. Structural model of vortex valve.

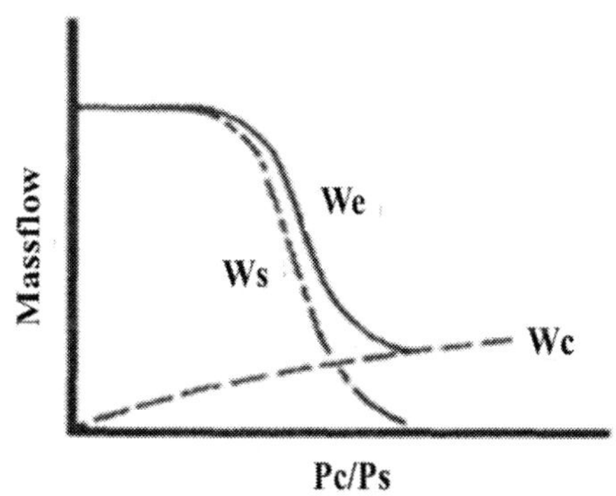

Figure 5. Flow regulation characteristics.

Figure 4 shows the basic components of vortex valve. The amplification effect of the vortex valve flow adjustment is reflected in that a larger supply flow rate could be adjusted by a smaller control mass flow. Among them, the supply flow enters the vortex chamber in a radial direction. When there is no control flow, the supply gas flow directly flows out through the outlet, while the vortex valve has the largest flow rate. With the continuous increase of the control flow rate, the angular momentum brought in induces rotation. A pressure gradient is generated in the radial direction of the vortex valve, which increases the flow resistance of the supply flow and reduces the flow of the supply, thereby generating the effect of flow regulation [8].

This paper uses the TD (Turndown Ratio) ratio to measure the effect of the flow regulating characteristics of the vortex valve. The TD ratio is defined as the ratio between the maximum mass flow rate and the minimum mass flow rate of the outlet of the vortex valve. Figure 5 shows the characteristic curve of vortex valve flow regulation under a typical working condition.

Based on the principle of vortex valve flow adjustment by establishing a pressure gradient, this paper mainly explores the characteristics of vortex valve flow adjustment when the pressure of the control flow changes, and also explores the effect of the area of control flow on the vortex valve flow adjustment characteristics.

This paper refers to the vortex valve structure designed by **R K Brodersen** [9] to design the physical model of the vortex valve. As shown in table 1. The design parameters are: D_o (Vortex Chamber Radius) = 20mm, D_e (Outlet Diameter) = 10mm, h (Vortex Chamber Height) = 10mm, A_s (Supply Flow Area) = 60mm^2, A_c (control flow area) = 2mm^2.

Table 1. Structural parameters of vortex valve.

Parameter	$\frac{D_o}{D_e}$	$\frac{A_s}{A_e}$	$\frac{A_c}{A_e}$	$\frac{h}{r_e}$
Range	3~4	2.546~9.42	0.021~0.254	2.0~5.3
This paper	4	3	0.1, 0.15, 0.2	4

2.2. Simulation Methods

In order to obtain accurate simulation results, this paper uses ICEM for structured mesh calculation, as shown in figure 6 and figure 7. The meshing quality is a key factor affecting the quality of calculation results. In order to improve the efficiency and quality of numerical calculations, according to the structure of the vortex valve, the calculation domain is divided into three sub-regions, and the meshes are generated and partially encrypted, which stitched by interface. The number of grid nodes of the vortex valve model is 32×10^4.

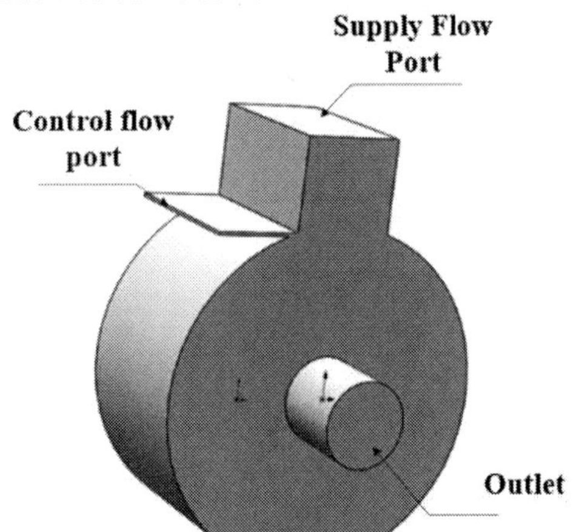

Figure 6. Simplified model of vortex valve.

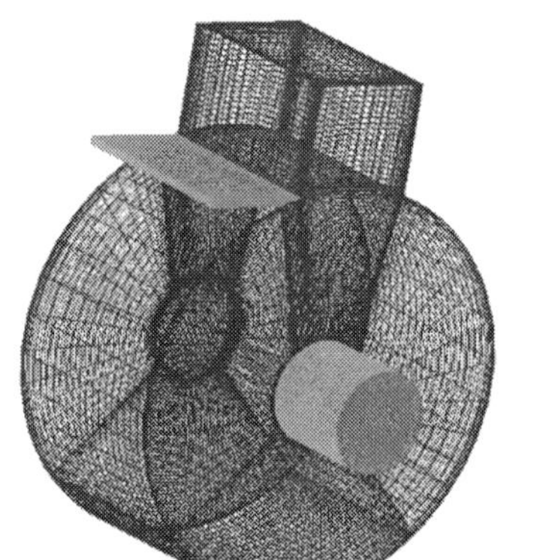

Figure 7. Computing meshing of vortex valve.

For the numerical simulation of the fluid condition inside the vortex valve, there are a variety of turbulence models to choose. According to the previous research [5], the lateral dissipation derivative term is added to the SST (Shear Stress Transport) model. The transport process of turbulent shear stress, so the SST turbulence model can better describe the eddy current calculation with backpressure gradient. The SST model combines the advantages of the model calculations in the near-wall region and the features of the model in the far-field calculation. It is calculated by multiplying the two models by the mixing function. In the near-wall area, the mixing function is equal to 1, so the near-wall area is equivalent to the $k-\omega$ model; the area far from the wall is automatically converted into the $k-\varepsilon$ model.

SST turbulence model transport equation:

$$\frac{\partial}{\partial t}(\rho k)+\frac{\partial}{\partial x_i}(\rho k u_i)=\frac{\partial}{\partial x_j}(\Gamma_k \frac{\partial k}{\partial x_j})+G_k-Y_k+S_k \tag{1}$$

$$\frac{\partial}{\partial t}(\rho \omega)+\frac{\partial}{\partial x_i}(\rho \omega u_j)=\frac{\partial}{\partial x_j}(\Gamma_\omega \frac{\partial \omega}{\partial x_j})+G_\omega-Y_\omega+S_\omega \tag{2}$$

In the formula, G_k represents the turbulent flow energy; G_ω is the ω generated amount; Γ_k, Γ_ω represents the effective diffusivity of k, ω, Y_k, Y_ω represents the dissipation due to k, ω.

In the calculation of this paper, gas is used as the working fluid for the specific vortex valve structure, the changes in the pressure of the combustion chamber caused by the mainstream leading from the combustion chamber are ignored, that is, the parameters of the supply pressure and temperature of the vortex valve do not change with the control flow adjustment process. The calculated gas parameters are as follows: $c_p = 1254 J\cdot kg^{-1}\cdot k^{-1}$ (Specific Heat Capacity), $\lambda_g = 0.3946 W\cdot m^{-1}\cdot k^{-1}$ (Thermal Conductivity of Gas), $\mu_g = 9.2\times 10^{-5} kg\cdot m^{-1}\cdot s^{-1}$ (Dynamic Viscosity), $R = 321.943 J\cdot kg^{-1}\cdot k^{-1}$ (Gas Constant).

Under the condition that the supply flow pressure and temperature of the vortex valve are constant, the change of the flow regulating characteristics of the vortex valve is explored by changing the control flow and the mainstream pressure ratio P_c / P_s. With reference to the research by **A Blatter** [4], the supply inlet is set to pressure inlet P_s = 16MPa (total pressure), T_s = 2500K (total temperature), the control flow adopts the pressure inlet boundary condition, and the wall surface is adiabatic and non-slip wall condition. Each set of calculations is performed until the supply flow is turned off.

3. Results and Analysis

3.1. Analysis of flow field inside vortex valve

Calculating the influence of the pressure change of the control flow on the flow regulation characteristics of the vortex valve to ensure that the supply pressure and temperature remain unchanged. The flow regulation characteristics of the vortex valve are explored by changing the pressure ratio of the control flow to the supply. Flow examples used for the numerical simulations are summarized in table 2.

Table 2. Examples of pressure change in control flow.

Examples	p_s(MPa)	T_s(K)	T_c(K)	A_s(mm^2)	A_c(mm^2)	p_c/p_s
1	16	2500	300	60	2	No control
2	16	2500	300	60	2	1
3	16	2500	300	60	2	1.05
4	16	2500	300	60	2	1.07
5	16	2500	300	60	2	1.1
6	16	2500	300	60	2	1.2
7	16	2500	300	60	2	1.4
8	16	2500	300	60	2	1.6
9	16	2500	300	60	2	1.8

IOP Conf. Series: Materials Science and Engineering **887** (2020) 012014 doi:10.1088/1757-899X/887/1/012014

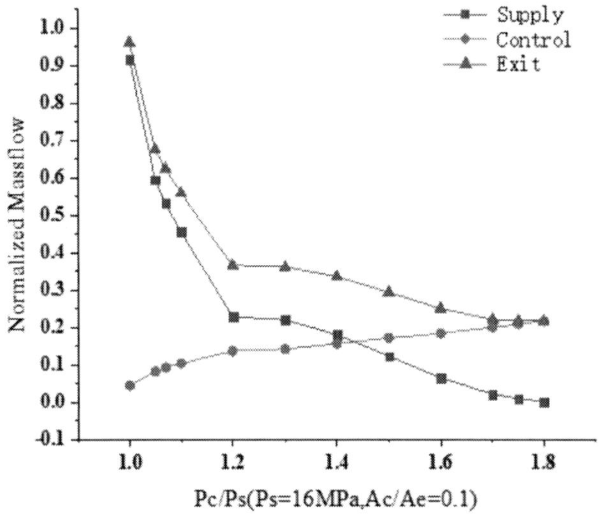

Figure 8. Flow characteristic curve of vortex valve.

It can be seen from figure 8 that under the condition that the supply pressure and temperature of the vortex valve remain unchanged, as the control flow pressure increases, the outlet flow of the vortex valve and the supply flow gradually decrease, finally the supply flow is completely cut off. This is because increasing the pressure and flow rate of the control flow increases the angular momentum of the control flow into the vortex chamber, enhances the rotational momentum of the vortex chamber, and increases the radial direction of the vortex chamber. The pressure gradient increasing resists the mainstream flow. It can be seen from figure 8 that when the control flow area ratio A_c / A_e = 0.1 and P_s = 16 MPa, the vortex valve flow adjustment ratio TD = 4.6. During this process, the control flow rate changed little, from 0.04 to 0.21, but the supply flow rate decreased from 0.92 to 0, and the vortex valve outlet flow rate decreased from 0.96 to 0.21, indicating that a small amount of control flow adjustment was used. The feasibility of the large flow main gas flow has verified the amplification effect of the vortex valve flow adjustment, which can use a small amount of control flow change to adjust the large vortex valve outlet flow.

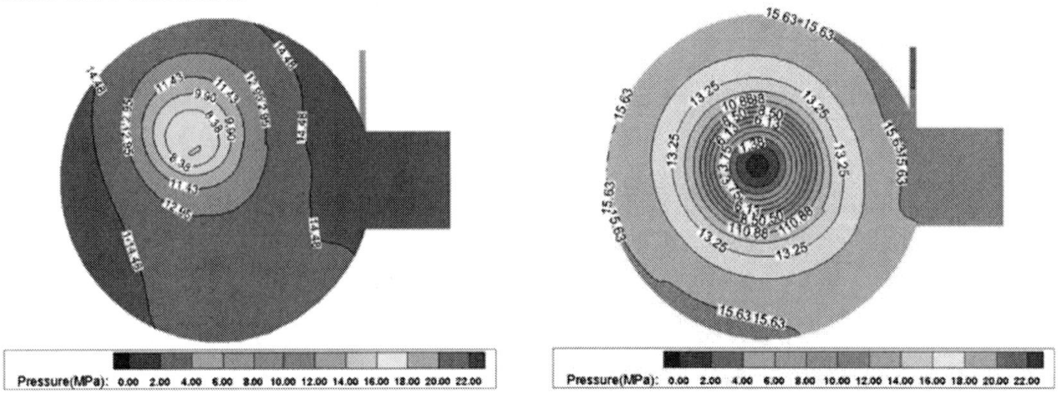

(a) No control (b) P_c/P_s = 1.2

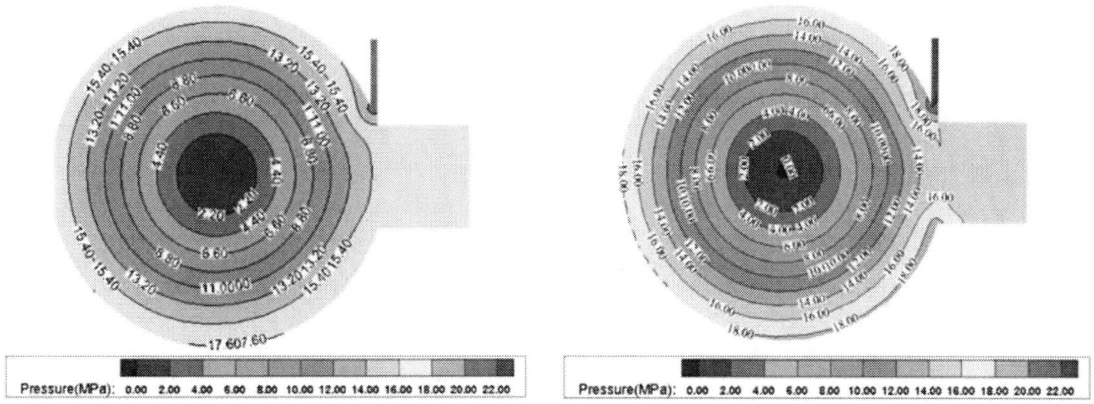

(c) $P_c/P_s = 1.6$ (d) $P_c/P_s = 1.8$

Figure 9. Cloud diagram of pressure distribution at X = 5mm section.

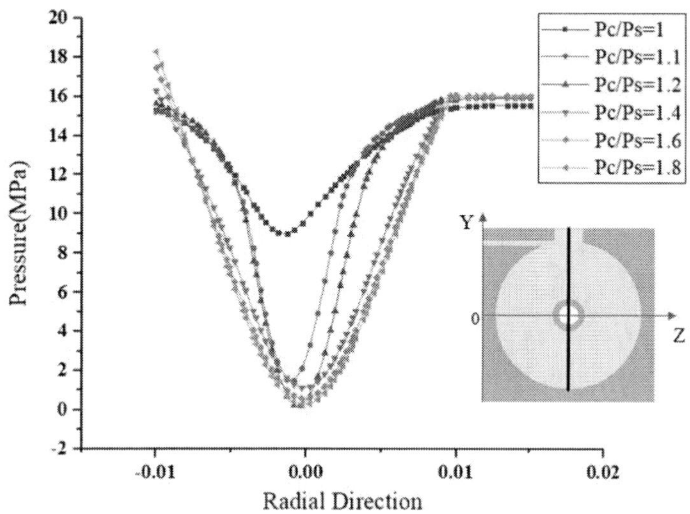

Figure 10. Y-direction pressure distribution at X = 5mm section.

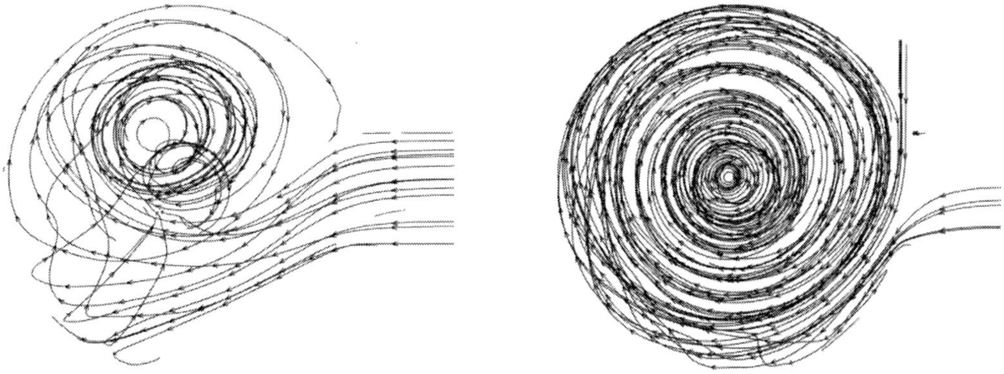

(a) No control (b) $P_c/P_s = 1.2$

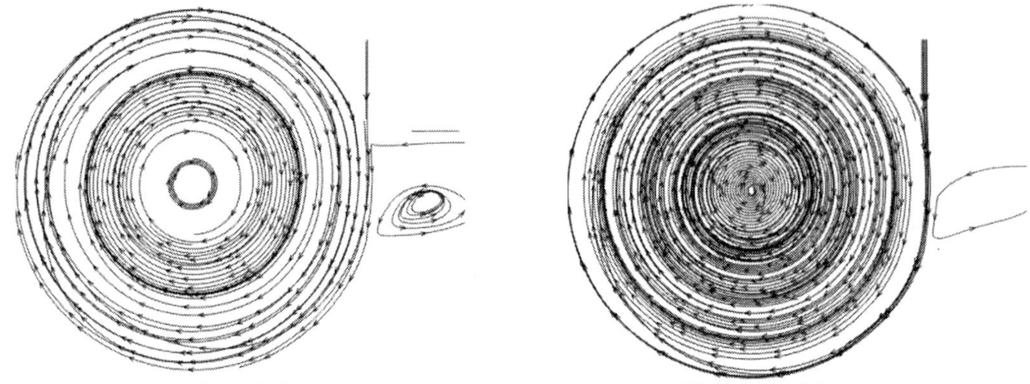

(c) $P_c/P_s = 1.6$ (d) $P_c/P_s = 1.8$

Figure 11. Streamline distribution of X = 5mm section

Figure 9 and figure 10 show the pressure distribution cloud diagram and the radial pressure distribution curve of the vortex valve X = 5mm section with different pressure ratios. It can be seen from figure 9 that the pressure of the flow field inside the vortex valve is annularly distributed due to the influence of the swirling flow, and the center of the annular pressure distribution gradually approaches the geometric center of the vortex valve with the increase of the force ratio 1.8. Figure 11 shows that the flow separation phenomenon occurs when the pressure ratio is relatively low inside the vortex field, that is, reverse flow occurs inside the vortex field. When the Cut-off state is reached(P_c / P_s = 1.8), the vortex field is almost in a vortex symmetrical state, and the pressure distribution center is almost at the geometric center of the vortex valve, the internal flow field tends to stabilize. Combining the pressure distribution cloud diagram and curve show that as the pressure ratio gradually increases, the pressure near the wall surface of the vortex chamber gradually increases, and the slope of the pressure distribution curve gradually increases. The pressure gradient inside the vortex field becomes larger, resulting in a smaller pressure drop between the supply inlet and the vortex field. Increased resistance to the supply causes mainstream traffic to gradually decrease. Therefore, by increasing the control flow pressure, the main flow has a certain inhibitory effect. When P_c / P_s = 1.8, the resistance caused by the radial pressure gradient is greater than the power of the supply flow to enter the vortex chamber. At this time, the main flow is completely cut off.

Combining figure 8 and figure 10, it is found that the pressure gradient of the radial pressure distribution of the vortex valve suddenly increases at a pressure ratio from 1 to 1.2, the pressure gradient of the process suddenly increases, and the sudden increase in the resistance of the vortex field to the mainstream is the main reason for the sudden decrease of supply flow in the range of 1-1.2. Within the range of pressure ratio 1.2-1.8, the pressure gradient changes little and the mainstream resistance changes slightly, so the supply flow decreases slowly in this range.

3.2. Vortex valve flow regulation performance changes when Ac / Ae changes

In order to investigate the influence of the control flow area on the flow regulating characteristics of the vortex valve, calculation conditions are shown in table 3.

Table 3. Structural parameters of vortex valve.

Conditions	A_c/A_e	A_e (mm^2)	D_o (mm)	h (mm)
1	0.1	20	20	10
2	0.15	20	20	10
3	0.2	20	20	10

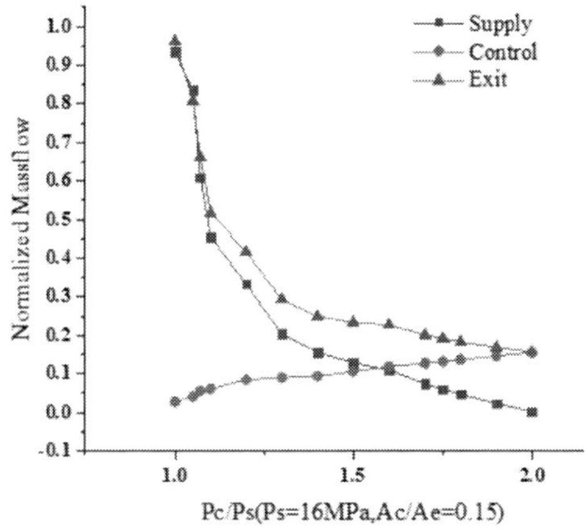

Figure 12. Result of vortex valve's flow regulation characteristics (P_s=16MPa, A_c/A_e=0.15).

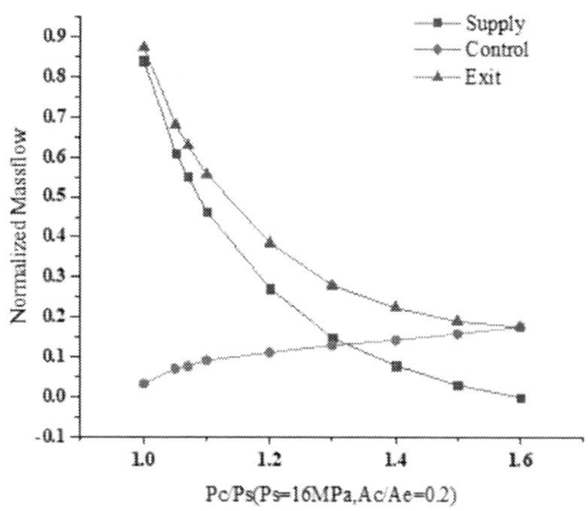

Figure 13. Result of vortex valve's flow regulation characteristics (P_s=16MPa, A_c/A_e=0.2).

Combining the flow adjustment characteristic curves with A_c / A_e values of 0.1, 0.15, and 0.2, it can be known that with A_c / A_e values of 0.1, 0.15, and 0.2, respectively, the vortex valve flow adjustment ratios TD are 4.6, 6.4, and 5.7, respectively. As the control flow area becomes larger, the TD ratio increases first and then decreases. It can be known from figure 12 that change the control flow area within a certain range can increase the vortex valve flow adjustment range. When A_c / A_e = 0.1, the cutting force ratio is 1.8; when A_c / A_e = 0.15, the cutting force ratio is 2.0; when A_c/A_e = 0.15, the cutting force ratio is 1.6. This is because as the area of the control flow increases, the tangential speed of the vortex valve becomes smaller and the angular momentum decreases when the same pressure ratio is used. When the cut-off effect is reached, the pressure ratio needs to be increased, and the vortex valve flow adjustment range is also increased. It can be seen from the curve in figure 13 that when A_c / A_e increases to a certain value, the pressure ratio required for cutting becomes smaller, and the flow adjustment process of the vortex valve is more stable. The reason is that as the area of the control flow increases, the flow of the control flow varies with the pressure increase changes relatively slowly, the angular momentum brought by the control flow becomes smaller, and the pressure gradient tends to be smooth, which makes the internal flow field more stable, so the adjustment process is smoother.

4. Conclusions

In this paper, the physical model and mathematical model of the vortex valve are established to mesh the internal flow field. Numerical analysis of the flow regulating characteristics of the vortex valve applied to the secondary jet of the nozzle shows that:

1) When the mainstream parameters of the high temperature jet vortex valve are constant, depending on the radial pressure gradient generated by changing the control flow ratio, the flow rate can be adjusted for the main stream and the outlet, thereby achieving the secondary jet flow adjustment effect of the nozzle, which proves that the high temperature jet is used. The vortex valve is a feasible solution to adjust the secondary jet flow of the nozzle. With the gradual increase of the ratio P_c/P_s, the outlet flow of the vortex valve and the main gas flow gradually decrease, which can realize the secondary jet flow of the nozzle. Continuous adjustment, meanwhile, the high-temperature jet vortex valve scheme has the amplification effect of a small amount of control flow to adjust a larger jet flow.

2) By changing the ratio of the area of the control port of the vortex valve to the area of the outlet of the vortex valve (A_c/A_e), the flow adjustment range of the vortex valve can be changed. When A_c/A_e is 0.1, 0.15, 0.2 respectively, the flow adjustment TD ratios are 4.6, 6.4, 5.7, and the pressure ratio when the main gas flow is cut off is 1.8, 2.0, 1.6, indicating that when the predetermined flow adjustment range of the vortex valve is reached, At this time, there is an optimal A_c/A_e area ratio. At this time, the

ratio of the required control flow to the mainstream pressure of the vortex valve is the smallest.

References

[1] Zhang J H and XIE K 2012 Vector Control Scheme for Secondary Flow of Fluid Throat Nozzle *Journal of aeronautical dynamics* **03** 309-8.
[2] Li B, Yu X Y and Xie K 2016 Vector Characteristics of Fluid Throat of Solid Rocket Motor *Journal of aeronautical dynamics* **12** 2996-03.
[3] Beckwith I E 1975 Development of a High Reynolds Number Quiet Tunnel for Transition Research *AIAA Journal* **13** 300-6.
[4] Blatter A and Keranen T W 1969 Research and development of a vortex valve for flow modulation of a 16-percent aluminized 550 deg F solid propellant gas *AIAA* vol 69 p 424
[5] Blatter A and Keranen T W 1970 A vortex valve for flow modulation of 5500 deg F gas *Journal of Spacecraft and Rockets* **7** 169-4.
[6] Woolhouse R J, Tippetts J R and Beck S B M 2001 A comparison of the experimental and computational modelling of the fluidic turn-up vortex amplifier at full and zero swirl conditions *ARCHIVE Proceedings of the Institution of Mechanical Engineers Part C Journal of Mechanical Engineering Science* **215** 893-3.
[7] Ali J and Georgios H V 2004 Vortex chamber flows *AIAA* p 5620.
[8] Sun B, Yang Z H and Quan W L 2014 Cold Flow Test Investigation on Modulation Property of Fluidic Vortex Valve *Journal of Projectiles, Rockets, Missiles, and Guidance* **34** 107-9.
[9] Brodersen R and Papadopoulos J 1981 Hot gas control system design and vortex valve tests *Automatic Control IEEE Transactions on* **26** 625-7.

Effect of Obstacle's Length-to-Height Ratio on Aerodynamic Quantities of Rarefied Hypersonic Flow

L WANG[1] and S FANG[1]

[1] School of Aerospace Engineering, Beijing Institute of Technology, Beijing, 100081, China

Email: 314475440@qq.com

Abstract. The DSMC method was used to calculate and analyse the flow field structure and the surface aerodynamic quantities when the hypersonic rarefied gas flow passing obstacle at a height of 70km. The changes of surface aerodynamic quantities such as skin friction coefficient, heat transfer coefficient, and pressure coefficient when the length-to-height ratio of the rectangular obstacle is 1/1, 1/2, 1/3, and 1/4 are analysed. The results show that under the above aspect ratio of the obstacle, a recirculation region is formed at each corner of the obstacle. A bow shock will be formed on the windward side of the obstacle, and with the increase of the obstacle's height, the bow shock will be superimposed on the separated shock wave in front. A strong expansion wave will be formed on the backside of the obstacle, which affects the downstream flow field structure and aerodynamic quantities. When the obstacle's height increases, the maximum value of the aerodynamic quantities on each surface of the obstacle will increase significantly. It is calculated that the quantities such as heat transfer coefficient and pressure coefficient reach the maximum at the convex point of the windward surface.

1. Introduction

The knowledge of the factors that affect the aerodynamic and thermal loads acting on the vehicle surface becomes significant while designing a hypersonic vehicle. And the hypersonic spacecraft flying in the upper atmosphere inevitably has geometric surface discontinuities, such as cavities, gaps, steps or obstacles, which occurs as a desired or undesired design feature in modern aerodynamic configurations. The obstacles, for example, may be formed due to sensor installations, fabrication tolerances, and differential expansion or ablation rates between different materials [1]. These discontinuities, due to the interaction with external flows, may constitute in a potential source in heat flux rise to the surface and may contribute to physical phenomena related to the problem of flow separation.

The flow over a cavity, gap or obstacle causes locally thermal and aerodynamic loads that may seriously exceed those of a smooth curve in hypersonic flight. For safety reasons, these loads must be predicted correctly. This can be done either by experiments, which are very expensive and time consuming for real flight conditions, or by numerical simulation, which is getting constantly increasing importance.

In this context, forward-facing steps positioned in a rarefied hypersonic flow has been studied by Leite and Santos [2] using the DSMC method. The study was inspired by the concern in investigating the impact of frontal-face height on the flow field structure and on the aerodynamic surface properties in the transition flow regime. The computational results suggested that the diversity of the step height

Content from this work may be used under the terms of the Creative Commons Attribution 3.0 licence. Any further distribution of this work must maintain attribution to the author(s) and the title of the work, journal citation and DOI.
Published under licence by IOP Publishing Ltd

led to significant changes in the flow field structure ahead of the step. High pressure and heating loads were noticed in the vicinity of the step fontal face because of the recirculation region ahead the step.

In the following, Paolicchi and Santos [3] worked on gaps positioned in a rarefied hypersonic flow also by using the DSMC method. The work attempt to investigate the impact of length-to-depth ratio (L/H) on the flow field structure. The primary highlight was to analyse the behavior of the primary properties, such as pressure, velocity, temperature and density when the gap L/H ratio changes. The results showed that the gap flow in the transition flow regime acts differently from that found in the continuum flow regime. It was observed only one vortex for the L/H ratio of 1, 1/2, 1/3 and 1/4. However, in the continuum flow regime, the number of vortices inside the gap is almost given by the amount H/L [4].

In the present report, numerous studies have attempted to investigate the previous work [2-3]. For example, Leite and Santos have studied on a surface discontinuity, a combined gap/step, in a rarefied hypersonic flow [5], and a back-facing step in the same condition [6]. However, few writers have been able to draw on any systematic research into the geometric discontinuity as an obstacle in the hypersonic rarefied flow. Therefore, the primary goal is to provide a comprehensive characterization of the effects on pressure and heating loads in the flow field structure and on the obstacle surfaces. Freestream flow conditions employed in the present calculations are those given in the previous work [2-3], and represent those experienced by a re-entry vehicle at an altitude of 70 km with freestream Mach number 25. In these conditions, the degree of molecular non-equilibrium is such that the Navier-Stokes equations are inapplicable. Therefore, the DSMC method will be employed to calculate the hypersonic flow on the obstacle configuration.

2. Computational method and procedure

The Direct Simulation Monte Carlo (DSMC) method, developed by Bird [7], has been proved very outstanding in simulating complex physical phenomena in rarefied non-equilibrium flows. The DSMC method using a computer to model real gas flows by simulating a great number of modeling particles, which represents a fixed number of real gas molecules. The states of simulated particles are stored and modified with each time step as the particles move, collide and exit boundary in the simulated domain, so that the position coordinates, velocities and other physical properties such as internal energy can be calculated. In this scenario, the DSMC method will be used in the present work in order to perform a parametric study of hypersonic rarefied flow over obstacles.

In the present work, the molecular collisions are modeled by using the variable soft sphere (VSS) molecular model, which is proved to be more accurate than variable hard sphere (VHS) [8], and the no time counter (NTC) collision sampling technique [9], and the Borgnakke–Larsen statistical model [10] for energy exchange between translational and internal modes of diatomic or polyatomic molecules. Simulations are operated by using a non-reacting gas model consisting of two chemical species, N_2 and O_2. Energy exchanges between the translation and internal modes, rotation and vibration, are considered. Relaxation collision numbers are obtained in a collision energy-based procedure as suggested by Boyd [11] for rotation and by Bird [12] for vibration. For a given collision, the probabilities are designated by the inverse of the relaxation numbers, which correspond to the number of collisions necessary, on average, for a molecule to relax.

In order to implement the collisions between a pair of particles, the flow field around the obstacle configuration is divided into an arbitrary number of regions, which are subdivided into computational cells. The cells are further subdivided into subcells, two subcells/cell in each coordinate direction. The cell provides a reference for the sampling of the macroscopic gas properties, such as density, velocity, pressure, temperature, etc., while the subcell is used in order to select collision partners for the establishment of the collision rate. Therefore, the physical space network is used to facilitate the choice of molecules for collisions and for the sampling of the macroscopic flow properties.

2.1. Geometry definition

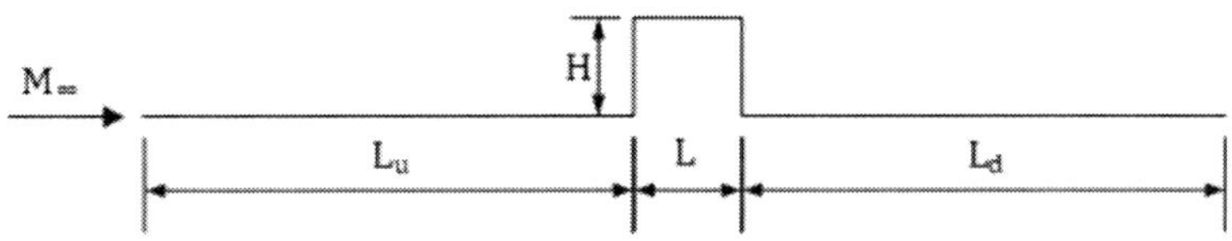

Figure 1. Geometry definition of the obstacle.

Figure 1 shows the definition of the geometric model of an obstacle. A zero-thickness flat plate with an obstacle of length L and depth H was assumed. Since the height H of the obstacle is much smaller than the nose radius R of a re-entry spacecraft, the geometric model can reasonably represent the state when the hypersonic spacecraft has a surface protrusion.

Referring to Figure 1, M_∞ stands for the free stream Mach number, H is the obstacle height, L is the obstacle length, L_u is the length of the obstacle upstream surface, and L_d is the length of the obstacle downstream surface. The following were assumed: L_u and L_d of 45.15 mm, a length L of 3 mm, and a height H of 3, 6, 9, and 12 mm. Therefore, the obstacles investigated correspond to a length-to-height ratio, L/H, of 1, 1/2, 1/3, and 1/4, respectively. In addition, L_u and L_d correspond to $50\lambda_\infty$, where $\lambda_\infty$is the freestream mean free path at an altitude of 70 km.

An understanding of the L/H ratio effects on the aerodynamic surface properties can be gained by comparing the flow field behavior of a flat plate with an obstacle to that without an obstacle. In this manner, a flat plate without an obstacle works as a benchmark for the cases with an obstacle and will be referred herein as the flat-plate case. For convenience, it was assumed a flat plate with a total length of $200\lambda_\infty$.

2.2. Freestream and flow conditions

Table 1. Freestream and flow conditions.

Parameter	Value	Unit
Altitude	70	km
Velocity (U_∞)	7456	m/s
Temperature (T_∞)	220.0	K
Pressure (p_∞)	5.582	N/m^2
Density (ρ_∞)	8.753×10-5	kg/m^3
Viscosity (μ_∞)	1.455×10-5	Ns/m^2
Number density (n_∞)	1.8209×1021	m^{-3}
Mean free path (λ_∞)	9.03×10-4	m

Freestream flow conditions used for the numerical simulations are summarized in Table 1. These freestream conditions represent those experienced by a spacecraft at an altitude of 70 km [13]. This altitude is associated with the rarefied flow regime, which is characterized by the overall Knudsen number the order of or larger than 10^{-2}. In addition, the gas properties, associated with the working fluid, N_2 and O_2, are given by Bird [7], and tabulated in Table 2. In this table, m, d, and ω stand for molecular mass, molecular diameter, and viscosity index, respectively.

Table 2. Gas properties.

	m(kg)	d(m)	ω
O_2	5.312×10^{-26}	4.07×10^{-10}	0.77
N_2	4.650×10^{-26}	4.17×10^{-10}	0.74

The freestream velocity U_∞, assumed to be constant at 7456 m/s, corresponds to a freestream Mach number M_∞ of 25. The wall temperature T_w was fixed at 880 K, and assumed to be uniform on the obstacle. This temperature is representative of the surface temperature near the stagnation point of a re-entry capsule.

The degree of rarefaction of the flow is usually expressed through the overall Knudsen number, $Kn = \lambda/l$, where λ is the mean free path in the freestream gas and l is a characteristic length of the flow field. By assuming the obstacle length L as the characteristic length, the overall Knudsen number Kn_H corresponds to 0.3095, 0.1548, 0.1032, and 0.0774 for height H of 3, 6, 9, and 12 mm, respectively. In addition, the Reynolds number Re_H is around 121, 243, 365, and 486 for height H of 3, 6, 9, and 12mm, respectively, also based on conditions in the undisturbed stream.

2.3. Computational flow domain and grid

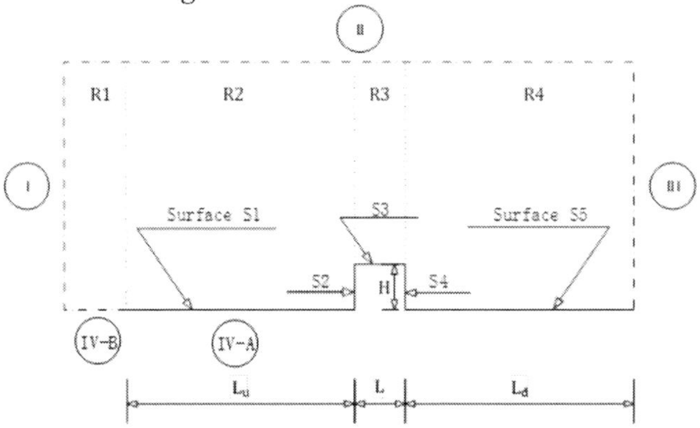

Figure 2. Schematic view of the computational domain.

The physical space around the obstacle was divided into four regions, which were subdivided into computational cells, in order to improve the choice of particles for collisions and for sampling of the macroscopic flow properties. According to previous studies [14-15], the size of each cell should be smaller than the local mean free path λ_∞, which provided a proper reference for the sampling of the macroscopic gas properties. Also, cells were further subdivided into subcells on the basis of the number of simulated particles, by two subcells per cell in each coordinate direction, in order to ensure physical accuracy in the collision process. In this case, the collision partners were chosen from the same subcell for the establishment of the collision rate. For a given pair of particles, the collision probabilities are inversely proportional to the relaxation numbers, which correspond to the number of collisions necessary, on average, for a molecule to achieve thermal equilibrium.

A schematic view of the computational domain is plotted in Figure 2. This computational domain was made large enough so that flow disturbances caused by the obstacle did not reach the computational boundaries, where freestream conditions were specified. As can be seen in this figure, side IV-A is defined by the obstacle surface. Diffuse reflection with complete thermal accommodation is applied to this side. Side IV-B is a plane of symmetry, where all flow gradients normal to the plane are zero. At the molecular level, this plane is equivalent to a specular reflecting boundary. Sides I and II are the freestream sides through which simulated particles enter and exit. Side I is positioned $5\lambda_\infty$ upstream of the flat-plate leading edge, and side II defined at $50\lambda_\infty$ in obstacle height H of 3 and 6mm cases, and $70\lambda_\infty$ in H of 9 and 12mm cases above the flat plate. Finally, at the downstream outflow boundary, side III, the flow is predominantly supersonic, and vacuum condition was specified [12]. Therefore, simulated molecules can only exit at this boundary.

3. Computational results and discussion

This section focuses on the effects that take place in the aerodynamic surface quantities due to variations on the obstacle L/H ratio. Aerodynamic surface quantities of particular interest in the transitional flow regime are heat flux, wall pressure, and shear stress. In this scenario, this section discusses and compares the differences of these quantities expressed in a coefficient form.

Before proceeding with the analysis of the aerodynamic surface quantities, it proves instructive to first present some features of the flow patterns around the obstacles. In doing so, particular attention is paid to the distribution of pressure coefficient in the entire computation domain.

Distribution of pressure coefficient around the obstacle is illustrated in Figure 3. In this plot, pressure is normalized by the freestream pressure, ρ_∞. According to Figure 3, it is observed that the maximum value for the pressure coefficient takes place in a region in the vicinity of the front side of the obstacle. It is obvious that there is a shock wave at the leading edge and a bow shock in front of the obstacle.

Figure 3. Pressure Coefficient C_p distribution in the vicinity of obstacles.

3.1. Heat transfer coefficient

The heat transfer coefficient C_h is defined as follows:

$$C_h = \frac{q_w}{(1/2)\rho_\infty U_\infty^3} \tag{1}$$

where $q_\mathrm{w}$is the heat flux to the body surface.

The heat flux q_w to the body surface is calculated by the net energy flux of the molecules impinging on the surface. The net heat flux q_w is related to the contribution of the translational, rotational, and vibrational energies of both incident and reflected molecules as defined by the following expression:

$$q_w = q_i - q_r = \frac{F_N}{A\Delta t}\left\{\sum_{j=1}^{N}\left[\phi_j\right]_i - \sum_{j=1}^{N}\left[\phi_j\right]_r\right\} \tag{2}$$

where $\phi_j = \frac{1}{2}m_j c_j^2 + e_{R_j} + e_{V_j}$, c is the velocity of the molecules, e_R and e_V stand for rotational and vibrational energies, respectively.

The dependence of the heat transfer coefficient C_h on the L/H ratio is demonstrated in Figure 4 for surfaces S1 and S5, and in Figure 5 for surfaces S2, S3 and S4. As shown in Figure 4, important features can be observed in the heat transfer coefficient C_h along surface S1 due to changes in L/H ratio. Of particular interest, it is observed that the heat transfer coefficient C_h for obstacles follows the same behavior presented by the flat-plate case close to the sharp leading edge, region unaffected by the presence of the obstacles. Further downstream along the surface S1, the heat transfer coefficient C_h significantly increases and reaches peak values close to the frontal face, then decreases to almost zero at the stagnation region. Along the frontal face S2, the heat transfer coefficient increases monotonically, from zero at the stagnation point to a maximum value near the S2/S3 junction, which depends on the frontal-face height H. Along the upper surface S3, the heat transfer coefficient presents a maximum value at the S2/S3 junction and then decreases downstream along the surface. It is quite apparent that this significant increase in the heat transfer coefficient is due to the flow reattachment zone. Along the back surface S4, the heat transfer coefficient is two orders of magnitude lower than those observed on the surfaces S1 and S5, which is due to the recirculation region formed in the vicinity of the back face. Along the surface S5, the minimum heat transfer coefficient occurs in the vicinity of the back face in the recirculation region. After that, the heat transfer coefficient increases to a maximum value after the flow reattachment. In the following, C_h almost decays and approaches the level observed for the flat plate without an obstacle.

IOP Conf. Series: Materials Science and Engineering **887** (2020) 012015 doi:10.1088/1757-899X/887/1/012015

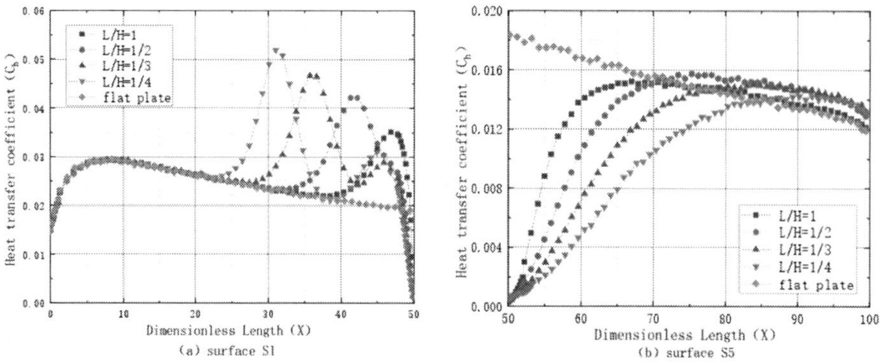

Figure 4. Heat transfer coefficient C_h distribution along surface S1 and S5

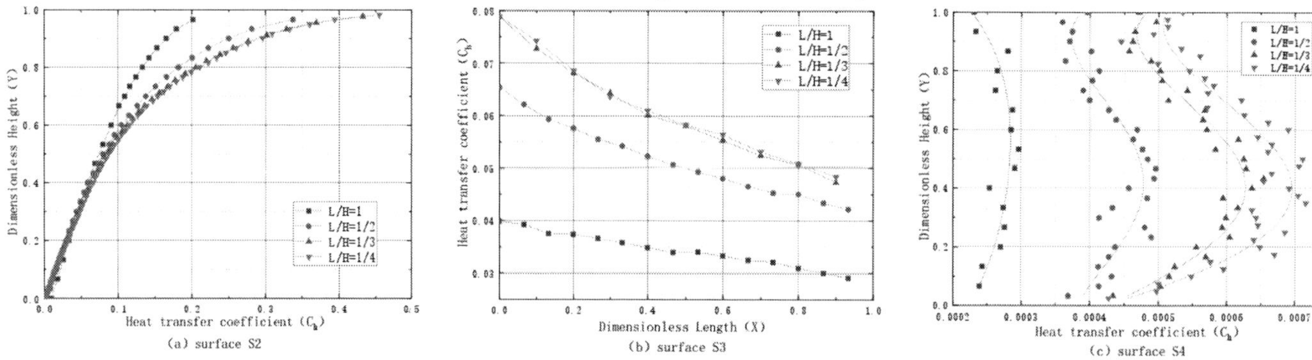

Figure 5. Heat transfer coefficient C_h distribution along obstacle surfaces S2, S3, and S4.

For comparative purpose, the maximum values for C_h are around 0.21, 0.34, 0.43 and 0.46 for height H of 3, 6, 9 and 12mm, respectively. In contrast, the C_h for the flat-plate case, i.e., a flat plate without obstacles, is around 0.0284 at section X = 9.11 in the surface. Therefore, C_h of 0.21, 0.34, 0.43 and 0.46 correspond respectively to 7.39, 11.97, 15.14 and 16.20 times the pick value for the flat-plate case. Furthermore, it is very encouraging to observe that, for H of 12 mm case, the amount of energy transferred to the obstacle corner represents around 50% of the total energy ($\frac{1}{2}\rho_\infty U_\infty^3$) of the gas coming from the freestream.

Another feature of particular interest in the heat transfer coefficient behavior is related to the peak values on the surface S1 due to the presence of the obstacles. As shown in Figure 4, the maximum values observed for the heat transfer coefficient C_h as well as the location in which they occur depend on the obstacle's height H. For height H of 3, 6, 9 and 12 the maximum values for the heat transfer coefficient C_h correspond to section X of 46.8, 42.2, 35.7 and 31.1, respectively. For the time being, it is worth taking a closer look at the peak values for the H = 9 and 12 case. Referring to Figure 4, it is clearly noticed a second point of maximum heat transfer coefficient at the vicinity of the obstacle, more precisely for section X = 45.9 and 45.00. This behavior differs from the other two cases, i.e., obstacle height H of 3 and 6 mm.

3.2. Pressure coefficient

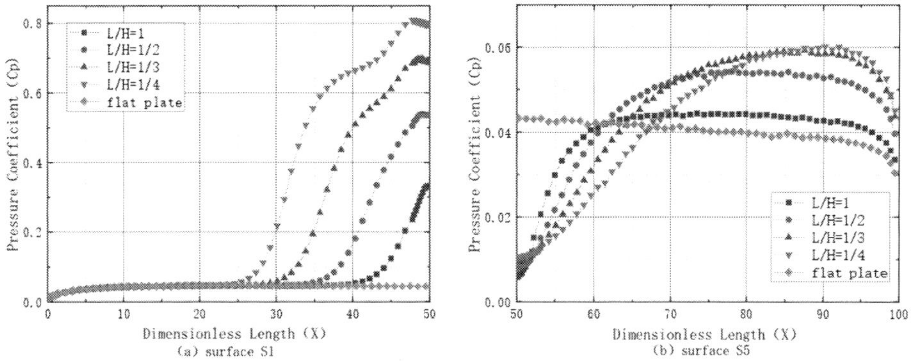

Figure 6. Pressure coefficient C_p distribution along surface S1 and S5.

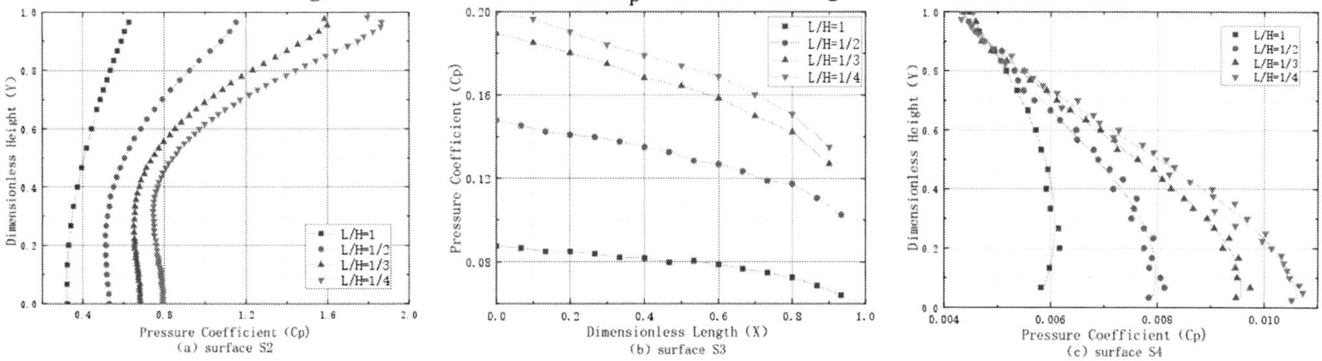

Figure 7. Pressure coefficient C_p distribution along obstacle surfaces S2, S3, and S4.

The pressure coefficient C_p is defined as follows:

$$C_p = \frac{p_w - p_\infty}{(1/2)\rho_\infty U_\infty^2} \quad (3)$$

where the pressure p_w on the body surface is calculated by the sum of the normal momentum fluxes of both incident and reflected molecules at each time step as follows,

$$p_w = p_i - p_r = \frac{F_N}{A\Delta t}\sum_{j=1}^{N}\left\{\left[(mv)_j\right]_i - \left[(mv)_j\right]_r\right\} \quad (4)$$

where F_N is the number of real molecules represented by a single simulated molecule, Δt is the time step, A is the area, N is the number of molecules colliding with the surface per unit time and unit area, m is the mass of the molecules, v is the velocity component of the molecule j in the surface normal direction. Subscripts i and r refer to incident and reflect molecules.

The impact on the pressure coefficient C_p caused by changes in the obstacle's height H is depicted in Figure 6 for surfaces S1 and S5, and in Figure 7 for surfaces S2, S3 and S4. According to this group of plots, it is noted that the pressure coefficient on the surface depends on the obstacle height, the pressure coefficient behavior along the obstacle surface is similar to that for the flat-plate case at the vicinity of the sharp leading edge. The upstream disturbances in the pressure coefficient C_p, due to the presence of the obstacle, were felt up to section X corresponding to approximately 28.27, 32.92, 38.49 and 44.07 for obstacles with height H of 12, 9, 6 and 3, respectively. From these sections up to the obstacle position, X = 50, the pressure coefficient C_p increases dramatically when compared to that for the flat-plate case. The maximum values for the pressure coefficient C_p along the surface S1 occur at the stagnation point, at the S1/S2 junction. Along the frontal surface S2, the peak value for C_p occurs close to the S2/S3 junction, similar to that peak value location for the heat transfer coefficient. Along the upper surface S3, the pressure coefficient presents a maximum value at the S2/S3 junction and then decreases downstream along the surface. Along the back surface S4, the smallest values for the pressure coefficient take place along the back face. And in this region, C_p is negative, this behavior is directly related to the recirculation region downstream of the back face. It should be mentioned in this context that the recirculation region in the vicinity of the back face is a region of low pressure.

Nevertheless, very close to the back corner, the pressure coefficient drops off due to the flow expansion around the back corner. After that, in the vicinity of the back face on surface S5, the pressure coefficient suddenly almost goes down and afterward increases significantly as the flow develops along surface S5. In addition, a significant reduction in the pressure coefficient is observed at the end of the lower surface. The reason for that is associated with the vacuum condition assumed for the downstream outflow boundary, side IV, as explained earlier.

For comparison purpose, the maximum values for C_p on the frontal face are around 0.63, 1.15, 1.59 and 1.80 for height H of 3, 6, 9 and 12, respectively. In contrast, the maximum value of C_p for the flat-plate case, i.e., a flat plate without obstacles is around 0.0393 at section X = 25.96 in the surface. Therefore, C_p correspond respectively to 16.03, 29.26, 40.45 and 45.80 times the pick value for the flat-plate case, which corresponds to a smooth surface.

It is noteworthy that the pressure coefficient rise at the vicinity of the frontal face is directly related to the recirculation zone that forms ahead of the obstacle. This pressure rise is explained by the fact that molecules, confined in the recirculation zone, collide with the surface S1 and with the face S2 of the obstacle, causing an increase in the normal momentum to surfaces of the obstacle. What stands out in Figure 6 is the peak value of pressure coefficient on surface S5, in L/H ratio of 1/2, 1/3 and 1/4 case, the pressure coefficient is 25%-50% higher than the flat-plate case.

3.3. Skin friction coefficient

The skin friction coefficient C_f is defined by,

$$C_f = \frac{\tau_w}{(1/2)\rho_\infty U_\infty^2} \tag{5}$$

where the shear stress τ_w on the body surface is calculated by the sum of the tangential momentum fluxes of both incident and reflected molecules impinging on the surface at each time step by the following expression,

$$\tau_w = \tau_i - \tau_r = \frac{F_N}{A\Delta t}\sum_{j=1}^{N}\left\{\left[(mu)_j\right]_i - \left[(mu)_j\right]_r\right\} \tag{6}$$

where u is the velocity component of the molecule j in the surface tangential direction.

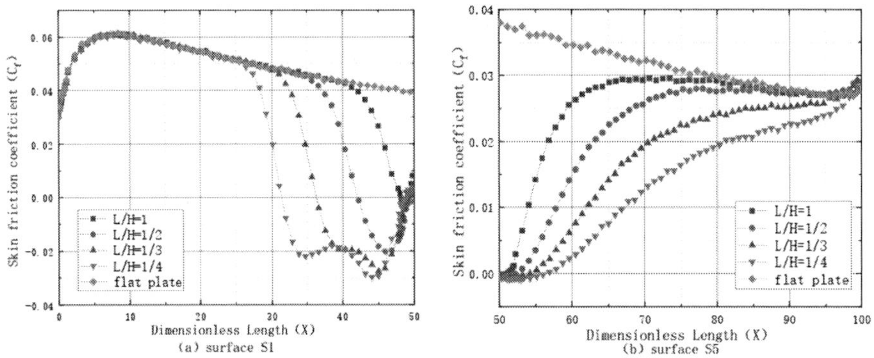

Figure 8. Skin friction coefficient C_f distribution along surface S1 and S5.

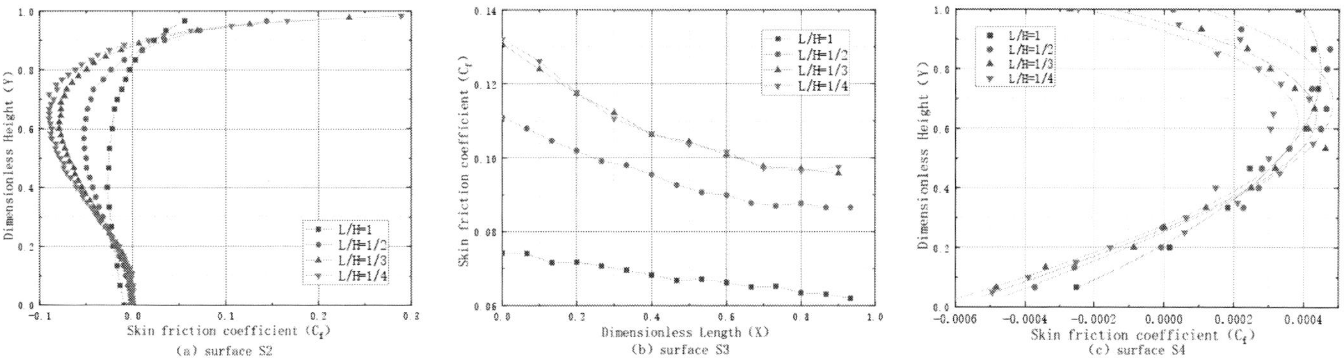

Figure 9. Skin friction coefficient C_f distribution along obstacle surfaces S2, S3, and S4.

The distribution of skin friction coefficient C_f along the surfaces S1 and S5 is displayed in Figure 8 and the obstacle surfaces S2, S3, and S4 in Figure 9 as a function of the obstacle height H. Looking first to Figure 8, it is observed that the upstream disturbances, due to the presence of the obstacles, are felt in the skin friction coefficient C_f approximately up to section X of 27.34, 31.06, 35.7, and 41.2, for height H of 12, 9, 6, and 3, respectively. From this position up to the obstacle position, X = 50, the skin friction coefficient C_f decreases, when compared to that for the flat-plate case, and reaches zero for section X of 31.99 , 36.63, 42.21, and 48.22 for height H of 12, 9, 6, and 3, respectively. After that, as a result of the recirculation region, the skin friction coefficient C_f continues to decrease up to a minimum point. After the minimum point, C_f increases again and reaches values close to zero at the stagnation point at the base of the obstacle. Along the upper surface S3, the skin friction coefficient C_f presents the maximum value at the obstacle shoulder, then drops off downstream and reaches the value observed for the flat-plate case. Along the surface S5, the skin friction coefficient is negative near to the face of the obstacle and becomes positive at the reattachment point, as shown in the magnified view. After that, the skin friction coefficient increases and reaches the values observed for the flat plate without obstacles.

Turning to Figure 9, along the frontal face S2, the skin friction coefficient C_f is basically zero at S1/S2 junction. After that, it is negative from the obstacle base up to the flow reattachment point. From this point up to the S2/S3 junction, the skin friction coefficient drastically increases, since this is basically a region exposed to a high speed flow. Afterwards, due to the flow expansion around the S1/S2 junction, the skin friction coefficient C_f diminishes by approximately 50% in comparison to the values observed at the beginning of the upper surface. Along the back surface S4, it is seen that the skin friction coefficient C_f along the back face is at least two orders of magnitude smaller than that observed along the upper surface. In addition, the peak values take place at the upper-most portion of the back face. Moreover, peak values decrease by increasing the obstacle height H. It should be mentioned that the section corresponding to the condition of $C_f = 0$ ($\tau_w = 0$) was used to define the separation point.

A more careful analysis in the distribution of the skin friction coefficient C_f reveals important flow peculiarities at the vicinity of the S1/S2 corner. As shown in Figure 8, the skin friction coefficient C_f on S1, after reaching the minimum negative value, it increases again and reaches values around zero at the stagnation point. A magnified view of the stagnation region indicates that the skin friction coefficient C_f becomes positive in this region, more precisely at section X of 49.33, 49.34, 49.00 and 48.89 for height H of 3, 6, 9 and 12, respectively, as illustrated in Figure 8. For the time being, it is worth taking a closer look at the peak values for the H = 9 and 12 cases. Referring to Figure 8, it is clearly noticed a second point of maximum skin friction coefficient at the vicinity of the obstacle, more precisely for section X = 40.35 and 34.78. This behavior differs from the other two cases, i.e., obstacle height H of 3 and 6 mm.

4. Conclusions

Computations of a rarefied hypersonic flow over obstacles have been performed by using the Direct Simulation Monte Carlo (DSMC) method. The simulations provided information concerning the nature of the aerodynamic properties acting on the surface of the obstacle. The effects of the obstacle height on the heat transfer, pressure and skin friction coefficients for a representative range of parameters were investigated. The obstacle height ranged from 3 to 12 mm, which corresponded to Knudsen numbers in the transitional flow regime.

The analysis showed that hypersonic flow past an obstacle is characterized by a strong bow-shock ahead of the frontal face, which influences the aerodynamics surface properties upstream the frontal face. And a strong expand wave was formed at the back face of the obstacle, which affects the properties downstream the back face. It was found that the disturbance at both upstream and downstream due to the presence of the obstacles increased with the obstacle height rise. Nevertheless, locally high heating rates were observed on the surface upstream and downstream the obstacle. The simulations showed that these high heating rates are a consequence of the molecules confined in the recirculation zones. As a result, for the conditions investigated in this work, in terms of heat and pressure loads, the presence of a discontinuity like an obstacle on the vehicle surface should be avoided in the vehicle design. And heat protection should be enhanced if an obstacle on the surface is inevitable.

References

[1] Bertram M H, WIGGS M M 1963 Effect of surface distortions on the heat transfer to a wing at hypersonic speeds *AIAA Journal* **1** 1313-1319
[2] Santos W, Leite P 2011 Computational analysis of rarefied hypersonic flow over a forward-facing step *42nd AIAA Thermophysics Conference* 2011-3310
[3] Paolicchi L T L C, Santos W F N 2018 Length-to-depth ratio effects on aerodynamic surface quantities of a hypersonic gap flow *AIAA Journal* **56** 780-792
[4] Everhart J, Alter S, Merski R, Wood W, Prabhu R 2006 Pressure gradient effects on hypersonic cavity flow heating *AIAA Paper* 2006-0185
[5] Leite P H M, Santos W F N 2014 Mach number impact on heat flux and pressure distributions of a hypersonic flow over combined gap/step geometries *AIP Conf. Proc.* **1628**, 176-184 (2014)
[6] Santos P H M L 2019 Computational analysis of a rarefied hypersonic flow over backward-facing steps *Journal of Thermophysics and Heat Transfer* 1533-6808
[7] Bird G A 1994 *Molecular gas dynamics and the direct simulation of gas flows* Oxford University Press
[8] Koura K, Matsumoto H 1991 Variable soft sphere molecular model for inverse-power-law or Lennard-Jones potential *Physics of Fluids A Fluid Dynamics* **3** 2459
[9] Bird G A 1989 Perception of numerical methods in rarefied gasdynamics *16th International Symposium on Rarefied Gas Dynamics: Theoretical and Computational Techniques* 211-226
[10] Borgnakke C, Larsen P S 1975 Statistical collision model for Monte Carlo simulation of polyatomic gas mixture *Journal of Computational Physics* **18** 405-420
[11] Jun E, Boyd I D 2018 Assessment of the LD-DSMC hybrid method for hypersonic rarefied flow *Computers & Fluids* **166** 123-138
[12] Bird G A 2009 A comparison of collision energy-based and temperature-based procedures in DSMC *Rarefied Gas Dynamics:26th International Symposium* 245-250
[13] Moss J, Boyles K, Greene F 2006 Orion aerodynamics for hypersonic free molecular to continuum conditions *14th AIAA/AHI International Space Planes and Hypersonic Systems and Technologies Conference* AIAA paper 2006-8081
[14] Alexander F J, Garcia A L, Alder B J 1998 Cell size dependence of transport coefficients in stochastic particle algorithms *Physics of Fluids* **10** 1540-1542
[15] Alexander F J, Garcia A L, Alder B J 2000 Erratum: "Cell size dependence of transport coefficients in stochastic particle algorithms" *Physics of Fluids* **12** 731-731

IOP Conf. Series: Materials Science and Engineering **887** (2020) 012016 doi:10.1088/1757-899X/887/1/012016

A Study on Towing Capacity with Deflated Tires of a Certain Type of Civil Aircraft

Zhenwei Zhang

Aircraft Integration Department, Shanghai Aircraft Design and Research Institute

zhangzhenwei@comac.cc

Abstract: It is a very important ability whether civil aircraft allows the towing operation when tires flat in actual operating. Based on a certain type of regional aircraft, associating with ground load, ground equipment, landing gear structure, weight balance and maintenance engineering specialty etc, this paper studies the related airworthiness clauses of towing with deflated tires, analyzes the towing load of a certain type of aircraft and defines the scene when towing with deflated tires occur. Procedures of the normal towing and deflation towing are analyzed and similarities and differences of competing types of aircraft are also compared. On this basis, the actual towing ability of a certain type of aircraft is evaluated and more research on weight, safety factor of towbar and other aspects is conducted. Finally, the conclusion is provided that this type of aircraft has towing ability with single deflated tire by exceeding 60% of load requirement of the clause and being confirmed by strength evaluation.
Keywords: Airworthiness clause; Deflated tires; Towing load; A certain type of regional aircraft

1. Introduction

Aircraft towing is an operation of moving an aircraft by a ground tractor. According to the different ways of traction, aircraft towing can be divided into two categories: conventional towbar towing and towbarless towing. For a certain type of civil regional aircraft, it adopts the way of towbar towing. Under normal towing conditions, the aircraft is on the paved rigid pavement surface, and it is pulled forward or backward by the tow bar connected to the nose landing gear buffer pillar. The maximum allowable towing angle of the nose wheels is 102 ° to the left and right. Towing operation must ensure normal tire pressure. When the tire is deflated, it must be replaced before the towing operation can be performed.

2. Questions raised

In 2015, when the aircraft was in the course of demonstration operation, the airline proposed that there are other aircraft that allows at least one deflated tire when towing, which can greatly improve the convenience of operation, especially in special scenarios where the aircraft is fully loaded. If the towing operation is allowed with deflated tires, it can avoid the consequences of closing the runway and reduce the operating burden of the airline. Therefore, airline companies hope that the main manufacturer can analyze whether this type of aircraft allows towing operation with deflated tires.

3. Evaluation of towing ability with deflated tires

Four aspects of the towing ability with deflated tires are analyzed and explored in this chapter, those are towing design and airworthiness clauses related to towing with deflated tires, towing scenarios with

Content from this work may be used under the terms of the Creative Commons Attribution 3.0 licence. Any further distribution of this work must maintain attribution to the author(s) and the title of the work, journal citation and DOI.
Published under licence by IOP Publishing Ltd

deflated tires that may occur, towing operation procedures with deflated tires of similar type of aircraft and towing ability of a certain type of aircraft.

3.1 Research on airworthiness clauses

a) 25.509 Towing loads

According to CCAR25.509(a)[1], specifies the calculation method of towing loads for different aircraft weights. These loads must be applied at the towing fittings and must act parallel to the ground.

For a certain type of regional aircraft, under normal tire inflation conditions, its towing load is given by

$$F_{TOW}=[(6W_T + 204,100)/70]*g=0.1526*W_T*g \quad (1)$$

Where WT is the design ramp weight.

According to CCAR25.509(d), specifies the towing loads for different towing points and towing directions. For this type of regional aircraft, the towing loads of the nose landing gear and the main landing gear in different directions are shown in Figure 1.

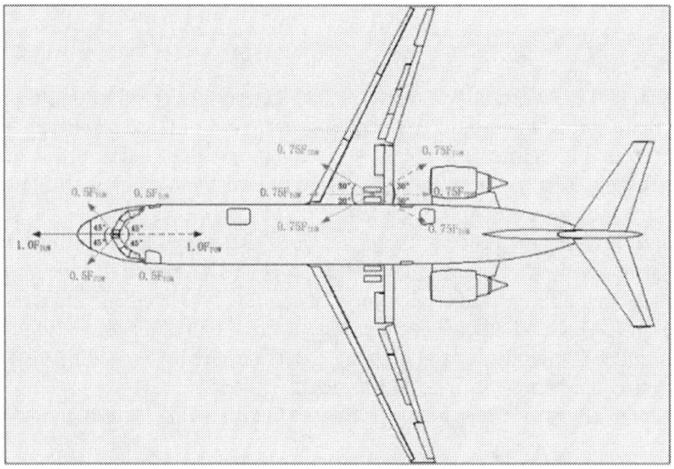

Figure 1. Schematic diagram of the towing loads of a certain type of aircraft

b) 25.511 Ground load: unsymmetrical loads on multiple-wheel units

According to CCAR25.511(b)(1), stipulates the relevant considerations for wheel arrangements with deflated tires. For a multiple-wheel landing gear device, one of the tires must be considered to be deflated. For a landing gear device with four or more wheels, two of the critical tires must be considered to be deflated.

For a certain type of regional aircraft, one deflated tire of its two-wheel landing gear device, that is, a single deflated tire should be considered.

CCAR25.511(f) considers the effect of deflated tires on the structure and specifies corresponding towing load limits. For the case of one and two deflated tires, the towing load F_{Tow} must be 60% and 50% of the load prescribed, respectively.

3.2 scenarios analysis

Towing scenarios with deflated tires are analyzed as follows.

a) When the aircraft is on the tarmac for pre-flight inspection, if the tire is deflated, it will not be released, and the tire should be replaced on the spot or towed to the nearby area.

b) When the aircraft is on the runway and taxiway before take-off and after landing, if the tore deflates, the aircraft needs to be towed away from the runway and taxiway because there is no deflation taxiing procedure.

c) For maintenance purposes, move the aircraft with deflated tires (for example, from the boarding gate to the maintenance hangar).

d) Other scenarios where the aircraft may need to be moved when the tire is deflated.

3.3 Comparison of other types of aircraft

After consulting the manuals of other types of aircraft[2~5], the towing procedures with deflated tires for each type are obtained, as shown in Table 1.

Table 1. The towing procedures with deflated tires of related types of aircraft

Aircraft type	Whether to allow towing operation with deflated tires	Tire deflation condition	Specific description of towing
A320	Yes	a) a maximum of one deflated tire on each landing gear.	Towing with nose landing gear and the maximum towing angle is +/- 10 °
		b) one deflated tire on a main landing gear.	Towing with nose landing gear and the maximum towing angle is +/- 40 °
		c) two deflated tires on one main landing gear (for four-wheel landing gear)	Towing with main landing gear and the towbar cannot be deflected
B737	Yes	a) one deflated tire on the landing gear.	a) normal towing process b) minimum towing speed c) avoid sharp turns
		b) more than one deflated tires on each landing gear.	towed by main landing gear with cable, and the maximum angle of cable is +/- 30 °
ERJ190	Yes	at most one deflated tire on each landing gear (for four-wheel landing gear)	normal traction process, while a) ensuring no one on the plane b) the aircraft is as light as possible to reduce the load on the landing gear c) the maximum towing speed is 7km/h walking speed
CRJ200	No	NA	All towing procedures require inflated tires.

3.4 Towbar design conditions

According to references[6] , the actual maximum towing load of a normal aircraft towing is obtained

$$T_{MAX} = (0.06+\gamma) *W_T*9.8=0.09W_T*9.8 \quad (2)$$

Where γ is the maximum slope of the airport taxiway, taking 3%.

The actual maximum towing load ratio to the specified towing load F_{TOW} is given by

$$T_{MAX}/F_{TOW}=0.09W_T*9.8/0.1526W_T*9.8=0.59 \quad (3)$$

In the design of the towbar of this type of aircraft, the shearing force of the safety bolt takes 0.8 times of the limiting towing load, that is, the safety factor of the towbar is designed to be 0.8. When there is starting acceleration, large turning or improper operation, the towing load for deflated tires is likely to exceed 60% of the prescribed towing load. If it is between 60% -70%, the towbar safety pin will not be disconnected, which may damage the landing gear structure.

3.5 Evaluation of towing ability with deflated tires

This paper attempts to reduce the towing load by unloading the aircraft and increasing the corner limit. This type of aircraft has a minimum passenger load ratio of 7%. After conservative calculation, the load factor can only be adjusted to 0.645 and the turning angle is limited to 41 ° by the way of unloading weight. But this method cannot cover the safety factor of the towbar which is 0.8. The towing load limit cannot be fully released.

The "N" working condition when the tire is normally inflated is defined in this paper, as well as the most severe working condition when the single wheel deflates, that is, the "F" working condition that all three tires on the left side of the landing gear deflate, and the "G" working condition that all three tires on the right side of the landing gear deflate, as shown in Figure 2 and Figure 3.

" **N** " - Nominal Tire:

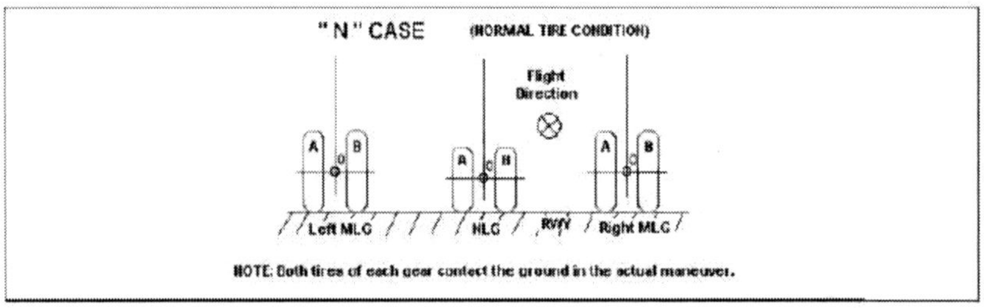

Figure 2. Schematic diagram of normal tire inflation conditions

" **F** " / " **G** " **- Tire Asymmetry (deflated tires) [25.511(c)]:**

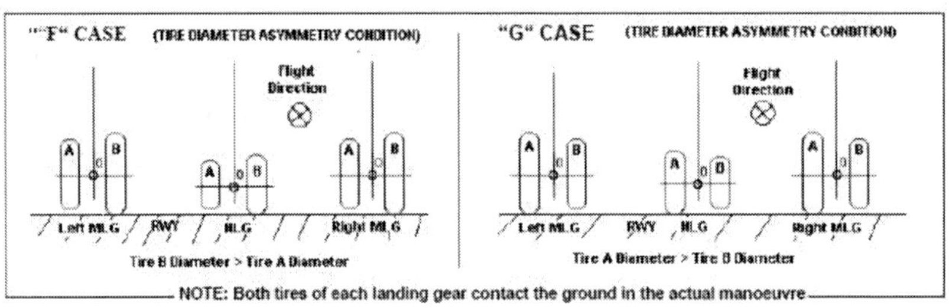

Figure 3. Schematic diagram of the most severe working condition of single-wheel deflation

Based on the above conditions, strength of the longitudinal (0 °, 180 °)and 45 ° of the nose and main landing gear is calculated when the towing load is 60% and 100% of fully inflated conditions applied to the nose landing gear with deflated tires, which demonstrates the actual carrying capacity of nose landing gear and main landing gear.

The nose landing gear was used as an example to define the main structural stress points of the nose landing gear. Among them, P1 and P2 are the connecting points of the main pillar of the landing gear and the fuselage, P3 and P4 are the connecting points of the landing gear resistance rods and the fuselage, and PA is the wheel pivot point, as shown in Figure 4.

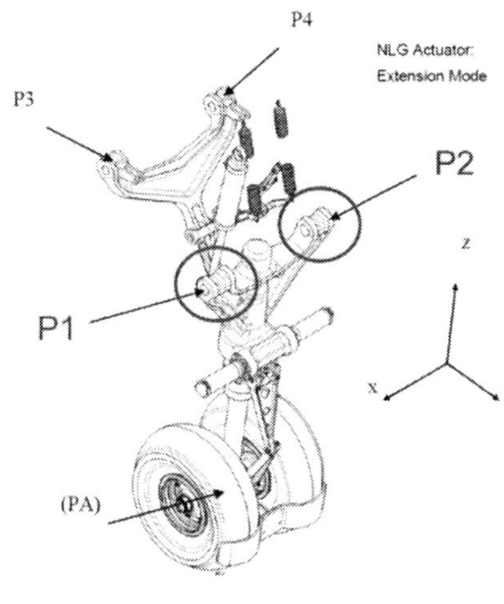

Figure 4 Schematic diagram of the stress points on the structure of the nose landing gear of a certain type of aircraft

The comparative analysis of the strength of the nose landing gear is shown in Figure 5.

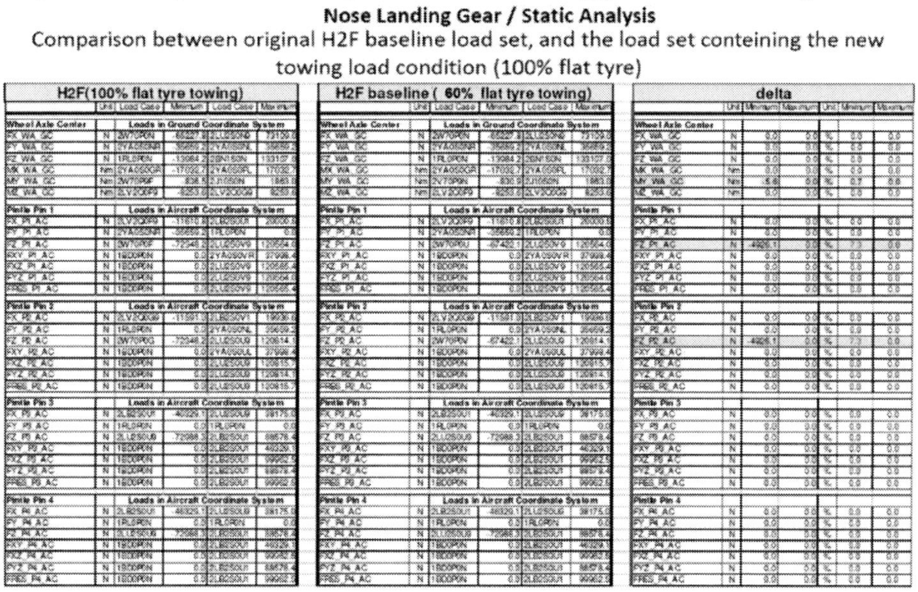

Figure 5. Comparison of the strength data of the nose landing gear of a certain type of aircraft (100% flat tire towing load and 60% flat tire towing load)

Through calculation and analysis, the result of the strength using 100% flat tire towing load with single deflated tire, comparing with the strength using 60% flat tire towing load ,is given: the force Fz of P1 and P2 structure of the nose landing gear has an increase of 4926N,which is 7.3%, the load of P3 and P4 structure has no change; the load of main structure has no change. After consulting the original strength data of the landing gear design, P1 and P2 have a strength margin of more than 10% in the structural design, which can completely cover the load increase caused by the tire deflation.

4. Conclusion

This type of aircraft can adopt the normal towing procedure in the case of single-wheel deflation, allowing towing operation of the nose landing gear with unlimited towing load. At the same time, the towing speed is required to be as low as possible and a constant speed should be kept when performing the towing operation. Also, sharp turns should be avoided.

References

[1] 25.509 Towing loads, and 25.511 Ground load: unsymmetrical loads on multiple-wheel units, 2011.11. Civil Aviation Regulation of China (CCAR), *Part 25-Airworthiness Standards: Transport Category Airplanes*. R4.
[2] 09-10-00 Towing with tires deflated, 2008.11. *A319/A320/A321 Aircraft Maintenance Manual*, 264-270.
[3] 09-11-00 Towing the airplane with flat tires, 2008.02. *737-600/700/800/900 Aircraft Maintenance Manual*, 201-208.
[4] 04-20-05 Towing the aircraft with flat tires, 2008.10. *ERJ200 Aircraft Maintenance Manual*, 265-267.
[5] 09-10-00 Towing-description and operation, 2004.04. *CRJ200 Aircraft Maintenance Manual*, 239-243.
[6] 5.6.2 Towing Equipment Design, 2000.11. Aircraft Design Manual, *Volume 21-Comprehensive Product Guarantee*, 105-110.

IOP Conf. Series: Materials Science and Engineering **887** (2020) 012017 doi:10.1088/1757-899X/887/1/012017

Compensation Measurement Study for Aspherical Primary Mirror of Space Camera with CGH

Xiaolei Liu*, **Xiaofeng Sun, Xiaomeng Dong, Dong Yang**

Beijing Institute of Spacecraft System Engineering, Beijing 100094 China

*xlliuhit@163.com

Abstract. With the development of modern optical technology, demands on the performance of optical systems are increasing. Aspherical surfaces are widely used in space optical system considering its excellent ability in correcting aberrations, reducing the optical system weight and simplifying the optical construction. However, the interferometry of this kind of aspherical surface is difficult. To overcome the difficulty in testing aspherical primary mirror of space camera system, we provide a compensation method to accomplish the null testing of the aspherical primary mirror with CGH (Computer generated hologram). Our proposed method is introduced in detail in the theory part. In addition, we provide a detailed design for an actual parabolic surface. The results show that our proposed method can accomplish the testing of aspherical primary mirror in large aperture with satisfactory accuracy.

1. Introduction

Interferometry is the mostly often used method in the optical mirror surface testing. For interferometers, it generally emits a stand plane or spherical wavefront and measures the relative wavefront deviation from planar or spherical wavefronts. When testing a plane surface or a spherical surface, it can be applied directly. However, when testing an aspherical surface or freeform surfaces, a null lens must be adopted to generate a relative wavefront which matches the test surface exactly.

For the null lens method, many researchers have made in-depth studies on this issue and many kinds of null lens such as Doll null lens and Offner null lens are designed [1-4]. Besides null lens, CGH (Computer generated hologram) can also be used to generate a desired wavefront matching the testing aspherical mirror surface exactly. For the CGH, it is a kind of diffractive optical element which is capable of producing wavefront in desired shape. With the developing of the micro-lithographic technology, the CGH can be produced with satisfactory precision and accuracy [5-9]. Many scholars have also made a lot of studies in this kind of testing method [10-14]. For example, Talha used CGH to test aspherical surfaces and the PV of relative wavefront aberration for the first diffraction order is less than 0.0479λ [15]. Li applied the CGH to testing of an off-axis aspherical surface and obtained good results [16].

In this paper, we focus on the CGH testing technique for aspherical primary mirror surface of space camera, including the design methods of aspherical surface and analysis of design results. This paper is organized as follows. In section 2, the basic theory of CGH compensation measurement is introduced. In section 3, we apply our proposed method to an actual aspherical primary mirror of space camera and analysed the design results. The relative conclusion is given in section 4.

Content from this work may be used under the terms of the Creative Commons Attribution 3.0 licence. Any further distribution of this work must maintain attribution to the author(s) and the title of the work, journal citation and DOI.
Published under licence by IOP Publishing Ltd

2. Theory

For the CGH, it can be treated as a collection of linear gratings with variable spatial frequencies. By adjusting the spatial frequency of relative linear gratings on the CGH, the wavefront of incident light can be transformed to an aspherical wavefront in a desired form.

To simplify the model, we take a linear grating as an example. Considering a binary grating model as shown in Fig.1, the period of the linear grating is D and the width of the etched area in the grating is qD. T_1 is the amplitude of the output wavefront from the unetched area of the grating, while T_0 is the amplitude of the output wavefront from the etched area of the grating. Ψ is the phase function which represents the phase difference between the peak and valley of the linear grating.

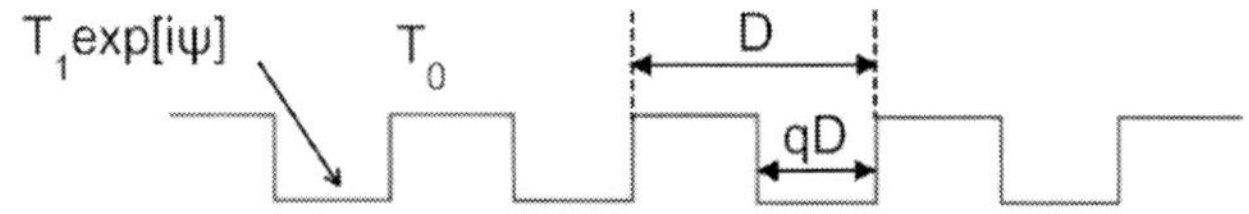

Figure 1. Linear grating model of CGH

Considering the wavelength of the incident light is much smaller than the period of the linear grating D, for a planar wavefront which comes into the grating at normal incidence, the amplitude of the output wavefront can be described as:

$$u(x) = T_0 + (T_1 e^{i\psi} - T_0) rect(x/qD) \cdot \frac{1}{D} comb(x/D) \tag{1}$$

According to the Fraunhofer diffraction theory, the far-field wavefront function of a planar wavefront which comes into the grating at normal incidence can be expressed as:

$$\begin{aligned} U(\xi) = & \left\{ T_0 \delta(\xi) + \left[T_1 \cos(\psi) - T_0 \right] \cdot q \cdot \sin c(qD\xi) \cdot \sum_{m=-\infty}^{\infty} \delta(\xi - m/D) \right\} \\ & + i \left\{ T_1 \sin(\psi) \cdot q \cdot \sin c(qD\xi) \cdot \sum_{m=-\infty}^{\infty} \delta(\xi - m/D) \right\} \end{aligned} \tag{2}$$

where m is the diffraction order, the ξ represents the spatial frequency.

For the one order,

$$U(\xi) = \begin{cases} \left\{ T_0 + \left[T_1 \cos(\psi) - T_0 \right] \cdot q \right\} + i \left\{ T_1 \sin(\psi) \cdot q \right\} & m = 0 \\ \left\{ \left[T_1 \cos(\psi) - T_0 \right] \cdot q \cdot \sin c(mq) \right\} + i \left\{ T_1 \sin(\psi) \cdot q \cdot \sin c(mq) \right\} & m \neq 0 \end{cases} . \tag{3}$$

As CGH is designed with variable fringe spacing, the design concept of the CGH relative to testing aspherical surface can be shown in Fig.2.

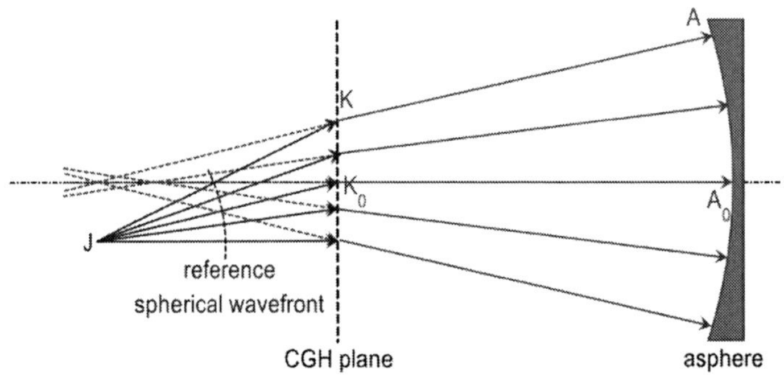

Figure 2. Phase function calculation model of CGH.

For the aspherical surface testing, the output wavefront is exactly the same with the aspherical equation, which can be used to guide the designing of CGH.

As shown in Fig.2, *KA* is a ray of light passes through any point of the CGH and it is also in the normal direction of the point *A* in the aspherical surface. Then the phase of point *K* can be expressed as the difference between reference light K_0A_0 and light *KA* as follows:

$$\phi(x,y)=\phi(K)=\frac{2\pi}{\lambda}(opl[K_0A_0]-opl[KA]) \tag{4}$$

where λ represents the wavelength of incident light and $opl[\bullet]$ represents the optical path of the corresponding rays. Considering that the phase function of the incident rays used to reconstruct the wavefront on the CGH plane is:

$$\varphi(x,y)=\varphi(K)=\frac{2\pi}{\lambda}(opl[JK]-opl[JK_0]) \tag{5}$$

Then the phase function that the CGH needs to encode can be expressed as:

$$\begin{aligned}\psi(x,y)&=\phi(x,y)-\varphi(x,y)\\&=\frac{2\pi}{\lambda}(opl[K_0A_0]-opl[KA])-\frac{2\pi}{\lambda}(opl[JK]-opl[JK_0])\\&=\frac{2\pi}{\lambda}(opl[JK_0A_0]-opl[JKA])\end{aligned} \tag{6}$$

The diffraction pattern design can be accomplished according to the above mathematical model.

3. CGH design of a parabolic mirror

In this part, combined with an actual engineering example, we designed a CGH for an aspherical primary mirror of space camera. The primary mirror is a parabolic surface. The clear aperture of it is 865mm and the vertex radius of curvature *R* is 2638.4mm.

The departure between the parabolic mirror and its best-fit sphere is shown in Fig.3.

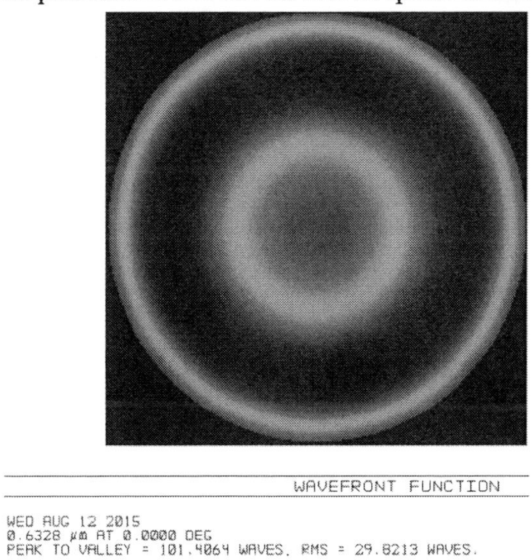

Figure 3. Departure of parabolic mirror.

It can be seen from Fig.3 that the PV (peak-to-valley) and RMS (Root mean square) of the departure is 101.4064λ and 29.8213λ respectively, where λ is 632.8nm.

According to our proposed model described in section.2, the optical path for CGH testing can be described as shown in Fig.4.

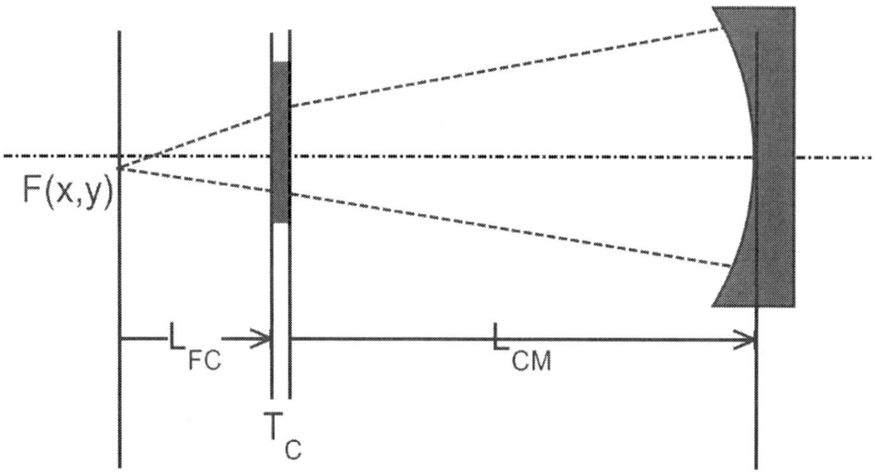

Figure 4. Testing optical path of parabolic mirror.

In the above optical path, the basic parameters are as follows:

- The carrier frequency of interferometer focus: (x,y)=(0,-5);
- The distance between the interferometer focus and CGH: L_{FC}=150mm;
- CGH substrate thickness: Tc=8mm;
- The distance between CGH to the parabolic mirror: L_{CM}=2500mm;
- Aperture of CGH: D=80mm;
- F number of standard lens in the interferometer: F#=1.5;

The residual map of the testing area is shown in Fig.5.

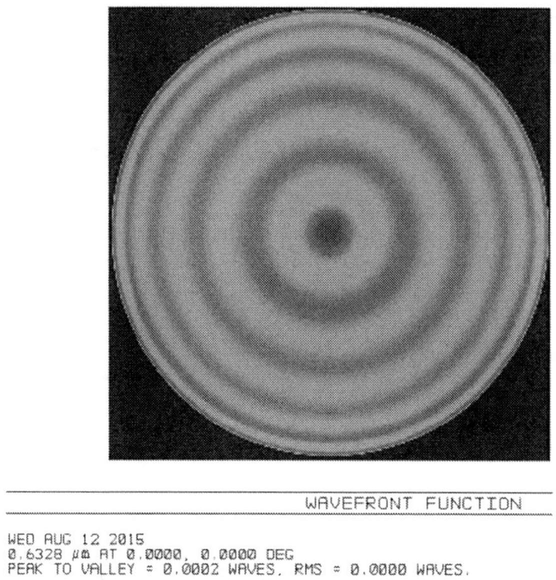

Figure 5. Residual map of testing area.

It can be seen from Fig.5 that the PV and RMS of the residual map is 0.0002λ and 0 respectively, which means the parabolic mirror can be tested in null with CGH in satisfactory accuracy.

In the designing of CGH, each diffraction order should also be separated from each other, which avoids the extra interference in the aspherical surface testing. The separation of diffraction orders for our parabolic mirror is shown in Fig.6.

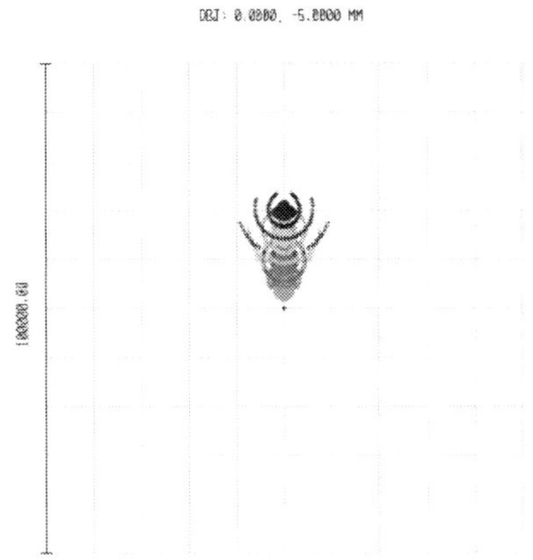

Figure 6. Separation of diffraction orders.

It can be seen from Fig.6 that each diffraction order is separated from the testing diffraction order very well.

To accomplish the alignment between CGH and testing interferometer, we also designed the special alignment area on CGH. The relative alignment optical path is shown in Fig.7 and the residual map of the alignment area is shown in Fig.8.

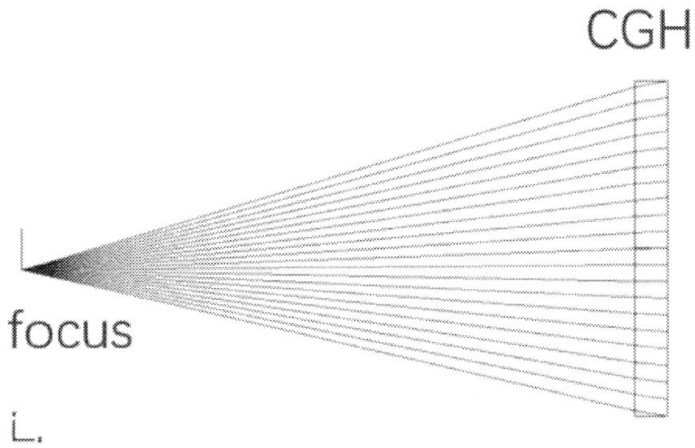

Figure 7. Optical path of alignment on CGH.

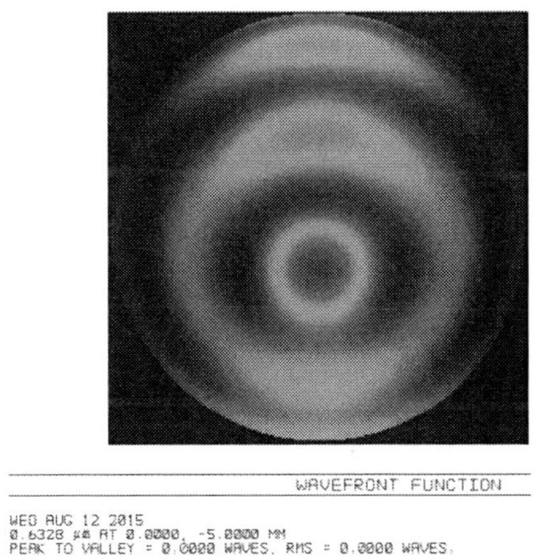

Figure 8. Residual map of alignment area on CGH.

In actual testing, considering the interferometer can be aligned with CGH accurately, each diffraction order is separated from the testing diffraction order and the residual map of the testing area is negligible, the parabolic mirror surface can be tested with our proposed method with satisfactory accuracy.

4. Conclusion

We have provided a CGH design method to accomplish the null testing of aspherical primary mirror of space camera. To evaluate the designing accuracy of our proposed method, an actual engineering example (parabolic mirror surface) is taken as a design goal. Combined with our design method and considering actual designing demands such as the accurate alignment between interferometer and CGH, the separation of diffraction orders, the design residual is negligible and the design result is satisfactory. As the design analysis is now only for aspherical surfaces, further theoretical and experimental studies for freeform surfaces will be taken.

References

[1] Malacara D. Optical Shop Testing 2007 Wiley.
[2] Noble R, Malacara D, Cornejo A. Multistep Hindle Test. Appl. Opt. 1974, 13(11): 2476-2476.
[3] Offner A. A Null Corrector for Paraboloidal Mirrors. Appl. Opt., 1963, 2(2): 153-155.
[4] P.J.Guo and J.H.Yu. Design and certification of a null corrector to test hyperboloid convex mirror SPIE,2006,6150.
[5] Lohmann A W, Paris D P. Binary Fraunhofer Holograms, Generated by Computer. Appl. Opt. 1967, 6(10): 1739-1748.
[6] Lee W-H. Binary Synthetic Holograms. Appl. Opt., 1974, 13(7): 1677-1682.
[7] Lee W-H. Circular-carrier holograms. J. Opt. Soc. Am., 1975, 65(5): 518-523.
[8] Lee W-H. Binary computer-generated holograms. Appl. Opt., 1979, 18(21): 3661-3669.
[9] Brown B R, Lohmann A W. Complex Spatial Filtering with Binary Masks. Appl. Opt. 1966, 5(6): 967-969.
[10] F.Z, X.Luo, J.L.Zhao, D.L.Xue, L.G.Zheng and X.J.Zhang, "Test of off-axis aspheric surfaces with CGH," Opt. Precis. Eng. 2011, 19:706-716
[11] J.H.Burge, L.B.Kot, H.M.Martin, R.Zehnder, and C.Zhao, "Design and analysis for interferometric measurements of the GMT primary mirror segments," Proc.SPIE 2006, 6237:62370M
[12] H.M.Martina, R.G.Allena, J.H.Burge, D.W.Kim, J.S.Kinsley, M.T.Tuell, S.C.West, C.Zhao and T.Zobrist, "Fabrication and testing of the first 8.4m off-axis segment for the Giant Magellan Telescope," Proc.SPIE, 2010, 7739:77390A
[13] Y.Xie, Q.Chen, F.Wu, C.K.Liu, X.Hou, and J.Zhang, "Design and fabrication of twin computer-generated holograms for testing concave aspherical surfaces," Opto-Electron.Eng,2008, 35:40-43
[14] S.Li, B.Liu,A.Tian, Z.Guo, P.Yang, and J.Zhang, "A practical mthod for determining the accuracy of computer-generated holograms for off-axis aspheric surfaces," Opt.Laser Eng, 2016, 77:154-161
[15] M.M. Talha, J.Chang, Y.Wang, D.Cheng, T.Zhang, M.Aslam and A.N.Khan, "Computer generated hologram null compensator for optical metrology of a freeform surface," Optik 2010, 121:2262-2265
[16] S.Li, F.Wu, Q.Chen, B.Fan, and L.Li, "Design and experiment of testing an off-axis aspheric surface by computer generated hologram," Proc.SPIE 2013, 8788:878833

Sounding Rockets Scattering Factors and their Influence on Motion Accuracy

Nickolay Zosimovych

Intelligent Manufacturing Key Laboratory of Ministry of Education, Shantou University, 243, Daxue Road, Shantou, Guangdong, 515063, China

nzosimovych@stu.edu.cn

Abstract. The increase of sounding rocket active part of trajectory compared to the "normal" can be due to the launches of ballistic flying vehicle for the maximum specified range. In this case, rockets should use propulsion systems with a relatively low fuel specific impulse. This effect also occurs in case of implementation of minimum scattering programs during launches at intermediate ranges (between the minimum and maximum mission ranges). In proposed research was developed methods for analysing the influencing of ballistic and navigation support the uncertain factors on the accuracy of sounding rocket motion with an extended active part of trajectory and was given a relevant assessment to the contribution of each component in relation to specific cases of their occurrence.

1. Introduction

The current level of rocket technology development is characterized by extremely high requirements for guaranteed mission accuracy. These requirements practically apply for all flying vehicles classes, but they are especially hard for ballistic type vehicles, or Sounding Rockets (SR). By sounding rockets we mean a wide class of ballistic vehicles, which, along with the active, controlled part, also has a passive, uncontrolled part of the trajectory [1].

Generally, active and passive parts of trajectory can be several and they can alternate with each other. The controlled movement of the centre of mass of this type of flying vehicle is usually considered as a guidance task, the solution of which should ensure that the vehicle could reach to a given point in inertial space, or to achieve a predetermined mission range [2, 3].

High precision Sounding Rocket mission could be possible only if available control algorithms, reliable mission's ballistic and navigation support (BNS). BNS based on the available primary and source navigation information, which are contains information about the physical environment, and directly during the motion control, in the structure of the a priori model of the SR dynamics motion, etc.

Development of methods and means of improving the reliability of the BNS has always been among the priorities of the SR dynamic design. Currently, a lot of publications have been published related to solving problems of increasing the reliability of measurement information [4-9].

In a much lesser extent, these issues were raised in relation to the assessment of the impact of ballistic motion errors. It should, however, bear in mind that whatever approaches are not used for the reliability and accuracy of the initial and primary navigation information always remain uncontrollable factors that affect the accuracy of controlled motion.

Content from this work may be used under the terms of the Creative Commons Attribution 3.0 licence. Any further distribution of this work must maintain attribution to the author(s) and the title of the work, journal citation and DOI.
Published under licence by IOP Publishing Ltd

There is a number of problems in which their importance is growing significantly. Such tasks include flight control problems with Sounding Rockets extended active part of trajectory (APT). Longer APT may be due to ballistic SR for a predetermined maximum mission range. In this case, launchers should use propulsion systems with a relatively low specific impulse.

This effect also occurs in case of implementation the minimum scattering programs [1, 10-13] during launches at intermediate ranges. The SR mission trajectory under the minimum scattering program is steeper than the maximum range mission trajectory. The implementation of this type of program requires a significantly higher thrust-to-weight ratio, and, consequently, a longer APT. If we using minimum range launch it is also interlinked with the discussed effect. This is explained by the fact that this removal trajectory leads to a reduction of the SR transit time through high-dense atmosphere, as well as with the possibility the increasing afterburning of fuel, compared to the nominal version.

There are quite few publications devoted to this kind of research. It should be noted that the BNS, as a whole, are often not objects of decomposition and require a comprehensive approach in assessing the influence in both the measurement information and the initial (ballistic) data. First of all, it has to do with the level of reliability of the used model of the Earth's gravitational field.

The above gives reason to believe relevant the topic of the study, which analyses the influence of BNS uncertain factors on the accuracy of SR motion with extended APTs.

The purpose of the study is to develop methods for analysing the influence of BNS uncertain factors on the accuracy of SR motion with an extended APT and to give a relevant assessment to the contribution of each component to specific cases of their occurrence.

In this research, we used methods of differential equations, probability and statistics methods, linear algebra, matrices, ballistics, rocket dynamics and SR navigation.

2. Scattering Components of Sounding Rockets with Inertial Control System

Scattering of the trajectory is caused by deviations from the nominal (calculated) values of the design characteristics of the complex (Sounding Rocket and its systems), as well as flight parameters.

Possible unforeseen factors connected with the launch of SR and its provision are the reasons for these deviations [2, 3, 11-17].

Scattering is a measure criteria of the rocket accuracy and its main characteristic functioning as an integral evidence of the accuracy of different types of SR systems. Due to the functional diversity, there are reasons to consider the scattering components as a statistical sum of some physical independent components. This should correspond to the stratified level of the target accuracy structure, which correspond to the SR at certain structurally aggregated levels [11, 18].

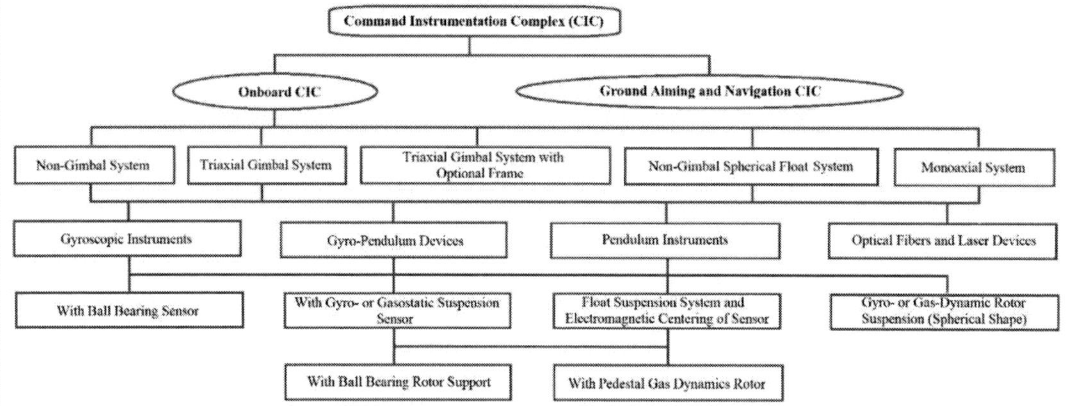

Figure 1. Command instrumentation complex (CIC) classification.

Command instrumentation complex (CIC) for Sounding Rockets classification is represents on Figure 1. The resulting scattering value shall be presented as an additive sum of three main components [12, 17, 18]:

$$\bar{\delta} = \bar{\delta}_{grav} + \bar{\delta}_{tech} + \bar{\delta}_{init}, \quad (1)$$

where $\bar{\delta}_{grav}$ – gravimetric and geodetic errors; $\bar{\delta}_{tech}$ – technical errors; $\bar{\delta}_{init}$ – errors in data preparation.

Gravimetric and geodetic are including next inaccuracies:

1. In true parameters of Earth shape.
2. In passive part of the trajectory (PPT).
3. In geodetic reference.

The above types of inaccuracies are characterized by the fact that they announced in total deviation of the SR from the target. That are because of errors in the initial conditions definition of motion and calculating some capture trajectory inaccuracies, which is a reference to determine are limit sizes of the scattering ellipse [4, 12, 19].

Generally, considered appropriate to subdivide technical preparation inaccuracies into the following components [20]:

1. Methodical control errors from the deviation point.
2. Instrumental errors in control system.
3. GPS orientation errors for platform INS.
4. GINS orientation errors.
5. Errors in the PPT.

The above errors systematization cause by technical scattering. They are explained by the peculiarities of their expression in the total deviation from the target of the point of incidence of the SR last stage.

For example, with the perfect operation of the SR angular stabilization system, the orientation errors in the axes of sensitive measuring systems are announced only through isochronous variations of the motion parameters. And they do not depend on DPI control action, and, consequently, on the methodical control errors. DPI caused by instrumental errors in CS. In the first approximation shall be determined only by the instrumental component of the switch-off time variation for the propulsion system (PS) arising due to errors in the all system elements [2, 21]. As for the methodological error, it practically does not depend on from any instrumental errors in DPI CS. And it does not depend on errors inside devices sensitivity axes orientation. For most control methods, this error is announced in DPI through the motion parameters isochronous deviation, as well as the methodological component of the variation of PS switch-off time.

Finally, the errors (perturbations) in PPT mainly at the stage of afterburning effect, separation phase and descend in the atmosphere, announced in DPI, can be considered independent from the groups of errors, which have been analysed. Whereby:

- the pulse scatter after thrust cut-off;
- separation perturbations;
- deviation of the velocity vector, as well as angle-of-attack and angle-of-slip deviations from nominal values, when entering the dense layers of the atmosphere, are usually classified as perturbations of the initial conditions of motion.

External perturbations:

1) perturbations of atmospheric thermodynamic parameters (deviations from normal conditions);
2) wind (speed and direction);
3) deviation of the mass distribution parameters in the last SR stage (centre of mass displacement, angular misalignment of the main inertia axes);
4) form deviation, and, consequently, deviation of aerodynamic force coefficients and moments of SR last stage.

Errors in data preparation for SR launch are directly related to errors during solving the inverse boundary problem, which is determine the ballistic control parameters. They are generating the following errors:

1. The capture trajectory calculation errors.
2. Errors in CS settings calculation and on-board entering data errors during its preparation for launch.

These errors occurrence may be depending from:

1. Methodical errors in calculation and formulae.
2. Calculation methods errors (interpolation errors, numerical integration, etc.).
3. Computational errors (finite presentation of numbers in the bit grid of the on-board computer, the use of rounding in the calculation process, etc.).
4. Input and storage data errors.

It is assumed that data preparation errors can be reduced to negligible values. This gives us a reason to confine to the first two components in the formula (1) while studying the scattering characteristics, i.e., to consider only the components due to gravimetric and geodetic errors, as well as technical errors.

3. The Capture Trajectory Calculation

Currently, there are many approaches and methods, which are used for calculation of capture trajectory, i.e. a nominal trajectory, which goes through the target point [12, 13, 15, 17, 18, 20, 22, 23]. This circumstance makes it impossible to describe in sufficient detail the possible ways to solve the problem within the framework of this study. In this regard, we should limit ourselves to the problem general mathematical formulation, and a brief description of its solution aspects.

We assume that the coordinates of the point of incidence are given in the form of nonlinear spherical coordinates. And they uniquely determined by the SR motion parameters at the time t_k command is sent to the last stage or head section (SP). These coordinates can be represented by the following functional dependencies [18]:

$$D_c = D_c[q_{a1}(t_k), q_{a2}(t_k), \dots, q_{a6}(t_k), t_k]$$
$$Z_c = Z_c[q_{a1}(t_k), q_{a2}(t_k), \dots, q_{a6}(t_k), t_k], \quad (2)$$

where $q_{ai}(t_k)$ – the absolute motion parameter values at the time moment $(i = 1, \dots, 6)$ t_k.

The motion parameters $q_{a1}(t_k), q_{a2}(t_k), \dots, q_{a6}(t_k), t_k$ for the given programs uniquely depend on the of the take-off point H_0, B_0, L_0 coordinates, targeting azimuth $A_0$and the time t_k. In this case, the dependences (2) can be presented as follows:

$$D_c = D_c[H_0, B_0, L_0, A_0, t_k]; Z_c = Z_c[H_0, B_0, L_0, A_0, t_k]. \quad (3)$$

For a known take-off point, the dependencies (3) may be written as follows:

$$D_c = D_c(A_0, t_k); Z_c = Z_c(A_0, t_k). \quad (4)$$

The coordinates determination under given mission parameters and time is a direct ballistic problem, which is solved by numerical integration of a system of differential equations of nominal motion [12, 18, 20]. Determination of the initial data for the FV launch is based on the inverse ballistic problem solution [12, 18].

Nevertheless, the calculation of the capture trajectory can be made by successive approximations during solving a direct ballistic problem. This method presented here as follows. Using simple, approximate dependencies, the first approximation of the A_0 and t_k parameters shall be determined. Then the direct problem, based on the obtained coordinates of the point of incidence D_c, Z_c shall be determined by the corrections δA_0 and δt_k, which shall be introduced into the values A_0 and t_k. The calculation will be repeated until the next condition is fulfilled a required accuracy as in (4).

The main input data for the calculation of the capture trajectory are:

1) geodetic coordinates of the take-off point H_0, B_0, L_0;
2) geodetic coordinates of the target H_T, B_T, L_T;
3) model of Earth shape parameters;
4) model of the Earth's gravitational field parameters;
5) the Earth's standard atmosphere parameters;

6) FV control system characteristics;
7) errors $\Delta D_{allowable}, \Delta Z_{allowable}$ in the calculation of the range and direction of the capture trajectory.

4. Discussion

The calculation of the capture trajectory can be made according to the following scheme.

1. A linear spherical range to the target D_T and a spherical targeting azimuth $A_{spheric}$ shall be determined from the known geodetic coordinates of the take-off point B_0, L_0 and the target B_T, L_T by the formulas:

$$D_T = R_N \Phi_N = R_N \left\{\frac{\pi}{2} - arcsin[sin\,\varphi_T\, sin\,\varphi_0 + cos\,\varphi_T\, cos\,\varphi_0\, cos(L_T - L_0)]\right\}, \quad (5)$$

$$A = \begin{Bmatrix} \frac{\pi}{2} - A_T^*; when \Rightarrow sin(L_T - L_0) > 0; \\ \frac{3\pi}{2} + A_T^*; when \Rightarrow sin(L_T - L_0) < 0; \end{Bmatrix}$$

where $A_T^* = arcsin\left(\frac{sin\,\varphi_T - sin\,\varphi_0\, cos\,\varphi_T}{cos\,\varphi_0\, cos\,\varphi_T}\right); sin\,\varphi_l = \frac{(1-e^2)\, sin\, B_l}{(1-e^2)^{0.5} sin^2 B_l}, (l = 0, T\, arg\, e\, t).$

2. The spherical azimuth $A_{spheric}$ accepted as the first approximate of the targeting azimuth $A_0 = A_{spheric}$ and approximate time for sending for a separation command of the last stage or a nose cone of FV. Knowing the time limits change $tk_{k_{max}k_{min}}$ and corresponding minimum and maximum range changes $Dmax$ *and* D_{min} it is possible to determine t_k, the appropriate ranges D_T by linear interpolation from the formula:

$$t_k = t_k^* + \frac{t_{k_{\max}} - t_{k_{\min}}}{D_{\max} - D_{\min}}\left(D_T - D^*\right), \quad (6)$$

where D^* – is the range corresponding to the time t_k^*;

$$t_k^* = \{tT(Dmin_{max}/2; |t(Dmin_{max}/2 \le D_T \le D_{max})_{k_{max}})min_{k_{min}}\}$$

3. Than we can calculate coordinates $x_r^{(c)}, y_r^{(c)}, z_r^{(c)}$ using the equations of nominal motion for example, the equations with the specified parameters A_0, t_k and the coordinates of the incidence point D_c, Z_c are determined from the next formulas

$$\left.\begin{aligned} D_c = R_N \Phi_c = R_N \left(\frac{\pi}{2} - arcsin\frac{x_r^{(0)}x_r^{(c)} + y_r^{(0)}y_r^{(c)}}{r_0 r_c}\right); \\ Z_c = -R_N E_c\, sin\,\Phi_c = -R_N(A_0 - A_c)\, sin\,\Phi_c, \end{aligned}\right\} \quad (7)$$

where A_c – I the azimuth of incidence point;

$$A_c = \begin{Bmatrix} \frac{\pi}{2} - A_c^*; when \Rightarrow z_r^{(c)} < 0; \\ \frac{3\pi}{2} + A_c^*; when \Rightarrow z_r^{(c)} > 0; \end{Bmatrix}$$

$$A_c = arcsin\frac{x_r^{(0)}x_r^{(c)} - x_r^{(c)}y_r^{(0)}}{\sqrt{(r_0 r_c)^2 - \left(x_r^{(0)}x_r^{(c)} + y_r^{(0)}y_r^{(c)}\right)}},$$

$$r_c = \sqrt{\left(x_r^{(c)}\right)^2 + \left(y_r^{(c)}\right)^2 + \left(z_r^{(c)}\right)^2}; \; r_0 = \sqrt{\left(x_r^{(0)}\right)^2 + \left(y_r^{(0)}\right)^2};$$

$$x_r^{(0)} = (N_{e0} + H_0)\, cos\, B_0; \quad y_r^{(0)} = [(N_{e0} + H_0) - e_e^2 N_{e0}]\, sin\, B_0;$$

$$N_{e0} = \frac{a_e}{\sqrt{1 - e_e^2\, sin\, B_0}}.$$

4. Now we may verify the conditions:

$$|D_c - D_T| \le \Delta D_{extra}; \; |Z_c - Z_T| \le \Delta Z_{extra}. \quad (8)$$

If these conditions are not satisfied, we shall improve A_0 and t_k, correcting by formulas:

$$t_k^{(i+1)} = t_k^{(i)} + \delta t_k^{(i)}; \quad A_0^{(i+1)} = A_0^{(i)} + \delta A_0^{(i)}. \tag{9}$$

Corrections δA_0 and δt_k shall be defined in different ways. Particularly, these corrections can be determined by solving a system of two linear equations obtained from decomposing the functions (4) up to terms of the 2nd order of smallness:

$$\delta D = \dot{D}_k \delta t_k + \frac{\partial D}{\partial A_0}\delta A_0; \quad \delta Z = \dot{Z}_k \delta t_k + \frac{\partial Z}{\partial A_0}\delta A_0, \tag{10}$$

where $\dot{D}_k$ and $\dot{Z}_k$ are the full derivatives from the point of incidence coordinates in APT flight; $\frac{\partial D}{\partial A_0}, \frac{\partial Z}{\partial A_0}$ - are partial derivatives from the point of incidence coordinates in targeting azimuth A_0.

Derivatives $\dot{D}_k$ and $\dot{Z}_k$ can be calculated, for example, using the following formulas:

$$\dot{D}_k = \sum_{j=1}^{6} \frac{\partial D}{\partial q_j^{(k)}} q_j^{(k)}; \quad \dot{Z}_k = \sum_{j=1}^{6} \frac{\partial Z}{\partial q_j^{(k)}} \dot{q}_j^{(k)}, \tag{11}$$

where $\frac{\partial D}{\partial q_j^{(k)}}, \frac{\partial D}{\partial q_j^{(k)}}$ - are partial derivatives of the coordinates D_c, Z_c in terms of the motion parameters $q_j^{(k)}$ in APT end (so-called ballistic derivatives); $\dot{q}_j^{(k)}$ – are full derivatives of the motion parameters $q_j^{(k)}$ in APT flight time.

The process of improving successive approximations in determining the input data t_k and A_0 would be continue until the conditions (8) are met.

The ballistic derivatives in the expression (11) and purposed to construct the capture trajectory are necessary for the scattering characteristics analysis as well. Within the solution of the problem under consideration, which are often called sensitivity functions, and have a physical meaning of weight functions. The weight functions determine how single deviations of different perturbing motion factors influence to the total range. The weigh function is expressed in terms of the point of incidence coordinates.

Among the sensitivity function methods determining, the following are the most common [2]:

- numerical methods (finite difference methods and variation methods);
- analytical method based on Kepler's model of FV motion along PPT.

In this publication, used the finite difference method in its simplified version in order to determine desired values of the sensitivity functions.

The method of variations based on a single integration of the system of differential equations of motion (SDEM), describing the SR nominal motion, and multiple integration of linearized SDEM under single initial conditions.

In contrast to numerical methods, the analytical method allows [8, 11, 18] to get a solution based on finite analytical relations. But there is possible to find the corresponding relations only for simple models, among which, is the Kepler model. This model is obtained under the following assumptions: 1) the Earth is taken as a fixed sphere of radius R_N; 2) the gravitational field of the Earth is determined by the Newtonian potential $U = \frac{\pi_0}{r}$, where $\pi_0 = \mu = fM = 3.986 \cdot 10^{14} \frac{m^3}{sec^2}$ is the coefficient 3) the atmosphere is absent; 4) SR is considered as a material point with a mass equal to the mass of FV and concentrated in its center of mass.

Under these assumptions the trajectory of SR motion on PPT, will be a curve lying in a plane, which passes through the center of the attracting body. In this case, it is completely determined by the motion parameters r_k, V_k, θ_{FV} and time t_k (see Figure 2) [20, 21].

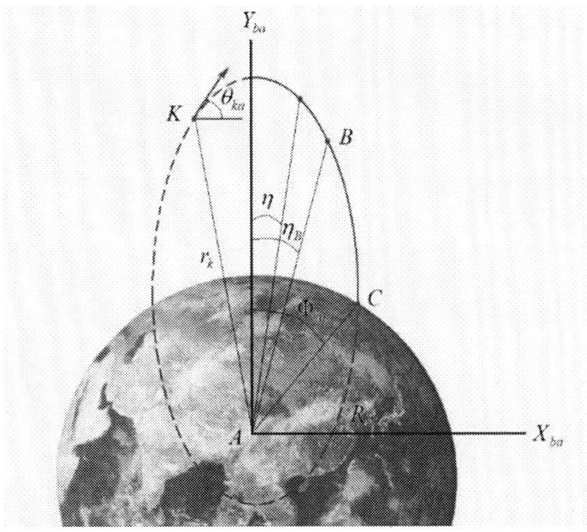

Figure 2. Main points of SR trajectory.

The main advantage of analytical method compared to the numerical one is in reduction of calculation time required to implement the algorithm for calculating ballistic derivatives. However, the analytical method has its drawbacks. And the essential drawback of this method are arising errors from the model of Kepler motion, which is not quite accurate [22, 23].

5. Conclusion

Proposed general estimation methogology for influence scattering factors of ballistic and navigation support uncertain factors on the accuracy of sounding rocket motion with an extended active part of trajectory and was given a relevant assessment to the contribution of each component in relation to specific cases of their occurrence.

References

[1] Jacques S Gansler, Ballistic Missile Defense: Past and Future. Center for Technology and National Security Policy National Defense University, 2010, Washington DC, USA, 88

[2] Mattew R, Tetlow, Commercial Launch Vehicle Design and Predictive Giudance Development, 2003, The University of Adelaide, A 254

[3] Zosimovych N, Commercial Launch Vehicle Design, 2016, LAP LAMBERT Academic Publishing, 184

[4] Dmitrievsky AA, Lysenko LN, Prikladnye zadachi teorii optimalnogo upravleniya dvizheniem bespilotnyh letatelnyh apparatov, 1978, Mashinostroeniye, M 328

[5] Jorge Piovesan, Chaouki Abdallah, Magnus Egerstedt, Herbert Tanner, Yorai Wardi. Statistical Learning for Optimal Control of Hybrid Systems, *American Control Conf.*, ACC '07, 9-13 July 2007, IEEE

[6] Smirnov IP, Optimal control of a dynamic system with random parameters under incomplete information: *Journ. of Applied Math. and Mech.*, 1979, V 43, No 4, 621–628

[7] Gustafson ED, Stochastic Optimal Control of Spacecraft. *PhD Dis.*, 2010, The University of Michigan

[8] Lebedev AA, Bobronnikov VT, Krasilstshikov MN, Statisticheskaya dinamika I optimizaciya upravleniya letatelnyh apparatov, 1985, M Mashinostroeniye, 280

[9] Huang, Kerson, Statistical Mechanics, 1987, Wiley

[10] Siharulidze YuG, Ballistika letatelnyh apparatov, 1982, M Nauka, 351

[11] Mishin VP, Bezverby VK, Pankratov BM, Osnovy proektirovaniya letatelnyh apparatov (Transportnye sistemy), 1985, M Mashinostroeniye, 260

[12] Appazov RF, Lavrov SS, Mishin VP, Ballistika upravlyaemyh raket dalnego deystviya, 1966, M Nauka, 308
[13] Spencer J, The Ballistic Missile Threat Handbook, 2002, The Heritage Fondation, USA, 104
[14] Fedosov EA, Bobronnikov EA, Krasilstshikov MN, Kuhtenko VI, Dinamicheskoye proektirovaniye system upravleniya avtomaticheskih manevrennyh letatelnyh apparatov, 1977, M Mashinostriyeniye, 336
[15] Andreev VD, Theory of Inertial Navigation (Aided Systems), 1967, M Nauka, (Available from the U.S. Department of Commerce, Springfield, Va. 22151), 409
[16] Zosimovych N, Integrated Guidance System of a Commercial Launch Vehicle, 2014, "Modern European Science – 2014" Proc. X Int. Scient. and Pract. Conf., Sheffield, 30 June – 7 July, Vol. 18. Tech. Sc., Science and education LTD, 10-22
[17] Mogilevsky VD, Navedeniye ballisticheskih letatelnyh apparatov, 1982, M Nauka, 352
[18] Krasilstshikov Mn, Sebryakova GG, Upravleniye I navedeniye bespilotnyh manevrennyh letatelnyh apparatov na osnove sovremennyh informacionnyh tehnologiy, 2003, Fizmatlit, 28
[19] Zhdanuk BF, Osnovy statisticheskoy obrabotki traektornyh izmereniy, 1978, M Mashinostroeniye, 384
[20] Volkov LI, Prokudin AI, Gavrilov VS, Tochnost mezhkontinentalnyh ballisticheskih raket, 1996, M Mashinostroyeniye, 304
[21] Chatfield AB, Fundamentals of High Accuracy Inertial Navigation, 1997, Progress in Astronautics and Aeronautics, Vol. 174
[22] Prati A, Calderara S, Cucchiara R, Using Circular Statistics for Trajectory Shape Analysis, 2008, Vol. 25, AIAA 978-1-4244-2243.
[23] Dorf RC, Bishop RH, Modern Control Systems, Pearson, 12th edition, 807

IOP Conf. Series: Materials Science and Engineering **887** (2020) 012019 doi:10.1088/1757-899X/887/1/012019

Research on Design Method of High Speed and High Dynamic Pressure Separation Structure

Jian XU and Siyi FAN

China Academy of Launch Vehicle Technology, Beijing, 100076

776773695@qq.com

Abstract. The inter-stage separation scheme has always been the focus and difficulty of low-altitude high-speed aircraft design. The separation reliability of low altitude high-speed aircraft is required to be higher. Experts and scholars have carried out design research on different forms of separation schemes. In this paper, two typical kind of inter-stage separation schemes are introduced. For aircraft A, a point cold separation scheme is adopted, which is connected by four explosive bolts, and two piston push rods provide separation impulse. For aircraft B, exhaust ports are opened on both sides of the rear of the transition section. The tail section is equipped with four "X" full-motion rudders for flight control. In addition, two horizontal stabilizers were added to improve stability. Based on the comparison and analysis of different inter-stage separation schemes for The separation reliability of low altitude high-speed aircraft is required to be higher. Experts and scholars have carried out design research on different forms of separation schemes. two different types of aircraft A and aircraft B, a new high reliability inter-stage separation scheme is proposed, the feasibility of the scheme is verified by simulation analysis, and the design of the stage separation structure under high dynamic pressure and high speed with compact layout is realized.

1. Introduction

The separation reliability of low altitude high-speed aircraft is required to be higher. Experts and scholars have carried out design research on different forms of separation schemes. In this paper, a new separation scheme is proposed based on the analysis of several separation schemes for aircraft, and the simulation analysis is carried out.

2. Analysis of separation scheme

2.1. Separation scheme of aircraft A

The aircraft A is 3.66m long and weighs 1270kg. The aircraft is connected to the carrier through an adapter. When the separation condition is reached, the aircraft is disconnected from the adapter.

A point cold separation scheme is adopted, which is connected by 4 explosive bolts, and 2 piston push rods provide separation impulse. The detailed separation structure is shown in figure 1. Two explosion bolts pass through the adapter compartment frame and are screwed into the rear bulkhead of the aircraft. They are installed horizontally to the lower ends of the two keels to connect the aircraft to the corresponding adapter bulkhead This structure effectively connects the keel of the test aircraft with the collet keel of the adapter.

Content from this work may be used under the terms of the Creative Commons Attribution 3.0 licence. Any further distribution of this work must maintain attribution to the author(s) and the title of the work, journal citation and DOI.
Published under licence by IOP Publishing Ltd

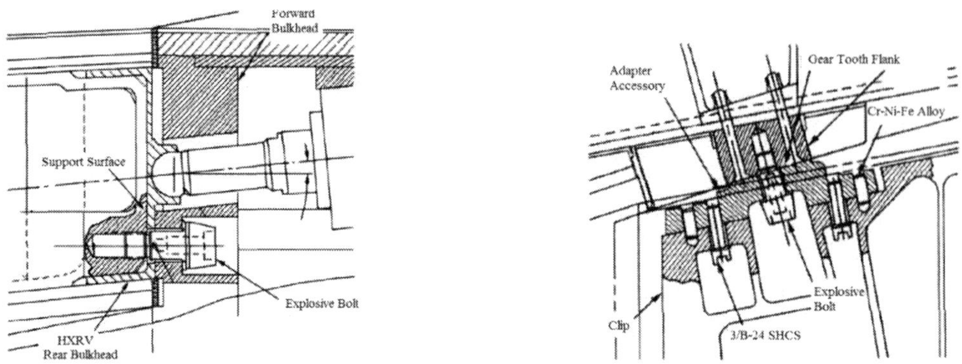

(a)Back-end connection structure (b) Front-end connection structure

Figure 1. Detailed connection structure

2.2. Flight test and separation scheme of aircraft B

The aircraft B verification machine consists of a booster stage, an intermediate transition section and a cruise stage. As shown in the figure 2, the verification machine has a total length of 7.6m, a total weight of 1780.4kg, and a maximum width of 0.6m. Exhaust ports are opened on both sides of the rear of the transition section. The tail section is equipped with four "X" full-motion rudders for flight control. In addition, two horizontal stabilizers were added to improve stability.

Before the booster was separated, the cruise stage slowly rotated 180 ° around the main axis, and the precursor compression surface changed from downward to upward. Subsequent booster separation and ignition of the super-combustion ram engine. After the engine is stable, the aircraft rotates 180 ° again to resume normal cruise. By flipping the aircraft B flight attitude into a negative angle of attack, this scheme improves the intake port lip Mach number, ensuring the start performance of the intake port at low flight Mach number conditions; while reducing the intake port. The upper surface structure design of the aircraft does not reduce the lift coefficient after the overturn, the aircraft can still maintain the flight altitude and cruise trajectory, and the control difficulty has not increased significantly.

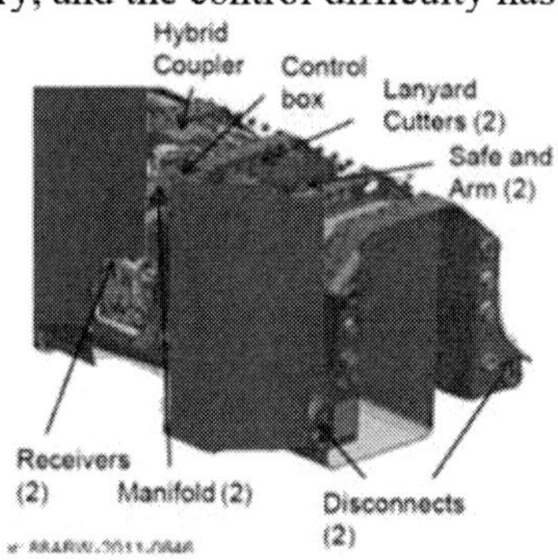

Figure 2 Aircraft B tail structure details

From the above figure and related information, it can be inferred that the aircraft B and the transition capsule are connected by 4 explosive bolts. A boss is provided at the installation position of the explosion bolt on the tail end of the aircraft. No use of guides such as guide pins was found. The separation momentum may come from the aerodynamic pressure of the inlet.

2.3. Connect the unlocking device

Commonly used connection unlocking devices are: explosion bolts, low-impact separation devices and cutting cables[1]. Explosive bolt impact is relatively large, but the structure is simple and the reliability is high. Compared with the explosion bolt, the impact of the low impact separation device is relatively small, but the structure is complicated, the installation operation is cumbersome, and the connection thickness is strict. The use of cutting ropes has requirements on the cabin structure and materials, and the ambient temperature.

2.3.1. Selection of connection and unlocking device. Limited by the aircraft, lower-level structure and layout. According to the loading conditions of the tail end of the aircraft, the tail end of the aircraft mainly bears the axial pressure and the bending moments in the Y and Z directions[2]. Therefore, the explosion bolt or the low-impact separation device should be installed along the Y and Z directions. May be remotely located. The initial design is four explosive bolts or low-impact separation devices, the layout is shown below, and the calculation of the diameter of the connecting bolts of the explosive bolts or low-impact separation devices is carried out. The connection structure between the aircraft and the lower stage is shown in figure 3.

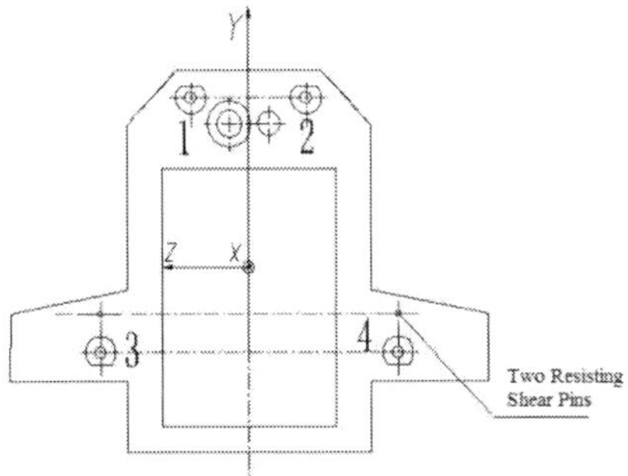

Figure 3. The connection structure between the aircraft and the lower stage

Use formula (1) to calculate the working tension:

$$P = \frac{2M}{nD} + \frac{T}{n} \tag{1}$$

Where M is the bending moment load on the abutment surface; D is the relative distance of the connection; T is the axial tensile load; n is the number of connections. Calculated under the worst working conditions, the working tension of connection point 4 is the largest, P_{max}=29748.3N.

Under this working condition, the total pull force of the connection is P0=1.8 Pmax=1.8×29748.3 =53546.94N. Design load P_{sj}=fP0=2×53546.94=107093.88N.

The total tensile force at the connection should be less than the tensile limit of the explosive bolt or low-impact separation device. Depending on the load capacity requirements, a low-impact separation device or an explosion bolt can be selected.

2.3.2. Comparative analysis. Depending on the carrying capacity, both the low-impact separation device and the explosion bolt can meet the separation requirements. From the perspective of structural layout and operability, the installation of both needs to open the operation port at the lower stage, but when installing the low-impact separation device, the connecting bolt needs to be pulled through the connection surface with a tow bar[3]. Tighten the release nut and apply a tightening torque on the release nut. After completion, remove the tow bar and install the pressure plate to fix the separation nut. The lower stage is required to have a larger opening, which adversely affects the strength and rigidity of the lower stage. Explosive bolts are easy to install. They only need to be installed on the lower stage side and apply a tightening torque. The requirements for the size of the opening are relatively low. Comprehensive consideration, choose the explosion bolt as the separation device.

2.4. Guide positioning device

Separation system guide positioning device generally uses positioning guide pins to resist shear and guide positioning, and its length is related to connection thickness and separation simulation results. Two types of shear guide structure types, positioning guide pin and cone surface + cone sleeve, were compared and demonstrated.

2.4.1. Locating Pin Scheme. A short positioning pin is provided on the butt surface of the flight test warhead body for shear support and installation positioning, which can prevent the explosive bolt from being sheared and affect normal separation[4]. There are 4 positioning pins installed near the explosion bolt mounting holes on the tail frame of the aircraft, as shown in the figure 4.

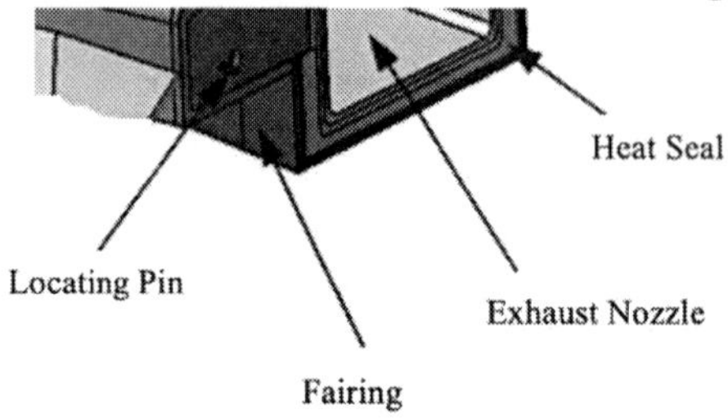

Figure 4. Schematic diagram of the rear end of the aircraft

According to the load conditions, the tail frame Mx of the aircraft is converted into shear force, and after combining with Fx and Fy, the worst case shear force of positioning pin 2 is 3937N.

Therefore, the diameter of a single positioning pin is:

$$D = 2\sqrt{\frac{fQ}{\pi\tau_b}} = 2\sqrt{\frac{2\times3927}{3.14\times640}}=3.95\text{mm} \quad (2)$$

Considering installation and other factors when designing the structure, the positioning pin diameter and other parameters are designed as follows: the diameter of the pin segment of the positioning pin is Φ10mm, the length of the column segment is 20mm; the thread M8 × 1, the length is 15mm; Plane out at the column section for easy clamping and tightening.

2.4.2. Conical surface + cone sleeve scheme. Considering that the positioning pin has a certain effect on the separation, a tapered sleeve (titanium alloy) is installed at the lower end front frame explosion bolt connection hole. The corresponding aircraft tail frame is provided with a tapered hole, and the contact between the tapered sleeve and the tapered hole is used for shearing. The structure of the connection between the aircraft and the lower stage is shown in the figure 5 and figure 6.

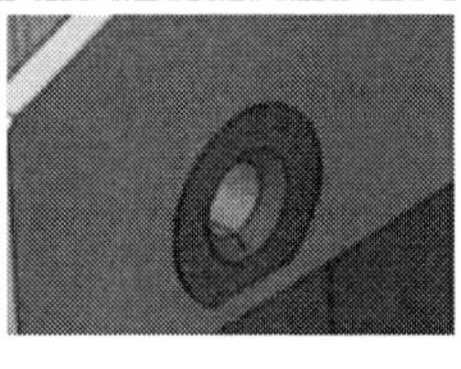

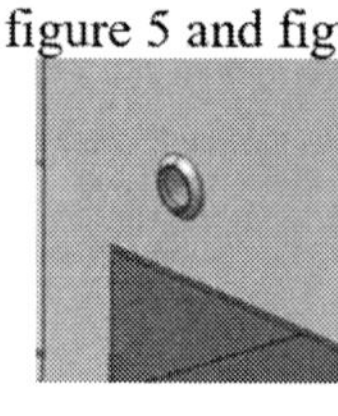

Figure 5. Part of the connecting boss **Figure 6.** A part of the cone sleeve connected

2.4.3. Summary. The aircraft is separated in the atmosphere, and there is strong aerodynamic interference during the separation process. The positioning pin may cause interference to the separation and even cause jams. Therefore, this solution uses a cone surface + cone sleeve solution.

2.5. Impact absorption device

The function of using the explosion bolt box as the shock absorbing device is mainly to store explosive separated nuts and screws, etc., to avoid causing harm to the equipment in the aircraft. The nut that cooperates with the explosion bolt in the aircraft is a homemade square nut. An explosion bolt box is designed on the outside of the square nut (connecting the corner box, as shown in figure 7). At the same time, a honeycomb structure is designed on the wall of the explosion bolt box to absorb the impact of the explosion bolt.

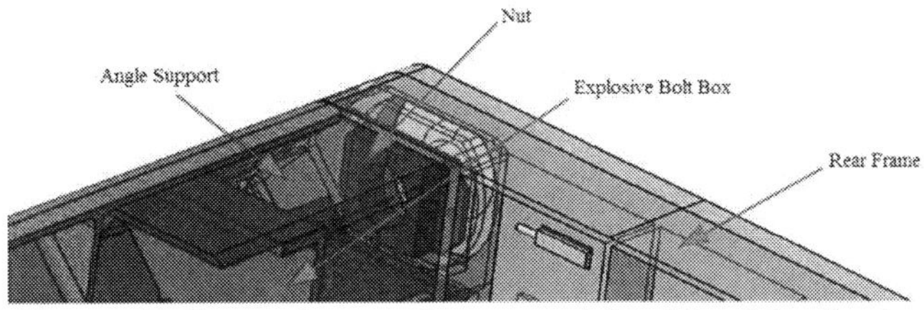

Figure 7. Corner box on the upper right side of the fuselage

When the explosion bolt is installed, place the square nut on the side directly opposite the square nut in the corner box, and apply a pre-tightening force to the explosion bolt with a tool from the lower stage side. The schematic diagram of the structure of the left side nut and corner box on the lower part of the aircraft fuselage is shown in figure 8, and the structure of the rest of the check is similar[5].

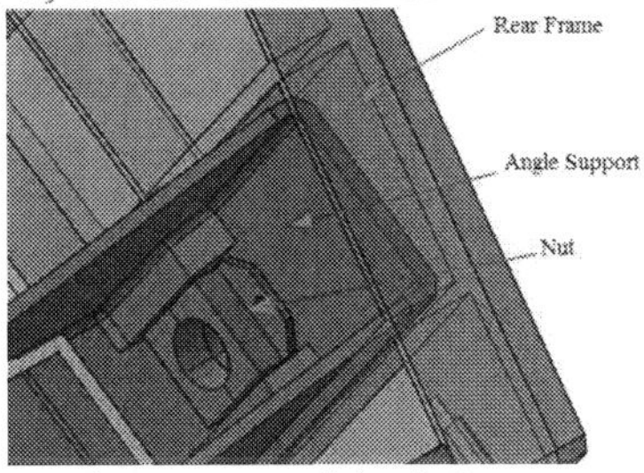

Figure 8. Internal structure of the left corner of the fuselage

2.6. Separation impulse device

The scheme uses mature separation impulse devices, namely reverse thrust rocket and thrust termination device to provide separation impulse.

2.6.1. Thrust termination device. The boost engine will activate the thrust termination device at the end of the flight, that is, open 4 ejection tubes. The comprehensive negative thrust curve of the boost engine is shown in figure 9. It can be seen from the figure that after the thrust termination device is turned on, the engine thrust is positive, and after a certain period of time, the engine thrust starts to become negative.

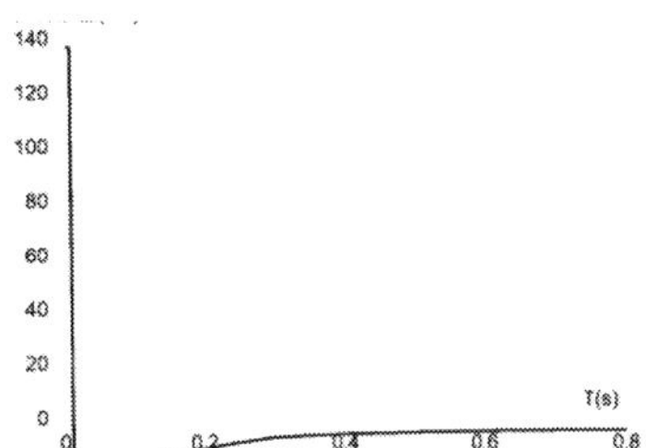

Figure 9. Negative thrust-time curve

2.6.2. Retro-Rocket. A total of two retro-rockets are located at the end of the booster stage. The line connecting the thrust action points of two reverse thrust engines and the intersection of the bullet axis is 1.145m away from the full bomb origin.

The average thrust of a single retro-rocket thrust is 10kN, and the included angle between the nozzle installation axis and the rotary axis of the booster engine is 25 °. Retro-rocket performance is shown in table 1. Thrust of retro-rocket is shown in figure 10. Position of retro-rocket thrust point is shown in figure 11.

Table 1. Capability of retro-rocket.

Parameter Name	Unit	Parameter Value
Average Thrust	kN	10
Average Thrust Deviation		≤15%
Average Working Time	s	≥0.35
Ignition Delay Time	ms	t80≤35ms
Asynchrony	ms	≤35ms
Back Thrust Engine Thrust Line Traverse	mm	±5
Deflection of Thrust Line of Reverse Thrust Engine	′	22

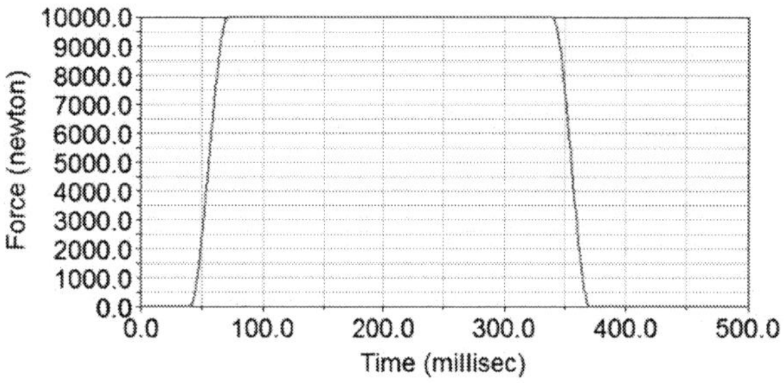

Figure 10. Thrust of retro-rocket

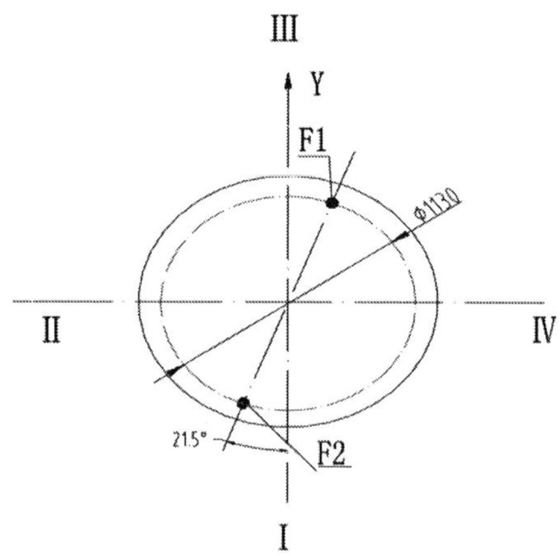

Figure 11. Position of retro-rocket thrust point

3. Conclusion

In this paper, a new separation scheme is proposed, which is composed of a connecting unlocking device, a guiding positioning device, a separating impulse device and a shock absorbing device. The feasibility of the scheme is verified by simulation analysis, and the design of the separation structure under high speed and high dynamic pressure is realized, which improves the reliability of the separation system.

References

[1] Zhao Shanshan, Comparative Analysis on the Aircraft Stage Separation Based on Rigid Guided-Contact Model, Missiles and Space Vehicles, 2013

[2] Jiang Ruigang, Discuss of Tactical Missile Stage Linear Separation, Spacecraft Recovery & Remote Sensing, 2000

[3] Yang Tao, Finite Analysis of Stage-separation Connecting Appliances of Missile, Journal of Projectiles, Rockets, Missiles and Guidance, 2006

[4] Qian Cheng, Simulation and Analysis of Rigid-Flexible Coupling for Stage Separation of Missile Based on ADAMS, Journal of Ordnance Equipment Engineering, 2017

[5] Zhou Yanhua, Type Synthesis and Optimization of Main Driving Mechanism for Servo-punch Press, Journal of Mechanical Engineering, 2015

Research on Command Allocation Method for Flying Wing Aircraft

Ning Zhang

China Academy of Aerospace Aerodynamics, Beijing 100074, China

soviet2005@sina.com

Abstract. Flying wing has nice aerodynamic efficiency, good loading and stealth characteristics, it is an ideal aerodynamic configuration for reconnaissance, transportation and bombing type aircraft. However, flying wing aircraft suffers from poor stability and control characteristics, which makes it difficult to design and control. Due to elimination of horizontal and vertical tails, flying wing aircraft has poor longitudinal and directional dynamic characteristics. Unlike most of the conventional aircraft, flying wing aircraft usually have a large number of control effectors, the redundant effectors are designed and required to provide adequate lift, pitch and yawing force or moments. There-fore, flying wing aircraft usually need to adopt reasonable control allocation algorithms to exploit the control capability of redundant effectors. Especially for large-aspect ratio flying wing with very high level of effector redundancy, the pursuit of performances, e.g., gust load alleviation control and maximum lift mode, makes control allocation extremely important and indispensable. This paper summarizes the typical characteristics of command allocation methods, and its applicability in different flight missions, providing a reference for the design of command allocation method for flying wing aircraft.

1. Introduction

When compared with conventional aircraft, the longitudinal and lateral static stability of flying wing aircraft is greatly degraded, the directional axis usually is weakly unstable, and the three-axis damping is significantly reduced. Hence the dynamics characteristics of flying wing aircraft is difficult to meet the flying quality requirements, and control augmentation system is necessity for flying wing. Since horizontal and vertical tails have been cancelled, flying wing aircraft usually adopts a combination of multiple sets of elevens and new type of drag rudder to meet the aircraft's requirements for control power.

The flying wing aircraft adopts redundant multi-control surface configuration to improve control power, but each control surface has multi-axis control capability and there is strong control coupling effect. In order to solve the problem of control surface redundancy and the axial function of the control surface is not clear, it is necessary to introduce a control allocation unit in the flight control system to map the control command required by the control law to the redundant control surfaces in real time, so that the aircraft can accurately complete the flight mission. Conventional aircraft usually manual decouples control surface function according to control characteristics of effectors, although this method can reduce the complexity of control allocation unit, which would limit the control surface's control effectiveness, especially for flying wing aircraft, and cannot achieve the maximum attainable moment set (AMS) of control surfaces. Therefore, it is necessary to design a reasonable control allocation unit for the allocation of control command in order to maximize the rudder surface control ability of the

Content from this work may be used under the terms of the Creative Commons Attribution 3.0 licence. Any further distribution of this work must maintain attribution to the author(s) and the title of the work, journal citation and DOI.
Published under licence by IOP Publishing Ltd

aircraft.

This paper studies and summarizes common control allocation methods for flying wing aircraft, including generalized inverse, daisy chain, direct allocation, and non-linear optimization allocation, etc., which aims at providing a reference for the design of command allocation algorithms for flying wing aircraft.

2 Command allocation problem description

2.1 Flying wing command allocation

Conventional aircraft has three typical effectors, elevator, aileron, and rudder. They are used to generate three-axis control moments control the flight attitude of aircraft. Therefore, within the range that the three-axis moment can reach, the control deflections that the aircraft needed for maneuvering is uniquely determined, that is, the traditional control allocation problem can be expressed as follows

$$\begin{cases} M = qS_w\bar{c}(M_0 + M_\alpha\alpha + M_q q + M_{\dot\alpha}\dot\alpha + M_{\delta_e}\delta e) \\ L = qS_w\bar{b}(L_\beta\beta + L_p p + L_r r + L_{\delta_a}\delta a + L_{\delta_r}\delta r) \\ N = qS_w\bar{b}(N_\beta\beta + N_p p + N_r r + N_{\delta_a}\delta a + N_{\delta_r}\delta r) \end{cases} \quad (1)$$

It can be seen that the desired moment $[L, M, N]_d$ corresponds to unique control vectors $[\delta_a, \delta_e, \delta_r]$ and the three control effectors have a clear functional axis, and there is almost no control coupling in the vertical and horizontal axes.

Flying wing aircraft adopts multiple sets of new type of control surface configurations. In order to solve the problem that redundant control surfaces are more than the desired control axes and have no explicit axial control function, it is necessary to adopt command allocation unit to allocate the desired control moments required by the control law to relevant control surfaces optimally.

Due to the usage of innovative control effector of flying wing aircraft, especially the drag based control effector for the directional axis, such as the all-moving wing tip used on tailless fighter aircraft, as shown in Figure 1, it is found that for flying wing aircraft not only the longitudinal and lateral-directional axes under different flight conditions have strong coupling effects, but also the generated large amount of drag cannot be ignored. Therefore, drag, lift and side force should also be considered in command control allocation.

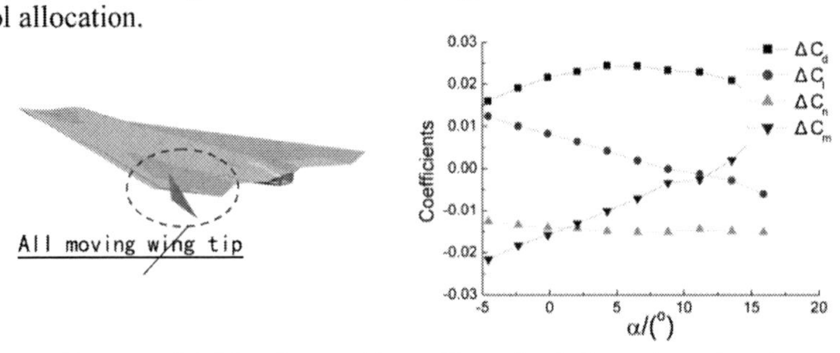

Figure 1. Multi-axis coupling effect of a full-motion wingtip

The control surface aerodynamic nonlinear characteristics of the flying wing aircraft are very strong, which leads to the traditional simple linear allocation method no longer applicable [1]. Besides, the drag based control effector usually adopts a single-sided large deflection strategy, such as the all-moving wing tip deflection range is [0, 60] degree, the split drag rudder is even in the range [0, 90] degree, and so the dynamics of the actuator cannot be ignored, including the frequency and rate of deflection. Here only the speed limit $\dot\delta_i$ of the actuator is considered, namely,

$$\rho_{i\min} < \dot\delta_i < \rho_{i\max} (i = 1 \quad ... \quad n) \quad (2)$$

Where $\rho_{i\text{mm}}$ and $\rho_{i\max}$ are the minimum and maximum deflection rate of each control surface. Therefore, the command allocation problem for tailless aircraft is described as follows,

$$\begin{gathered} g(u(t)) = M_d(t) \\ s.t.\ \ \bar{\delta}_{i\min} \le u_i \le \bar{\delta}_{i\max} \end{gathered} \tag{3}$$

Where $g(\cdot)$ represents the nonlinear function of mapping control surface deflection $u(t)$ into control moments, $M_d(t)$ are the desired three-axis control moments, $\bar{\delta}_{i\min}$ and $\bar{\delta}_{i\max}$ are the lower and upper position limits for the i_{th} control surface. For a multi-control surface flying wing aircraft usually has n control surface. And u is,

$$u = [\delta_1, \delta_2, \ldots, \delta_i, \ldots, \delta_n]\,(n > 3) \tag{4}$$

The control surface position and rate limit can be approximated as follows, the upper and lower limits of the position within a single sampling period,

$$\begin{cases} \bar{\delta}_{i\max} = \max(\delta_{i\max}, \delta_i(t-T) + T\rho_{i\max}) \\ \bar{\delta}_{i\min} = \min(\delta_{i\min}, \delta_i(t-T) - T\rho_{i\max}) \end{cases} \tag{5}$$

2.2 Attainable moment set

Under different control effectiveness matrix and control effector constraints, the maximum attainable moment set $\{M\}$ are quite different. Large AMS means that the aircraft has good controllability and may achieve maneuverability.

In the three dimensional vector space $\mathbb{R}^3$ composed of aircraft three-axis moment coefficients, the AMS of conventional aircraft is a space hexahedron, which corresponds to the three control surfaces elevator, aileron and rudder, respectively. But the AMS of flying wing aircraft usually is space polyhedron (greater than six), and it is a spatial convex body if nonlinear aerodynamic effects are considered. As shown in Figure 2, the AMS of flying wing aircraft has strong influence on aircraft total drag.

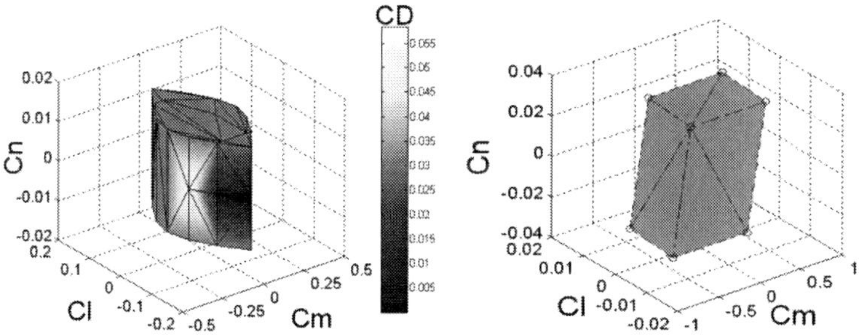

(a) A large aspect ratio flying wing (b) Boeing747 aircraft
H = 20000m, Ma = 0.6 cruise H = 6000m, Ma = 0.65 cruise

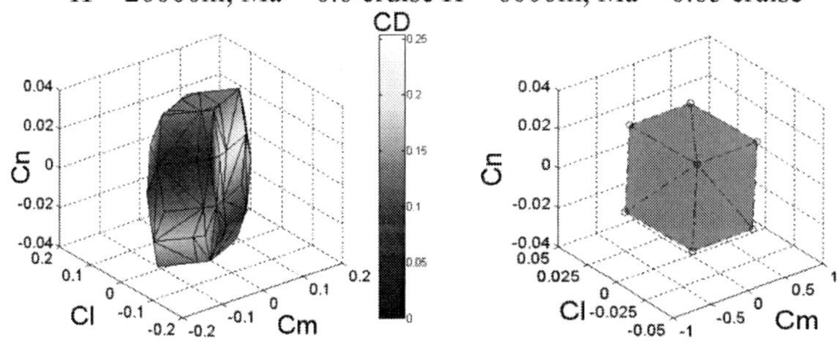

(c) A small aspect ratio flying wing (d) F-4 aircraft
H = 4572m, Ma = 0.6 cruise H = 10668m, Ma = 0.6 cruise

Figure 2. Attainable moment set of several aircraft [2]

IOP Conf. Series: Materials Science and Engineering **887** (2020) 012020 doi:10.1088/1757-899X/887/1/012020

2.3 Solution situation

Control allocation problem can be treated as find solution to nonlinear equations $g(u) = M_d$ with constraints $\bar{u}_{\min} \le u \le \bar{u}_{\max}$. Because the number of control surfaces is more than the expected control variables, the nonlinear equation of the command allocation problem is underdetermined. As shown in Figure 3, there are two cases for the solution of this problem.

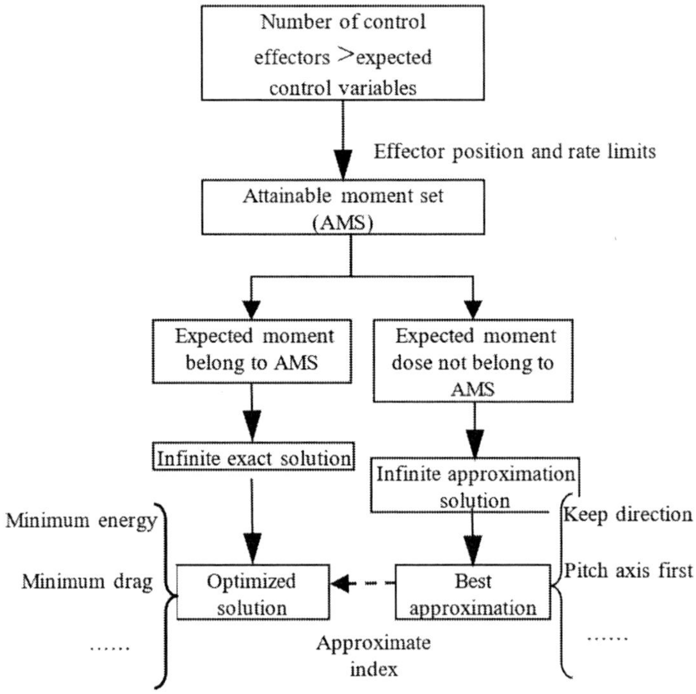

Figure 3. Solution to the instruction allocation problem

When the expected moment exceeds the AMS range, in order to accurately control the difference between the magnitude and direction of the realized moment and the expected moment, the expected moment must first be converted into attainable moment according to performance index, and then convert it into optimal exact solution.

3. Control allocation method

3.1 Generalized inverse

The generalized inverse method is the most classic and simple control allocation method, which is only suitable for linear control allocation problem. By solving the pseudo inverse of the control effectiveness matrix, we can obtain the minimum norm solution u of the control vectors, i.e.

$$u = B^T[BB^T]^{-1}, M_d \Rightarrow \min \|u\|_2^2 \tag{6}$$

Where M_d is the desired moment, B is the control effectiveness matrix.

Considering the authority of different control axes, a weighted generalized inverse method is proposed [3]. As is shown in equation (7), this method can prevent the control surface with high control efficiency from being saturated in advance.

$$u = N(BN)^T[BN(BN)^T]^{-1} \Rightarrow \min N^{-1} \|u\|_2^2 \tag{7}$$

This method has been developed for a very long time since it is simple and easy to implement. However, this method cannot reach all the values in AMS. More importantly, if using this method the deflection of the control surface and the rate constraint may cause great errors and pose uncertain risks to the control system.

In order to reduce the above-mentioned risk, a redistributed pseudo-inverse algorithm is proposed, that is, re-allocating the control surface that is not saturated in previous allocation and redistribute command to it until it is fully saturated [4]. But this improved method based on generalized inverse still cannot reach all the values in AMS.

3.2 Daisy chain

In daisy chain method [5], before the allocation routine is started, the control surfaces are divided into several groups based on their functions, such as the primary control surface and the secondary control surface combination (or the conventional control surface and the non-conventional control surface). Then the control allocation routine will first use the primary control surfaces, when primary control surfaces are saturated, the secondary control surfaces will be used. This method can maximize the use of conventional control surfaces and avoid using non-conventional control surfaces.

$$M = \begin{bmatrix} g_1(u_1) & g_2(u_2) \end{bmatrix} \begin{bmatrix} u_1 \\ u_2 \end{bmatrix} \to M_d \tag{8}$$

Where B_1 and u_1 are the control effectiveness matrix and control vector of the primary control surfaces, B_2 and u_2 are the control effectiveness matrix and control vector of the secondary control surfaces.

The daisy chain allocation algorithm is to find the pseudo inverse of each group of the control matrices in turn, as described below,

$$\begin{cases} M_i = B_i u_i \\ u_i = B_i^T [B_i B_i^T]^T (M_d - \sum_1^{i-1} M_j) \end{cases} \tag{9}$$

Different groups of control surfaces can also use different algorithms to allocate control command. This allocation method is suitable for aircraft that performs simple tasks during level flight states.

3.3 Direct allocation

Direct allocation method is first proposed by Durham [6], it is a geometric space method based on the maximum usable AMS. This method searches for the boundary element of AMS to find the intersection point of the desired moment, that is,

$$\hat{M} = M_d / |M_d| \tag{10}$$

$$M = \rho \hat{M} \tag{11}$$

In case $\rho \geq 1$, the expected moment is reachable, otherwise, the expected moment is unreachable, reducing the moment M proportionally to intersect AMS boundary, because we assume the aircraft dynamics as linear system, the amount of control deflection is reduced in proportion, which ensures that the realized moment is in the same direction as the desired moment.

Because the direct allocation method can reach all the values in AMS and has clear physical meaning, there are many control allocation algorithms are developed based on this method. Like for example, Durham has proposed a face search algorithm[7], edge search algorithm [8] and adjoint surface search algorithm[9]. Besides, direct allocation method can also be realized by linear programming method[10,11]. Currently direct allocation methods are limited to linear control allocation problems.

3.4 Optimization based (nonlinear programming) method

Due to the emergence of advanced aerodynamic configuration and innovative control effectors, the problem of non-linear control allocation is becoming increasingly prominent. At the same time, in order to effectively solve the control allocation problem when the expected moment exceeds the AMS range, control allocation method based on mathematical programming have been rapidly developed.

Optimization based allocation method can be simply described as,

$$\min f(u) \tag{12}$$

$$\text{S.t}\begin{cases} g(u) = M_d \\ \bar{u}_{\min} \le u \le \bar{u}_{\max} \end{cases} \tag{13}$$

Where $f(u)$ is optimization goals, $g(u)$ is the control effectiveness function of control surfaces, $[u_{\min} \quad u_{\max}]$ is the position limit of control surfaces.

Assume equation (13) is in the feasible region of optimization problem, if the feasible solution does not exist, that is, when the expected moment is outside the AMS, then the equation constraint will be relaxed and the unreachable moment need to be constrained in the objective function.

Optimization based control allocation method can also achieve multi-target, variable-target control allocation, such as minimal drag during cruise[12], maximum lift during take-off and landing phase or the smallest radar reflection cross-sectional area during combat phase[13]. Nelson et al also proposes to maximize the priority of the pitching moment under any maneuver[14] to solve the problem of the expected moment beyond the maximum AMS under optimization based control allocation, which has gain positive supports from pilots. But at present, the classic quadratic programming method can only solve the case where the control effector model is linear.

3.5 Dynamic allocation method

In an actual control system, control allocation does not statically allocate the desired moment at current time, but needs to allocate the expected moment command that changes over a period of time required by the control law. Härkegård et al. considered this issue as dynamic control allocation [15], this method has been initially applied on a tailless aircraft [16] and developed the generalized method for dynamic control allocation,

$$\min J = \left\| W_1 \left(u(t) - u_p(t) \right) \right\|_2^2 + \left\| W_2 \left(u(t) - u(t-T) \right) \right\|_2^2 \tag{14}$$

Where W_1 and W_2 is the weighting factor, and $W = \left(W_1^2 + W_2^2 \right)^{1/2}$, then

$$u(t) = Eu_p(t) + Fu(t-T) + GM_d(t) \tag{15}$$

Where E, F and G satisfy

$$\begin{cases} E = (I - GB)W^{-2}W_1^2 \\ F = (I - GB)W^{-2}W_2^2 \\ G = W^{-1}(BW^{-1}) \end{cases} \tag{16}$$

Dynamic allocation considers the relationship between control allocation command of current time and the allocation result of previous time step $u(t) = f(M_d(t), u(t-T))$ to prevent the abrupt jump of control surfaces. It can be seen that when $W_2 = 0$, dynamic control allocation is the same as generalized inverse method.

Considering the dynamic characteristics of multiple control surfaces as second-order system, only damping ratio ξ_j and frequency ω_j are modeled, then we will have simplified control surface model,

$$\delta_j = \frac{\omega_j^2}{s^2 + 2\xi_j \omega_j s + \omega_j^2} \delta_{j.des} \tag{17}$$

Since the dynamic characteristics of each control surface may be different, Venkataraman et al. adopt a Fourier transform to the optimization target (moment error, etc.) to consider the constraints and optimization problem in frequency domain [17]. The optimization result uses the low-frequency control effectors to complete the low-frequency moment command and improve the tracking accuracy of the commanded moment.

Table 1. Comparison of typical allocation methods

Allocation method	Integration	Control model		Implementing moment sets	Allocation target	Consider dynamic allocation	When expected moment ∉ AMS	Algorithm	Applicable flight phase
	Full participation on the rudder surface	Linear	Non-linear	AMS reachable	Achieve other goals				
Generalized inverse method	√	√				√	Uncertainty risk	Simple (weighted) pseudo-inverse method, redistribution pseudo-inverse method	Difficult to apply directly
Daisy chain		√	√		√		Depends on different algorithms	Multi-level generalized inverse	cruise
Direct allocation	√	√		√			Holding torque direction	Area search, edge search, linear programming, etc.	Maneuver
Optimization based allocation	√	√	√	√	√	√	Achieving optimization goals	Linear programming, quadratic programming, penalty function method, etc.	Field / Cruise / Maneuver

It can be seen from Table 1 that in different flight phase and tasks of the aircraft, it is not possible to rely on a single command allocation method, and the actual aircraft command allocation has become an optimization problem. Therefore, the command allocation scheme for flying wing aircraft should consider the different flight phases, from the perspective of optimization, it can be converted to achieve different expected performance indicators under different flight phases.

4. Conclusion

This paper studies and summarizes common command control allocation algorithms, including generalized inverse allocation, daisy chain allocation, direct allocation, and optimization based allocation for flying wing aircraft. The generalized inverse method is simple to implement, but the saturation of the control surface may bring uncertainty risks to control system. Daisy chain allocation can be treated as an improved method of the generalized inverse method, which can deal with control surface saturation, but cannot reasonably allocate command based on the dynamic characteristics of the control surface. The direct allocation method is based on the geometric space method, which is relatively simple to implement, but can only be applied to linear control effectiveness model. The optimization based allocation method is a more ideal method for the command allocation of flying wing aircraft in the future. It can achieve the command allocation that meet specific optimization goals according to different mission requirements, but the nonlinear optimization algorithm is currently difficult to solve in real time. Hence, according to the control effector characteristics of flying wing aircraft, we should select reasonable command allocation algorithms to maximize the control power of redundant effectors for flying wing aircraft.

References

[1] Yang Enquan, Research on Key Technologies of Flight Control System for Unmanned Combat Aircraft. PhD dissertation of Beijing University of Aeronautics and Astronautics. 2006.5.

[2] Wang Lei, Wang Lixin, Jia Chongren.Characteristics and operating characteristics of cracked rudder of flying wing aircraft. Acta Aeronautical Sinica, 2011, 32 (8), 1392-1399.
[3] Kenneth A. Bordignon, Wayne C. Durham. Closed-Form solution to the constrained control allocation problem. Virginia Polytechnic Institute and State University, AIAA-94-3552-CP.
[4] Zhan Zhengyong, Liu Lin. Research on Multi-Control Surface Advanced Layout Aircraft Control Distribution Technology. Flight Mechanics. March 2006, Volume 24, Issue 1.
[5] Teel A. R, Buffington J M. Anti-Windup for an F-16's daisy chain control allocation. Journal of Guidance, Control and Dynamics, 1996, 19(6): 1226-1230.
[6] Wayne C. Durham, Constrained control allocation. Virginia Polytechnic Institute and State University, AIAA-92-4550-CP.
[7] Wayne C. Durham. Computationally efficient control allocation. Journal of Guidance, control and Dynamics. 1994, 17(6): 1371~1373.
[8] Durham W C. Attainable moments for the constrained control allocation problem. Journal of Guidance, control and Dynamics. 2001, 24(3): 519~524.
[9] Li Weiqi, Wei Chen, Chen Zongji. A New Direct Assignment Algorithm for Restricted Control. Journal of Beijing University of Aeronautics and Astronautics. November 2005, Volume 31, Issue 11.
[10] Shen Jinglai, Gao Jinyuan. Optimal Allocation of Limited Control Variables Linear Programming for Flight Control Systems. Flight Mechanics. June 1999, Volume 17, Issue 2
[11] Marc Bodson. Evaluation of optimization methods for control allocation. University of Utah, Electrical Engineering. AIAA-2001-4223.
[12] Wayne C. Durham, John G. Bolling, Kenneth A. Bordignon. Minimum drag control allocation. Virginia Polytechnic Institute and State University, AIAA-96-3410-CP.
[13] Wang Meixian, Li Ming. Review of Control Assignment Methods for Advanced Fighters. Aircraft Design. 1673-4599 (2006) 03-0017-03.
[14] Mark D. Nelson, Wayne C. Durham. A comparison of two methods used to deal with saturation of multiple, redundant aircraft control effectors. Virginia Polytechnic Institute and State University, AIAA-2002-4498.
[15] Ola Härkegård. Dynamic control allocation using constrained quadratic programming. AIAA-2002-4761.
[16] Liu Yanbin, Lu Yuping. Application of Nonlinear Adaptive Control in Tailless Flight Control System. Chinese Journal of Aeronautics. September 2006, Volume 27, Issue 5.
[17] Ram Venkataraman, Michael Oppenheimer, and David Doman. A new control allocation method that accounts for effector dynamics. IEEE Aerospace Conference, Big sky, MT, March 6-13 2004.

Study on the Computation Method for the Generalized Force and Mass of Flexible Aerial Rudder in Space Vehicle

Guo Peng-fei, Wu Jian-hui, Zhang Wei-qiang, Zhang Yong and Yang Yang

China Academy of Launch Vehicle Technology, Beijing 100076, China

Corresponding author's e-mail address: gpfmail@126.com

Abstract. Aerial rudders are significant actuators in the spacecraft control systems and are responsible for the stability and control of different motions such as pitching, yawing and rolling around mass center. With the development of new generation of space vehicles, high speed flight demonstration, light and thin aerial rudders with large span-chord ratio have been widely used. In the design of control strategies in pitching, yawing and rolling channels and control feedback loop parameters, the influence caused by aerial rudder elastic deformation should be considered particularly. In this paper, elastic motion equations of the spacecraft flexible rudder are studied firstly. Then, according to the modal vibration features, aerodynamic force derivative distribution characteristics and generalized force expressions, the rudder station division and computation method of generalized rudder aerodynamic force derivatives are proposed in the pitching, yawing and rolling channels based on the computational fluid dynamics simulation. By comparison the results of the proposed method with these of the traditional computation methods, the accuracy and validity of the proposed method are verified which can provide considerable references and basis in the control design of flexible rudder in the spacecraft.

1. Introduction

The motion of spacecraft during flight is the coupling of the overall rigid movement and the elastic oscillation. Under the loadings including aerodynamic force, power system thrust and servo force, the rigid motion of the spacecraft could lead to structural oscillation which will also affect overall rigid motion simultaneously [1]. Therefore, the spacecraft considering structural elasticity challenges the traditional flight control which is based on rigid body dynamics, and makes the control of real spacecrafts more complicated.

Traditional spacecrafts represented by launch vehicles and missile weapons hold symmetric aerodynamic shapes and structures with relatively large rudder stiffness. The corresponding pitching, yawing and rolling channels which are affected by servo mechanism, control force and aerodynamic force are coupled in a low degree. Thus, tri-channel control system with decoupled strategy is employed generally [2]. However, with the increasing weight, thrust and slenderness ratio of rocket and missile as well as extensive applications of bundled propellants, the new generation spacecraft presents the characteristics of concentrated low frequency mode, strong coupling vibrations and complex local deformations. In this case, Wu et al [3] established elastic vibration equations of spacecrafts with spatial mode based on the finite element method, and demonstrated the influence of lateral, longitudinal, torsional vibrations and booster local mode on the attitude control system.

Aerial rudders are significant actuators in the spacecraft control systems and are responsible for the

Content from this work may be used under the terms of the Creative Commons Attribution 3.0 licence. Any further distribution of this work must maintain attribution to the author(s) and the title of the work, journal citation and DOI.
Published under licence by IOP Publishing Ltd

stability and control of different motions such as pitching, yawing and rolling around mass center, which are the vital basis to ensure the predetermined trajectory flight [3-4]. In traditional space vehicle design, aerial rudders hold large structural stiffness and connection stiffness and thus can be regarded as rigid body in the control system [5-7]. With the development of new generation of space vehicles represented by USA X-37B and X-51 high speed flight demonstration, light and thin aerial rudders with large span-chord ratio have been widely used. These dimension changes cause a dramatic decline of rudder stiffness comparing with those in traditional launch vehicles and missile weapons, even make local modal frequency of the rudders lower than the bending modal frequency of the structure in directions of pitching and yawing. Besides, these changes increase the difficulty in the design of control systems which guarantee space vehicle stability against attitude divergence and out of control. So in the design of control strategies in pitching, yawing and rolling channels and control feedback loop parameters, the influence caused by aerial rudder elastic deformation should be considered particularly.

In this paper, rudder station division and computation method of generalized rudder aerodynamic force derivatives are proposed according to the modal vibration features, aerodynamic force derivative distribution characteristics and generalized force expressions. By comparing with the results of traditional computation method, the accuracy and validity of the proposed method are verified which can provide considerable references and basis in the control design of the flexible rudder.

2. Elastic motion equation and corresponding parameters

Elastic vibration modal shapes of the spacecraft structure mainly include motions in longitudinal, pitching, yawing and rolling directions with pitching axis perpendicular to trajectory plane and yawing axis in the trajectory plane perpendicular to structural longitudinal axis. When elastic vibrations in pitching and yawing directions are analyzed as shown in Figure 1, computational fluid method and finite element method should be employed to obtain characteristics of the aerodynamic distribution and each order modal shape respectively.

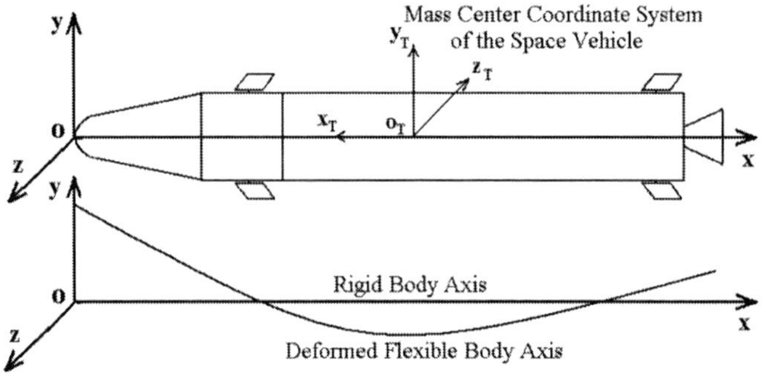

Figure 1. Elastic deformation of space vehicle structure

In modal analysis, the finite element method is widely adopted and the spacecraft structural elastic vibration equation can be expressed by,

$$M\ddot{X}(\mathrm{t}) + C\dot{X}(t) + KX(t) = F(t) \tag{1}$$

where M, C and K are the structural system mass matrix, damping matrix and stiffness matrix respectively; $\mathrm{X(t)}$, $\dot{X}(t)$ and $\ddot{X}(t)$ are the finite element node displacement, velocity and acceleration vector respectively; $F(t)$ is the loading vector on finite element node. Mode-superposition method is exploited to solve the equation, and it is assumed that

$$X(t) = \sum_{i=1}^{n} \Phi_i q_i(t) \tag{2}$$

where Φ_i is the ith-order modal shape of the structural system, then Equation (1) can be decoupled as:

$$\ddot{q}_i(t) + 2\zeta_i\omega_i\dot{q}_i(t) + \omega_i^2 q_i(t) = F_i(t)/M_i \quad i = 1,2,\dots,n \tag{3}$$

where $q_i(t)$, $\dot{q}_i(t)$ and $\ddot{q}_i(t)$ are the ith-order mode displacement, velocity and acceleration in generalized coordinates respectively; ω_i , ζ_i , $F_i(t)$ and M_i are the ith-order mode natural frequency, damping coefficient, generalized force and generalized mass of the structural system. M_i and $F_i(x,y,z,t)$ can be expressed as:

$$M_i = \Phi_i^{\mathrm{T}} M \Phi_i = m(x,y,z)\cdot\Phi_{i,X}^2(x,y,z) \tag{4}$$

$$\begin{aligned} F_i(x,y,z,t) = \Phi_i^{\mathrm{T}} F(x,y,z,t) = \Phi_{i,X}(x,y,z)F_X(x,y,z,t) + \\ \Phi_{i,Y}(x,y,z)F_Y(x,y,z,t) + \\ \Phi_{i,Z}(x,y,z)F_Z(x,y,z,t) \end{aligned} \tag{5}$$

where $F(x,y,z,t)$ and $\Phi_i(x,y,z)$ are the external excitation and modal shape in corresponding position respectively.

When generalized mass and generalized force are substituted into Equation (3), spacecraft lateral elastic vibration equation can be described by

$$\ddot{q}_i(t) + 2\zeta_i\omega_i\dot{q}_i(t) + \omega_i^2 q_i(t) = D_{1i}\Delta\dot{\varphi} + D_{2i}\Delta\alpha + D_{3i}\Delta\delta_\varphi + D_{3i}''\Delta\ddot{\delta}_\varphi + Q_{iy} \tag{6}$$

$$D_{1i} = -\frac{180\cdot qS_M}{\pi M_i V}\sum_{s=1}^{K} C_{nn}^{\alpha}\Phi_i(X)(X_{\mathrm{T}} - X_s)dX \tag{7}$$

$$D_{2i} = \frac{180\cdot qS_M}{\pi M_i}\sum_{s=1}^{K} C_{nn}^{\alpha}\Phi_i(X)dX \tag{8}$$

If the stiffness of the aerial rudder is large enough, the modal shape $\Phi_i(X_K,Y_K,Z_K)$ and the modal shape slope $\Phi_i'(X_K,Y_K,Z_K)$ at any point on rudder are equal, and then the following can be derived.

$$D_{3i} = \frac{N_K H_K}{M_i}[F_{nn}^K\Phi_i(X_K) + m_K a_X l_K \Phi_i'(X_K)] \tag{9}$$

$$D_{3i}'' = \frac{N_K H_K}{M_i}[m_K l_K \Phi_i(X_K) + J_K\Phi_i'(X_K)] \tag{10}$$

where q is the dynamic pressure; S_M is the reference area; V is the flight velocity; M_i is generalized mass of the ith-order mode; Φ_i and Φ_i' are the modal shape and corresponding slope respectively; $\Delta\dot{\varphi}$ is the pitching(or yawing) rotational angular velocity; $\Delta\alpha$ is the attack angle (or sideslip angle); $\Delta\delta_\varphi$ is the angle of aerial rudder rotation; $\Delta\ddot{\delta}_\varphi$ is the angular acceleration of aerial rudder rotation; Q_{iy} is the generalized force corresponding to the disturbing force in the direction of main shaft vibration; C_{nn}^{α} is the coefficient derivative of transverse(in the direction of pitching or yawing) distribution of aerodynamic force; F_{nn}^K is the coefficient of the transverse(in the direction of pitching or yawing) aerodynamic force of aerial rudder; X_T is the longitudinal coordinate of the mass center of space vehicle; X_s, X_R are the longitudinal coordinates of all rockets(missiles) body stations and connection location of the aerial rudder respectively; a_X is the longitudinal acceleration; m_K is the mass of aerial rudder; l_K is the distance between mass center of the aerial rudder and corresponding spinning shaft; J_K is the moment of inertia of aerial rudder; N_K is the number of aerial rudders; H_K is the distribution coefficient of aerial rudder; $F_{nn}^K\Phi_i(X_K)$ is the derivative of the generalized aerodynamic force on rudder.

However, when the stiffness of aerial rudder is small, vibration of the rudder under lateral vibration mode exists different patterns as shown in Figure 2. The modal shape $\Phi_i(X_K,Y_K,Z_K)$ and the corresponding slope $\Phi_i'(X_K,Y_K,Z_K)$ at any point on rudder are quite different. When effect of flexibility is considered, rudder generalized aerodynamic force derivative $F_{nn}^K\Phi_i(X_K)$ and generalized mass M_i in Equations (9) and (10) according to overall spacecraft station computation are not suitable any more.

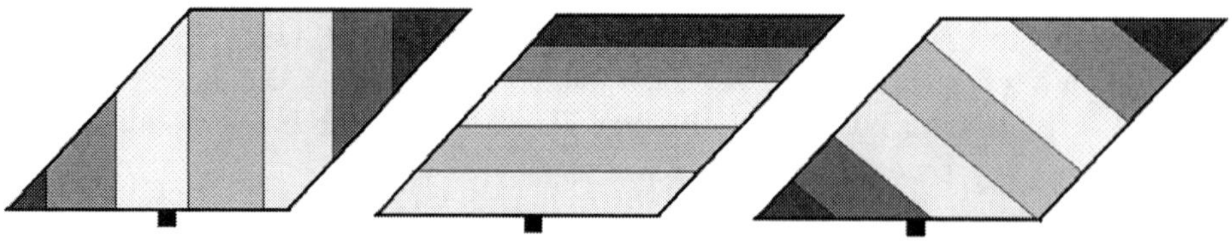

(a) Pattern 1 (b) Pattern 2 (c) Pattern 3

Figure 2. Rudder vibration shape patterns in lateral vibration mode

3. Computation method on generalized force and generalized mass of flexible aerial rudder

According to the expressions of generalized mass in Equation (4) and generalized force in Equation (5), generalized mass of the aerial rudder and generalized aerodynamic force on the rudder can be generated as follows:

$$M_i = \int\int\int \begin{bmatrix} \Phi_{i,X}^2(x,y,z)m(x,y,z) + \\ \Phi_{i,Y}^2(x,y,z)m(x,y,z) + \\ \Phi_{i,Z}^2(x,y,z)m(x,y,z) \end{bmatrix} \mathrm{d}x\mathrm{d}y\mathrm{d}z \tag{11}$$

$$F_i = \int\int\int \begin{bmatrix} \Phi_{i,X}(x,y,z)F_X(x,y,z,t) + \\ \Phi_{i,Y}(x,y,z)F_Y(x,y,z,t) + \\ \Phi_{i,Z}(x,y,z)F_Z(x,y,z,t) \end{bmatrix} \mathrm{d}x\mathrm{d}y\mathrm{d}z \tag{12}$$

Because aerial rudder is a continuous structure, each vibration modal shape is distributed continuously in space, namely, there are equal modal shape perpendicular lines as shown in Figures 3(a)-3(c), which means these modal shapes perpendicular to vertical plane are equal.

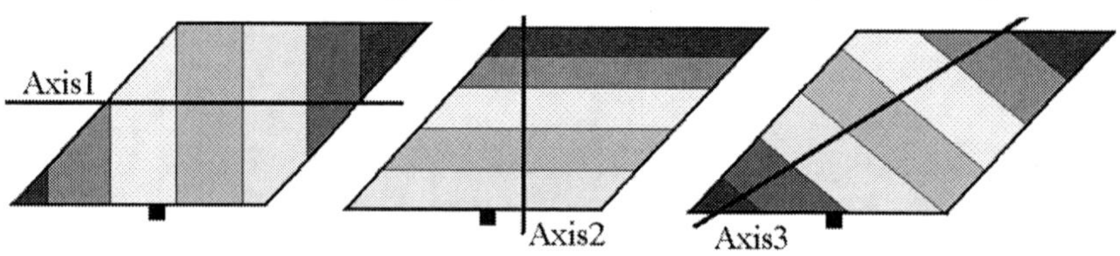

(a) Pattern 1 (b) Pattern 2 (c) Pattern 3

Figure 3. Equal vibration shape axis of rudder vibration shape patterns under lateral vibration mode of spacecraft

When the vertical direction of equal vibration shape is considered as the station division direction of flexible rudder, the rudder can be divided into s stations, then the station mass $m^{(j)}$ and aerodynamic force $F^{(j)}, j = 1,2,\dots,s$ can be calculated. The generalized mas and the generalized aerodynamic force can be defined as follows.

$$M_i^* = \sum_{j=1}^{s}\left[\left(\Phi_{i,X}^{(j)}\right)^2 m^{(j)} + \left(\Phi_{i,Y}^{(j)}\right)^2 m^{(j)} + \left(\Phi_{i,Z}^{(j)}\right)^2 m^{(j)}\right] \tag{13}$$

$$F_i^* = \sum_{j=1}^{s}\left[\Phi_{i,X}^{(j)}F_X^{(j)} + \Phi_{i,Y}^{(j)}F_Y^{(j)} + \Phi_{i,Z}^{(j)}F_Z^{(j)}\right] \tag{14}$$

When vibration shape pattern in Figure 3(a) is taken as an example, the rudder vibration shape $\Phi_i(x,y,z)$ varies along X axis, while has no obvious change along the other two axis. Therefore, the vibration shape of the rudder can be simplified as $\Phi_i(x,y,z) = \Phi_i(x)$. When this simplification is substituted to Eq.(11) and Eq.(12), and the following can be derived,

$$M_i = \int\{m(x)[\Phi_{i,X}^2(x) + \Phi_{i,Z}^2(x) + \Phi_{i,Y}^2(x)]\}\mathrm{d}x \tag{15}$$

$$F_i = \int[\Phi_{i,X}(x)F_X(x) + \Phi_{i,Y}(x)F_Y(x) + \Phi_{i,Z}(x)F_Z(x)]\mathrm{d}x \tag{16}$$

where
$m(x) = \int\int m(x,y,z)\mathrm{d}y\mathrm{d}z$, $F_X(x) = \int\int F_X(x,y,z)\mathrm{d}y\mathrm{d}z$, $F_Y(x) = \int\int F_Y(x,y,z)\mathrm{d}y\mathrm{d}z$ and $F_Z(x) = \int\int F_Z(x,y,z)\mathrm{d}y\mathrm{d}z$.

When the stations of aerial rudder are divided densely, the mass, vibration shape and derivatives of aerodynamic force among successive stations will vary moderately and linearly. It is assumed that the aerodynamic force is $F_X(x) = Ax + \mathrm{B}$, vibration shape is $\Phi_{i,X}(x) = Cx + \mathrm{D}$, distance between two stations is L, left and right station masses are m_L and m_R respectively, as a result, generalized mass and generalized force can be expressed by

$$M_{i,X} = \int_0^L (m_\mathrm{L} + (m_\mathrm{R} - m_\mathrm{L})/L \cdot x)/L \cdot (Cx + D)^2\mathrm{d}x \tag{17}$$

$$F_{i,X} = \int_0^\mathrm{L} (Ax + B)(Cx + D)\mathrm{d}x \tag{18}$$

Similarly, the similar results could be obtained in Y and Z directions.

When stations are divided in the direction of equal vibration shape vertical line, generalized mass and generalized force can be expressed by

$$M_{i,X}^* = m_\mathrm{L}D^2 + m_\mathrm{R}(CL + D)^2 \tag{19}$$

$$F_{i,X}^* = D\int_0^{L/2}(Ax + B)\mathrm{d}x + (CL + D)\int_{L/2}^{L}(Ax + B)\mathrm{d}x \tag{20}$$

Similarly, the similar results could also be obtained in *Y* and *Z* direction.

It can be proved that when the stations of flexible aerial rudder are divided in the vertical direction of equal vibration shape and detailed enough, the generalized mass and the generalized force with high accuracy of the flexible rudder can be calculated, where $M_{i,X} = M_{i,X}^*$, and $F_{i,X} = F_{i,X}^*$.

4. Simulation case studies

In order to verify the effectiveness of the proposed method, the aerial rudder shown in Figure 4 is taken as an example. The thickness of the aerial rudder is 30mm and the mass is m_d(kg) and distributes uniformly, the other dimensions are shown in Figure 4. Simulation cases are listed in Table 1.

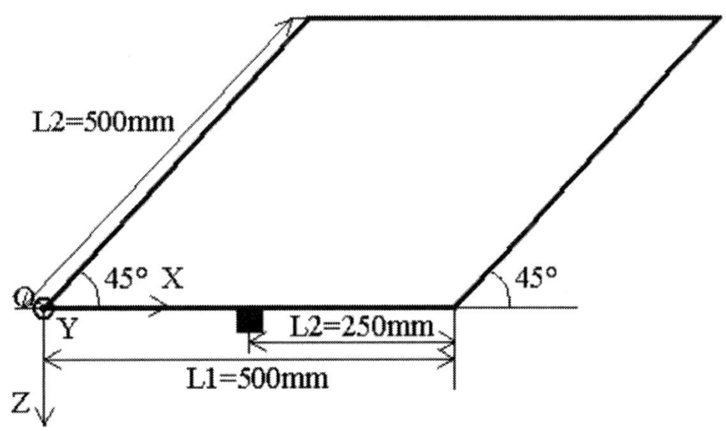

Figure 4. Structural dimensions of an aerial rudder

IOP Conf. Series: Materials Science and Engineering **887** (2020) 012021 doi:10.1088/1757-899X/887/1/012021

Table 1. Simulation cases

Mach Number	height(km)	Flight deviation (°)	Dynamic pressure(N/m^2)
2.5	12	0	84872
2.5	12	1	84872

The aerodynamic pressure distribution of the aerial rudder can be obtained in conditions of rudder deflection 0 degree and 1 degree respectively by using aerodynamic CFD simulation, and the distribution characteristics of its derivatives of aerodynamic force in different directions can be obtained by subtraction, as shown in Figure 5.

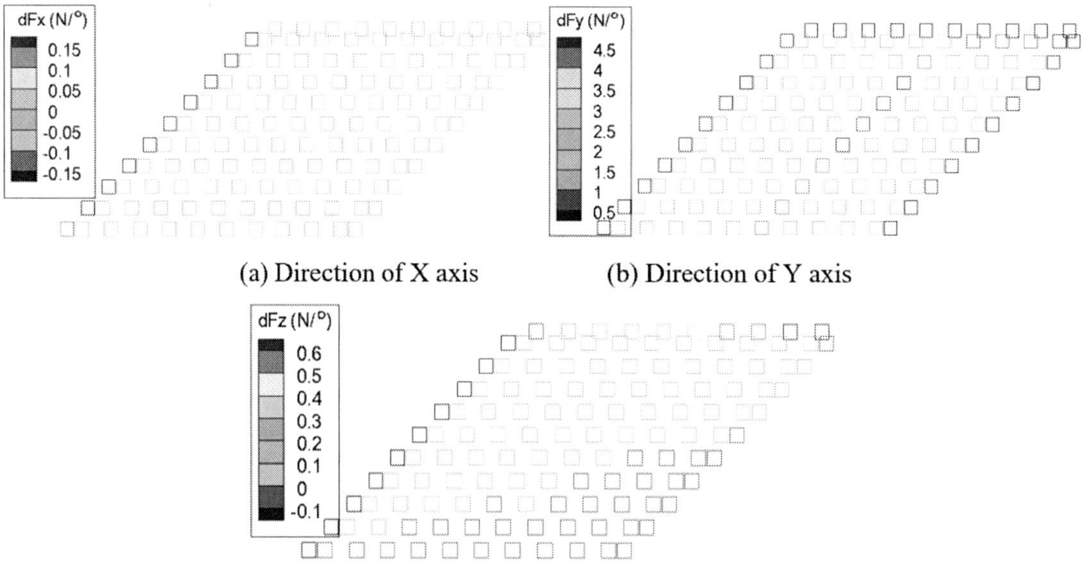

(a) Direction of X axis (b) Direction of Y axis

(c) Direction of Z axis

Figure 5. Distribution of certain aerial rudder grid aerodynamic force derivative

The aerial rudder is connected with spacecraft structure through the rudder shaft. However, there can be many different modes for the rudder ($\Phi_{i,X}(x,y,z)$,$\Phi_{i,Y}(x,y,z)$ and$\Phi_{i,Z}(x,y,z)$) under the modes of pitching, yawing and rolling of the spacecraft. When three modes shown in Figure 2 are considered as examples, the mode functions can be assumed as follows:

(1) pattern1: $\Phi_{i,X}(x,y,z) = 0$, $\Phi_{i,Y}(x,y,z) = -0.4x + 0.1$ and $\Phi_{i,Z}(x,y,z) = 0$;

(2) pattern2: $\Phi_{i,X}(x,y,z) = 0$, $\Phi_{i,Y}(x,y,z) = 0.4z$ and $\Phi_{i,Z}(x,y,z) = 0$;

(3) pattern3: $\Phi_{i,X}(x,y,z) = 0$, $\Phi_{i,Y}(x,y,z) = 0.5x - 0.5z - 0.5$ and $\Phi_{i,Z}(x,y,z) = 0$.

According to the different vibration modes of the aerial rudder, the rudder is divided into 10 stations by using three different station division methods as shown in Fig. 3. The average value of the section vibration mode data of each station is taken, and the corresponding generalized mass and generalized force derivative are calculated, and compared with the definitions in Equations (11) and (12) of the generalized mass and generalized force derivative, the results are listed in Tables 2 and 3.

Table 2. Generalized mass with different station divisions in different vibration patterns (kgmm^2)

Shape pattern	Generalized mass definition	Axis1 station division	Axis2 station division	Axis3 station division
1	0.012m_d	0.014m_d	0.031m_d	0.022m_d
2	0.0066m_d	0.0031m_d	0.0070m_d	0.0042m_d
3	0.055m_d	0.042m_d	0.062m_d	0.052m_d

Table 3. Generalized force derivative with different station divisions in different vibration patterns (N)

Shape pattern	Generalized force derivative definition	Axis1 station division	Axis2 station division	Axis3 station division
1	-25.69	-24.21	-41.22	-33.75
2	-26.75	-10.92	-27.11	-17.85
3	-82.19	-65.26	-97.23	-83.21

It can be seen from Table 3 that for the flexible aerial rudder with obvious local modes, the equal modal shape vertical direction (Axis1 station in pattern 1, Axis2 station in pattern 2, and Axis3 station in pattern 3) should be taken as the direction of station division, and then the generalized mass, generalized force derivative can be calculated which holds higher engineering calculation accuracy.

5. Conclusions

In this paper, influence of the flexible rudder used in the control system design of new generation spacecraft is analysed and the elastic motion equation is studied. Rudder station division and computation method of generalized aerodynamic force are proposed according to the modal vibration features, derivative distribution characteristics of aerodynamic force and generalized force expressions of modal shape so as to determine generalized mass and force of flexible aerial rudder. The research results can provide an important theoretical basis for the wide application of thin and light rudder with large span-chord ratio, and it is also an important technical basis for the realization of high precision and high reliability control of flexible rudder.

References

[1] Pan Z W. The dynamical modeling and modal slope indicate technique of launch vehicle [J]. *Science in China, Series E: Technological Sciences*, 2009, **39**(3): 469—473.
[2] Zhang W D, Liu Y X, Liu H B, Ding X F, Zhang K B. Development trend and prospect of attitude control technologies of launch vehicle. *Aerospace Control*, 2017, **3**(35): 85-89.
[3] Wu Y S, He L S. Attitude control technology of new generation launch vehicles [J]. *Journal of Beijing University of Aeronautics and Astronautics,* 2009, **35**(11): 1294-1297.
[4] Wei C Z, Ju X Z, Zhang L, Cui N G. The modeling and controller design for aeroelastic rocket launcher [J]. *Astronautical Systems Engineering Technology*, 2017, **1**(1): 21-26.
[5] Yin Y Y. Loads design foundation for solid rockets [B]. *China Aerospace Publishing House*, 2007.
[6] Deng W Y. Research on the methods of bending modes data for launch vehicle [J]. *Aerospace Control*, 2010, **28**(2): 7-11.
[7] Feng X X, Xia G Q, Han X L, Han X M. Analysis of stability for launch vehicle attitude control system [J]. *Journal of Dalian University of Technology*, 2015, **5**(55): 542-547.

Research on the Operating Mechanicals of the Helicopter Robot Pilot

Zibo Jin, Daochun Li* **and Ziyu Wang**

School of Aeronautic Science and Engineering, Beihang University, 100191, Beijing, China

*Corresponding author: lidc@buaa.edu.cn

Abstract. The requirements for the quantity and quality of pilots are getting more and more demanding. The helicopter robot pilot was put forward as a new autonomous driving scheme which can achieve fast switch between pilot driving and autonomous driving without modifying helicopters. The overall design of the robot was determined and the actuators of the robot was demonstrated in detail according to the handling characteristics of helicopter manipulation mechanism. Based on the detailed design scheme, the robot pilot was processed and assembled. In order to test the control performance of the robot pilot, the robot was installed in the flight simulator to carry out helicopter flight simulations. The test results illustrated that the robot pilot is able to accomplish basic flight control of helicopter.

1. Introduction

With the development of aircraft technology, the number of aircrafts will be further improved, which puts forward higher requirements for the quantity and quality of pilots. It costs extremely high on training a human pilot, while pilots still take many risks when flying missions in extreme environments such as radiation areas and plateau regions. Furthermore, a large number of newly developed and modified aircraft are in the stage of development and test flight. It takes great risks and may cause great losses for human pilots carrying out these flight tests. Advances in unmanned aerial vehicle (UAV) technology can solve the problem of pilot safety to some extent. However, due to the limited development of artificial intelligence, as well as the low carrying load of UAVs, it can be assumed that manned aircraft will remain the majority in the foreseeable future. Meanwhile, the existing technology which converts manned aircrafts into UAVs is irreversible and costly as well.

Based on the above background, the helicopter robot pilot is proposed as a new automatic driving solution. The helicopter robot driver refers to an autonomous robot system, which can be installed in the helicopter cockpit. The actuators of the robot is connected with the helicopter manipulation mechanism, and the robot drives the helicopter through controlling the terminal movement of each actuator. In the future, the helicopter robot driver can not only carry out flight missions in extreme environments or test flights of new aircraft independently, but also assist the human pilot to complete complicated flight missions as the co-pilot, reducing the number of crew members.

Helicopter robot pilot belongs to the emerging field of aviation and robotics. Relevant research has just started and is being actively carried out. The Korea institute of science and technology (KIST) developed PIBOT in 2014[1][2][3]. The name combines the meanings of "Pilot" and "Robot". PIBOT is shaped like a human, including a head, a torso, arms and legs. The robot's head is equipped with a camera, positioned in the same position as a human eye. The arms and legs can operate actions like a

Content from this work may be used under the terms of the Creative Commons Attribution 3.0 licence. Any further distribution of this work must maintain attribution to the author(s) and the title of the work, journal citation and DOI.
Published under licence by IOP Publishing Ltd

human. The first generation of PIBOT was a scaled-down version of the commercially available BioLoid enhanced robot suite (BioLoid Premium) and it carried out simulated flight tests of fixed-wing aircraft. The second generation of PIBOT adopted humanoid joint mechanism to control the arms. The kinematic and dynamic characteristics of the manipulator mechanism were studied and a multi-loop PID control system was used for the robot to control the aircraft in the flight simulator. In 2015, Aurora Flight Science in USA developed the "Alias co-pilot" robot. The Alias pilot consists of a robotic arm, a visual head, and a sensor system that monitors the flight instrument data through a camera. The project grew out of DARPA's early requirements to support autopilot in the crew cabin and was designed to create a closer connection between human pilot and increasingly powerful computers and sensors. The "Alias co-pilot" robot successfully manipulated a small Cessna airplane under the supervision of a human pilot and completed the flight test. In December 2017, the Alias system was installed on UH-1 helicopter and tested for resupply missions.

In this research, a detailed scheme of helicopter robot pilot for typical helicopters was proposed. The position and movement space of manipulation mechanism in the helicopter cabin were measured and simulated. Considering the response speed, control complexity and operation accuracy, the actuators of the helicopter robot pilot were determined. The prototype of the robot pilot was built and installed in the flight simulator to test the actual control performance.

2. Design scheme of the robot pilot

2.1 Helicopter manipulation mechanism

The helicopter robot driver is designed to be mounted in the helicopter cockpit and the actuators should be connected to the helicopter operators. Therefore, it is necessary to accurately measure the position of helicopter manipulation mechanism and establish the movement space of each operator to serve as the design basis for the actuators of helicopter robot pilot.

The helicopter controlled by the robot pilot given in this research is a typical light helicopter SVH4, whose main manipulation mechanisms are a control stick, a total moment throttle and rudders. The position information of operators and seat size in the cabin were measured on the real helicopter and the cockpit environment was established, as shown in Figure 1. The trajectory of the end of the total moment throttle and rudders is in a plane, while the trajectory of the end of the control stick is in a curved surface. Figure 2. shows the motion space at the end of the control stick based on the established cockpit model.

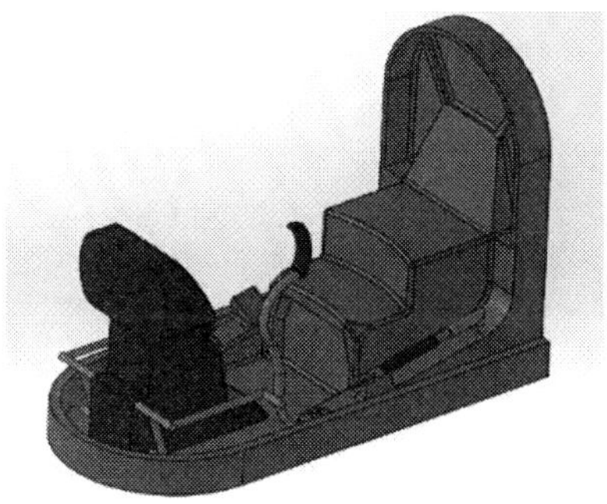

Figure 1. Helicopter cockpit model

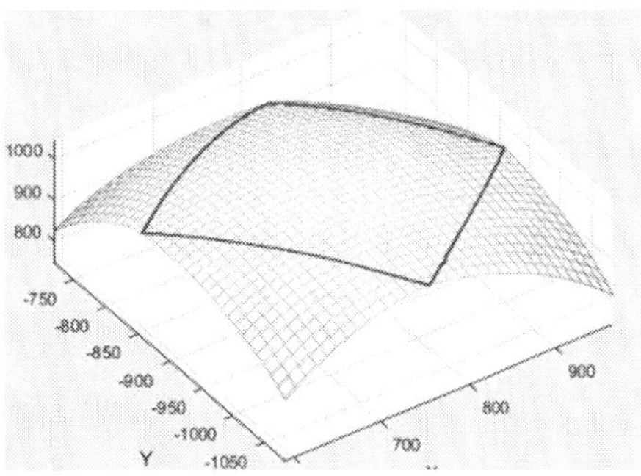

Figure 2. Motion space of the control stick

2.2 Actuator of the control stick

The control stick controls the pitch and roll of the helicopter, and the helicopter needs to adjust its attitude through the control stick at all times during flight. Therefore, the actuator of the control stick needs to have high manoeuvrability and high accuracy. Considering the operation requirement and that the end of the control stick is shifted in three directions, the 3-dof delta parallel robot was adopted as the actuator of the helicopter control stick, as shown in Figure 3.

The delta parallel robot is regarded as a typical 3-dof parallel mechanism in space because of its compact structure, high efficiency and accurate positioning[4][5]. The 3-dof parallel robot is a spatial mechanism that connects the fixed platform and the moving platform with three identical motion chains at an angle of 120 degrees. In each motion chain, there is a parallelogram closed loop composed of four ball hinges and rod groups. One end of the driving arm is connected to the closed loop, and the other end is connected to the fixed platform by a rotating pair. The three groups of parallelogram mechanisms ensure that the fixed platform and the moving platform are always parallel, eliminating the rotational degrees of freedom of the moving platform and preserving three plane degrees of freedom of space.

Figure 3. 3-dof delta parallel robot

2.3 Actuator of the rudders

The helicopter rudders change the pull force of the tail rotor through the coordinated control of the variable pitch of the tail rotor to control the helicopter course. The human pilots control rudders mainly by stepping on legs. Therefore, the actuators of the rudders takes the form of mechanical legs to achieve human-like operations. The length of the mechanical leg should be adjustable to make it suitable for a variety of helicopters. The end speed of the mechanical leg should have a wide range, which can correct course slowly or change course quickly to fly the helicopter.

The mechanical leg is based on the design concept of imitating human leg to design the executive parts, mainly including thigh, hip joint, knee joint, foot pedal, pushrod, encoder and so on. The overall 3d model of the mechanical legs are shown in figure 4. The mechanical leg uses the servo motor as the power source. In the actual working condition, the operation of the helicopter rudder does not need a particularly fast speed and a large torque. Therefore, the speed of the output rotation of the servo motor at the hip joint was slowed down. Because of the small space in the helicopter cockpit, a relatively compact deceleration device should be chosen. In this research, the harmonic reducer was adopted and the dc motor was selected.

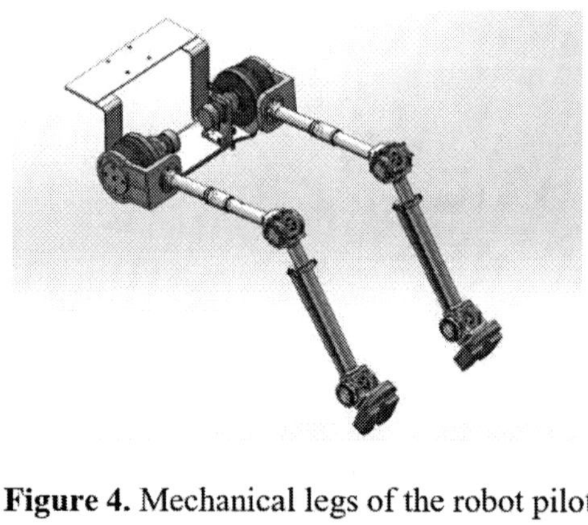

Figure 4. Mechanical legs of the robot pilot

Since the two rudders in a helicopter move in tandem, with one side pushing down and the other side pushing up, the robot was designed to adapt to the movement of the rudders. An electromagnetic clutch was installed between the motor and the control thigh to solve this problem. When one side of the mechanical leg drops, the other side controls the break of the clutch to realize the adaptive motion of the rudder. The rotation of the foot pedal was achieved by a push motor fixed to the pushrod. The extension of the pushrod corresponds to the raise of the pedal.

2.4 Actuator of the total moment throttle

The total moment throttle is arranged on the left side of the driver's seat and rotates along the support. When raising the total moment throttle, the automatic tiller rises and increases the total pitch of the rotor blade to increase the tension of the rotor, so as to control the lifting movement of the helicopter. Therefore, the actuator of the total moment throttle was designed as a mechanical arm to realize the lifting function similar to that of the human arm to complete corresponding action.

The total moment throttle only needs to perform a pull up and down action, it has one degree of freedom to operate the arm. The mechanical arm mainly consists of the upper arm, the forearm and the end-effector of the forearm. The upper arm is responsible for the structural support of the manipulator, while the forearm is driven by the dc reduction motor. Through the positive and negative rotation of the motor, the pulling and pressing of the end of the mechanical arm can be obtained.

In order to make the control volume of the mechanical arm have more margin to adapt to different types of helicopters, the upper arm was designed as an angle-adjustable structure. When installing the robot driver, the tilt angle of the upper arm can be adjusted according to the actual position of the total moment throttle. A deceleration motor and a synchronous belt were selected to ensure the stability of the end movement of the mechanical arm and the brushless dc motor was selected.

2.5 Trunk and head structure

The helicopter robot pilot is considered to have image and sensor information acquisition capability to realize remote control function based on visual transmission. For this purpose, the trunk and head of the robot were designed, as shown in figure 5. The trunk of the robot pilot is mainly used to place the power supply, controllers and driver of actuators. Therefore, the trunk part adopted a hollow design and was provided with an external power jack for power inlet and cut-off. At the bottom of the trunk, there was a navigation connector to facilitate the insertion and unplugging of cables. The head of the robot driver is mounted above the trunk and driven by two motors to achieve left-right rotation and up-down pitching. Two sets of cameras were installed inside the head, which can monitor the helicopter indicator panel images and output to the ground station.

The mechanical structure of helicopter robot pilot can be obtained by integrating the above actuators and robot trunk into the SVH4 cockpit, as shown in Figure 5.

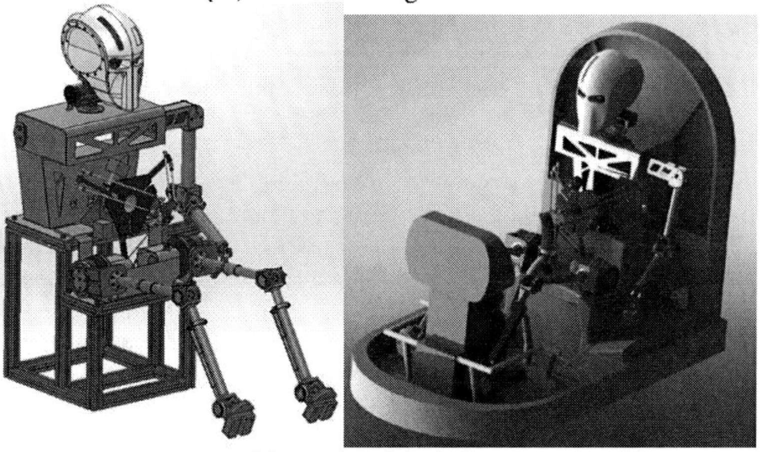

Figure 5. Assembly model of the helicopter robot driver

3. Manufacture and flight simulation test of the robot pilot

3.1 Manufacture of the helicopter robot pilot

Through the above dynamic analysis, the preliminary dynamic performance and manoeuvring performance of the helicopter robot pilot have been studied. Based on the analysis results, the selection of motors and sensors was determined. The equipment selection, detailed design of the mechanical system and control system of the robot pilot was studied. The parallel robot, mechanical arm and the mechanical legs were processed by aluminium alloy, the robot trunk was processed by 1mm stainless steel, and the robot head was processed by 3D printing. On the basis of the parts processing and the arrangement of electronic components, the assembly and testing of each part were conducted and the lines were connected according to the designed wiring scheme, so as to complete the whole prototype assembly of helicopter robot pilot, as shown in Figure 6. The prototype helicopter robot pilot has compact structure and complete functions. The robot pilot was also installed in the helicopter cockpit to test the control performance of the actuators of the robot, as shown in Figure 7. It has been confirmed that the actuators of the robot pilot is well connected with the helicopter manipulation mechanism and can complete quick and sensitive motion control.

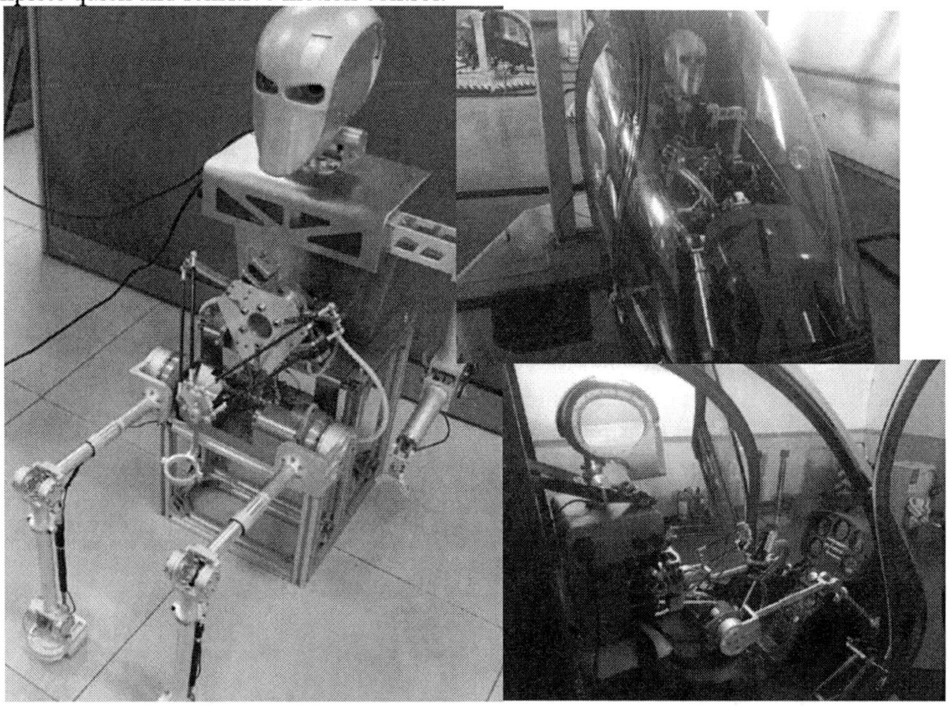

Figure 6. Manufacture of the helicopter robot pilot and the installing test

3.2 Flight simulation test of the helicopter robot pilot

In order to verify the control performance of the robot pilot, the robot pilot was installed in the flight simulator to conduct flight simulation test. The flight simulator adopted the X-plane flight simulation platform. The robot pilot fly the helicopter through the joystick that connected the computer and output signals to X-plane. The joysticks were fixed with the actuator terminals of the robot pilot, as shown in Figure 7.

The flight test mainly simulates the helicopter hovering, rolling, yawing and landing performance. During the test, the mechanical arm was executed through the upper control software to perform the pull up action on the total moment throttle and then it can be observed that the helicopter slowly raised in the computer screen. When the helicopter climbed to a certain height, the helicopter pitching and rolling action can be realized through controlling the joystick forward, backward, left and right through the parallel robot. At the same time, the helicopter can also yaw by controlling the mechanical leg. After the simulation of all these actions, the total moment throttle was forced to pull down gradually and the helicopter slowly landed to the ground, so as to complete the flight simulation test.

Figure 7. Flight simulation of the helicopter robot pilot

4. Conclusions

In this research, the helicopter robot pilot was put forward as a brand new autonomous driving scheme based on the thought of modularity, which can achieve fast switch between the pilot driving and autonomous driving without modifying the helicopter. The measurement of the helicopter cockpit was used to establish the working space for the helicopter robot pilot. Then, the overall scheme of the helicopter robot pilot was discussed and designed. The actuators of the robot to control the helicopter manipulation mechanism were investigated in detail and the mechanical structure and sensor placement of various parts were determined.

Based on the selection of robot equipment, the detailed design of the mechanical system and the control system, the prototype of the robot pilot was processed. Accordingly, a flight simulation platform was built to carry out flight simulations using the robot pilot. It was demonstrated that the robot pilot is able to fly the helicopter in the flight simulator and complete taking off and landing, hovering, rolling and yawing actions.

Reference

[1] Jeong H , Kim J , Shim D H . [American Institute of Aeronautics and Astronautics 52nd Aerospace Sciences Meeting - National Harbor, Maryland] 52nd Aerospace Sciences Meeting - Development of an Optionally Piloted Vehicle using a Humanoid Robot.

[2] H. Jeong, D. H. Shim and S. Cho, "A Robot-Machine Interface for full-functionality automation using a humanoid," 2014 IEEE/RSJ International Conference on Intelligent Robots and Systems, Chicago, IL, 2014, pp. 4044-4049.

[3] Hanjun Song, Heemin Shin, Haram You, Jun Hong, & David Hyunchul Shim. (2016). Toward autonomous aircraft piloting by a humanoid robot: Hardware and control algorithm design. 2016 IEEE/RSJ International Conference on Intelligent Robots and Systems (IROS). IEEE.

[4] Wang Bo, Zhang Dapeng, Li Xiaoman. Accuracy analysis of a 3-dof planar parallel robot. Journal of Jiangsu University of Science & Technology, 2009, 131(1):889-894.

[5] Staicu S . Inverse dynamics of the 3-PRR planar parallel robot. Robotics and Autonomous Systems, 2009, 57(5):556-563.

Experimental Study on Hysteresis Characteristics of Fluorosilicone Rubber Damper

Qiang Yang[1,a] and Yunlai Zhou[2,b]

[1]Aircraft strength Research Institute of China, Aviation Key Laboratory of Science and Technology on Structures Impact Dynamics, Xi'an, China,710065
[2]Faculty of Engineering, Lusofona University, Campo Grande, 376 1749-024 Lisboa – Portugal

[a] yqiang1230@163.com; [b] yunlai.zhou@ulusofona.pt

Abstract. This study tested the hysteresis characteristics of the damper for determining the damping characteristics. Firstly, dampers with four kinds of hardness, such as 20*HA*, 30*HA*, 40*HA* and 50*HA*, were obtained by vulcanization. Then hysteresis loops of the dampers were tested by cyclic loading method with displacement of ±0.5mm, ±1.0mm and ±1.2mm. The relationship between stiffness of dampers and hardness was obtained, and it was found that the stiffness of dampers changed quadratically with rubber hardness. The ellipse method is used to calculate the damping factor of the damper. The damping factor of the damper varies with the change of loading displacement and rubber hardness.

Key words: Fluorosilicone rubber; Hysteresis loop test; Damping factor

1. Introduction

Fluorosilicone, as a polymer damping material, was originally developed by the US Air Force and Dow Corning in 1951. Fluorosilicone rubber is an elastomer with trifluoropropyl side chain introduced into the main chain of methyl vinyl siloxane. This kind of rubber has not only the resilience, high and low temperature resistance and weatherability of silicone rubber, but also the oil resistance and solubility resistance of fluorine rubber [1,2].

Damping behavior of rubber macromolecule material is a process of transforming mechanical energy into heat energy dissipation by relative movement of macromolecule chain in rubber under cyclic loading. This kind of damping behavior results in the deformation hysteresis load in the loading process, which forms hysteresis loop under cyclic loading. The envelope area of hysteresis loop represents the energy consumed by the relative motion of molecular chains. A large amount of investigations were conducted on hysteresis loop features of rubber damping materials, for instance, Hu Zhenxian [3] et al. tested the hysteretic loops of natural rubber connectors in automotive shock absorbers with different amplitudes. It was found that the damping and stiffness characteristics of the shock absorber vary nonlinearly with the loading amplitude. Zhang Lixia [4] carried out loading tests at different frequencies for rubber dampers in bogies of fast freight cars. The effects of loading frequencies on rubber hysteresis loops were obtained, and the dynamic stiffness and loss factors of rubber materials were obtained as a function of the excitation frequencies of dynamic loads. Zeng Cheng [5] et al. acquired hysteresis loops of shock response for JXD rubber isolator with low frequency and large displacement on ship equipment by drop hammer impact testing machine at different impact speed and pulse width, and analyzed the

Content from this work may be used under the terms of the Creative Commons Attribution 3.0 licence. Any further distribution of this work must maintain attribution to the author(s) and the title of the work, journal citation and DOI.
Published under licence by IOP Publishing Ltd

influence of loading parameters on stiffness and loss factor of rubber isolator. Liu Wenwu [6] used simulation analysis method to study the response of rubber products under harmonic load, obtained the hysteresis loop of rubber products, and calculated the dynamic stiffness and loss factor of rubber products. Lei Gang [7] et al. carried out dynamic characteristic test of rubber bushing in automobile by constant amplitude frequency conversion loading method, obtained hysteresis loops of rubber bushing at different frequencies, and then calculated the dynamic stiffness and loss factor of rubber bushing.

J.B. Le Cam [8] studied the hysteresis of natural rubber by experimental method and found that the hysteresis loop of natural rubber is not caused by internal and heat dissipation, but is used to change the microstructure of the material. It is proved that natural rubber can store mechanical energy without converting mechanical energy into thermal energy, which is a realistic method to explain the excellent crack resistance of natural rubber.

T. Bhave [9] et al. studied the effect of nonlinear viscoelastic behavior of automotive synthetic rubber tires on its hysteresis coefficient. The analytical model was first used to predict the hysteresis friction of rubber tires. Then the model results are compared with the test results of rubber tires, and the friction performance evaluation method based on the nonlinear viscoelasticity of rubber tires is obtained.

Lin Song [10] and others tested different temperature, loading frequency and dynamic displacement of butyl rubber viscoelastic damping materials. An M-RT model considering the effects of temperature, frequency and displacement on the dynamic mechanical behavior of materials is proposed.

Fluorosilicone rubber has excellent damping characteristics and environmental resistance, and is suitable for vibration-damping installation of aerospace high-precision electronic equipment. At present, there are few studies on the mechanical properties of fluorosilicone rubber, especially the hysteretic behavior related to material damping. In this paper, for the four hardness fluorosilicone rubber dampers, the material testing machine is used to carry out three kinds of displacement cyclic loading, and the hysteresis loops with different hardness and different loading displacements are obtained. The damping loss factor of fluorosilicone rubber is calculated by elliptic method. The variation of the damping factor of the material with rubber hardness and loading displacement is obtained, which provides technical support for fluorosilicone rubber in vibration isolation installation of aerospace equipment.

2. Fluorosilicone Rubber Damper

The damper structure is illustrated in Figure 1. Four kinds of rubber with hardness of 20 *HA*, 30 *HA*, 40 *HA* and 50 *HA* were obtained through the formulation design of fluorosilicone rubber. According to the structure size of the damper, the damper was made by molding method, and the curing mold was designed. The shrinkage of rubber after vulcanization was considered in the design process of the vulcanization mould. The vulcanization process was designed and conducted hereinafter. Fluorosilicone rubber is a kind of resistant high temperature rubber. In this paper, the vulcanization process of fluorosilicone rubber is as follows: the temperature is 160 °C, the pressure is 20 MPa, and the time of vulcanization is 15 min. The second stage vulcanization temperature is 200 °C and vulcanization time is 4 h. According to the vulcanization process above, the specimens of fluorosilicone rubber damper was made. Four damper specimens were made from each hardness rubber. The corresponding specimen numbers of 20 *HA*, 30 *HA*, 40 *HA* and 50 *HA* were 11#-14#, 21#-24#, 31#-34#, 41#-44#, respectively, as shown in Figure 2.

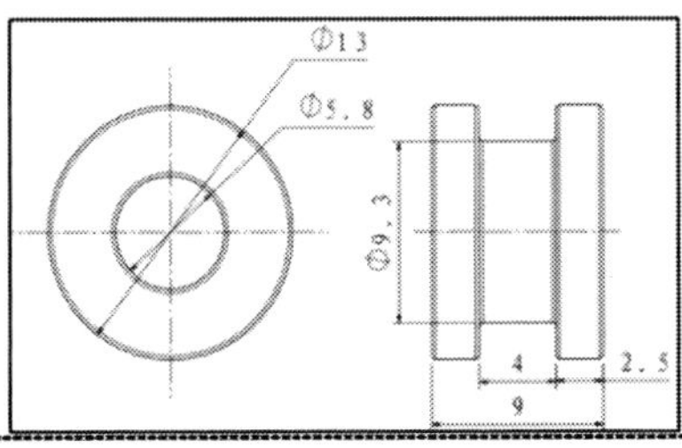

Figure 1. Shape and dimension of damper structure.

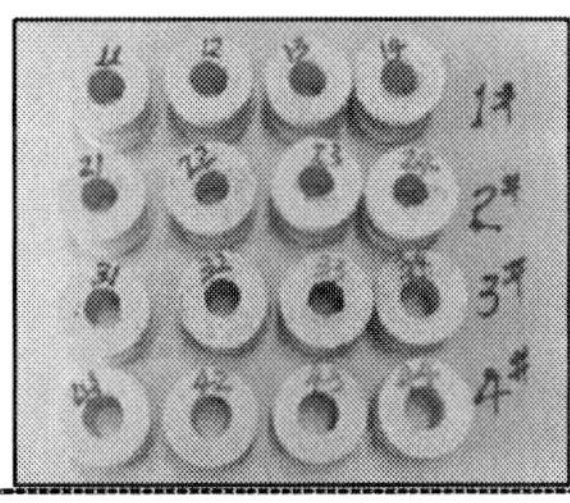

Figure 2. Specimens of damper.

3. Hysteresis Characteristic Test

The hysteresis characteristic test of the fluorosilicone damper was carried out by WDW-500 material testing machine, and the displacement control method was used to load at a speed of 1 mm/min. The stiffness and hysteresis characteristics of the damper were tested, respectively. The test system of the hysteresis characteristic of damper was established which shows in Figure 3. In the stiffness characteristic test, in order to avoid damaged the damper, the compression dimension of each sample was controlled to be 0.5 mm. In the hysteresis characteristic test, each sample was loaded with ±0.5mm, ±1.0mm and ±1.2mm, respectively. At least three hysteresis loops for complete cycles are guaranteed. The matrix of hysteresis characteristic is shown in Table 1.

Figure 3. Hysteresis characteristic test system of damper.

Table 1. Test matrix of hysteresis characteristic.

Rubber No.	Sample No.	Rubber hardness (*HA*)	Load displacement (mm)	Loading speed ($mm \cdot min^{-1}$)
1#	11#-14#	20	0.5/1.0/1.2	1
2#	21#-24#	30		
3#	31#-34#	40		
4#	41#-44#	50		

4. Test Results and Analysis

4.1. Stiffness test results and analysis

To prevent the damage of the damper caused by excessive compression during the stiffness characteristic test, the compression of each sample was 0.5mm. Figures 4, 5, 6, and 7 demonstrate the load-displacement curves of the four hardness dampers. The stiffness of the damper with four kinds of hardness was obtained by fitting the load displacement curve. The stiffness of the damper is 62.6 N/mm for the hardness of 20 *HA*; the stiffness of the damaper is 73.67 N/mm for the hardness of 30 *HA*; the stiffness of the damaper is 118.7 N/mm for the hardness of 40 *HA*; the stiffness of the damaper is 187.6 N/mm for the hardness of 50 *HA*. Therefore, the relationship between the stiffness K_s of the damper and the rubber hardness *HA* is as follows

$$K_s=0.0422(HA)^2+1.5069(HA),\ R^2=0.976 \tag{1}$$

4.2. Hysteresis characteristics test results and analysis

Figure 8 is a load-displacement curve obtained when the 11# specimen was loaded at ±0.5 mm, ±1.0 mm, and ±1.2 mm. When the loading displacement was 0.5mm, the maximum and minimum loads were 32.975N and -21.8N, respectively; when the loading displacement was 1.0mm, the maximum and minimum loads were 75.225N and -50.1N, respectively; when the loading displacement was 1.2mm, the maximum and minimum loads were 84.625N and-67.625N, respectively. The maximum load of the hysteresis loop increases as the loading displacement increases. With the increase of loading displacement, the envelope area of hysteresis loop increases, indicating that the more energy the damper consumes under one cycle loading. The maximum damping force of the damper under three kinds of loading displacement conditions, that is, the value of the intersection of the hysteresis loop and the Y-axis, respectively $F_{D11\text{-}0.5}$=6.125N, $F_{D11\text{-}1.0}$=9.325N, $F_{D11\text{-}1.2}$=9.175N. According to the standard GB/T 15168-2013 named as vibration and shock isolators measuring method for its static and dynamic characteristics, the loss factor η was calculated using the elastic force F_T and the damping force F_D in the hysteresis loop as follows

$$\eta = F_D / F_T \tag{2}$$

The damping loss factors under the three loading displacements can be calculated from equation (2): $\eta_{0.5}$=0.186, $\eta_{1.0}$=0.124, $\eta_{1.2}$=0.108. The damping loss factor of the fluorosilicone damper decreases with the increase of the loading displacement.

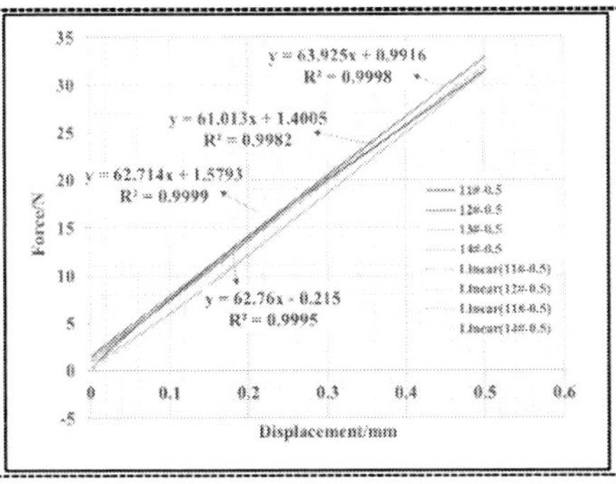

Figure 4. Load-displacement curve with 20*HA*.

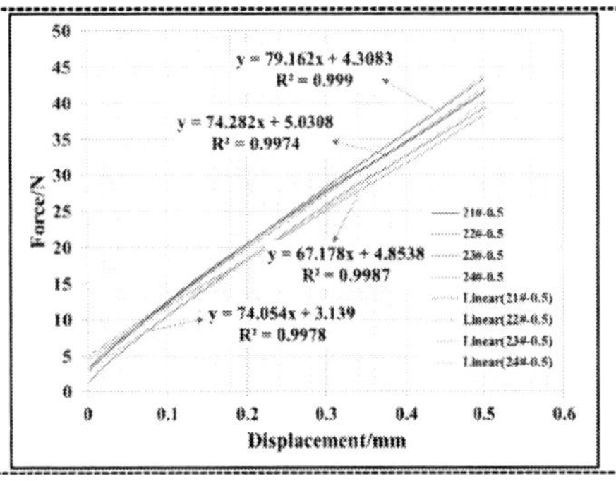

Figure 5. Load-displacement curve with 30*HA*.

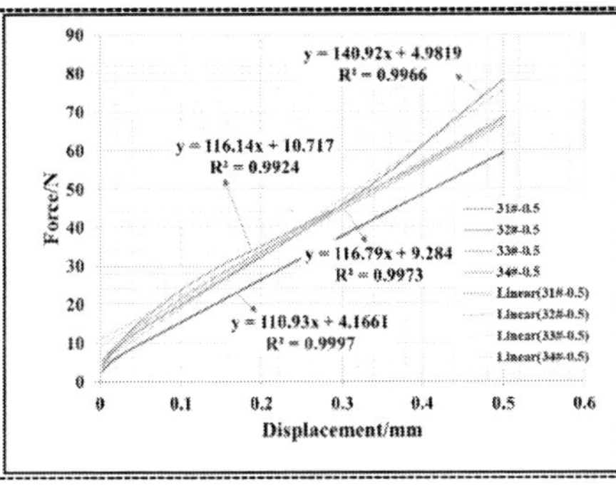

Figure 6. Load-displacement curve with 40*HA*.

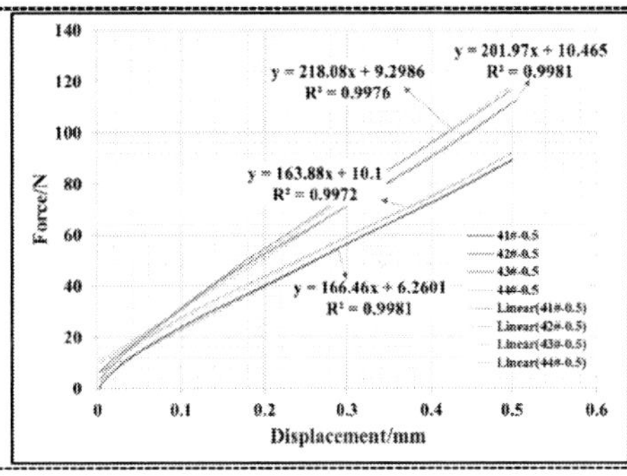

Figure 7. Load-displacement curve with 50*HA*.

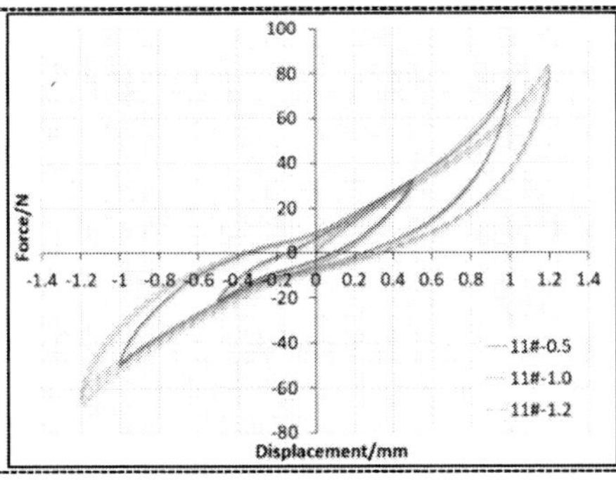

Figure 8. Hysteresis loop of 11 # damper under different loading displacements.

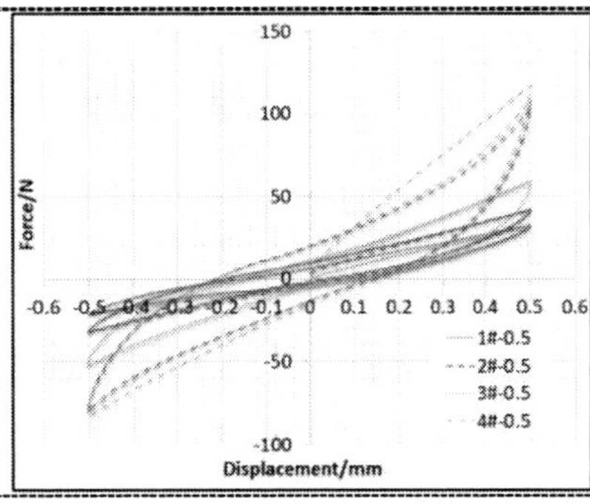

Figure 9. Hysteresis loops of four hardness dampers under ±0.5mm loading displacement.

Figure 9 is the load displacement hysteresis loop obtained by the four hardness dampers under ±0.5mm loading displacement. The maximum load of hysteresis loop increases as the rubber hardness increases; the damping force of the damper increases as the rubber hardness increases; With the increase of rubber hardness, the envelope area of hysteresis loop increases, which indicates that the high hardness damper consumes more energy under one cycle loading. Detailed results can refer to Table 2.

Table 2 Statistical of elastic force, damping force and damping factor of Hysteresis Loop of Dampers

Sample No.	$F_{T0.5}$(N)	$F_{T1.0}$(N)	$F_{T1.2}$(N)	$F_{D0.5}$(N)	$F_{D1.0}$(N)	$F_{D1.2}$(N)	$\eta_{0.5}$	$\eta_{1.0}$	$\eta_{1.2}$
11#	32.975	75.225	84.625	6.125	9.325	9.175	0.186	0.124	0.108
12#	32.250	60.175	70.725	4.625	11.138	12.600	0.143	0.185	0.178
13#	32.825	72.100	103.275	6.800	11.725	15.300	0.207	0.163	0.148
14#	31.350	64.775	73.45	4.563	8.263	10.975	0.146	0.128	0.149
1#Avg	32.350	68.069	83.019	5.528	10.113	12.013	0.171	0.149	0.145
21#	40.600	83.725	89.475	9.525	12.850	15.850	0.235	0.153	0.177
22#	43.375	89.125	108.00	9.100	12.275	13.925	0.210	0.138	0.129
23#	38.200	82.450	109.600	10.100	9.638	14.450	0.264	0.117	0.132
24#	38.325	82.625	99.890	8.125	11.800	15.150	0.212	0.143	0.152
2#Avg	40.125	84.481	101.741	9.213	11.641	14.844	0.230	0.138	0.146
31#	55.300	123.675	136.225	11.650	14.275	18.100	0.211	0.115	0.133
32#	73.625	153.375	159.175	21.525	22.425	19.200	0.292	0.146	0.121
33#	64.250	135.575	136.013	14.738	20.925	19.775	0.229	0.154	0.145
34#	66.013	131.100	182.792	15.400	16.600	21.550	0.233	0.127	0.118
3#Avg	64.797	135.931	153.551	15.828	18.556	19.656	0.244	0.137	0.128
41#	84.000	214.975	289.975	14.388	17.850	22.000	0.171	0.083	0.076
42#	106.625	241.125	319.100	17.363	14.175	17.750	0.163	0.059	0.056
43#	113.225	225.975	254.625	20.600	22.400	23.875	0.182	0.099	0.094
44#	86.975	83.800	189.15	15.825	10.425	17.150	0.182	0.124	0.091
4#Avg	97.706	191.469	263.213	17.044	16.213	20.194	0.174	0.085	0.077

According to GB/T 15168-2013 standard, the maximum elastic force and corresponding damping force of each specimen under three loading displacements were calculated. The damping loss factors of the damper were determined via Equation (2). The results can refer to Table 2. The average values of the test results of four specimens under each hardness were used to express the maximum elastic and damping forces of the damper under this hardness, and their loss factors were calculated. Taking the hardness of the damper as the transverse axis and the calculated loss factor as the longitudinal axis, the hardness-loss factor curves of the damper under different loading displacements were drawn, as shown in Figure 10. When the loading displacement was 0.5 mm, as the hardness of the rubber increases, the damping loss factor of the damper first increases and then decreases; when the loading displacement was 1.0 mm, as the hardness of the rubber increases, the damping loss factor of the damper decreases, but the middle section decreases very slowly; when the loading displacement was 1.2 mm, as the rubber hardness increases, the damping loss factor of the damper increases slightly and then decreases rapidly. However, when the three curves were fitted by quadratic polynomials, the quadratic coefficients of the fitted curves were all negative, indicating that the damping loss factor of the rubber has a maximum value in a certain range of hardness, which can guide the design of the damper. Under the same hardness, the damping loss factor at 0.5 mm loading displacement is greater than the loss factor of 1.0 mm and 1.2 mm, while the loss factors of 1.0 mm and 1.2 mm loading displacement are not much different. The fitting curves are also very close. It shows that the damper should be designed to control the maximum amount of deformation within the appropriate range to obtain the best damping loss factor.

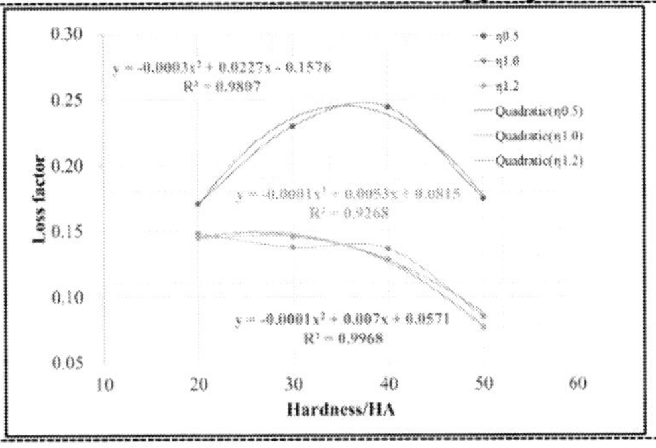

Figure 10. Curve of damping factor of damper with rubber hardness and loading displacement.

5. Conclusion

In this paper, the stiffness and hysteresis characteristics of fluorosilicone rubber under different hardness are studied. The relation between the stiffness of the damper and the hardness of the rubber, and the relation between the loss factor variation of the damper and the rubber hardness, and loading displacement are determined. The conclusions are as follows

(1) The stiffness Ks of the damper increases quadratically with the increase of rubber hardness *HA*; the maximum load of the damper also increases with the increase of rubber hardness;

(2) The single loading loop energy consumption of the damper increases as the hardness of the rubber increases;

(3) The damping loss factor of the damper increases first and then decreases with the increase of the rubber hardness, and the loss factor of the damper has a maximum value within a certain hardness range;

(4) The damping loss factor of damper decreases with the increase of loading displacement, but it remains unchanged when the loading displacement is large, indicating that the optimum damping loss factor can be obtained only when the maximum deformation is controlled in a proper range in the design of cushion.

References

[1] FLITNEY B. Extending the application of fluorosilicone elastomers[J]. *Sealing Technology*, 2005, (**2**):6-11

[2] Ye Xiwang, Guo Jianhua, Zeng Xingrong. Research progress of fluorosilicone rubber[J]. *China Elastomerics*, 2017, **27(2)**:60-66

[3] Hu Zhenxian, Gu Liang. Shocker Mount Dynamic test and property analysis[J]. *Transactions of Beijing Institute of Technology*, 2010, **30(4)**:400-414

[4] Zhang Lixia. Static and dynamic analysis of rubber isolators on rapid wagon[D]. Chengdu: Southwest Jiaotong University, 2013

[5] Zeng Cheng, Hua Hongxing. Study on shock response characteristics of nonlinear rubber isolator[J]. *Noise and Vibration Control*, 2012, **4**:20-24

[6] Liu Wenwu, Weng Xuetao, Lou Jingjun, et al. Study on the harmonic response method to analyse the dynamic characteristics of rubber products based on ANSYS[J]. *Journal of Wuhan University of Technology (Transportation Science&Engineering)*, 2010, **34(5)**: 966-968

[7] Lei Gang, Liu Ying, Hu Peng, et al. A study on dynamic characteristics of rubber bushing[J]. *Science Technology and Engineering*, 2015, **15(7)**:74-78

[8] J.B. Le Cam. Energy storage due to strain-induced crystallization in natural rubber: The physical origin of the mechanical hysteresis[J]. *Polymer*, 2017, **127(3)**:166-173

[9] T. Bhave, M. Tehrani, M. Ali, et al. Hysteresis friction and nonlinear viscoelasticity of rubber composites[J]. *Composites Communications*, 2018, **9**:92–97

[10] Lin Song, Gao Qing. Dynamic Constitutive study for visoelastic damping material[J]. *Atomic Energy Science and Technology*, 2008, **42S**:584-587

IOP Conf. Series: Materials Science and Engineering **887** (2020) 012024 doi:10.1088/1757-899X/887/1/012024

Optimization Design of an Aircraft Wing Structure based on Response Surface Method

A A G Hanif[1], H S Li[1], M A Raza[2], M Kamran[2]and M Abdullah[2]

[1] College of Aerospace Engineering, Nanjing University of Aeronautics and Astronautics, Nanjing 210016, China.
[2] Centres of Excellence in Science & Applied Technologies (CESAT), No. 2801, Islamabad, Pakistan.

E-mail address: ahmad@outlook.com

Abstract. Wing is the most critical part of an aircraft whose structural weight holds prime importance. It is always desirable to reduce the weight of an aircraft. In this work, a detailed analysis of wing is carried out and results pertaining to stress distribution are recorded. Commercial software ANSYS is used to perform simulations related to stress, strain and displacement distributions. Moreover, the optimization design of an aircraft wing was conducted using design of experiments (DOE) and response surface method (RSM) technique for the reduction of computational cost. The aircraft wing is efficiently optimized by changing thickness under the same loading and flight conditions. Results show that this optimization technique is useful to reduce stresses as well as overall weight of the structure with minor variation in thickness of components.

1. Introduction

An aircraft wing is mainly used to produce required lift while drag force resists the forward motion. Therefore, wing is the most critical part of an aircraft and its failure is unaffordable. The wing structuresof aircraft are generally made up of alloys and metals, thus ensuring a cantilever design where no exterior support is necessary. Mostly the skin of wings also serves as a component of wing frame thus allowing it to bear flight loads and generated stresses in flight. The wing consists of ribs, spars and stringers. The ribs are located chord wise while the spars are positioned span wise. Spars are key structural members of the aircraft wing also known "beams".

Usually, the wing consists of front and rear spars alongside the span. The two spars used in the structure of wing are identified as "two-spar" construction. Spar is very crucial and vital structural component of the wing which starts inside the fuselage up to the wing tip. The spar is mainly responsible to cater for the bending strength in the wing. Ribs are positioned crossways in between the spars, which allow maintaining the airfoil shape of the wing. First of all the flight load is experienced by the wing skin which is shifted to the ribs and then the ribs transfer some load to the spars. In this way the loads are distributed among each structural component.

It is observed that the analysis can be improved or refined by including ribs and spars in geometric model as performed by Kavya and Reddy [1].

The analysis of aircraft wing is a versatile, challenging and time demanding process. Optimization of wing structure is one approach to cope up with this intricacy which can be successfully attained by numerical method technique such as finite element methods. Numerical analysis is conducted in a

Content from this work may be used under the terms of the Creative Commons Attribution 3.0 licence. Any further distribution of this work must maintain attribution to the author(s) and the title of the work, journal citation and DOI.
Published under licence by IOP Publishing Ltd

commercially available tool named as ANSYS, as used by Sureka and Meher [2] in their study regarding A300 aircraft wing consisting of ribs and spars. Standard airfoil NACA 64215 was modelled for analysis. Senthilkumar et.al [3] also conducted structural analysis in ANSYS. FE model of the problem used shell elements. Kumar et.al [4] studied stress distribution, using shell elements for surface modelling. They also performed optimization technique on FE model.

To maximize or minimize a required objective function in the presence of existing constraints is termed as an optimization procedure [5]. Mazhar and Khan [6] optimized stiffness and strength of UAV wing using FEA ANSYS software. They used Artificial Neural Networks approach to decide the best possible arrangement through optimum stiffness and minimized cost and weight.

As far as optimization is concerned, it has always been a quest for designers to improve any desired parameter but not at the cost of other aspects. In 1951, Box and Wilson introduced RSM for the first time. It is characterized as a scheme to build global estimates to the system performance based on outcome considered at diversified points in the design space [7]. RSM reduces the computational costs acquired in solving such scenarios. Structural optimization using ANSYS and RSM was performed by Ajay Kumar Menon [8]. The stress reduction was successfully achieved in his work.

In this study we used ribs and spars for geometric modelling to emulate near practical situation. Shell elements were employed in FE model. Boundary and loading conditions are applied to perform static structural analysis. RSM is subsequently used to optimize the maximum stress value over the wing.

2. The Aircraft Wing Model

In this paper, we used, twin-spar wing geometry consisting of ribs and skin at top and bottom surface of the wing. In order to select the material structural steel, stainless steel and aluminium alloy were considered as shown in Table 1.

Table 1. Material selection for the wing.

Material	Density (kg/m^3)	Young's Modulus (MPa)	Poisson's Ratio
Structural steel	7.85 x 10^3	2.00 x 10^5	0.30
Stainless steel	7.75 x 10^3	1.93 x 10^5	0.31
Aluminium alloy	2.77 x 10^3	7.10 x 10^4	0.33

The size of aircraft wing corresponds to that of Boeing 767, a twin-aisle commercial aircraft. A supercritical airfoil Boeing BACXXX, was selected for this analysis, as used by other researchers too. [9]-[11] The dimensions of this aerofoil were recorded from the UIUC database (University of Illinois at Urbana-Champaign). The normalized dimensions are shown in Figure 1. [12] The waterline to span for the top and bottom surface of wing is given in Figure 2. In addition, the wing model design calculations are in agreement with the method, as recommended by Torenbeek. [13]

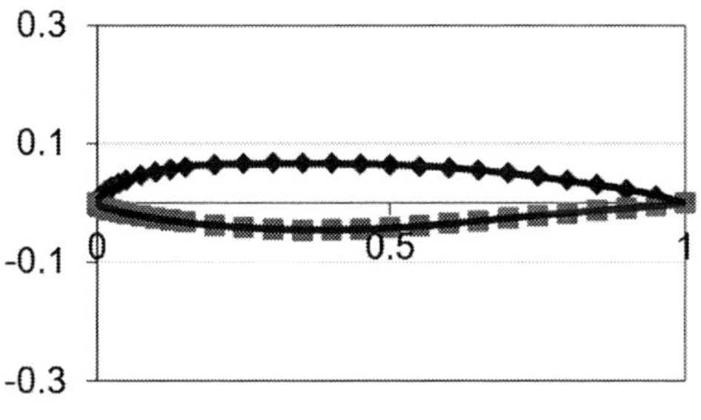

Figure 1. Supercritical Airfoil BACXXX (Normalized Coordinates).

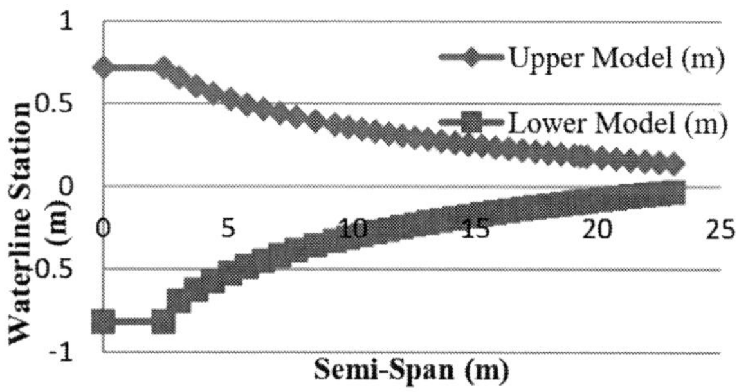

Figure 2. Waterline to span for the top and bottom surfaces of aircraft wing.

The ".igs" file of 3D model created in CAD software is then imported in Ansys 15.0 software. The geometry was "cleaned up" and load condition was applied. 1G-load was considered as a load (self-weight of the wing). FE model is then imposed by a boundary condition as side constraint, thus cantilevering the wing. The meshing was performed using shell elements. In total, there are 3902 shell elements and 10676 nodes. The weight of current wing model is 6266 kg. The FE model is prepared to perform static structural analysis to calculate stresses and deformation produced as shown in Figure 3 below.

Static structural analysis was performed to calculate the stress, strain and the displacement in the wing with all the three suggested materials. The results of all the three materials are tabulated in Table 2. Hence, for the purpose of this study, we opted Aluminium alloy, as the material of the wing structure.

Table 2. Static structural analysis.

Material	Max. Stress (Pa)	Total Deformation (mm)	Total Weight(kg)
Structural steel	84.91×10^6	67	6266.04
Stainless steel	82.57×10^6	65	5612.26
Aluminium alloy	30.20×10^6	65	1411.30

3. Optimization Design Of Aircraft Wing

The finite element process is a mathematical method for the analysis of structures. [14] After obtaining the results from structural analysis, DOE and RSM are used for the reduction of computational costs. In the field of design optimization, widespread research is being conducted using simulation tool of FEA (finite element analysis). Nevertheless, structural optimization usually engages costly function evaluations. For example a simulation of a traveller automobile for crash acquires around 27 hours with an probable computational cost of about US Dollars 5,200/- [15]. As a result, alternate technique of function evaluations; for instance, DOE and RSM are generally engaged in design engineering to minimize the computational costs.

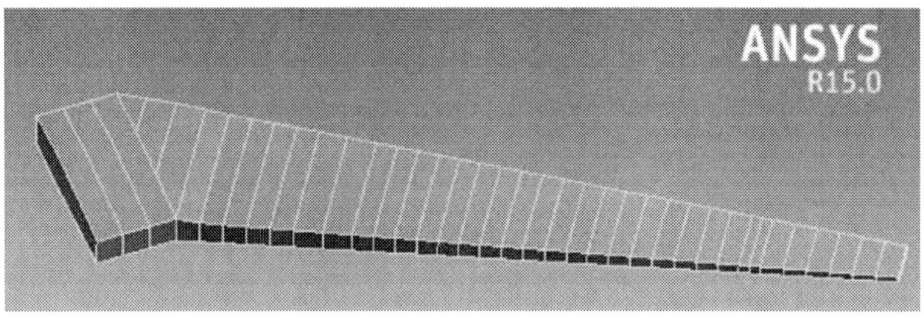

Figure 3. Finite element model of the wing for structural analysis.

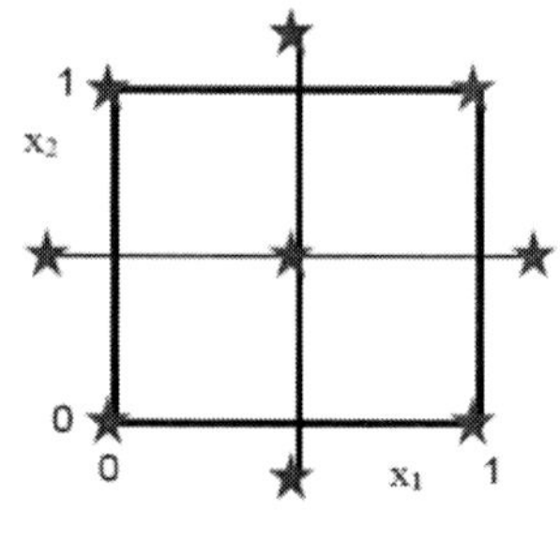

Figure 4. CCD representation for 2 design variables

3.1. Design of experiments (DOE)

The method to select a group of samples in a given design space, with the aim to maximize the amount of information gained from an inadequate number of samples is referred as Design of experiments (DOE) [16]. The objective of DOE study is to approximate and envisage the trend in the response data. In this method, the design variables are considered to be uniformly distributed along the upper and lower limits. The measured response quantity in DOE is expressed as:

$$y_m(x) = y_t(x) + \varepsilon \tag{1}$$

where $y_m(x)$ is the measured response, $y_t(x)$ is the true response and ε represents error term.

Central composite design (CCD) is a DOE technique which has been used in this study. In CCD the numbers of samples are given as 2^n+2n+1. n represents number of design variables, whereas 2^nsamples shows corner points in the design space and $2n$ samples shows the points outside the design space. For a two dimensional CCD problem, the design boundaries are scaled from 0 to +1, 8 out of the 9 samples are lying on or outer surface of the border of the design space, while only the center point, lies inside the design space as presented in Figure 4, above.

3.2. Response Surface Method (RSM)

RSM is a compilation of procedures used in mathematics and statistics which are helpful for various processes such as improving, developing and optimizing. In the industrial world the use of RSM is very common and efficient. In general, the relationship of a method linking a response u, that involves the input parameters ξ_1, ξ_2, … ξ_n is written as follows:

$$u = f(\xi_1, \xi_2, ..., \xi_n) + \varepsilon \tag{2}$$

whereas f is the response function and the term ε represents error(other sources of variability):

$$E(u) \equiv \eta = E[f(\xi_1, \xi_2, ..., \xi_n)] + E[\varepsilon] = f(\xi_1, \xi_2, ..., \xi_n) \tag{3}$$

It is handy to change these variables to the coded variables $\mathbf{v_1, v_2, \ldots v_n}$. The coded variables are usually dimensionless with standard normal distribution and it can be denoted as:

$$\eta = f(v_1, v_2, ..., v_n) \tag{4}$$

Since f is unknown, therefore, an approximation is needed. We often use first or second order polynomial when the small zone of independent variable space is available. The first-order model expressed in the form of coded variables is expressed as:

$$\eta = \beta_o + \beta_1 v_1 + \beta_2 v_2 \tag{5}$$

If there is an interaction between the parameters, the main effects model is used which can be written as:

$$\eta = \beta_o + \beta_1 v_1 + \beta_2 v_2 + \beta_{12} v_1 v_2 \tag{6}$$

Sometimes in order to define the response function, we need the second-order model. The following equation describes interaction between the variables:

$$\eta = \beta_o + \beta_1 v_1 + \beta_2 v_2 + \beta_1 v_{12} + \beta_2 v_{22} + \beta_{12} v_1 v_2 \tag{7}$$

The second-order model is flexible and has the ability to implement the estimation of the β values conveniently [17].

4. Results And Discussion

In this study, the design objective is to curtail the weight of aircraft wing, restricted to stress and displacement constraints. The weight has a pivotal role in the structure of any aircraft. It is a well known fact that that the incremental factor is 4.525 for any weight bearing part; meaning thereby, 1 lbs (pound) decrement in the weight of a structure represents 4.525 lbs reduction in the gross take-off weight of an aircraft [18]. In the presented work, stress limits are 30 MPa in tension and compression, while; allowable z-displacement component within 0.065 m. The thickness of upper and lower skin, thickness of ribs and thickness of spars are considered as design variables. DOE was performed to fix and ascertain the thicknesses of spars and ribs. The optimum thickness of spars and ribs is found 27.3 mm and 8.5 mm respectively. The equivalent stress, equivalent strain and total deformation in the wing are determined using the static structural analysis as shown in Figures. 5, 6 and 7, correspondingly.

IOP Conf. Series: Materials Science and Engineering **887** (2020) 012024 doi:10.1088/1757-899X/887/1/012024

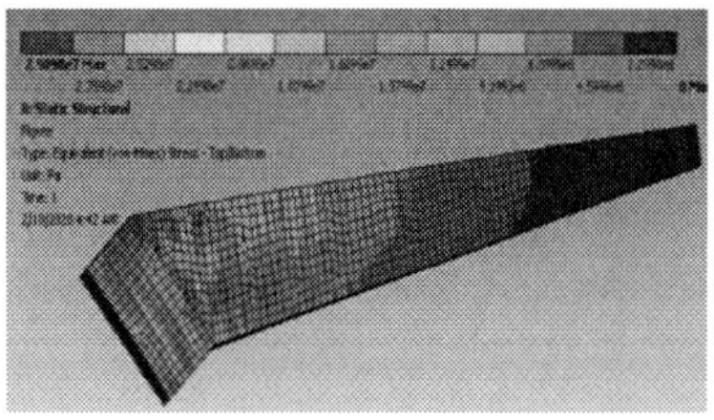

Figure 5. Equivalent stress of the wing model.

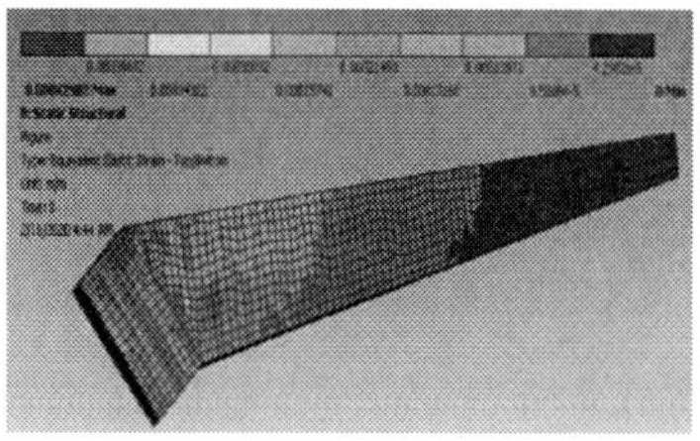

Figure 6. Equivalent elastic strain of the wing model.

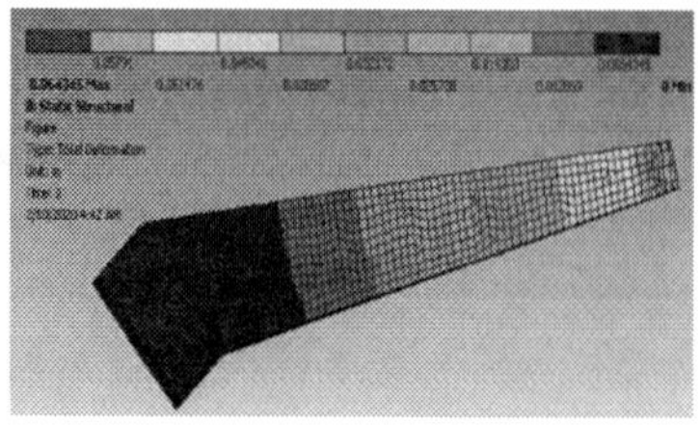

Figure 7. Total deformation of the wing model.

The trade-off chart of response surface optimization is presented in Figure 8; where the feasible area is indicated in green colour. In this technique, the behaviour of stress over the wing in response to thickness variation of wing skin is observed. It reveals that the value of stress can be reduced from 30.2 MPa to 29.7 MPa by optimizing the thickness of wing. Results were optimized so that the maximum stress reduction is achieved with minimum increment in thickness. Figure. 9 depicts the response chart showing the behaviour of maximum stresses against thickness of wing skin. The optimum value of design variable i.e. wing skin is found to be 3.1 mm.

In our work, it is revealed that optimized values of design variables that is; thickness of spar, thickness of rib, upper skin thickness and lower skin thickness for least weight should be 27.3 mm, 8.5 mm, 3.1 mm and 3.1 mm respectively, subjected to the stresses within the allowable limit. It is pragmatic here that with the presented optimization method, the procedure converged to the minimum value of weight = 1031.1 kg having the above quoted optimum values of design variables in altogether eleven iterations. The function convergence plot is presented in Figure. 10.

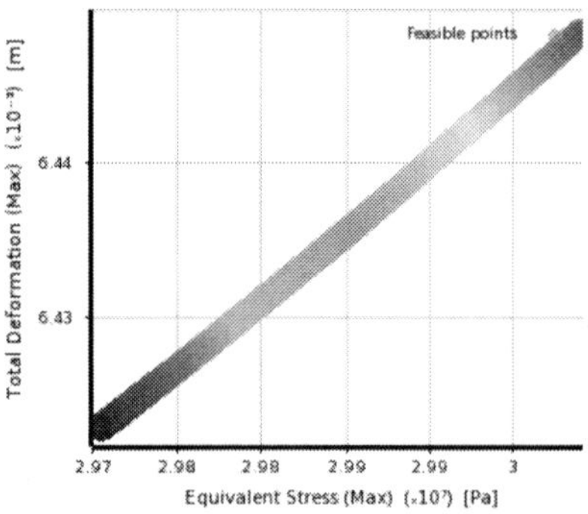

Figure 8. Trade-off chart of response surface optimization.

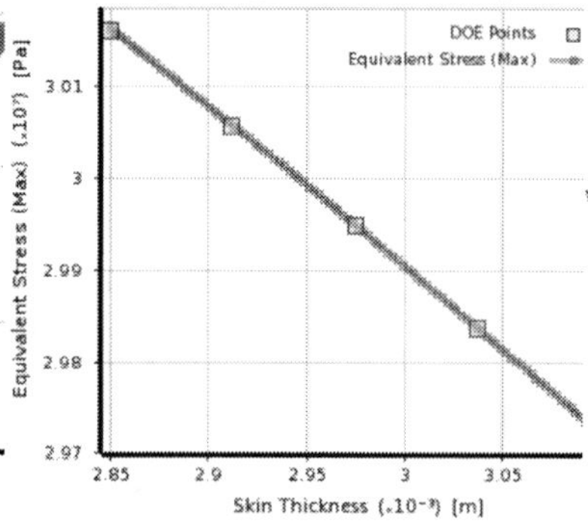

Figure 9. Equivalent maximum stress w.r.t skin thickness.

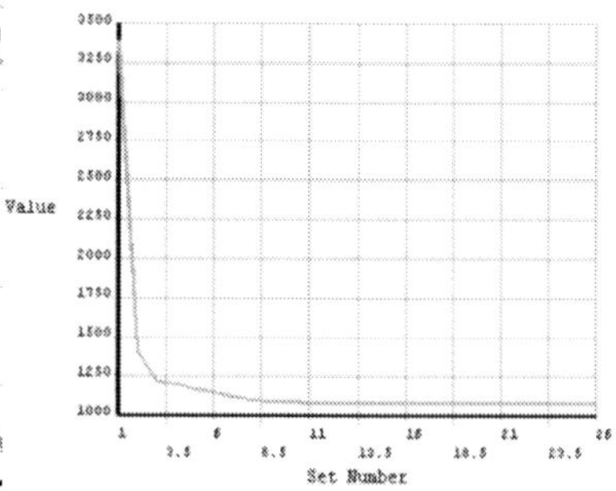

Figure 10. Convergence plot in ANSYS.

5. Conclusion

In this study, an aircraft wing was optimized considering the stress and displacement constraints. The thickness of the wing ribs, spars and skins were selected as design variable. Commercial software ANSYS was utilized to calculate the structural response of the wing, i.e. stress, strain and displacement distributions. The objective of optimization design is set to reduce the maximum stress as well as overall weight of the wing. For the purpose of reduction in the computational cost, response surface method technique has been employed to build up surrogate for optimization problem. After the process of optimization, the equivalent maximum stress is successfully reduced. The structural weight is also reduced by 27% using DOE and RSM technique, which is a remarkable weight saving that requires very little effort. The weight reduction could be very beneficial in terms of carrying more payloads. Furthermore, the amount of fuel to power up the aircraft will also be reduced, thus plummeting the running costs of the aircraft.

References

[1] Kavya G and Reddy B C R, 2015. Design and finite element analysis of aircraft wing using ribs and spars. *International Journal and Magazine of Engineering, Technology, Management and Research.***2**, Issue No: 11 (November).

[2] Sureka K and Meher R S, 2015. Modeling and structural analysis on A300 flight wing by using ANSYS. *International Journal of Mechanical Engineering and Robotics Research.***4**, IssueNo:2 (April).

[3] Senthilkumar S, Velayudham A, and Maniarasan P, 2013. Dynamic structural response of an aircraft wing using Ansys. *International Journal of Engineering Research & Technology (IJERT).***2** Issue No: 6 (June).

[4] Kumar T S V, Basha A W, Pavithra M and Srilekha V, 2008. Static and dynamic analysis of atypical aircraft wing structure using MSC Nastran. *International Journal of Research in Aeronautical and Mechanical Engineering.* **3** Issue No: 8 (August).

[5] Belligundu A D and Chandrupatla T R, 2011. *Optimization Concepts and Application inEngineering, 2nd ed.* (New York: Cambridge University Press)

[6] Mazhar F and Khan A M, 2010. Structural design of a UAV wing using finite element method. *51st AIAA/ASME/ ASCE/AHS/ASC Structures, Structural Dynamics, and Materials Conf
18th 12 - 15 April 2010* (Orlando, Florida) AIAA 2010-3099

[7] Roux W J, Stander N and Haftka R T, 1998. Response surface approximation for structural optimization. *International Journal of Numerical Methods in Engineering.***42** Issue No: 3, pp 517-34.

[8] Menon A K, 2005. Structural optimization using ANSYS andregulated multiquadric response surface model. *MS thesis at The University of Texas at Arlington* (December)

[9] Brown D A 1970, Advanced airfoils studied for transports. *Aviation Week and SpaceTechnology (New York)* vol. 92, No. 25, June 22, 1970, pp.55-57;60.

[10] Anderson D 1971, Dateline America: supercritical wing.*Flight*vol. 100, No. 3251, July 1, 1971, pp. 23-26.

[11] Albers R G, Tuttle M E andAvery W B, 2008. Subsonic transport composite wing model for optimization studies.*Proc. of the 12th AIAA/ISSMO Multidisciplinary Analysisand Optimization Conf, 10-12 September2008* (Victoria, British Columbia, Canada) AIAA 2008-6016.

[12] Seligh M, UIUC Airfoil Coordinates Database, Version 2.0, URL: https://m-selig.ae.illinois.Edu/ads.html

[13] Torenbeek E, 1992 *Development and application of a comprehensive, design-sensitive weight prediction method for wing structures for transport category aircraft* (The Netherlands: Delft University of Technology), Report LR-693, Sep., 1992.

[14] Cook R D, 1995 *Finite Element Modeling for Stress Analysis* (NY: John Wiley &Sons, Inc.)

[15] Yang R J, Kodialam S, Gu L and Tho C H, 2004. Multidisciplinary Design Optimization in aScalable, High Performance Computing Environment.*Structural and Multidisciplinary Optimization.* **26**, pp. 256-63 (February).

[16] Giunta A A,S F Wojtkiewicz Jr. and Eldred M S, 2003. Overview of modern design of experiment methods for computational simulations. *41st Aerospace Sciences Meeting and Exhibit, 6-9 January 2003*(Reno Nevada) AIAA 2003-649.

[17] Myers R H, Montgomery D C, Anderson-Cook C M, 2016. *Response surface methodology – Process and Product Optimization Using Designed Experiments, 4^{th}ed,* (New Jersey: John Wiley &Sons, Inc.)

[18] Ampofo J and Ferguson F, 2002. Optimal design of aircraft wing structures: a computer aided design method.*Proc of the 5th Biannual World Automation Congress* (Orlando, FL)pp. 471-480.

Analysis of Gas State in Venus Surface Environment Simulation System

Gao Wen, Yang Xiaoning, Wang Jing, Guo Qinliang, Lin Boying, Bi Yanqiang and Wu Yue

Beijing Institute of Spacecraft Environment Engineering, Beijing, 100094

18811772351@163.com

Abstract. Venus, as a terrestrial planet, not only has the closest distance to Earth in the solar system but also has similar size and mass to Earth, the detection on it has extremely high reference value for Earth and future research. While due to the high temperature, high pressure and weak acid environment on the surface of the planet, the landing detection is hindered. Based on comprehensive analysis and computation on the Venus surface environment simulation process and diffusion state of single-phase binary gas under high-temperature and high-pressure conditions, this article provides an optimization scheme, which would provide support for the establishment of future environmental simulation equipment.

1. Introduction

Venus is the closest planet in the solar system to Earth. The difference in radius and average density of Venus from Earth is only 5%, which indicates that the two planets have similarities in the main structure and composition. In contrast, in terms of atmospheric environment, the surface pressure of Venus is about 90 times that of Earth, and its main component is carbon dioxide, while the component of carbon dioxide in Earth's atmosphere is relatively low. According to the Venus observations, there are more active signs of volcanic activity on the surface of Venus, which indicates that convective motion exists within Venus, but there are still no signs that plate movements will be caused by this, and then planetary magnetic fields will be generated[1].

Venus is so close to Earth in mass and size, so probing Venus has important scientific significance for understanding the historical evolution of Earth. However, due to the extremely harsh environment of Venus, the Venus lander will face great challenges[2]. Before the Venus lander is launched, it needs to undergo a full environmental assessment to ensure the reliability of the mission. Therefore, simulating the Venus environment to evaluate the Venus lander is very important. The gas of the final state of the Venus environmental simulation system is high-temperature and high-pressure, and it is heated from normal temperature to high temperature to reach the required ambient temperature and pressure. Based on the actual two-component equation of state (EOS), this article analysed the gas state change process in the Venus environment simulation system.

2.Venus typical atmospheric environment

Pressure profile of Venus atmosphere [3]is shown in figure 1. The pressure of Venus atmosphere increases dramatically with decreasing altitude. The atmospheric pressure on the surface of Venus is about 90atm [4].

Content from this work may be used under the terms of the Creative Commons Attribution 3.0 licence. Any further distribution of this work must maintain attribution to the author(s) and the title of the work, journal citation and DOI.
Published under licence by IOP Publishing Ltd

IOP Conf. Series: Materials Science and Engineering **887** (2020) 012025 doi:10.1088/1757-899X/887/1/012025

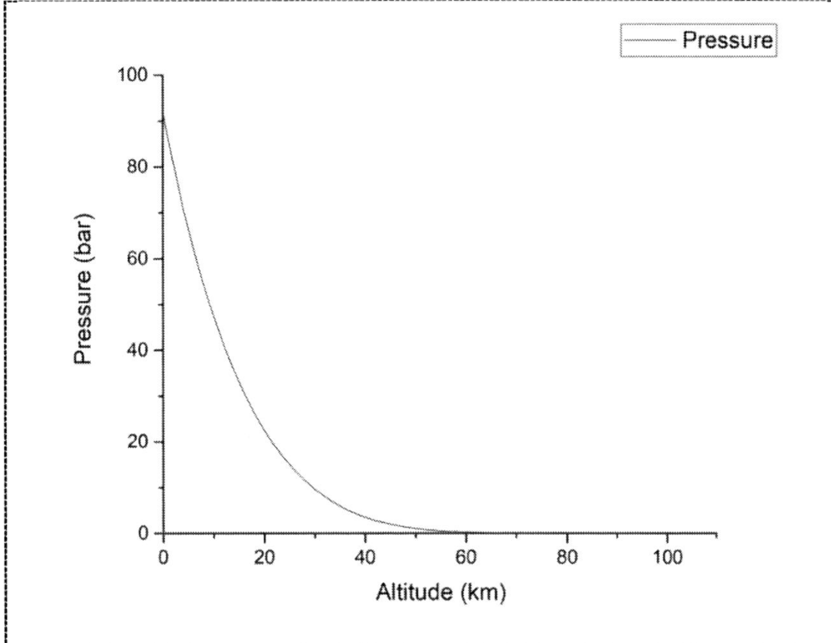

Figure 1. Pressure Profile of Venus Atmosphere.

Temperature profile of Venus atmosphere [4] is shown in figure 2, and the temperature of Venus surface is about 740K. Therefore, the temperature and pressure, required to simulate the Venus atmospheric environment, are about 773K and 92bar and due to the complexity of the atmospheric composition, it is so difficult to simulate.

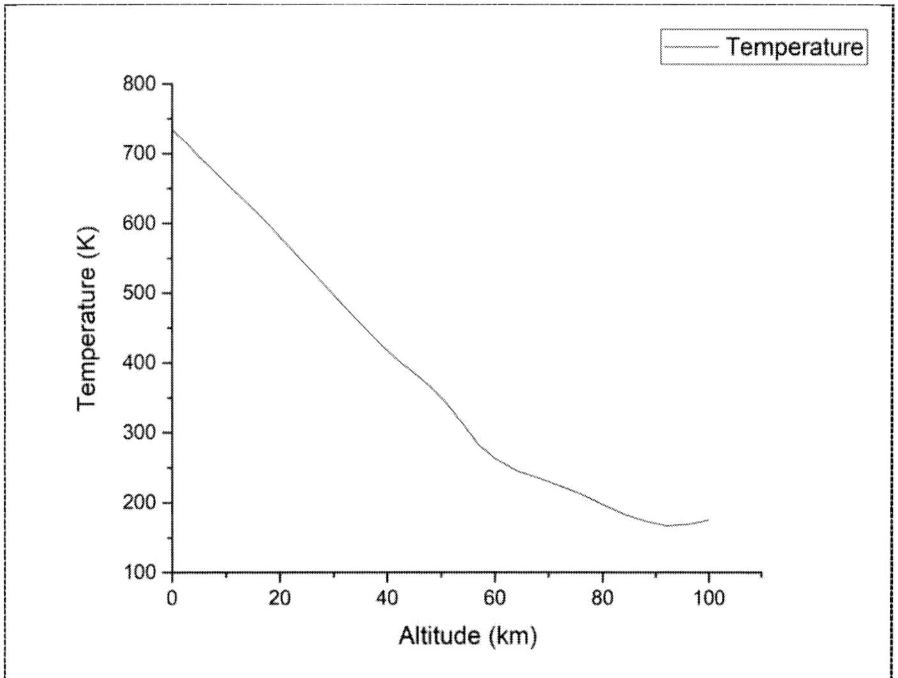

Figure 2. Temperature Profile of Venus Atmosphere.

The atmospheric composition table of Venus [1] is shown in table 1. In Venus atmosphere, CO_2 and N_2 account for 99.5%. When performing state analysis, CO_2 and N_2 are considered as the main factors for calculation.

Table 1. Composition of the atmospheres of Venus [1].

Species	Venus
Carbon dioxide	0.96
Nitrogen	0.035
Argon	0.00007
Neon	0.000005
Water vapor	30 parts per million
Heavy water (HDO)	3 parts per million
Sulphur dioxide	150 parts per million
Carbonyl sulfide	4 parts per million
Carbon monoxide	0.00004
Hydrogen chloride	0.5 parts per million
Hydrogen fluoride	0.005 parts per million
Atomic oxygen	trace
Hydroxyl	trace
Atomic hydrogen	trace

3. Typical Venus Surface Environment Simulation System Process

In order to reach the temperature and pressure of Venus surface, the container needs to have a certain pressure P_1 when the temperature is raised, so that it can reach the temperature and pressure after the temperature is raised. To reach P_1 from normal pressure conditions, it is necessary to use a pump to pump premix container to the container to achieve the purpose of compressing the volume and boosting the pressure. Simple system process is shown in figure 3.

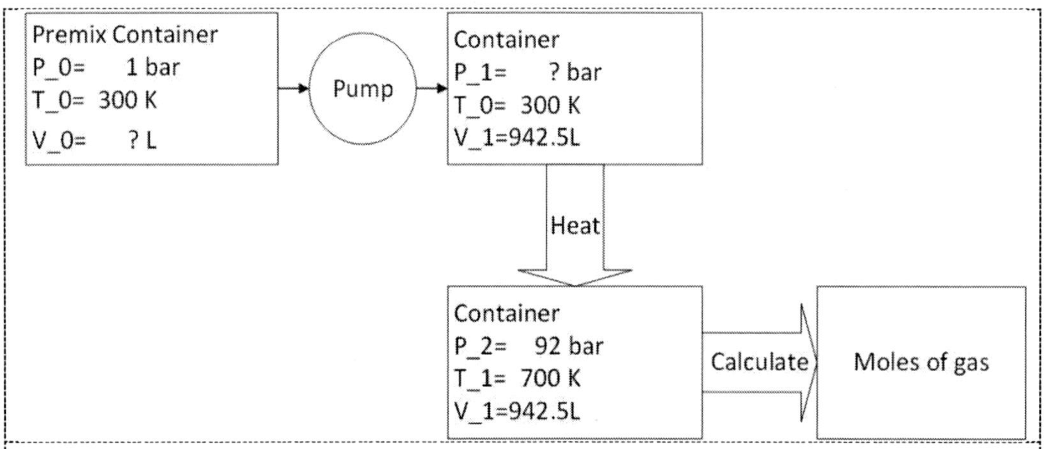

Figure 3. Simple system process.

4. Gas state analysis

4.1. Theory of actual gas state calculation

The gas composition can be regarded as a mixed gas of 96.5% CO_2 and 3.5% N_2 [5], and due to the large temperature and pressure, the gas is not ideal gas.

Research shows that there is a certain similarity in the thermodynamic properties of various substances, and the contrast equation of state (EOS) is a general equation of state for various substances expressed using dimensionless contrast parameters [6].

$$p_r = \frac{p}{p_C} \tag{1}$$

$$T_r = \frac{T}{T_c} \tag{2}$$

$$v_r = \frac{v}{v_c} = \frac{V_m}{V_{m,C}} \tag{3}$$

$$z = \frac{pV_m}{RT} \tag{4}$$

Among them, the comparison parameters p_r,T_r and $v_r$are the ratios of the parameters at each point and the same parameters at the critical point, and the compression factor is the ratio of the actual gas volume to the ideal gas volume. After experiments on various gases, a general compression factor graph is obtained, so in the calculation, reading graph or iterative method is used to obtain the compression factor.

The calculation of the actual mixture needs to be simplified, and the virtual critical constant method is used to treat the mixture as a pure substance. This method was proposed by W.B. Kay [7], and the following is the Kay rule:

$$T_{P_c} = y_1 T_{c1} + y_2 T_{c2} + y_2 T_{c2} + \cdots = \sum y_i T_{ci} \tag{5}$$

$$P_{P_c} = y_1 P_{c1} + y_2 P_{c2} + y_2 P_{c2} + \cdots = \sum y_i P_{ci} \tag{6}$$

Virtual contrast parameter [7]:

$$T_{P_r} = \frac{T}{T_{P_c}} \tag{7}$$

$$P_{P_r} = \frac{P}{P_{P_c}} \tag{8}$$

4.2. Gas state analysis

4.2.1. *Calculate the total amount of gas.* Calculate the mixed gas compression factor Z_m in this state by the virtual critical constant method to calculate the overall amount of mixed gas and obtain the amount of each component gas [7].

$$Z = \frac{1}{1-h} - \frac{A}{B}\left(\frac{h}{1+h}\right) \tag{9}$$

$$h = \frac{b}{V} = \frac{B}{Z} \tag{10}$$

where A and B can be calculated by:

$$\frac{A}{B} = \frac{4.934}{T_{P_r}^{\ 1.5}} \tag{11}$$

$$B = \frac{0.08664 P_{P_r}}{T_{P_r}} \tag{12}$$

The iterative process is: preset the initial value of the compression factor Z_0, substitute it into equation (9) to calculate h_0, and then pass equation (10) to obtain Z_1, and calculate the difference $\Delta Z = |Z_{k+1} - Z_k|$ to verify whether it is less than the setting ε, if it is smaller, then output $Z = Z_{k+1}$, otherwise continue to iterate until the result meets the requirement. The process of iteratively calculate is shown in figure 4.

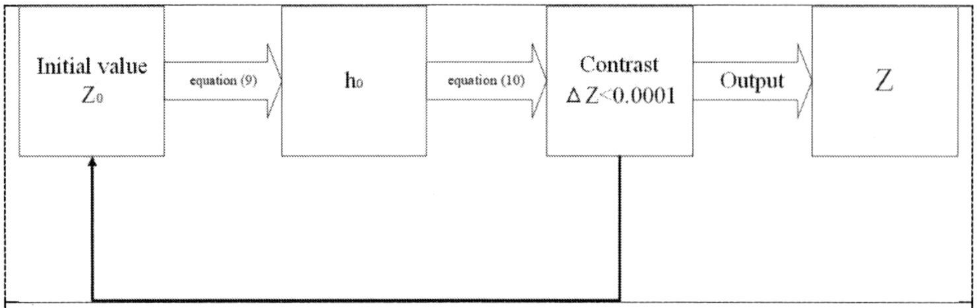

Figure 4. Iteratively calculate the compression factor Z_m.

4.2.2. *Iteratively calculate P_1 by the molar amount of gas.* Consistent with the above, when calculating P_1, the calculation of the relative volume is relatively tedious, so here we use the preset Zm_0, calculate Pm_0, and then calculate Tr_0 and Pr_0 to obtain $\Delta Z = |Zm_{\_k+1} - Zm_{\_k}|$, verify whether it is less than the setting ε, if it is smaller, output the result $Z = Zm_{\_k+1}$, otherwise continue iterating until the result meets the requirement. The process of iteratively calculate is shown in figure 5.

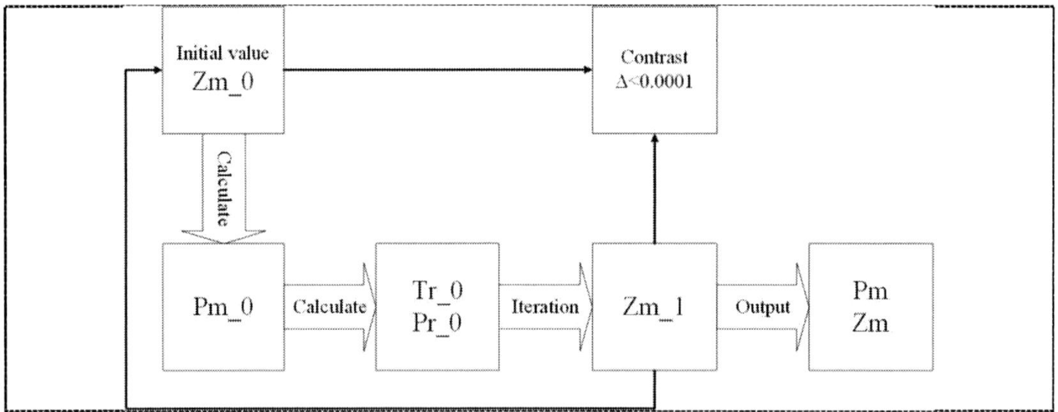

Figure 5. Iteratively calculate P_1.

4.2.3. Calculate the volume of the premixed container. Iteratively calculate Z_m similarly to the first step, combining N_m can be obtained the volume of Premix Container V_0.

4.3. Results and analysis

The amount and mass of each gas component substance [5] after calculation is shown in table 2.

Table 2. the amount and mass of each gas component.

No	Composition	mol	Molar mass	Mass(g)
1	CO_2	1307	44	57508
2	N_2	47.40	28	1327.2
3	SO_2	0.2438	64	15.6032
4	COS	0.0691	60	4.146
5	H_2O	0.0406	18	0.7308
6	CO	0.0163	28	0.4564
7	H_2S	0.0027	34	0.0918
8	HCl	0.000677	36.5	0.024711
9	HF	0.00000339	20	0.0000678
	Gas mixture	1354.77318	/	58856.25298

The calculation result Z_1=0.8451, P_1=3.0289MPa=30.289bar, and the reserve pressure before the temperature rise of the simulation process in the reference [5] is 500psi=3.45Mpa, and the error rate is 12.2%

Calculation result Z_2=0.9966, V_0=33.6261m^3, which is about 35 times of the container.

5. Conclusion

Venus environment simulation technology is of great significance to the success of the Venus landing mission. Based on the actual two-component equation of state, this article calculates the gas state at each stage of the Venus environment simulation system and obtains the temperature and pressure parameters of the gas at each stage. It provides a basis for the typical process design of the Venus simulation system.

References

[1] Fredric W. Taylor et al.2018 Venus: The Atmosphere, Climate, Surface, Interior and Near-Space Environment of an Earth-Like Planet. *J Space Sci Rev* **214** 35
[2] JIAO Weixin.2018 Future deep space exploration and related space environmental simulation. *J Spacecraft Environment Engineering* **35** 103
[3] Hou jianwen.2015 *Deep space Exploration-Venus Exploration* (Beijing National Defence Industry Press)
[4] Vladimir A. Krasnopolsky.2012 A photochemical model for the Venus atmosphere at 47–112 km. *J Icarus* **218** 230
[5] Kremic T et al. Extreme Environment Simulation - Current and New Capabilities to Simulate Venus and other Planetary Bodies,2014 IEEE Aerospace, 2014 (https: // ntrs. nasa. gov/ archive/ nasa/ casi.ntrs.nasa.gov/20140013390.pdf)
[6] Zhou yan.2014 *Engineering Thermodynamics* (Beijing Chemical Industry Press)
[7] Yujiazhi.2016 *Chemical Thermodynamics* (Beijing Chemical Industry Press)

Dynamical Analysis for the INS Vibration Control System Used in UAV

Yuxing Duan[1], Xiao Li[2], Bo Su[1], Xin Wang[1] and Qiang Yang[1]

[1] Aircraft Strength Research Institute of China, Xi'an 710065, CHN
[2] Beijing Zhonghangzhi Technology Co., Ltd, Beijing 100176, CHN

Email: yuhsing@163.com

Abstract. A six-degree-of-freedom mechanical dynamic model and its equations of the damping system are established for the offset installation environment of the UAV. The influence parameters of the damping system on the vibration coupling state is studied based on the model. And the optimal parameters of the damping system are obtained by doing the mechanical dynamic calculation. And then the vibration isolator of the UAV strap-down inertial measurement unit is designed and manufactured. Finally, the damping system was tested and verified on the shaking table and in the flight environment of the UAV. It is shown that the damping system could implement the tri-axial equal rigid design when the shaft-to-radius stiffness of the isolator is about 0.8. And the experiments also prove this point. The damping system weakens the coupling problem of the strap-down inertial measurement unit effectively, and it reaches the quasi-decoupled state. The amplitude of the acceleration decreases by more than 50% after vibration isolation.

1. Introduction

The strap-down inertial navigation system (INS) is the key equipment in aeronautics and astronautics, that mainly used for positioning and attitude measurement of carriers [1]. Its performance and reliability could directly affect the flight safety and working state of the carriers. However, its measurement accuracy could be reduced by the severe vibration environments. The vibration decrease has been the key problems in the safety and stability of the INS.

Yu et al [2] developed a hybrid mode tri-axial magnetorheological elastomer isolators and the phase-based control to suppress the vibration of the IMU in three directions. Chen et al [3] proposed a vibration isolation system (VIS) structural and established the decoupled six-degree-of-freedom dynamic model to eliminate the motion coupling errors. Tuo et al [4] studied a method for designing a IMU damping system and made the modal analysis of dithered ring laser gyroscope IMU damping system using the finite element analysis. Tu et al [5] proposed a novel dynamic analysis and experimental verification method for the deviation angles in dual-axis rotational INS in order to accurately analyse the deviation angles caused by rubber dampers deformation. The proposed model and dynamic analysis method were validated by comparing it against the real deviation angles measured by rotation experiments for a turntable.

Based on the research of Doctor Tu et al [5], a six-degree-of-freedom mechanical dynamic model and its equations of the damping system would be established for the offset installation environment of the UAV. The influence parameters of the damping system on the vibration coupling state was studied

Content from this work may be used under the terms of the Creative Commons Attribution 3.0 licence. Any further distribution of this work must maintain attribution to the author(s) and the title of the work, journal citation and DOI.
Published under licence by IOP Publishing Ltd

using the model. The optimal parameters of the damping system were obtained by doing the mechanical dynamic calculation.

2. Establishment of the damping system model

In order to simplify the model, the vibration isolator could be considered as a joint with a certain stiffness and damping. Then the damping system which is composed of the isolators can be regard as control by four matrixes: the translational stiffness matrix **K**, the translational damping matrix **C**, the rotational stiffness matrix **T** and the rotational stiffness matrix **Ξ**. And each isolator could be simplified and decomposed to one stiffness damping system which is shown in figure 1 along x, y, and z axes. It is assumed that the axial stiffness and damping of the isolator are K_a and C_a, and the radial stiffness and damping are K_r and C_r. Considering that the vibration isolator is a structure that is symmetrical along the axis rather than the radial direction, and the installation direction of the isolators in the damping system is parallel to the x direction. Thus

$$\begin{aligned} K_x = K_a,\ K_y = K_z = K_r \\ C_x = C_a,\ C_y = C_z = C_r \end{aligned}. \tag{1}$$

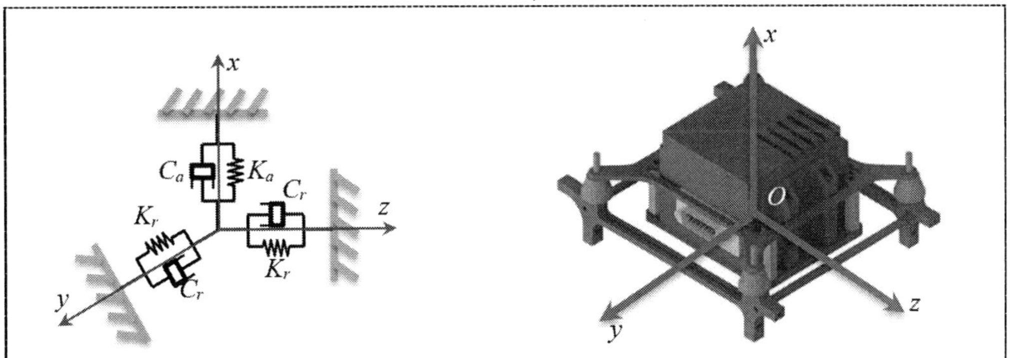

(a) Voigt model [6] (b) 3-D model of the damping system of IMU

Figure 1. Model of isolator of the damping system.

The dynamic state of the whole IMU damping system is shown in figure 1. The mass centre of the IMU is at the geometric centre of the damping system. If the damping system is completely symmetrical in three directions, the stiffness centre of the damping system can be coincided with the mass centre that we could decouple the whole dynamic system.

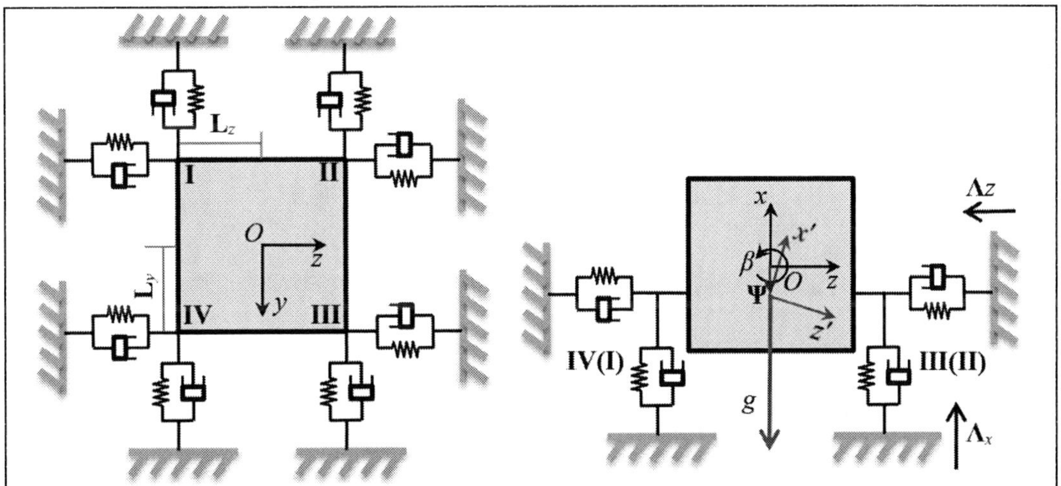

(a) For heading & side direction (b) For vertical direction

Figure 2. Dynamical mechanical coupling state of the model.

In practice, the heading direction (y) and lateral direction (z) are absolutely identical and symmetrical that the vibration state of the two directions is exactly the same. But vibration state of the vertical direction (x) is different from the two other direction since the stiffness centre was not coincided with the mass centre along axial direction of the isolator (x, namely vertical direction). We assume $\mathbf{\Psi}$ is the offset vector between the stiffness centre of the IMU damping system and mass centre. Analysis above shown that the components along y and z direction are both zero and we could take $\mathbf{\Psi}=[\psi, 0, 0]$.

The figure 2 shows that the components of $\mathbf{\Psi}$ along y and z direction are both zero so that the damping system of the IMU is decoupled along x direction when the system was vibrated along vertical direction (namely $\mathbf{\Lambda}_x$). Nevertheless, the x components of $\mathbf{\Psi}$ is not zero so that the linear vibration along y direction was coupling with the angular vibration along **z** direction and the linear vibration along z direction was also coupling with the angular vibration along y direction when the system was vibrated alongside direction ($\mathbf{\Lambda}_y$ or $\mathbf{\Lambda}_z$).

We assume that the IMU is one isotropic cube with a side length of $2l$, that is, $\mathbf{L}_y=\mathbf{L}_z=l$. Moreover, it's known that $\mathbf{L}_x=\psi$. Thus

$$\begin{gathered}
m(\ddot{x}-g)+4C_a\dot{x}+4K_a x=0 \\
m\ddot{y}+4C_r\dot{y}+4K_r y+4C_r\psi\dot{\gamma}+4K_r\psi\gamma=0 \\
m\ddot{z}+4C_r\dot{z}+4K_r z-4C_r\psi\dot{\beta}-4K_r\psi\beta=0 \\
\frac{1}{6}ml^2\ddot{\alpha}+8C_r l^2\dot{\alpha}+8K_r l^2\alpha=0 \\
\frac{1}{6}ml^2\ddot{\beta}+4(C_a l^2+C_r\psi^2)\dot{\beta}+4(K_a l^2+K_r\psi^2)\beta-4C_r\psi\dot{z}-4K_r\psi z=0 \\
\frac{1}{6}ml^2\ddot{\gamma}+4(C_a l^2+C_r\psi^2)\dot{\gamma}+4(K_a l^2+K_r\psi^2)\gamma+4C_r\psi\dot{y}+4K_r\psi y=0
\end{gathered} \tag{2}$$

3. Decoupling the damping system model

3.1. Coupling analysis of x direction

Firstly, one linear sine excitation $\mathbf{\Lambda}_x$ was exerted along x direction. Thus

$$m(\ddot{x}-g)+4C_a(\dot{x}-\dot{\Lambda}_x)+4K_a(x-\Lambda_x)=0. \tag{3}$$

Then the linear and angular transfer function (namely H_L) of the damping system along x direction are

$$H_L=(1+(\frac{C_a\omega}{K_a})^2)^{\frac{1}{2}}((1-\frac{m\omega}{4K_a})^2+(\frac{C_a\omega}{K_a})^2)^{-\frac{1}{2}}. \tag{4}$$

3.2. Coupling analysis of y direction

Firstly, one linear sine excitation $\mathbf{\Lambda}_y$ was exerted along y direction. Thus

$$\begin{gathered}
m\ddot{y}+4C_r(\dot{y}-\dot{\Lambda}_y)+4K_r(y-\Lambda_y)+4C_r\psi\dot{\gamma}+4K_r\psi\gamma=0 \\
\frac{1}{6}ml^2\ddot{\gamma}+4(C_a l^2+C_r\psi^2)\dot{\gamma}+4(K_a l^2+K_r\psi^2)\gamma+4C_r\psi(\dot{y}-\dot{\Lambda}_y)+4K_r\psi(y-\Lambda_y)=0
\end{gathered} \tag{5}$$

Transforming them into the matrix forms, one obtains

$$[\mathbf{M}_L]\begin{bmatrix}\ddot{y}\\ \ddot{\gamma}\end{bmatrix}+[\mathbf{C}_L]\begin{bmatrix}\dot{y}\\ \dot{\gamma}\end{bmatrix}+[\mathbf{K}_L]\begin{bmatrix}y\\ \gamma\end{bmatrix}=[\mathbf{K}_L]\begin{bmatrix}\mathbf{\Lambda}_y\\ \dot{\mathbf{\Lambda}}_y\end{bmatrix}. \tag{6}$$

Where

$$[\mathbf{M}_L] = \begin{bmatrix} m & 0 \\ 0 & \frac{1}{6}ml^2 \end{bmatrix}$$

$$[\mathbf{C}_L] = \begin{bmatrix} 4C_r & 4C_r\psi \\ 4C_r\psi & 4(C_a l^2 + C_r\psi^2) \end{bmatrix}$$

$$[\mathbf{K}_L] = \begin{bmatrix} 4K_r & 4K_r\psi \\ 4K_r\psi & 4(K_a l^2 + K_r\psi^2) \end{bmatrix}$$

$$[\mathbf{K}_L] = \begin{bmatrix} 4K_r & 4C_r \\ 4K_r\psi & 4C_r\psi \end{bmatrix}$$

Thus the impedance $[\mathbf{Z}_L]$ [7] is

$$[\mathbf{Z}_L] = [\mathbf{K}_L] - \omega^2[\mathbf{M}_L] + i\omega[\mathbf{C}_L]. \tag{7}$$

$$[\mathbf{U}_L] = [\mathbf{K}_L]\begin{bmatrix} 1 \\ i\omega \end{bmatrix}$$

Then the transfer function $[\mathbf{H}_L]$ is

$$[\mathbf{H}_L] = \begin{bmatrix} \mathbf{H}_{L11} \\ \mathbf{H}_{L21} \end{bmatrix} = [\mathbf{Z}_L]^{-1}[\mathbf{U}_L] = \frac{\mathrm{Adj}[\mathbf{Z}_L][\mathbf{U}_L]}{\det[\mathbf{Z}_L]}. \tag{8}$$

4. Design of the damping system and Experiments

The 1st passage frequency of the UAV in this article is 12 Hz. In order to avoid the coupling resonance between the IMU and engine of the UAV, the resonance frequency of the IMU damping system should be placed in the middle of the first two passage frequencies. So the resonance frequency of the damping system is designed as 30 Hz. Four identical isolators were installed on the flange of the UAV. The structure of the system was shown in figure 1(c) and parameters of the IMU shown in table 1. The IMU had the offset vector $\mathbf{\Psi}$ since the installation environment restrictions.

Table 1. Parameters of the IMU.

Parameters	Value
Mass of IMU(m)	0.58 kg
Installation distance of isolators(l)	120 mm
Offset vector($\mathbf{\Psi}$)	[50, 0, 0]

Figure 3 shows the influence of axial damping C_a and radial damping C_r on coupling vibration of the damping system when the resonance frequency is about 30 Hz. Therefore, we could conclude that the main contribution of C_r is to reduce the amplification of the linear resonance frequency and C_a can effectively eliminate the influence of coupling vibration. When C_a=7 kg s^{-1}, the influence on coupling vibration was eliminated basically. And C_r=7 kg s^{-1}, the amplification of the system is 4.8 which could also meet the requirement. So that we are setting C_a and C_r both equal 7 kg s^{-1} in the preliminary calculation.

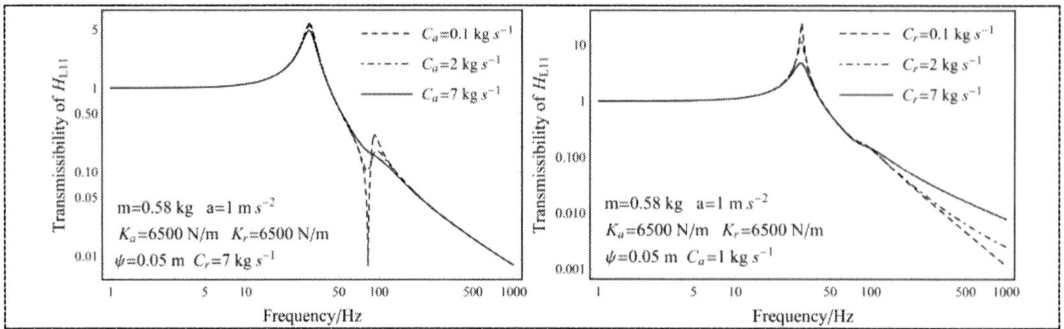

(a) The axial damping ratio of isolator (b) The radial damping ratio of isolator

Figure 3. The transmissibility of the damping system at y axis under Λ_y.

Substituting the values of C_a and C_r into equation (4), one obtains that the axial stiffness K_a equals 5320 N/m when the resonance frequency was 30 Hz. Then substituting the values of C_a, C_r and K_a into equation (8), one obtains that the radial stiffness K_r equals 6650 N/m.

In a word, the ratio of axial stiffness to radial stiffness is

$$K_a / K_r = \frac{5320\ \text{N/m}}{6650\ \text{N/m}} = 0.8 . \tag{9}$$

5. Results and discussions

Figure 4 shows the test results of the damping system on shaking table, that the solid line is the test result and the dashed line is calculation result. Figure 5 shows the test results of the damping system on the UAV, that the grey part is the vibration response on the flange of the UAV and the pure black part is the vibration response on the IMU.

One observes that the damping system was both decoupled along axial direction (x) and side direction (y). The 2nd resonance frequency peak of the angular coupling vibration had been largely eliminated. And the axial and side resonance frequencies were basically the same, so the damping system could be considered to be in the state of tri-axial equal stiffness.

Furthermore, the acceleration amplitude on the IMU decreased significantly. The peak-to-peak value was reduced by more than 50 percent compared with that before the vibration isolation, and the rotor frequency amplitude reduced to one third of the original value.

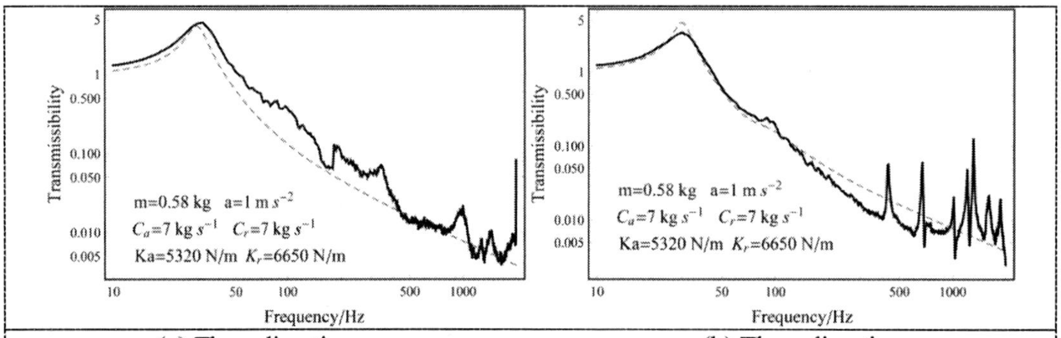

(a) The x direction (b) The y direction

Figure 4. Results of the vibration test.

IOP Conf. Series: Materials Science and Engineering **887** (2020) 012026 doi:10.1088/1757-899X/887/1/012026

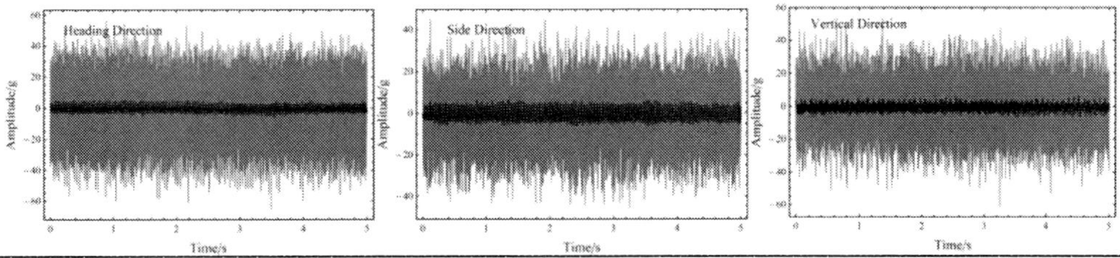

(a) The x direction (b) The y direction (c) The z direction

Figure 5. Results of the vibration test.

6. Conclusion

We studied the dynamic coupling state under the offset installation of the damping system, and concluded that

- The offset installation of the damping system will cause the coupling between the angular vibration and the linear vibration, and could be designed to achieve the quasi-decoupling state. When the ratio of axial stiffness to radial stiffness is around 0.8 and damping ratio is above 0.125, the 2nd resonance frequency peak caused by angular vibration could be eliminated basically and the damping system would be in the state of tri-axial equal stiffness.
- According to the design and calculation of the damping system, the vibration problems had been solved effectively. And the working state of the system is consistent with the theoretical analysis. The peak-to-peak value decreased by more than 50 percent compared with before the vibration isolation after filtering the output data.

References

[1] Tazartes D. An historical perspective on inertial navigation systems 2014 *International Symposium on Inertial Sensors and Systems* IEEE pp 1-5
[2] Yu Tao, Xiaoting Rui, Fufeng Yang, et al. Development of a MRE isolation system for strapdown inertial measurement unit 2019 *Mechanical systems and signal processing* **117** pp 553-568
[3] Junchao Cheng, Jiancheng Fang, Weiren Wu, et al. Optimized design method of vibration isolation system in mechanically dithered RLG POS based on motion decoupling 2014 *Measurement* **48** pp 314-324
[4] Tuo Zhouhui, Hu Dewen, Li Ruhua, et al. Damping design of strap-down inertial navigation system 2009 *Journal of Chinese Inertial Technology* Vol.**17** No.**6** pp 648-650
[5] Yongqiang Tu, Gongliu Yang, et al. Dynamical analysis and experimental verification of deviation angles caused by rubber dampers deformation in high precision mechanically dithered RLG dual-axis RINS 2019 *Mechanical Systems and Signal Processing* **126** pp 553-567
[6] S. Kaul. Maxwell-Voigt and Maxwell ladder models for multi-degree-of-freedom elastomeric isolation systems2015 *Journal of Vibration and Acoustics* **137** pp 021021-021029
[7] R. E. D. Bishop. *The Matrix Analysis of Vibration* 2008 Cambridge University Press pp 70-94

FEA and Optimization of a Door Used in Large Vacuum Chamber

Yi Zhang [a], **Shizeng Lv, Xiao Han and Jie Gong**

Beijing Institute of Spacecraft Environment Engineering, 100094 Beijing, China

[a] E-mail: spcomyth@139.com

Abstract. The door is the access for the test piece and personnel, and it is one of the important parts of the vacuum chamber. This paper introduces a kind of door of a large vacuum chamber, which moves radially along the ground guide rail. Because of the risk of overturning, we use the finite element software to create a finite element model, define the material properties, and apply the load. On this basis, we carry out stress analysis, displacement analysis and overturning analysis. According to the analysis results, we give two optimization methods for anti-overturning. Finally, we analyse the overturning resistance of the optimized model.

1. Foreword

In recent years, the demand for comfort testing in cockpit has gradually increased in aviation field. The large vacuum chamber with diameter of 10m is the principal part of the environmental simulation equipment used in the experiment, and the door is the key part of the vacuum chamber [1].

The door moves radially along the ground rail [2,3]. At present, there are many analyses on its airtightness, stress and displacement, but there is no analysis on its overturning [4,5]. However, there is a risk of overturning for this door, especially for the large vacuum chamber, the impact of the door's collapse is unimaginable. Therefore, we use the finite element software not only to analyse the stress and displacement of the door structure, but also to analyse its overturning risk [6]. According to the analysis results, the optimization method of anti-overturning is given to ensure the safety and reliability of the door.

2. Structure of the Door

The structure of the door is shown in Figure 1, which is composed of head, flange and supporting frame. It is the passageway for the specimen and personnel, and is one of the important parts of the vacuum chamber [7]. The support frame is equipped with wheels and moves radially along the ground guide rails. The head is spherical crown shape, the inner diameter is φ10m, and the size of support frame is 12000mm (L) × 3000mm (W) × 5300mm (H).

Content from this work may be used under the terms of the Creative Commons Attribution 3.0 licence. Any further distribution of this work must maintain attribution to the author(s) and the title of the work, journal citation and DOI.
Published under licence by IOP Publishing Ltd

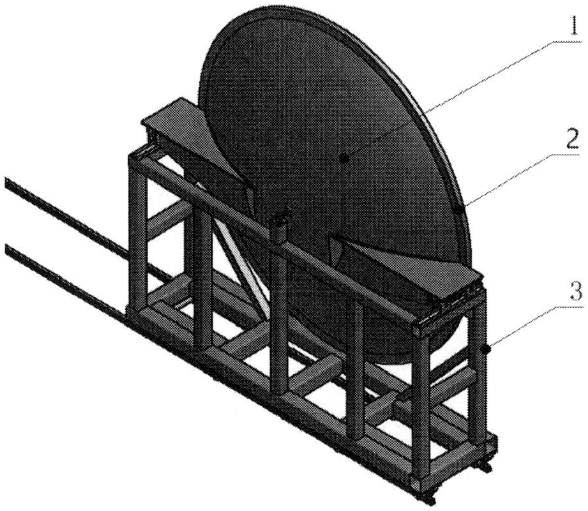

1. Head;
2. Flange;
3. Supporting frame;

Figure 1. Structural of door.

3. Finite element analysis

The structure of the door is complex, so the traditional mechanical analysis method can't be used to analyse it. However, the finite element method has its unique advantages in stress, displacement and overturning analysis [7].

We create the finite element model of the door, and carry out stress analysis, displacement analysis and overturning analysis.

3.1. Pre-processing

3.1.1. Create a finite element model. Because there are some fine structures in the model, in order to obtain the simulation results more quickly and accurately, we use CATIA software to simplify the model, the simplified model is shown in Figure 2. We use HyperMesh software to mesh the model, as shown in Figure 3, the mesh size is 80mm, the total number is 43775. The mesh is evenly distributed, appropriately dense and of good quality.

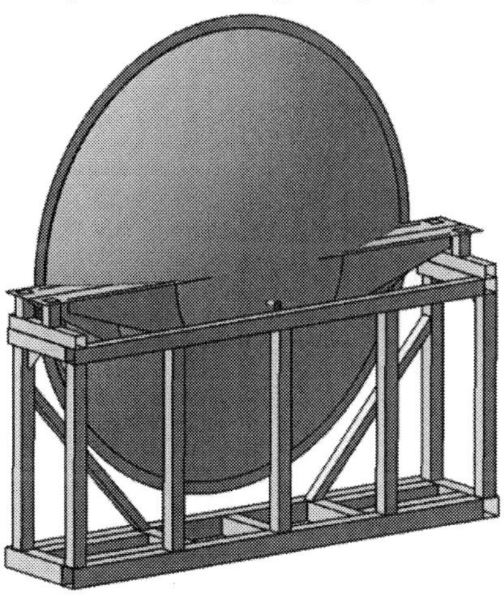

Figure 2. Simplified model.

Figure 3. Model after meshing.

3.1.2. Material properties. The door material is Q345R. See table 1 for the properties of materials [8].

Table 1. Material properties

Material	Elastic Modulus(MPa)	Poisson ratio	Density (kg/m^3)	Yield strength (MPa)	Factor of safety	Allowable stress (MPa)
Q345R	210000	0.3	7850	345	1.7	202

3.1.3. Applied load. The geometric boundary condition of the finite element model is mainly gravity load. According to the actual working condition, all units bear gravity (g = 9.8m / S2), and a horizontal support surface is built at the bottom, as shown in Figure 4.

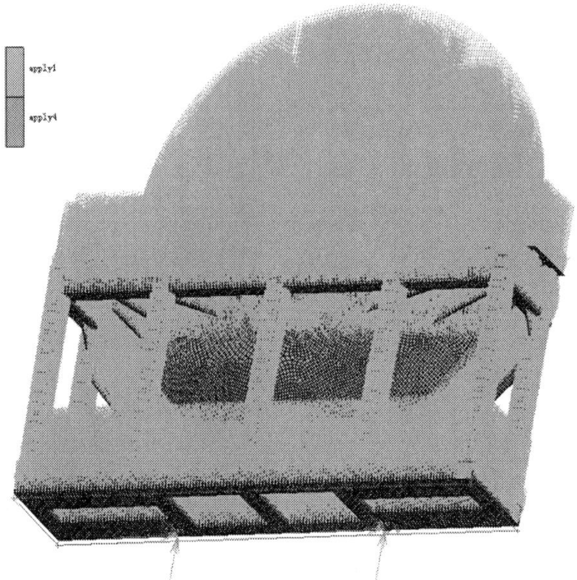

Figure 4. Boundary conditions.

3.2. Stress analysis results

The overall stress distribution of the door and the location of the maximum stress are shown in Figure 5. The area with high stress is mainly concentrated at the joint of inclined square pipe and vertical square pipe of frame. The maximum stress is 171.4 MPa, which is located at the bottom support point, which is lower than the allowable stress of material by 202MPa.

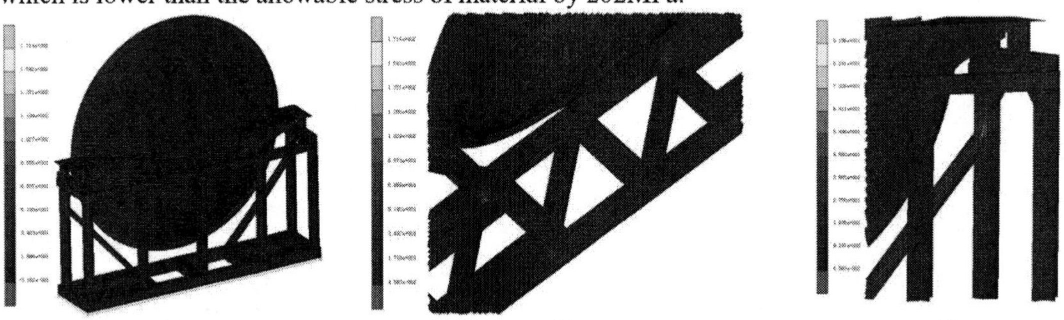

(a) (b) (c)

Figure 5. Stress distribution of door.

3.3. Displacement analysis results

The overall displacement distribution and the maximum displacement position of the door are shown in Figure 6. The maximum deformation displacement is 15.03mm, which is located on the top of the spherical crown head. The area with large displacement of the supporting frame is mainly distributed in the vertical square tube connected with the inclined support, as shown in Figure 7.

Figure 6. Displacement distribution of door.

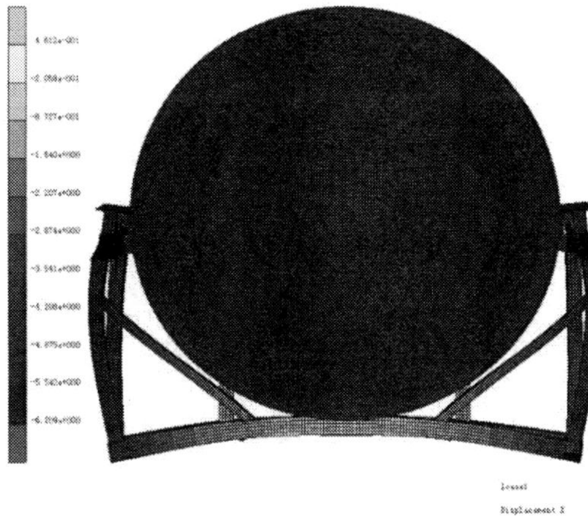

Figure 7. Displacement distribution along the direction of gravity.

3.4. Overturn analysis results

As shown in Figure 8, the support point at the bottom far away from the door has an upward displacement of 0.4mm, which indicates that the support point far away from the door has been suspended, so we can see that the door has a tendency of toppling forward.

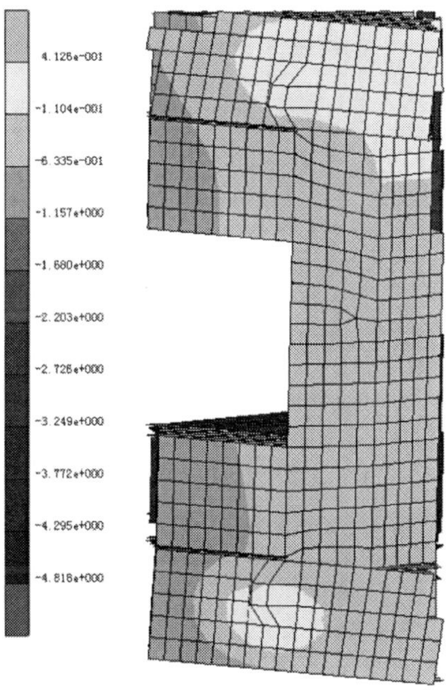

Figure 8. Displacement distribution of bottom support point in the direction of gravity.

4. Anti-overturning optimization

4.1. Optimization method 1

In order to avoid the risk of overturning, it is necessary to optimize the structure of the door. The first method of optimization is to widen the base, extend the square pipe with the mark of 1-5 by 100mm along the direction of red arrow. The comparison between the pre optimization and post optimization is shown in Figure 9.

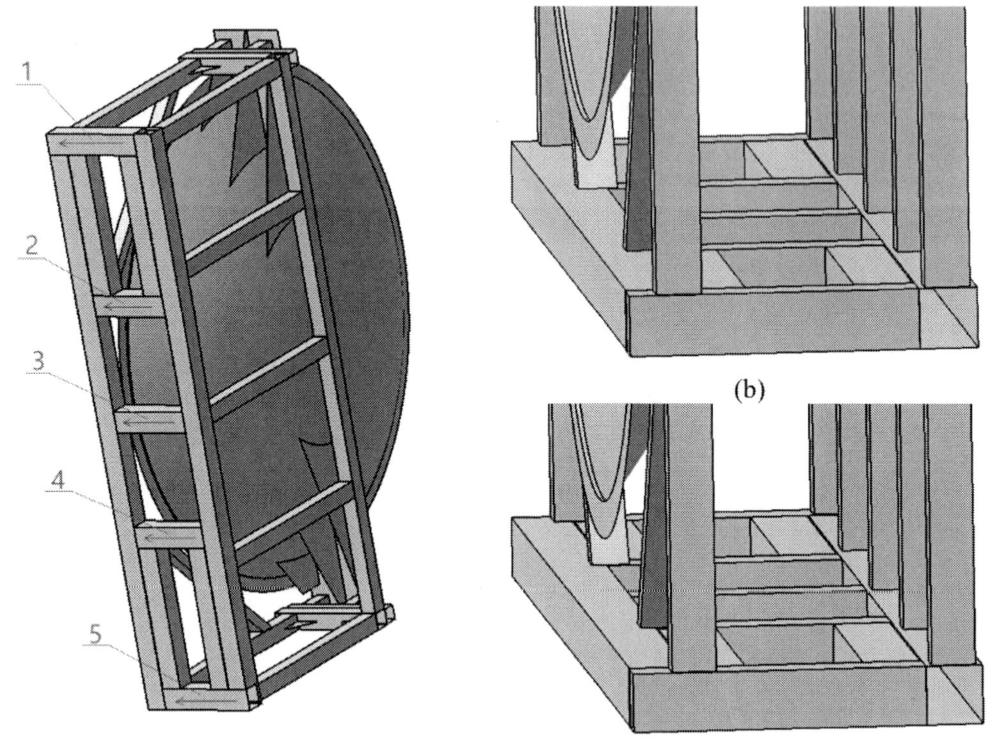

Figure 9. Comparison before and after optimization.

The finite element simulation results after widening the base are shown in Figure 10. The four support points are all in contact with the support surface, and the trend of the optimized model dumping disappears.

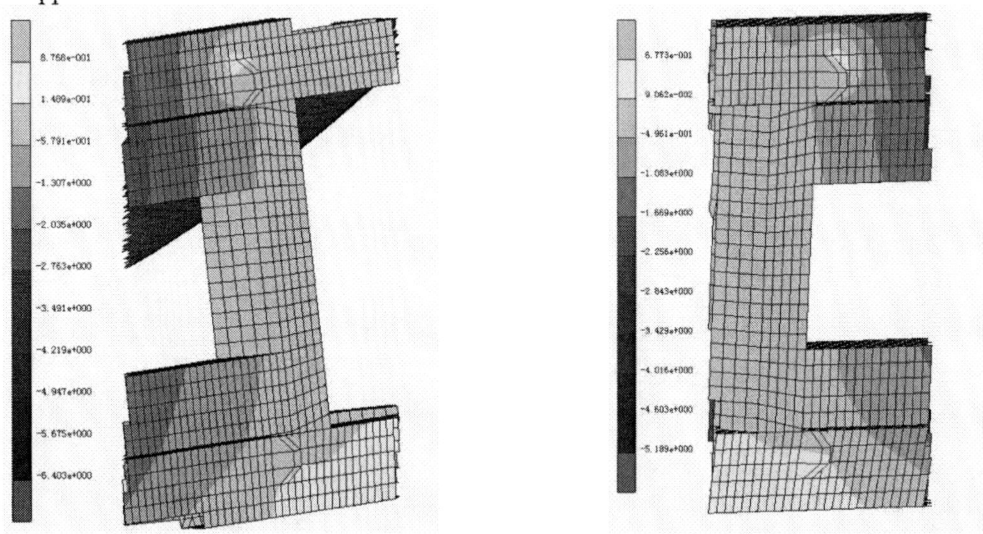

Figure 10. Displacement distribution of optimized bottom support point.

4.2. Optimization method 2

The second optimization method is to place a rectangular iron bar with a mass of 10t on the beam of the supporting frame to play the role of counterweight, as shown in Figure 11.

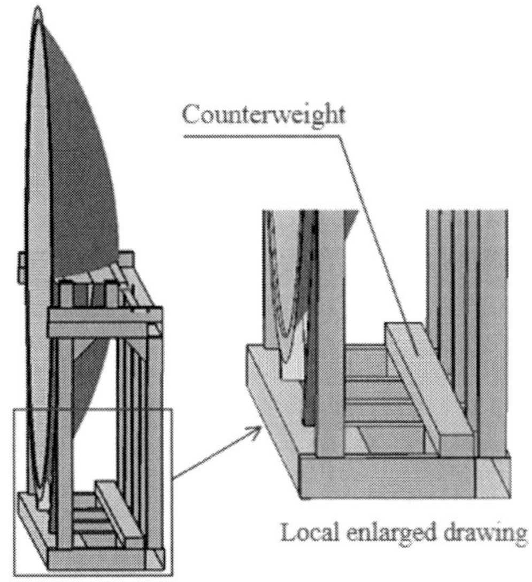

Figure 11. Optimization method 2.

After adding 10t counterweight to the supporting frame, the results of the finite element simulation are shown in Figure 12. All four supporting points are in contact with the supporting surface, and the trend of the optimized model dumping disappears.

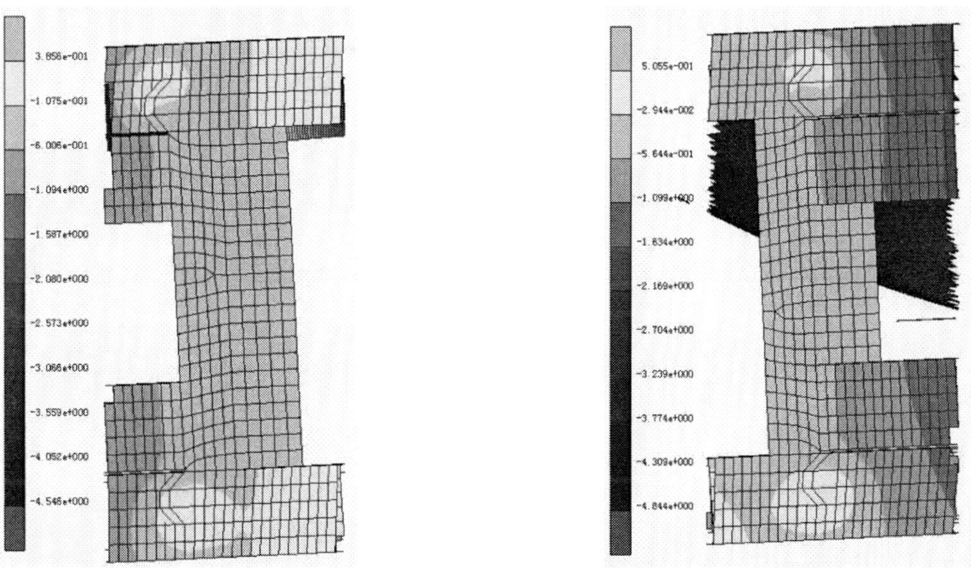

(a) (b)

Figure 12. Displacement distribution of optimized bottom support point.

4.3. Analysis of door overturning resistance

The first optimization method is adopted in this paper. In order to inspect its overturning resistance after optimization, the door is tilted forward 1° during pre-processing, as shown in Figure 13. It can be seen that the front support point of the door is in contact with the ground, while the back support point is far away from the ground.

The simulation result is shown in Figure 14, the back support point of the door contacts the ground. This shows that the door can return to its normal position without overturning under the condition of 1° inclination.

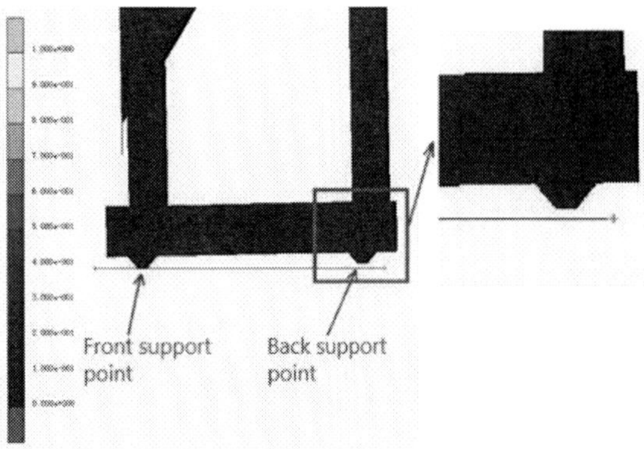

Figure 13. The diagram of overturning resistance test method.

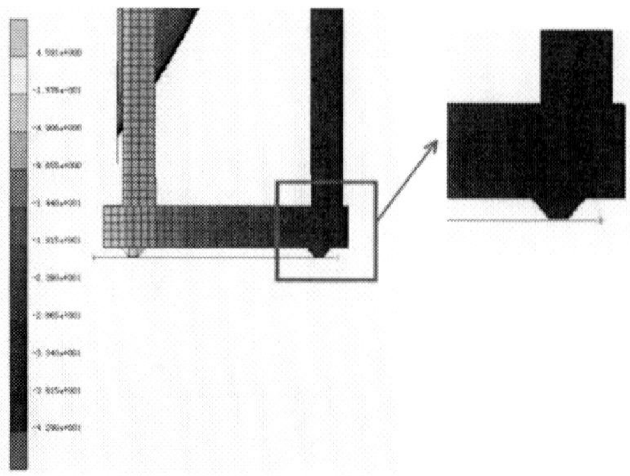

Figure 14. Simulation results of overturning resistance.

5. Conclusion

In this paper, a set of door used in large vacuum chamber is introduced, and its structure is analysed by finite element method. The analysis contents include: stress analysis, displacement analysis, overturn analysis. The analysis results are as follows:

- The maximum stress of the door is 171.4MPa, which is located at the bottom wheel support point, lower than the allowable stress of the material;
- The maximum displacement of the door is 15.03mm, which is located on the top of the head;
- The support point at the bottom away from the door has an upward displacement of 0.4mm, indicating that the support point away from the door has been suspended, and the door has a tendency of forward overturning.

In order to avoid the risk of door overturning, two kinds of optimization methods for anti-overturning and the analysis results after optimization are given:

- The base of the supporting frame is widened by 100 mm. The four support points of the optimized model are all in contact with the support surface, the tendency of dumping disappears, and can resist at least 1° of overturning;
- Place a weight of 10t above the beam under the supporting frame. The four support points of the optimized model are all in contact with the support surface, and the tendency of dumping disappears.

At present, the optimized door is in use with stable and reliable operation.

References

[1] Xiao Han, Yan Qi and Botao Liu 2010 *Structural design and optimization for door flange of a large vacuum vessel* (*Spacecraft Environment Engineering* vol 4) p 493-495

[2] Hengguo Mao, Guoming Xu, Hangsheng Zhou and et al 2012 *Three new designs of the motion mechanism for the vacuum vessel gate* (*Spacecraft Engineering* vol 6) p 671-673

[3] Shouquan Ke, Bencheng Huang, Chuanda He and et al 1993 *Spacecraft environmental engineering and simulation test* (Beijing: Astronautic Press) p 116-118

[4] Rongda Mao, Bo Li, YuanLiang Liu, and et al 2018 *ANSYS-based Structure Analysis and Improvement of a Special-shaped Vacuum Vessel* (*Vacuum & Cryogenics* vol 5) p 348-352

[5] Long Wen, Guilong Ling, Guobiao Cai 2011 *Deformation Simulation of Single—Axis Double-Pin-Hinged Doors* (*Chinese Journal of Vacuum Science and Technology* vol 1) p 50-56

[6] Shizeng Lv, Xiao Han and Yan Qi 2016 *Reliability optimization design of vacuum chamber for extra-large space environmental simulation equipment* (*Chinese Journal of Vacuum Science and Technology* vol 1) p 38-43

[7] Bencheng Huang and Jinming Chen 2005 *Space Vacuum Environment & Vacuum Technology* (Beijing: National Defense Industry Press) p 38-68

[8] GB 150.2-2011 2012 *Pressure vessels-Part 2: Materials* (Beijing: Standards Press of China)

IOP Conf. Series: Materials Science and Engineering **887** (2020) 012028 doi:10.1088/1757-899X/887/1/012028

Finite Element Analysis of Antenna Array of Large Meter Wave Height Mobile Dual Polarization Radar based on NASTRAN

JinWei Wang, YeQing Gu and BaoFu Tang

Nanjing Research Institute of Electronics Technology, Nanjing 210039, China

13913365261@139.com

Abstract. A large-scale meter wave high maneuvering dual polarization radar is designed with a cantilever folded beam. In order to realize the high tolerance spread of its dual polarization antenna, it is necessary to ensure that its folded spread antenna array has high plane stiffness. Firstly, the finite element model is established according to the geometrical model of the antenna array. Secondly, the corresponding load conditions are determined according to the material parameters of the antenna array, different wind loads and temperature fields. Then, the mechanical calculation and analysis of the antenna array are carried out by using NASTRAN software, and the deformation, stress and deformation of the antenna array element and folded beam under different conditions are obtained Square root value. The results show that the array can meet the requirements of stiffness and strength.

1. Introduction

With the gradual deployment of F-22, F-35 and other aircrafts around China, the demand of high mobility meter wave radar for long-range, medium and high altitude surveillance of stealth aircrafts in the region, mobile supplementary network, strong network and emergency air situation support tasks is becoming more and more urgent [1]. Due to the limitation of wavelength and warning task, the scale of meter wave radar antenna is large. In order to ensure the stable operation of its overall radar system, the mechanical structure of its antenna array surface needs to have sufficient rigidity and strength under complex working conditions [2].Therefore, during the design of radar antenna array, it is usually carried out simulation calculation [3]. At present, in view of the radar antenna array design and analysis, the main of antenna array the whole skeleton model of finite element model is established, and the corresponding just strength calculation by finite element software, but in this way the calculation is often the result of the reaction is the stress level and change trend of the antenna array [4], there's no other response matrix surface to the stress of the key components and the deformation tendency.

In this paper, a reasonable finite element analysis model is established for a large-scale meter wave high maneuvering dual polarization radar. Based on this, NASTRAN is used to analyze the antenna array, antenna element, folding beam and other key components of the radar under multiple working conditions, check the deformation and stress values, calculate the corresponding deformation root mean square value, and verify the whole array and related components Structural safety. The results show that the stiffness and strength of the antenna array can meet the corresponding requirements

Content from this work may be used under the terms of the Creative Commons Attribution 3.0 licence. Any further distribution of this work must maintain attribution to the author(s) and the title of the work, journal citation and DOI.
Published under licence by IOP Publishing Ltd

under complex working conditions, which can provide reference for the design and calculation of similar structure antenna array.

2. Array structure and establishment of finite element model

2.1. Array structure

The radar antenna array studied in this paper is mainly composed of antenna beam, folding beam, antenna unit blade, reflection net, roller and connecting bolt. For the finite element analysis of the whole structure of the antenna array, it is necessary to establish the corresponding structural model of the whole antenna array. However, due to the complexity of the actual structure, it takes a lot of time to model and calculate. The actual model is simplified into a reasonable structure model for calculation to ensure that the calculation results are similar to the actual test results. Figure 1 shows the simplified antenna structure model

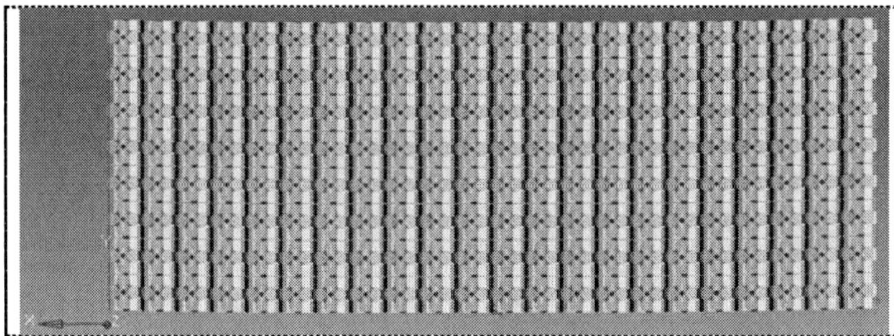

Figure 1. antenna structure model

2.2. Establishment of finite element model

In order to ensure the accuracy and reliability of the finite element analysis results, this paper first focuses on the basic composition of the radar array structure, the stress characteristics of key components, material performance, boundary conditions and external loads and other major factors, to determine the reasonable element type, material characteristics, constraints, connection relationship and load distribution, etc., to reduce the calculation knot as much as possible The error between the result and the actual result. The finite element model of the whole radar antenna array mainly includes

1) element type: due to the complex structure of the antenna array, in order to reduce the finite element model and calculation amount of the whole array, the parts on the crossbeam are represented by solid elements, and the other plate and shell structures, including crossbeam, reflection net, antenna element blade, are represented by shell elements

2) unit property: according to the stress characteristics, material properties and use requirements of different structural parts in the radar antenna array, the materials used for the antenna beam, folding beam, reflection net, roller, sleeve and connecting bolt are steel, and the materials used for the antenna unit blade are aluminum. The elastic modulus of steel is 210000 MPa, Poisson's ratio is 0.3, and density is 7.9e-9 (t/mm^3). The elastic modulus of aluminum is 72000 MPa, Poisson's ratio is 0.33, density is 2.7e-9 ((t/mm^3).

3) grid division: Figure 2 is the grid model of the antenna, which is based on the geometric model to ensure the quality and centroid remain unchanged, and the grid size is 30mm. The parts on the crossbeam are modeled by solid elements. The other plate and shell structures are expressed by shell element, and the grid treatment is carried out by the middle surface. In figure 2, X: antenna deployment direction; Y: gravity direction; Z: vertical antenna blade direction. figure 3 is the mesh model of antenna element. Based on the geometric model, the middle plane is extracted for mesh processing to ensure the quality and centroid remain unchanged, and the mesh size is 15 mm.

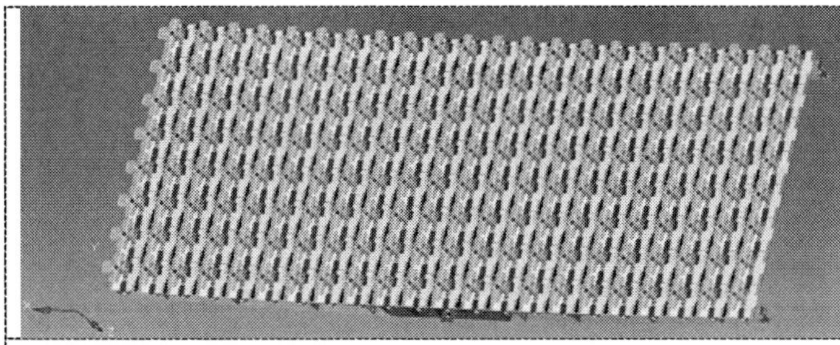

Figure 2. Model of antenna element

Figure 3. Model of antenna element

2.3. Connection problem

According to the connection relationship of the main components on the antenna array, the connection relationship of the main components in the process of finite element modeling is as follows: the reflection network on the adjacent folding beam creates an Rb2 on both sides respectively, so that the Rb2 center points of the reflection network on the adjacent side coincide, creates an Rb2 at the center point, connects two coincident nodes, and releases the degree of freedom around the Y axis, as shown in figure 4 It is shown. The antenna crossbeam and the reflection network are connected by hinges. Rb2 connection is established at the hinge rotation axis, and the freedom of rotation around the Y axis is released. Other freedom constraints are shown in figure 5. Rb2 is set up through bolt holes to connect the reflector support and beam, and fixed connection is shown in figure 6. The blade is bolted to the epoxy plate (the blade and the epoxy plate share the joint), and the blade and the hinge are bolted (the blade and the hinge share the joint), as shown in figure 7.

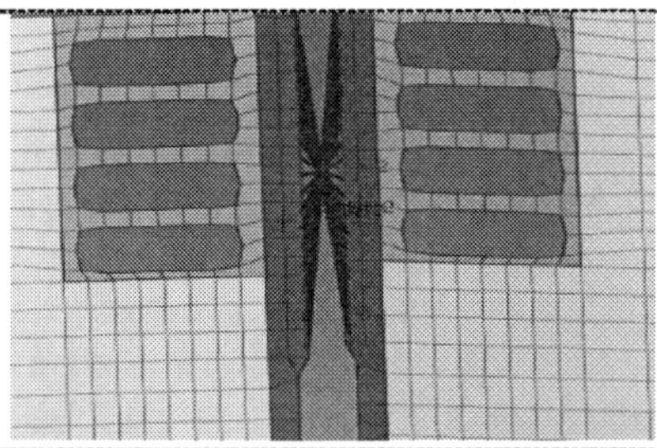

Figure 4. Connection of reflection network

Figure 5. Connection between reflection net and beam

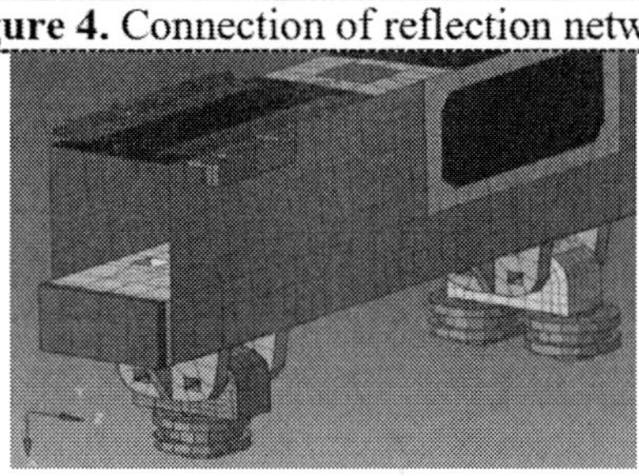

Figure 6. Connection between support and beam of reflection net

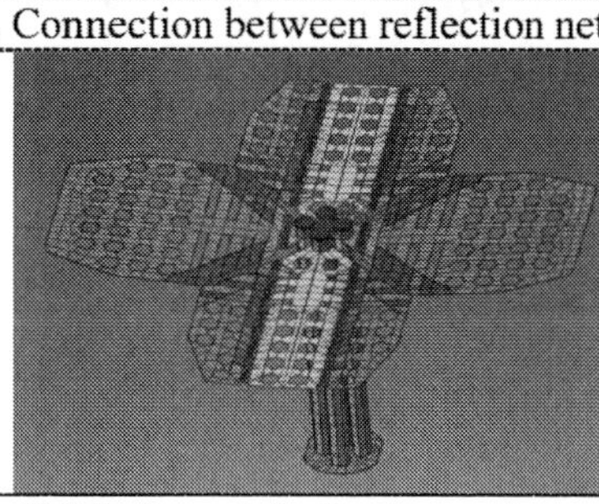

Figure 7. Mesh model of antenna element blade

3. Calculation condition and load

There are many loads on the radar antenna array in the outdoor working environment, including dead weight, wind load, ice and snow load, inertial load and temperature load when the antenna moves. Considering the actual working environment, this paper focuses on the analysis of the stiffness and strength of the antenna array under the combined action of gravity field and wind load, as well as the analysis of the stiffness and strength under the combined action of gravity field and temperature field. The basic equation of structural analysis is shown in formula (1) [5]:

$$[K]\{U\}=\{F\} \tag{1}$$

Where, [K] is the overall stiffness matrix of the structure; {U} is node displacement array; {F} is the node load array.

According to the working status and load bearing of antenna array situation faced by working condition, to calculate the wind load is divided into front and back to the antenna array antenna is surface, which is opposite with his back to the antenna array surface wind velocity of 25 m/s respectively, 30 m/s and 35 m/s, consider the corresponding wind speed at the same time the antenna surface of the centrifugal force, were set to 0.1 r/s, 0.05 r/s, 0 r/s, when the angular velocity is defined to antenna actual center of rotation axis rotation.

The effects of gravity field and temperature field on the antenna array face are set as two working conditions, namely, working condition 1 is considering the dead weight, and the ambient temperature is 20°C, reducing to -40°C. In working condition 2, considering the dead weight, the ambient temperature is 20°C and rises to 60°C.

4. Loading and constraint

4.1. Connection problem

Because the wind speed does not change regularly at any time, the wind load is a kind of dynamic load. This paper mainly considers that the structure of the antenna array is a large rigid structure with a large overall mass and the wind will not cause great vibration to the structure. Therefore, the wind load is considered as static load in the simulation calculation. The calculation formula of wind load is shown in (2)

$$F = C_{Fq} A \tag{2}$$

Where, C_{Fq} is the wind pressure value and A is the characteristic area of the object.

When the wind load is divided into the front face of the antenna, the wind load is evenly distributed on the reflective network, the beam and the parts on the surface of the beam. After calculation, the wind pressure corresponding to the wind speed of 25m/s, 30m /s and 35m /s is 0.0003949 MPa, 0.0005924 MPa and 0.0007899 MPa respectively. When the wind load is divided into the array facing the back of the antenna, the wind load is evenly distributed on the surface of the reflective network and the beam. The wind speed is 25m/s, the wind pressure corresponding to 30m /s and 35m /s is respectively 0.0003686 MPa, 0.0005528 MPa and 0.0007371 MPa.

For the gravity load of the antenna array, only the value of gravity acceleration needs to be specified, and the load can be completed through NASTRAN.

4.2. Constraints

According to the actual test situation, 6 degrees of freedom are constrained at the binding of the folded beam and the inner sleeve before calculation, as shown in figure 8

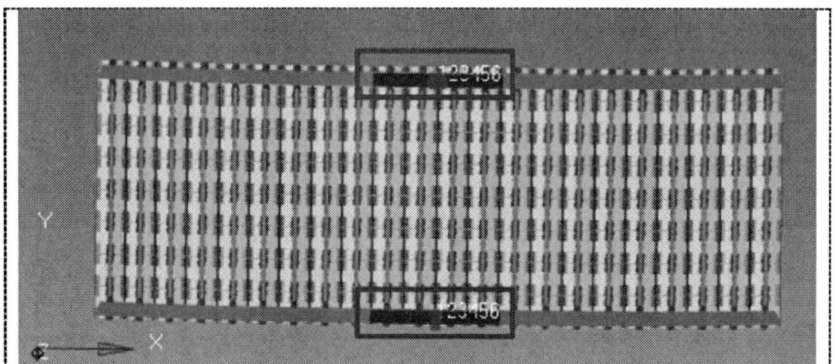

Figure 8. Antenna constraints

5. Calculation results and analysis

One of the main mechanical indexes of the antenna array surface is the precision of the reflection surface [6], and what affects the performance of the antenna array is the error condition of the whole reflection surface, which is measured by the root mean square value of the normal phase error of each node on the reflection surface. In this paper, the stress of the antenna array under different working conditions and the related deformation results are obtained through calculation and analysis, and then the RMS value of the reflection network model is obtained through calculation based on the obtained deformation results. Among them, the antenna array surface under the corresponding working conditions under the combined action of wind load and gravity field, the stress cloud diagram of folded beam and the displacement cloud diagram are extracted, as shown in figure 9, figure 10, figure 11 and figure 12 respectively. Through calculation, the deformation, stress and corresponding root mean square planting of antenna element and folded beam in antenna array face under different wind load conditions are obtained, as shown in table 1.

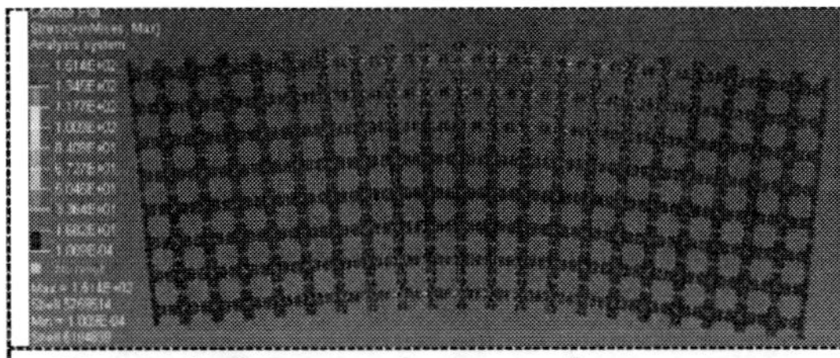

Figure 9. Stress nephogram of antenna array

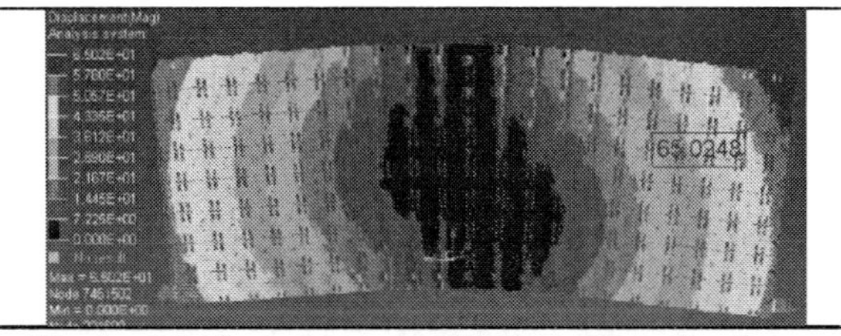

Figure 10. Antenna array displacement cloud map

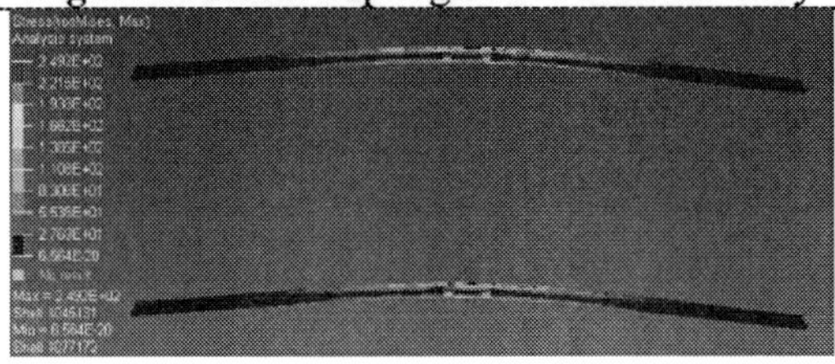

Figure 11. Stress nephogram of folded beam

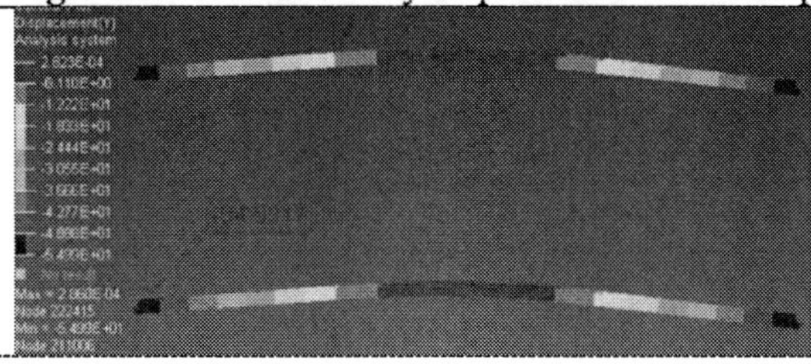

Figure 12. Displacement cloud diagram of folded beam

Table 1. The deformation, stress and RMS of antenna element and folded beam in different wind load conditions

		stress (MPa)			Displacement (mm)			Root mean square value	
ind eed	The direction of the wind	Antenna unit blade (aluminum)	Folding beam (steel)	Antenna array displacement	Antenna element displacement in Z direction	Z direction displacement of blade	Y direction displacement of folded beam	Reflection network	Antenna element blade
m/s	Facing the antenna array	151.37	249.23	65.02	32.86	29.98	55	3.8294	3.1201
	Back to antenna array	135.68	266.33	61.3	29.31	26.8	54.92	3.9610	3.2111
m/s	Facing the antenna array	138.06	222.83	61.82	27.45	27.45	55.2	3.8285	3.3409
	Back to antenna array	126.49	235.88	60.41	25.14	25.14	55.1	3.8753	3.4712
m/s	Facing the antenna array	136.49	216.39	62.55	29.77	27.18	55.41	4.0590	3.5832
	Back to antenna array	131.61	230.47	60.68	26.15	26.15	55.27	4.0614	3.7432

The displacement cloud images of the antenna array surface extracted under working conditions 1 and 2 under the action of temperature field and gravity field are shown in figure 13 and figure 14 respectively. The mean square deformation values of the corresponding reflection network and blade are calculated as shown in table 2.

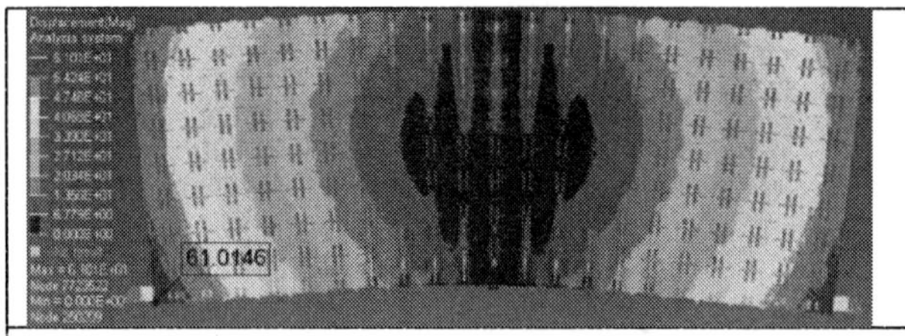

Figure 13. Antenna array displacement cloud map of working condition 1

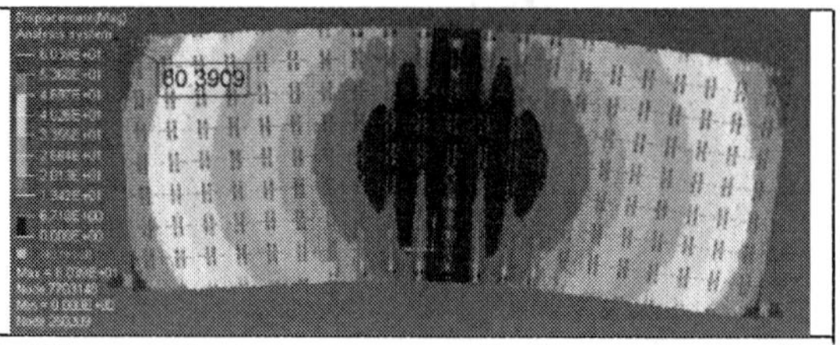

Figure 14. Antenna array displacement cloud map of working condition 2

Table 2. Root mean square value of blade deformation under the action of gravity field and temperature field

Working condition	RMS value of reflection netdeformation	Root mean square value of antenna element blade deformation
Working condition 1	4.9132	4.4808
Working condition 2	4.2566	4.0990

Combined with the calculation results, it can be seen that under the action of gravity field and wind load, the larger the wind load is in the positive direction, the smaller the centrifugal force is, and the smaller the z-direction displacement of the blade is. When the wind speed is 25M/S and the wind direction is positive, the displacement of the antenna array is the largest in the Z direction, and the largest displacement is located on the reflection network, and the displacement reaches 32.86mm. When the wind speed is 25M/S and the wind direction is in the lee direction, the stress value of the folded beam is the maximum. The maximum stress is located at the connection between the roller and the folded beam, and the maximum stress is 266.33mpa, which is less than the yield limit of the material, and the structure is safe. The working state of the antenna, in the combination of gravitational field and temperature field of conditions, when the environment temperature to reduce the 60 degrees, the antenna unit maximum displacement is 60.01 mm, deformation of blade root mean square value is 4.4808, reflecting mesh deformation root mean square value is 4.9132. The environment temperature 40 degrees, antenna unit maximum displacement is 60.39 mm, deformation of blade root mean square value is 4.0990, reflecting mesh deformation root mean square value is 4.2566.

6. Conclusion

1) In this paper the design of large wave high motor dual polarization radar antenna array face reasonable finite element analysis model is established, and according to the antenna array the material parameters and different wind load, temperature load cases, determine the corresponding application NASTRAN again in the gravity field of the antenna array surface and just under the action of wind load strength analysis, and in just strength under the action of gravity field and temperature field analysis, checking on the surface of the antenna array deformation and stress, verify the front of the whole structure of security.

2) The calculation results show that the larger the wind load is, the smaller the centrifugal force is and the smaller the z-direction displacement of the blade is. When the wind speed is 25M/S and the wind direction is opposite, the stress value of the folded beam is the maximum. Among them, the maximum stress is at the connection between the roller and the folded beam, and the maximum stress

is 266.33mpa, which does not exceed the yield strength of the material, and the structure is safe. In addition, under the action of gravity field and wind load, the maximum root mean square value of the reflection network of the antenna array face is 4.0614, and the maximum root mean square value of the blade is 3.7432, which meet the corresponding requirements.

3) The working state of the antenna, the environment temperature by 60 degrees, the maximum displacement is 60.01 mm, antenna unit deformation of blade root mean square value is 4.4808, reflecting mesh deformation root mean square value is 4.9132. The environment temperature 40 degrees, the maximum displacement is 60.39 mm, antenna unit deformation of blade root mean square value is 4.0990, reflective mesh deformation root mean square value is 4.2566, meet the corresponding requirements.

References

[1] Zhang qiang. Mi bo Three-coordinate radar: making stealth fighters have nowhere to hide [J]. Invention and innovation • big technology, 2017(3).
[2] Zhang gen heng, Cha jin shui, Hu jin song. Design of large phased array radar array structure [J]. Radar science and technology, 2016, 14(3):337-342.
[3] Zha jinshui. Structural design of a meter wave radar antenna system [J]. Electronic mechanical engineering, 2012(02):30-33.
[4] Huang xiao, tan gui hong, et al. Simulation and test analysis of a radar antenna skeleton [J]. Electronic mechanical engineering, 2015(6):47-49.
[5] Guo yi mu, Tao wei ming, Zhuang zhuo. Linear and nonlinear finite element and its application [M]. 2004.
[6] Su duan, Li zhi, Liu ji peng, et al. Accuracy analysis and control of large phased array radar antenna array structure [J]. Fire control radar technology (2):78-82.

Study on Unlocking Failure Mode for Mechanical Gyroscope in Spacecraft

Xiaohui Song[1,2], Liang Guo[1], Ming Shan[1], Qiang Zhang[1,2], Lili Wei[1], Pei Wang[1], Xiaochen Wang[1] and Tongchun Feng[1]

[1]Beijing Institute of Spacecraft Environment Engineering, Beijing, China
[2]Beijing Engineering Research Centre of the Intelligent Assembly Technology and Equipment for Aerospace Product, Beijing, China

songhsiaohui@126.com

Abstract. Mechanical gyroscope is a single-degree-of-freedom gyroscope used to measure the angular velocity of the carrier. In this paper, in accordance with the problem in the process of producing, based on the analysis of factors that influence the gyroscope's assembly& debugging precision, through the experiment and analysis, the cause of in the process of producing is found, the technique to solving the problem is presented. The problem of unable to unlock normally is solved. This technique can be used in the producing of gyroscope and can increase passing rate of mechanical gyroscope. The experimentations result indicate that the scheme is in reason. The efficiency of gyroscope assembly is greatly enhanced.

1. Introduction

Mechanical gyroscope is a single-degree-of-freedom gyroscope used to measure the angular velocity of the carrier. The Mechanical gyroscope is mainly composed of gyro motor, a locking mechanism, a float assembly, a housing assembly, an angle sensor, a permanent magnet torque device, etc. The appearance of a gyroscope is shown in Figure 1. The system block diagram of mechanical gyroscope in spacecraft is shown in Figure 2.

Figure 1. Schematic diagram of rate integration gyroscope

Content from this work may be used under the terms of the Creative Commons Attribution 3.0 licence. Any further distribution of this work must maintain attribution to the author(s) and the title of the work, journal citation and DOI.
Published under licence by IOP Publishing Ltd

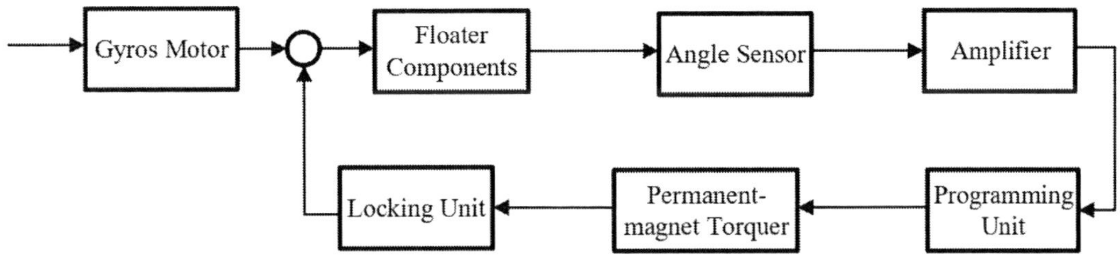

Figure 2. Block diagram of gyroscope

The locking mechanism is the core component of the gyroscope, and its performance determines the accuracy of the gyroscope. Unlocking accuracy is the main indicator for measuring the gyroscope. The stability of the working performance of the locking mechanism is directly related to the reliability and service life of the gyroscope.

During the production process, the launch site and the spacecraft flight, this index must be tested to determine the stability of the locking mechanism and confirm the normality of the mechanical gyroscope function. It must be tested during the production process, launch site and spacecraft flight to determine the stability of the motor and confirm the normality of the rate gyroscope function [1]. As shown in Figure 3, it is a schematic diagram of working principle of locking mechanism

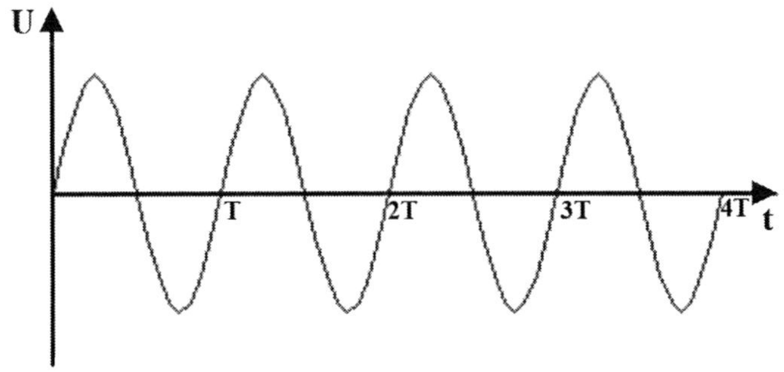

Figure 3. Working principle of locking mechanism

During the overall commissioning phase, this type of gyroscope has repeatedly failed to pass inspection and cannot be successfully delivered [2].

Aiming at this problem, through theoretical analysis and a large number of experiments, a set of process methods capable of improving unlocking reliability during the production process have been developed to ensure the quality and reliability of the product.

2. The problem of unlocking

According to the product structure, the assembly process structure and process flow are divided, and the workload of the assembly unit is evenly distributed by taking workload of each unit into consideration [3].

When a gyroscope performs a lock accuracy check outside the factory, the lock is not unlocked properly. Repeated 7 times, 5 of which were unlocked and 2 unlocked normally.

By analyzing the open-close loop of the product, the problem fault tree can be obtained as shown in Figure 4.

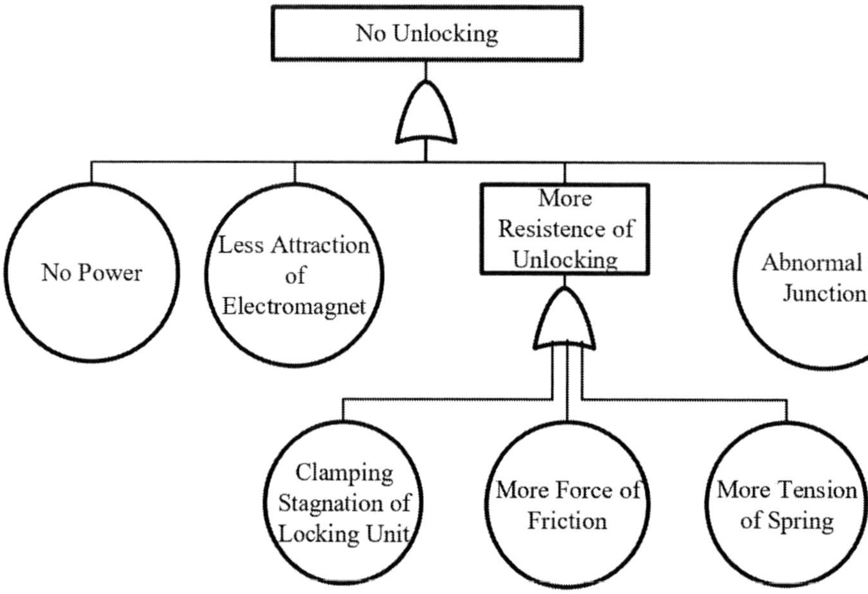

Figure 4. The fault tree of the problem of unlocking

3. Failure mode analysis

As shown in Figure 5, it is a schematic diagram of the unlocking operation of the locking mechanism. Under fault tree in Figure 4, unlock failure mode includes four aspects. Contact reed and contact subassembly are used in kinds of mechanical inertial apparatus, and they are is the media between electric signal and mechanical movement. The gyroscope performs lid opening inspection. It focuses on checking the contact gap of the contact components of the program mechanism.

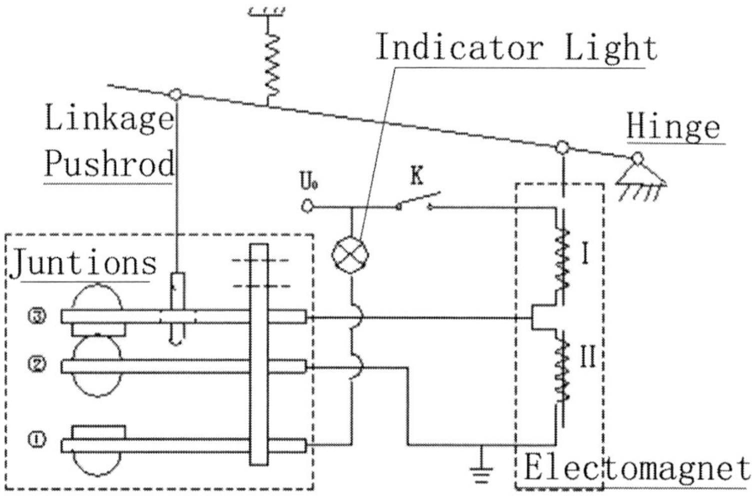

Figure 5. Schematic of unlocking work of locking mechanism

1) Power is not applied

When opening and closing the gyroscope and performing the open-close test, according to the on-site video records, it was clearly observed that the electromagnet has a pull-in action, which proves that the power of the locking mechanism is turned on, and this failure mode can be eliminated.

2) Unlocking resistance increases

This model can be divided into the following two factors:

First, Stuck and increased friction of the locking structure: Open the gyroscope cover and observe the state of pulling when the lock is unlocked. The moving parts are flexible, and no stuck phenomenon is found. Further inspection of the gyroscope with a magnifying glass, no extras were found, and it also proved that there were no stuck and increased friction problems in the lock structure, and this failure mode could be eliminated.

Second, the spring tension is increased: using a digital spring scale, the test spring tension is 1.75Kg, which meets the design technical requirements of 0.8Kg to 1.96Kg. According to the characteristics of the spring, during use, the spring force can only be gradually reduced, and will not be increased. This failure mode can be eliminated[4].

3) Small electromagnet suction

Re-test the resistance of the electromagnet winding coil I, which is 4.1Ω, and the design requirement is 4.1Ω ± 10%, which meets the requirements;

Re-test the resistance of the electromagnet winding coil II, which is 80Ω, and the design requirement is 80Ω ± 10%, which meets the requirements;

The re-testing of the electromagnet's suction force was 8 Kg, and the technical requirements were 6 Kg to 12 Kg, which met the requirements.

These parameters are within the range specified by the technical conditions, which proves that the electromagnet can work normally, and this failure mode can be eliminated.

4) Contact component abnormal

Observation and testing of the contact group found that:

① Contact component surface: ①② There are indentations on the reeds, and there is deformation, and there is ablation on the contacts.

② Due to pressure injury, the pressure of the normally closed contact is 14g, which is significantly less than the specified value of 20 to 30g.

③ The distance between normally open contacts is 0.35mm, which is less than the specified value of 0.5 + 0.1mm.

It can be seen that this contact assembly has not met the technical requirements. Replace the new contact assembly, and then perform the unlocking test. It is found that the unlocking is normal. Therefore, this failure mode cannot be ruled out, so it is thought that the reason that the gyroscope is unlocked abnormally is that the contact component is abnormal.

4. Unlocking work process

When the gyro coil is not powered, the contact reeds ② and ③ form a normally closed contact. When the switch K is closed, the electromagnet coil I works, which generates electromagnetic suction and overcomes the tension of the spring to drive the rocker arm to perform the unlocking action. At this time, the linkage push rod also moves with it. After a certain distance (that is, the distance between the top of the linkage push rod and the contact reed ②), the reed ② is pushed out of the reed ③.

At this time, the solenoid coil II is added to the circuit. The coils I and II work at the same time, so that even if the holding circuit produces a suction force greater than the elastic restoring force, and while ensuring the unlocking, the working current of the coil is also reduced. The linkage push rod pushes the reed ② and then passes a certain distance (the distance between the reeds ② and ①) until the reeds ② and ① are closed, so that the unlocking indicator light is connected to the circuit, the unlocking light is on, and the unlocking process is completed[5].

As shown in Figure 6, the moment when the push rod opens the normally closed contact ② in the unlocking action, it is the switching point of the suction characteristic of the electromagnet. By adjusting the position of the contact assembly relative to the push rod, it is ensured that the switching point is above point b in the figure, so as to ensure that the electromagnetic suction force is greater than the elastic restoring force, and the lock can be unlocked.

At point a, $F_m > f$, the lock can be unlocked normally.

At point b, $F_m = f$, the unlocking mechanism oscillates, and the contact assembly will generate sparks and ablate the contacts.

At point c, $F_m < f$, cannot be unlocked.

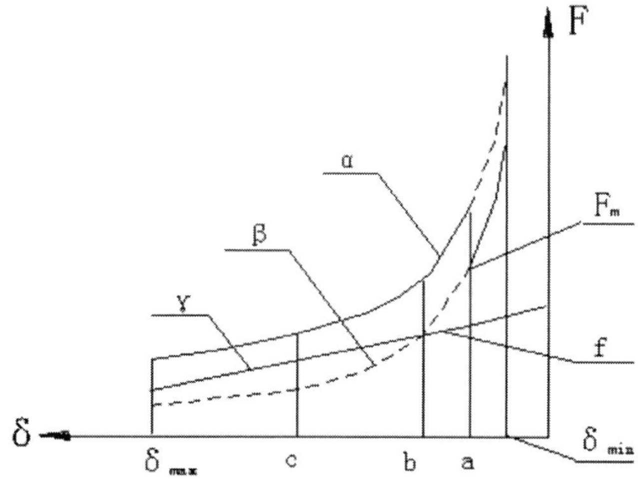

Figure 6. Schematic diagram of electromagnetic suction and spring restoring force

5. Analysis of causes

From the analysis of raw materials, the material of the contact reed is tin bronze. When the parts are processed, they have been heat treated and the elastic properties have reached the requirements. Judging from the retesting of the gyroscope contact assembly, the appearance of the contact assembly was flawed, and both the contact pressure and the distance between the contacts changed: the measured contact pressure was 14g and the distance was 0.35mm. It is thought that the contact damaged the reed during the assembly process. Although the contact assembly met the requirements of contact pressure and spacing at that time, it could be unlocked after adjustment on the instrument base.

Large internal stress, as the stress is released, the reed will produce a large amount of deformation and rebound, which causes the adjusted relative position and the contact pressure of the reed to change (becomes smaller), resulting in the shown in Figure 6. The switching points of α and β are advanced, and the electromagnetic suction is not enough to overcome the spring tension and cannot be unlocked[6].

When the amount of elastic jitter is close to or greater than the gap between two contacts, multi-pulse phenomenon will occur. When the amount of elastic jitter is less than the gap between the two contacts, there will be no sudden occurrence of two contacts, that is, no multi-pulse. Although the elastic vibration of the reed cannot be controlled, the contact gap between the two contacts can be prefabricated.

From the analysis of the manufacturing process, during the locking assembly, due to the pollution of the formula glue during the curing process, the local glue solution cannot be cured, which caused the partially incompletely cured glue solution in the mechanism to precipitate out of the gap when it was working on the instrument, and it was covered in a thin film[7]. On the outer circle of the locking mechanism, thereby preventing the contact between the brush and the conductive ring. When there is an excess on the locking mechanism, the contact between the corresponding brush wire and the brush will be affected, resulting in poor contact and increased contact resistance. The contact resistance increases, sparks occur between the contacts, which accelerates the ablation of the contacts, further worsening the working state of the contacts, and eventually leading to unlock failure.

6. Application effects

Based on the experimental analysis conclusions, combined with the actual situation of production, the following measure are proposed: Before the meter is assembled with the contact assembly, check the appearance of the contact assembly with a magnifying glass, and re-measure the contact pressure and contact gap to ensure that there are no abnormalities. After the meter is assembled with the contact assembly, check the appearance of the contact assembly again with a magnifier to ensure that it meets the technical requirements.

After taking measures, practical test shows the gyro meter inside and outside the plant did not have similar problems.

After adopting the above-mentioned technological measures, stricter process management and quality control have been carried out in each stage of production and use of contact reeds, which strictly guarantees and improves the pass rate of mechanical gyroscope.

7. Conclusion

Aiming at the actual problems in the generation process of the mechanical gyroscope, through theoretical analysis and experimental research, the mode that caused the failure of the unlocking mechanism was found, and a clear, effective and feasible process improvement measure was formulated to avoid similar problems fundamentally Recurrence.

This technique can be used in the producing of gyroscope and can increase passing rate of unlocking normally. The experimentations result indicate that the scheme is in reason. The efficiency of gyroscope assembly is greatly enhanced.

References

[1] Jihua Z 2009 Aircraft Digitized Measurement Aided Assembly Technology and Application. Aeronautical Manufacturing Technology 2009(24):49-52.
[2] Wang Yingyan, Xu Yeping, "The low frequency electric cable network design for satellite," Spacecraft Environment Engineering, 2013, 30 (2) :200～202.
[3] Xu Fuxiang, satellite Engineering, China Aerospace Press, 2002:69.
[4] Tan Weichi, Hu Jingang, Spacectaft System Engineering, China Science& Technology Press, 2009:131.
[5] Jihua Z 2009 Aircraft Digitized Measurement Aided Assembly Technology and Application. ERONAUTICAL MANUFACTURING TECHNOLOGY 2009(24):49-52.
[6] Robert D T, Michael E. Nonlinear finite elment evaluation of marman clamp structural capacity. AIAA 294213462 CP, 1994:320-330.
[7] Zhongyi M 2010 Digital Aircraft Assembly Technology Based on MBD AERONAUTICAL MANUFACTURING TECHNOLOGY 2010(18):42-45.

Experimental Study and Application Prospect Analysis of Truss Heat Pipes for Space Optical Cameras

Shen Chunmei[1,2], LUO Shikui[1,2] , CUI Jin[1,2], YANG Tao[1,2], Liu Zhao[1,2], ZHAO Zhenming[1,2] and YU Feng[1,2]

[1] Beijing Institute of Space Mechanics and Electricity, Beijing
[2] Key Laboratory for Advanced Optical Remote Sensing Technology of Beijing

123855964@qq.com

Abstract. To meet the application requirements of space optical cameras, the concept of a truss heat pipe is proposed in this paper. This truss heat pipe is composed of multiple (two or more) heat pipe shells with inner capillary wicks connected to each other by special joints with additional inner capillary wicks. It can be assembled into a variety of truss structures according to the layout requirements of the structure. Truss heat pipes can be used as the supporting structure of space optical cameras and can simultaneously efficiently transfer heat and maintain an even temperature, conserving the thermal control resources of space optical cameras. The function and performance of truss heat pipes were verified through tests of a single T-joint prototype and a multi T-joint prototype. Two truss heat pipes were then applied to the thermal management of a space optical camera. The thermal vacuum test results show that as the supporting structure of the front lens barrel heat shields, the two truss heat pipes can transfer heat from the space optical camera heat source to the front lens barrel heat shield both efficiently and evenly. This indicates that truss heat pipes can meet the requirements of the structure-thermal integrated design and thermal management of space optical cameras. As a new thermal control technology and product, truss heat pipes have the potential for broad applications in many aerospace fields in the future.

1. Introduction

A satellite in orbit is alternately heated and cooled by orbital heat flow (including solar radiation heat flow, Earth albedo heat flow, and Earth infrared heat flow) and the 4 K background of cold black space. The space environment in which satellites operate is thus very harsh[1-2]. As a main satellite payload, space optical cameras are also affected by the space environment. Meanwhile, space optical cameras are also affected by thermal disturbances due to their internal heat sources, such as electronic equipments, CCD components, cryocoolers, etc. To ensure imaging performance in orbit, the precision of temperature control for the optical lens and its supporting structure must be high. The temperature control accuracy of a typical high-resolution camera is up to ±0.3°C[3]. This makes thermal design one of the decisive factors for the performance of space optical cameras and also makes thermal design a key problem to be solved by thermal designers at home and abroad[3-14].

There are several traditional thermal control methods for the optical lens and corresponding supporting structure in space optical cameras. Components that are exposed to space are covered with multilayer insulation (MLI) to reduce the adverse impact of orbital heat flow and the 4 K background cold black space. Meanwhile, active temperature control heating circuits are arranged around the space

Content from this work may be used under the terms of the Creative Commons Attribution 3.0 licence. Any further distribution of this work must maintain attribution to the author(s) and the title of the work, journal citation and DOI.
Published under licence by IOP Publishing Ltd

optical camera to achieve accurate temperature control of the optical lens. Traditionally, internal heat sources are heat-insulated from the optical structures of the camera. The heat produced by the heat sources is transferred to an external radiator through an axially grooved heat pipe and is discharged to space. Simultaneously, additional active temperature control heating circuits are arranged on the internal heat sources to ensure that the temperatures of these components do not fall below the required lower limit when they are not working.

The traditional thermal control methods for space optical cameras are resource-intensive, requiring power, weight, layout considerations, installation space for the heat pipe and radiator, and so on. As long as the thermal control resources are sufficient, traditional thermal design ideas can keep the temperature of space optical camera components within the requirements. However, with the development of space optical camera technology, space optical camera thermal control is facing more and more challenges. The most prominent issue is that the number of internal heat sources with large power consumption and long working time is large. For example, the heat sources in one particular space spectrometer include three video processors and two pulse tube cryocoolers, whose total peak consumption can be up to 362 W[15]. In this case, using traditional thermal control measures would consume a large number of thermal resources, such as needing a large radiator and a large number of heat pipes, as well as having high power consumption.

Satellites cannot meet unlimited thermal control resource demands. Thermal engineers need new design ideas or methods to effectively solve thermal control challenges with limited resources. For instance, in the thermal design of the space spectrometer mentioned above[15], the thermal engineers used the power consumed by the electronic devices for thermal control by using multiple axially grooved heat pipes to connect the electronic devices with the bottom plate and cover shell of the space spectrometer, forming a thermal control network. This allowed them to simultaneously conserve both the thermal control power consumption and the radiator for the electronic devices. However, in the thermal control network, there are overlapping thermal resistances between multiple heat pipes, which leads to lower heat transfer efficiency. This is therefore not the best solution.

In this paper, truss heat pipe (THP) technology was proposed to better solve the above problems. Tests of the prototype were carried out. Additionally, two THPs were applied to a space optical camera and tested. Finally, the application prospects of the truss heat pipe were analyzed.

2. Overview of THP

A THP is a special new type of heat pipe that is composed of multiple (two or more) heat pipe tube shells with inner capillary wicks connected to each other by special joints with additional inner capillary wicks. The basic elements of a THP include the joint, heat pipe tube shell, sealed end cover, filling end cover, and filling tube. Different types of joints can be used, including L-type, T-type, cross-type, and others. According to the layout requirements, a THP can be assembled into a variety of truss structures. As a heat transfer element, a THP also has a mechanical bearing capacity and can be used as a supporting structure. The capillary wicks and vapor passages of the THP's different heat pipe tube shells can be connected to each other through joints. The working medium of the vapor-liquid phase change heat transfer can then circulate in the THP, just as in a single common heat pipe.

Figure 1 shows diagrams of plane single T-joint and multi T-joint truss heat pipe structures. The THP described in this paper is composed of an aluminum alloy, and the working medium for the vapor-liquid phase change heat transfer is high purity ammonia. Other compatible materials and working fluids can also be used according to requirements.

3. Functional verification of THPs

In this paper, prototypes of a single T-joint truss heat pipe and a multi T-joint truss heat pipe, shown in Figure 1, were manufactured and tested.

IOP Conf. Series: Materials Science and Engineering **887** (2020) 012030 doi:10.1088/1757-899X/887/1/012030

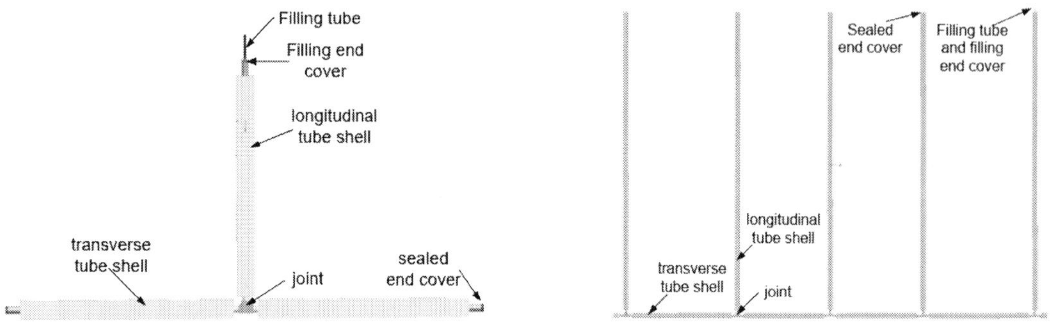

(a) single T-joint THP (b)multi T-joint THP

Figure 1. Diagrams of plane single T-joint and multi T-joint THP structures

3.1. Tests and results of the single T-joint prototype

The diagram and photo of the single T-joint prototype and its test system are shown in Figure 2. The size of the single T-joint prototype can be seen in Figure 2. During tests, 14 thermocouples connected to temperature measuring equipment were arranged on the single T-joint prototype for real-time temperature recording. A heating circuit was affixed on the evaporation section of the single T-joint prototype. The condensation section of the prototype was installed on the cold plate of the refrigerator. The prototype was covered with insulating cotton. The test cases considered are shown in Table 1.

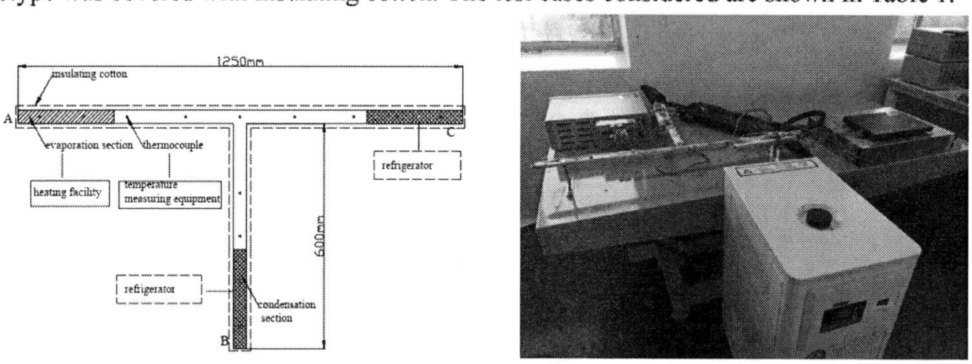

Figure 2. Diagram and photo of the single T-joint prototype and its test system

Table 1. Test cases considered for the single T-joint prototype

Evaporation section and condensation section	Thermal load on evaporation section(W)				Temperature of refrigerator cold plate(°C)
A is evaporation, B and C are condensation	60	90	120	150	10
B is evaporationn, A and C are condensation	60	90	120	150	10

Curves of the single T-joint prototype temperature over time under different test conditions are given in Figure 3. The maximum difference and standard deviation of all temperature measuring points on the single T-joint prototype are shown in Table 2.

For test cases in which A is the evaporation section and B and C are condensation sections (case A) and the thermal load on A is within 150 W, the maximum difference between all of the temperature measuring points on the single T-joint prototype varied between 1.3°C and 3.7°C with a standard deviation of about 0.4°C to 1°C, showing that the single T-joint prototype has good isothermal properties.

For cases in which B is the evaporation section and A and C are condensation sections (case B) and the thermal load on B is within 120 W, the maximum difference between all of the temperature measuring points varied between 1.1°C and 3.8°C with a standard deviation varying from about 0.3°C to 1°C, showing that the single T-joint prototype has good isothermal properties in these conditions as well. When the thermal load on B is increased to 150 W, however, the maximum difference and standard deviation are 8.8°C and 2.2°C, respectively. Both the maximum temperature difference and standard deviation are larger than those for the same heat load in case

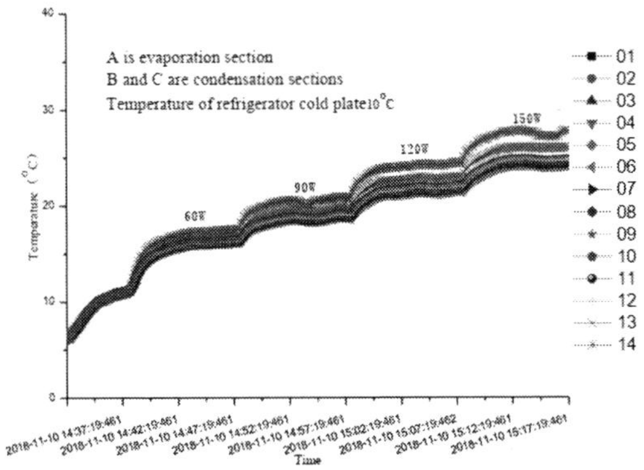

(a)

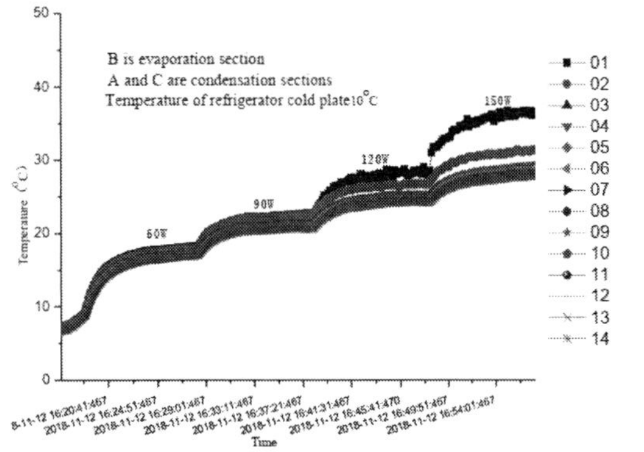

(b)

Figure 3. Curves of the single T-joint prototype temperature over time under different conditions

Table 2. Maximum difference and standard deviation of all temperature measuring points on the single T-joint prototype

-	Maximum difference(°C)				standard deviation(°C)			
Thermal load on evaporation (W)	60	90	120	150	60	90	120	150
A is evaporation, B and C are condensations	1.3	2.0	3.5	3.7	0.4	0.5	0.8	1.0
B is evaporation, A and C are condensation	1.1	1.9	3.8	8.8	0.3	0.5	1.0	2.2

This is because, compared with case A, both the backflow resistance of liquid in the capillary core and the flow resistance of vapor in the vapor channel at the joint are larger in case B. Thus, theoretically, the maximum heat transport capability at capillary limit is larger in case A than in case B. With the same 150 W load, the maximum temperature difference and temperature standard deviation of all temperature measuring points are larger in case B than in case A, indicating that 150 W is close to the maximum heat transport capability at capillary limit in case B. In contrast, the gap between 150 W and the maximum heat transport capability at capillary limit in case A is relatively large.

The shell wall temperature of the evaporation section is determined by the temperature of the backflow supercooled liquid, local steam temperature, and ambient temperature. With a fixed ambient temperature, when the actual heat transfer load is close to the maximum heat transport capability at capillary limit, the reflux rate of the supercooled liquid under capillary suction is slightly smaller than its evaporation rate. This will cause the end of the evaporation section to be partially dried or have little supercooled liquid. As a result, the cooling effect of the cold liquid on the shell wall of the evaporation section is weakened. On the other hand, when the heat load on the evaporation section is large, more steam will be produced and the steam flow speed will be large. The flow resistance of vapor in the vapor channel will be large. The difference in steam pressure between the evaporation section and the condensation section is then large, which means that the temperature difference between the steam in the evaporation and condensation sections is large. The above two phenomena lead to the large temperature difference between the evaporation section and the condensation section of the heat pipe.

3.2. Tests and results of the multi T-joint prototype

In this study, the multi T-joint prototype shown in Figure 1(b) was designed and manufactured. Its transverse and longitudinal dimensions were 2463 mm and 1670 mm, respectively. The multi T-joint prototype was tested in a low-temperature vacuum environment simulation room. The test system diagram is shown in Figure 4. A test support and test fixture were also designed. The test support was used to support the multi T-joint prototype and its test fixture. The test fixture includes a radiator and a cabin. The multi T-joint prototype was installed on the radiator thermally conductively. The radiator was then installed on the test support thermally non-conductively, and the side of the radiator facing the low-temperature heat sink was painted black to radiate heat. The cabin was installed on the radiator thermally non-conductively. The inner surface of the cabin was covered with multilayer insulation to ensure that the radiation heat exchange between the multi T-joint prototype and the cabin was as small as possible. Figure 5 shows the assembly diagram of the multi T-joint prototype, test fixture, and test support.

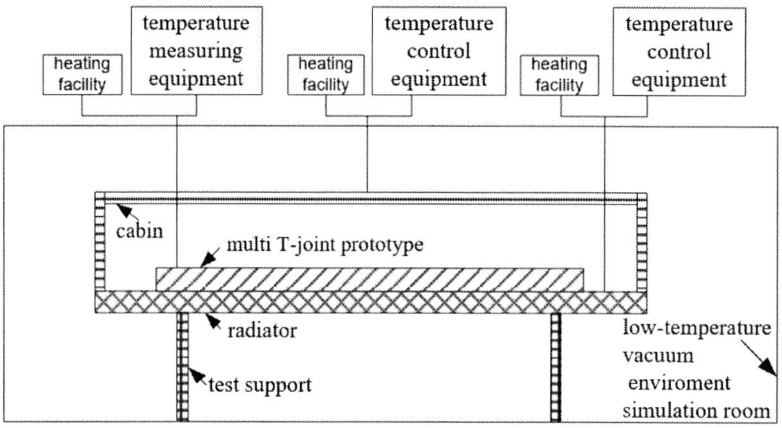

Figure 4. Test system diagram of the multi T-joint prototype

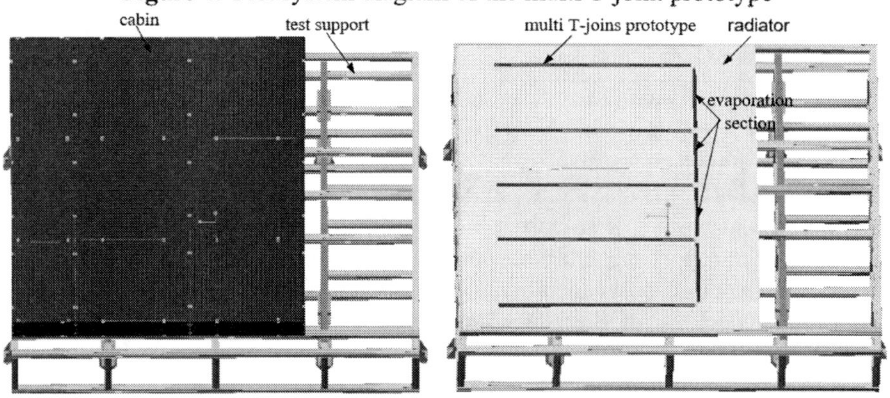

Figure 5. Assembly diagram of the multi T-join prototype, test fixture and test support

The heating circuit was located at the evaporation section of the multi T-joint prototype. Thermocouples were arranged on the shell to measure the temperature distribution on the prototype. The temperature control heating circuit and temperature control thermocouples were arranged on the radiator and cabin. During tests, the temperature of the environment simulation room was 100 K and the pressure was below 1.3×10^{-3} Pa. The working temperature distribution of the multi T-joint prototype was tested under the condition that when the evaporation section was exposed to a certain heat load, the radiator and cabin were controlled at a certain temperature. Specific test cases are given in Table 3.

Table 3. Test cases of the multi T-joint prototype

Thermal load on evaporation section(W)	Temperature of radiator and cabin(°C)	Temperature and pressure of the low-temperature vacuum environment simulation room
90	-29	
110	-28	Temperature:100K
130	-29	Pressure:≤1.3×10^{-3}Pa
150	-31	

The temperature curves of the multi T-joint prototype over time under different test conditions are given in Figure 6. The maximum difference and standard deviation of all temperature measuring points are shown in Table 4. From the test results, we observe the following. When the thermal load on the evaporation section was between 90 W and 150 W, the maximum temperature difference between all of the temperature measuring points on the multi T-joint prototype varied between 4.0°C and 8.7°C and the standard deviation varied between about 1.1°Cand 2.1°C, indicating that the multi T-joint prototype can work normally as a heat pipe. The temperature standard deviation is smaller than the maximum temperature difference, reflecting the fact that only a few data points have a slightly larger temperature difference, as can be seen in the temperature curves shown in Figure 6.

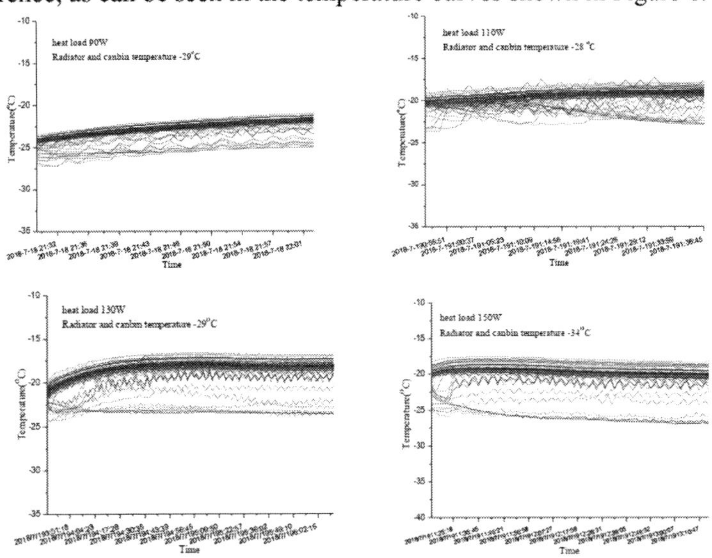

Figure 6. Temperature curves of the multi T-joint prototype over time under different test conditions

Table 4. Maximum difference and standard deviation of all temperature measuring points

Thermal load on evaporation section (W)	Maximum difference(°C)	standard deviation(°C)
90	4.0	1.1
110	6.1	1.3
130	7.0	1.7
150	8.7	2.1

4. Application and verification of the THP on a space optical camera

4.1. Application mode and THP structure

A diagram of the structure of the space optical camera is given in Figure 7. The main structure includes the baffle, front lens barrel, front optical lens assembly (main mirror assembly and secondary mirror assembly), rear optical system, focal plane assemblies, and two cryocoolers for the focal plane assemblies. According to traditional thermal design, the front lens barrel temperature should be controlled with temperature control heating circuits to provide a good temperature environment for the front optical lens assembly. Heat consumed by the two cryocoolers should be transferred to a space radiator to dump the heat to cold black space. Alternatively, to use the thermal control resources more efficiently, heat consumed by the two cryocoolers can be transferred to the front lens barrel. This can not only reduce the power consumption of the temperature control in the front lens barrel but can also eliminate the radiator of the cryocoolers. The heat transfer path from the cryocoolers to the front lens barrel using an ordinary axially grooved heat pipe is shown in Figure 8. In this heat transfer path, there are multiple overlapping thermal resistances. Therefore, the total thermal resistance of the heat transfer path is large. To reduce the thermal resistance of the heat transfer path as much as possible, the bottom circumferential heat pipes and the optical axis axial heat pipes were combined into one THP to greatly improve the efficiency with which the heat consumed by the cryocoolers was utilized.

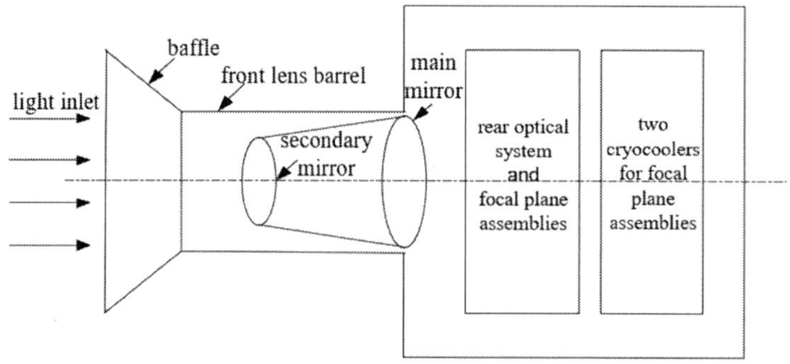

Figure 7. Diagram of the structure of the space optical camera

The Pictures of the THPs are given in Figure 9. Two sets of semicircular truss heat pipes, which can be seen in Figure 9(a) and 9(b), respectively, make up a cylindrical frame. The front lens heat shields are installed on the cylindrical frame of the THP, as shown in Figure 9(c). The inner diameter and height of the cylindrical frame of the truss heat pipe are 1602 mm and 1685 mm, respectively. The THP tube shell is an axially grooved heat pipe shell made of an aluminum alloy. The T-joint material is an aluminum alloy. The working medium is high purity ammonia.

Heat consumed by the cryocooler is first transferred to the circumferential tube shell evaporation section. The heat at the evaporation section is then transferred to other parts of the THP through the vapor-liquid phase transformation of the working medium in the THP. Finally, the THP transmits the heat to the front lens barrel heat shields through heat conduction. The evaporation section of THP A receives the heat consumed by one cryocooler. The evaporation section of truss heat pipe B receives the heat consumed by the other cryocooler.

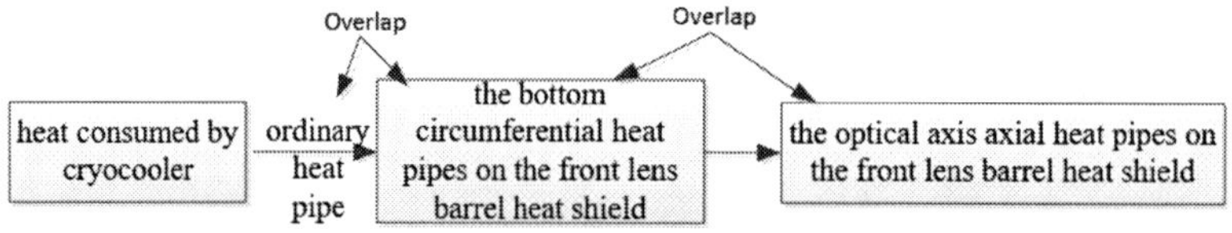

Figure 8. Heat transfer path from the cryocoolers to the front lens barrel using ordinary heat pipe

4.2. Vacuum thermal test results

During the vacuum thermal test, the two truss heat pipes were installed on the space optical camera as the supporting structure of the front lens barrel heat shields. The front lens barrel heat shields were installed on the two truss heat pipes thermally conductively. The space optical camera was vertically placed in the simulation room with the baffle vertically up. In this arrangement, the circumferential pipe shells of the truss heat pipe were at the bottom and the axial pipe shells of the truss heat pipe were on top.

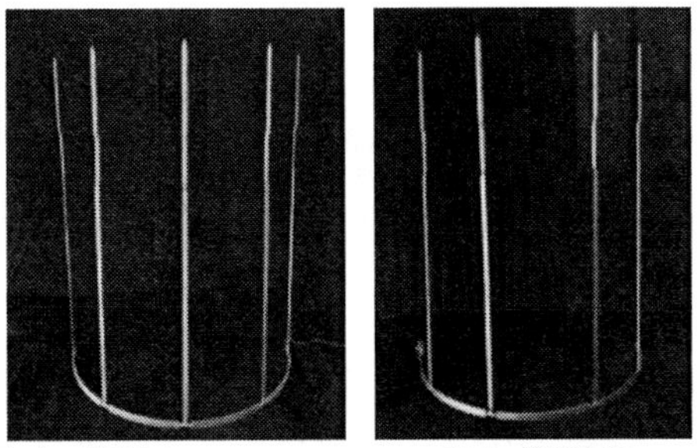

(a) THP A (b) THP B

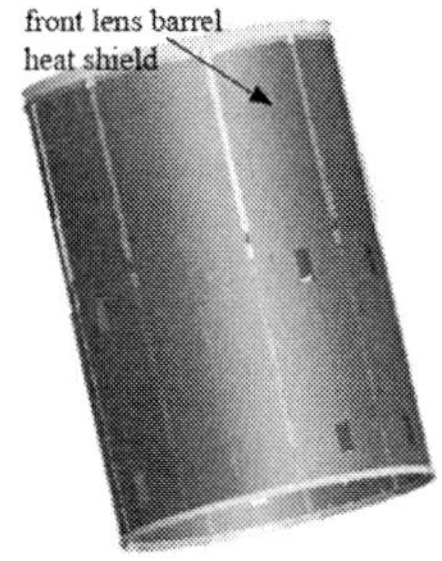

(c)

Figure 9. Pictures of two THPs and their assembly diagram with front lens barrel heat shield

During the tests, thermocouples connected to temperature measuring equipment were arranged on the truss heat pipe for real-time temperature recording. The temperature of the simulation room was 100 K and the pressure was below 1.3×10^{-3} Pa. The two THPs and the front lens barrel heat shields faced the 100 K low-temperature heat sink directly through the light inlet. The working heat consumption of the cryocooler thermally connected with truss heat pipe A was 140 W, while the working heat consumption of the cryocooler thermally connected with truss heat pipe B was 100 W.

The temperature curves of the two THPs over time are given in Figure 10. The maximum difference and standard deviation of all of the temperature measuring points on the two THPs are shown in Table 5. From the test results, we can see the following. The maximum difference among all of the temperature measuring points on THP A is 8.2°C with a standard deviation of 2.3°C, while the maximum difference of all temperature measuring points on THP B is 6.9°C with a standard deviation of 2.2°C. These results indicate that the two THPs can transfer the heat consumed by the two cryocoolers to the front lens barrel heat shield efficiently and evenly. The power consumption of the front lens barrel temperature control is saved. The radiator of the cryocoolers is eliminated. The uniform temperature applied to the front lens barrel heat shield by the THP also provides a good temperature environment for high precision temperature control of the front optical lens.

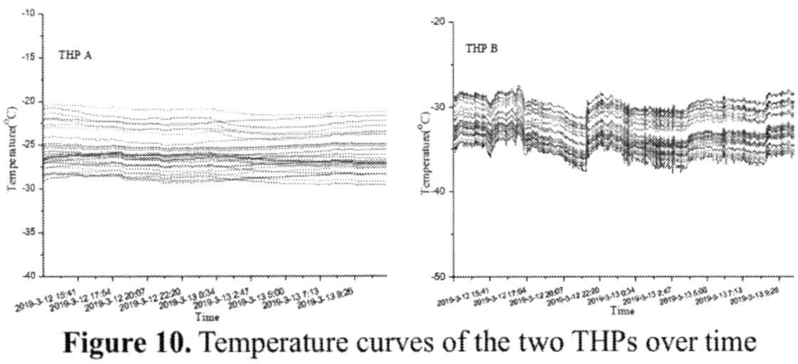

Figure 10. Temperature curves of the two THPs over time

Table 5. Maximum difference and standard deviation of all of the temperature measuring points on the two THPs

Truss heat pipe	Thermal load	Maximum difference(°C)	standard deviation(°C)
Truss heat pipe a	140	8.2	2.3
Truss heat pipe b	100	6.9	2.2

5. Application Prospect Analysis of THP

5.1. Structure-thermal integrated design and thermal management of space optical cameras

The THP proposed in this paper can be used as the supporting structure of a space optical camera and can also efficiently transfer heat and contribute to even temperature distributions. In addition, when bending an ordinary heat pipe, the bending radius generally must remain at least 5 times larger than the outer diameter in order to avoid damaging the internal axially grooved channel. In contrast, the joint design of the THP is not limited by this minimum bending radius, which can significantly conserve space in the heat pipe layout and reduce the difficulty of designing the heat pipe layout. Therefore, the THP has a wide application prospect in the structure-thermal integrated design and thermal management of space optical cameras.

5.2. Thermal control of on-board equipment with high power and high temperature stability requirements

With the development of large-capacity communication satellites, high-resolution remote sensing satellites, new scientific exploration satellites, and other civil spacecraft in China, more and more devices with high power and high stability requirements are used. For example, there are many TWTs with high power dispersed across the decks of communication satellites. In another example, there are many SAR antenna T/R modules on remote sensing satellites. The arrangement of these T/R modules is compact and the heat consumption of the T/R module array is large, while the temperature uniformity requirements of the T/R modules are high. If the heat cannot be drawn away, the performance of the feeder unit will be affected. At present, the traditional method is to use the outer surface of the satellite deck as a radiator surface to disperse the working heat consumption of TWTs and TR modules into cold black space. However, because the satellite deck is a sandwich structure with a honeycomb core between the composite panels of both sides, the thermal conductivity in both the plane direction and the thickness direction of the satellite deck is limited. In order to reduce the thermal resistance between heat sources on the inner panel and the outer panel as much as possible, and to avoid the uneven temperature of the deck caused by the uneven heat consumption of heat sources and the uneven orbit heat flow incident on the outer surface of the deck, an orthogonal lap heat pipe system is arranged on the satellite deck to strengthen the heat conduction in the thickness direction of the cabin and to promote uniform temperature in the plane direction. At present, there are three types of orthogonal arrangements of heat pipes, i.e. the fully embedded type, the hybrid type, and the fully externally protruding type, as shown in Figure 11. These three arrangements all have the problem of overlapping thermal resistance. There are also other problems in the layout of the 2-layer overlap fully embedded type, such as the thick deck and the complex fabrication process of the deck. The externally protruding type, meanwhile, affects the layout of the equipment on the deck.

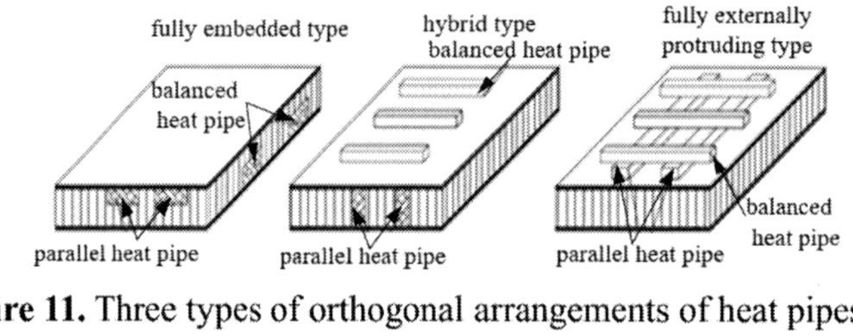

Figure 11. Three types of orthogonal arrangements of heat pipes on board

In view of the shortcomings of the traditional thermal control measures mentioned above, if the single-layer plane orthogonal integral truss heat pipe is embedded in the satellite deck, the overall heat transfer efficiency can be increased and the temperature equalization effect can be improved greatly . Meanwhile, both the difficulty of the manufacturing process and the thickness of the satellite deck can be greatly reduced. Therefore, truss heat pipes have a wide application prospect in the thermal control of equipment with high power and temperature stability requirements.

6. Conclusion

Based on the application requirements of space optical cameras, the concept of a THP composed of multiple heat pipe tube shells connected with each other by special joints was proposed in this paper. According to the layout requirements, the THP can be assembled into a variety of truss structures. THPs can be used as the supporting structure of a space optical camera and can also efficiently transfer heat and help maintain an even temperature, efficiently conserving the space optical camera's thermal control resources. Prototypes of a single T-joint THP and a multi T-joint THP were manufactured and tested. The test results fully verified the function and performance of THPs as heat pipes. Two truss heat pipes were then applied to the thermal management of a space optical camera. The thermal vacuum test results show that as the supporting structure of the front lens barrel heat shields, the two THPs can transfer heat from the space optical camera heat source to the front lens barrel heat shield both efficiently and evenly. The function and performance of the THP were also verified.

THPs can be used as the supporting structure of space optical cameras while efficiently transferring heat and maintaining an even temperature. In addition, the joint design of truss heat pipes is not limited by a minimum bending radius, which can significantly reduce the space needed by the heat pipe layout and reduce the difficulty of the layout's design. The plane orthogonal truss heat pipe can replace the traditional orthogonal lap heat pipe system and solve problems associated with current orthogonal lap heat pipe systems, such as large lap thermal resistance, thick satellite decks, and the complex fabrication process of the satellite deck. Therefore, as a new thermal control technology and product, THPs require further research and optimization but have the potential for broad application in many aerospace fields in the future.

References

[1] MIN G R. Satellite thermal control technology[M]. Beijing: China Astronautic publishing House, 1991.(in Chinese)

[2] David G. Spacecraft thermal control handbook volume I: fundamental technologies[M]. EI California: The Aerospace Press,2002.

[3] LI Chun-lin. Research on Space Optical Remote Sensor Thermal Control Technique[J]. Journal of Astronautics ,2014,35(8):863-870. (in Chinese)

[4] Bell J. L., Pavlov M. Design and analysis of a beryllium three—mirror anastigmat telescope for the Japanese advanced Meteorological imager (JAMI)[C]. International Asia—Pacific Environmental Remote Sensing Symposium, 4th: Remote Sensing of the Atmosphere, Ocean, Environment, and Space, Honolulu, Hawaii, Nov.8-11, 2004.

[5] Li Qiang, Chen Liheng. Thermal design of infrared detector components in complex heat flux[J]. Infrared and Laser Engineering, 2016, 45(9). (in Chinese)

[6] Chen Liheng, Li Yanchun, Luo Zhitao, et a1. Thermal design and testing of CCD for space camera[J]. Optics and Precision Engineering, 2011, 19(9):2117-2122. (in Chinese)

[7] Li Ming, Wu Qingwen, Jiang Fan. Design of thermal control system for three-linear array mapping cameras [J]. Optics and Precision Engineering, 2010, 18(6): 1367-1373. (in Chinese)

[8] Zhang Yue, Wang Chao, Su Yun Jiao, Jianchao. Thermal control scheme for ultrahigh resolution imaging system on geosynchronous orbit[J]. Infrared and Laser Engineering, 2014, 43(9):3116-3121. (in Chinese)

[9] ZHAO ZH M, LU P, SONG X Y. Themal design and test for high resolution space space optical camera on GF-2 satellite[J]. Spacecraft Recovery & Remote Sensing, 2015,36(4):34-40. (in Chinese)

[10] YU F, XU N N, ZHAO Y, et al. Thermal design and test for space space optical camera on GF-4 satellite[J]. Spacecraft Recovery & Remote Sensing,2016, 37(4):72-79. (in Chinese).

[11] ZHANG Y, WANG CH, SU Y, et al. Thermal control scheme for Ultrahigh Resolution Imaging System on Geosynchronous Orbit[J]. Infrared and Laser Engineering, 2014,43(9):3116-3121. (in Chinese).

[12] CHEN W CH, WANG H X. Verification of thermal design and in-orbit flight for large off-axis triple-mirror anastigmatic space optical camera[J]. OPTICAL INSTRUMENTS, 2015, 37(2):116-131. (in Chinese)

[13] LIU Wei-yi, Ding Ya-lin, Wu Qing-wen, et al. Thermal analysis and design of the aerial space optical camera's primary optical system components[J]. Applied Thermal Engineering, 2012,38(2012):40-47.

[14] RDESON R L, SHAUGHNESSY B M, WHALLEY M S, et al... The mechanical and thermal design and analysis of the VISTA infrared space optical camera[J]. SPIE,2004,5947:508-519.

[15] SHEN CH M, LIN CH L, GAO CH CH. Thermal management of Calorigenic Eqiupments in space spectral imager[J]. Spacecraft Recovery & Remote Sensing, 2012,33(6):80-85. (in Chinese)

IOP Conf. Series: Materials Science and Engineering **887** (2020) 012031 doi:10.1088/1757-899X/887/1/012031

Numerical Study of Shock Train in the Isolator of Scramjet Engine

Jing Fang, Xiao Xian Yao and Guan Lin Fang

School of Aerospace Engineering, Beijing Institute of Technology, Beijing, 100081, China

Corresponding author Email address: bitfang@163.com

Abstract: The isolator of scramjet engine has to separate the combustor from the inlet so that the scramjet can keep working stably. The steady-state flow field structure, the shock train shape and the starting position of the symmetrical model of isolator under different backpressure ratio are obtained by numerical simulation of commercial software fluent. Using the user-defined function to simulate the pressure fluctuation caused by the irregular heat release in the combustion chamber, the motion of the shock train in the isolation section is obtained. It is found that the start position of the shock train is larger than that of the steady state simulation, which provides a reference for the design of the combustion chamber.

1. Nomenclature

P = pressure
T = temperature
C_p = pressure coefficient
L = length
H = height
ρ = density
subscripts
b = background
in = inlet
0 = total

2. Introduction

The isolator is an important part of the scramjet engine, which is located between the inlet and the this document, combustion chamber. The isolator mitigates the mutual interference between the inlet and the combustion chamber and prevents the inlet does not start. It is very important for the stability of engine. However, there are complex flow phenomena such as the interference of the shock wave boundary and the formation of the shock train in the isolation section under the back pressure[1]. Due to the complexity of the internal flow in the isolation section, it is still not possible to accurately calculate the length of the shock train according to the conditions of flow and back pressure.

Many researchers have done a lot of research in this area. Billig and Waltrup [2],[3]experimentally study the shock wave structure in a circular tube and give the fitting formula between shock train length L and back pressure Pb, the Mach number of flow and the thickness loss θ of the momentum

Content from this work may be used under the terms of the Creative Commons Attribution 3.0 licence. Any further distribution of this work must maintain attribution to the author(s) and the title of the work, journal citation and DOI.
Published under licence by IOP Publishing Ltd

loss near the wall. Hoeger [4] used VULCAN software to provide back pressure gradually or instantaneously at the outlet section of two-dimensional supersonic flow, and analyzed the relationship between the final position of the shock train and the back pressure ratio [5]. Peek used a ramp device to artificially provide backpressure and studied the effect of backpressure on the leading edge position of the shock train [6]. Hut-zel established a dynamics model that correlates the leading edge position of the shock train in the isolation section with backpressure [7].

Therefore, it is very important to study the influence of the back pressure on the shock train structure of the isolation section. And it is of great significance for the study of the engine components and their overall performance.

In this paper, the flow field of the two-dimensional symmetrical model of isolated section is simulated by the commercial software Fluent. And it obtained the change of the starting position of the shock train in the isolator under different backpressure.

3. Numerical simulation Method

3.1 Physical model.

The mesh of the model shown in Figure 1. The length to height (L/H) is 6[8], and the height is 35mm. The inlet Mach number considered at the entry of the isolator was 2.5. The total pressure P_0=310000Pa, static pressure P_{in}=16535Pa, total temperature T_0=289K. And the outlet pressure of the isolator is 3 to 7 times of the static pressure of the inlet. The mesh of the model adopts ICEM professional grid generation software, which adopts structured meshing and refines the grid near the wall to improve the accuracy. The total number of grids is about 100,000.

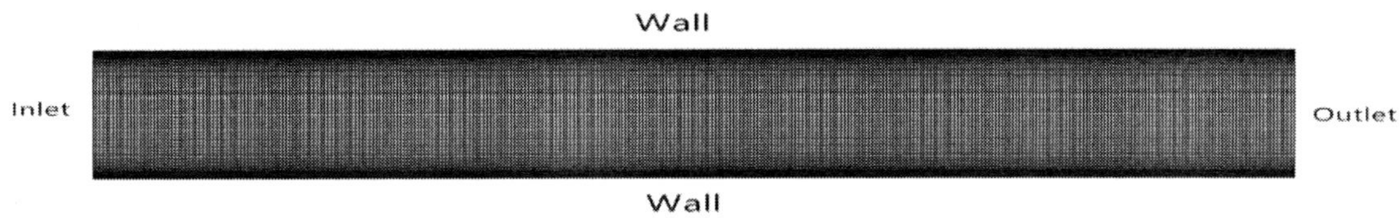

Figure 1. The model of isolator

3.2 Calculate model

In this paper, the commercial software Fluent is used, which uses the finite volume method to solve the Reynolds-averaged Navier-Stokes equations. The solver to choose the second order implicit scheme and the Gauss-Seidel iteration method. The turbulence model to choose the RNG k-ε model and the wall using unbalanced wall function method. The details of the numerical method are given in [9]. The solution is considered to have converged when the following conditions are met:

- Residuals of each control equation are less than 0.00001;
- With iteration, there is no change or little change in the distribution of wall static pressure;
- The quality of inlet and outlet is less than 0.0001;

And there is some basic assumptions:

- Gas is the ideal gas, subject to the ideal gas state equation;
- Do not consider the impact of gravity and other body force;
- Wall is no slip and adiabatic, and the entire flow field with the outside without heat exchange;

4. Results and Discussion

4.1 Shock train under different backpressure ratio

The outlet pressure (P_b) is set to 4 to 7 times the static pressure (P_{in}) of the inlet to study the evolution of the shock train and the variation of the exit parameters of the isolator with the backpressure ratio increase. Figure 2 shows the pressure cloud at different backpressure ratios. It can be seen from Figure 2 that when the backpressure ratio (P_b/P_{in}) is 4.0, a more obvious shock train structure has been formed in the flow field. As the backpressure ratio increases from 4.0 to 6.0, the starting position of the shock train continues to move forward, the length increases, and the number of shock waves increases. When the backpressure ratio rises to 7.0, the shape and shock wave number of the shock train are not changed obviously, but the length of the mixing zone is longer and the starting position of the shock train has a large forward, close to the entrance. It can be seen that the rise of the backpressure ratio will gradually move forward position of the shock train, the area affected by the backpressure in the isolator is expanded.

Figure 3 shows the distribution of the Mach number of the axis of the isolator. It can be seen that when the backpressure ratio is 4, the Mach number of the outlet (combustor inlet) of the isolation section is still greater than 1. This shows that the engine is in supersonic combustion mode. When the pressure ratio is greater than 5, the Mach number of the outlet of the isolator (combustor inlet) is less than 1, the outlet gradually transitions from partial subsonic areas to the complete subsonic area, this shows that the engine is in the subsonic combustion mode. It can be seen that with the increase of backpressure, it is possible to realize the modal conversion of the scramjet engine. It can be further analyzed that when the outlet pressure is high enough, the structure of the shock train will no longer change. With the further increase of the outlet pressure, the intensity of the shock strain is gradually increased, and the oblique wave is transformed into the positive shock wave strain, λ positive shock wave, and the position of the shock strain continues to move forward until the isolation section enters the inlet. It can affect the normal operation of the inlet, and even cause the inlet does not work. Captions should be typed in 9-point Times. They should be centred above the tables and flush left beneath the figures.

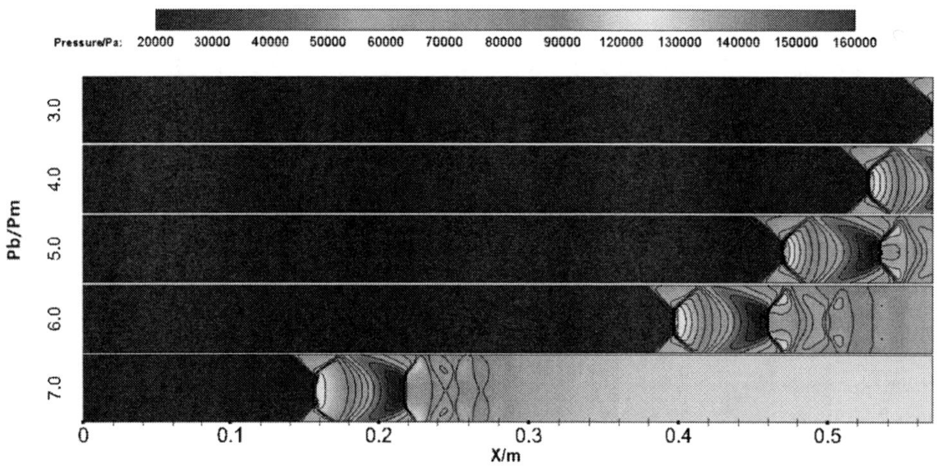

Figure 2. The pressure cloud at different backpressure ratios

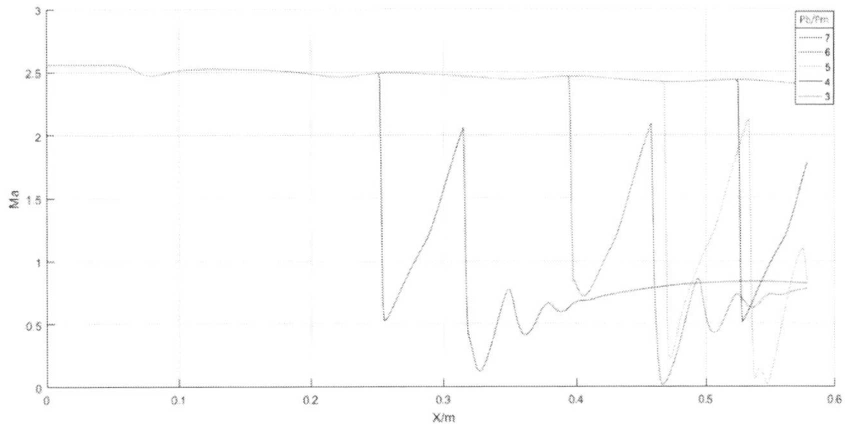

Figure 3. Mach number of the axis of the isolator

4.2 Shock train under outlet pressure of sine wave.

The outlet pressure P_b is set to change according to the sine wave law to ensure that the back pressure ratio also has the same variation. The method is that the user-defined function is written into the Fluent. Set the calculation of the combustion environment pressure is 8.9MPa, the pressure ratio is 5. The time at which the pressure disturbance begins to be added is 0.01 ms; and the dimensionless pressure is defined as 0.2. Figure 4 shows a periodic of the pressure distribution at different time under a sine wave. It can be seen from the figure that the length and the location of the shock train in the tube will also change periodically. In contrast to the pressure cloud at backpressure ratio of 7.0 in Figure 2 and the pressure cloud at the 1/4T in Figure 4, it can be seen that the starting position range of the shock train is larger than that of the steady state simulation due to the hysteresis. When it is at 3/4T, the starting position of the shock train is large, close to the outlet of the isolation section.

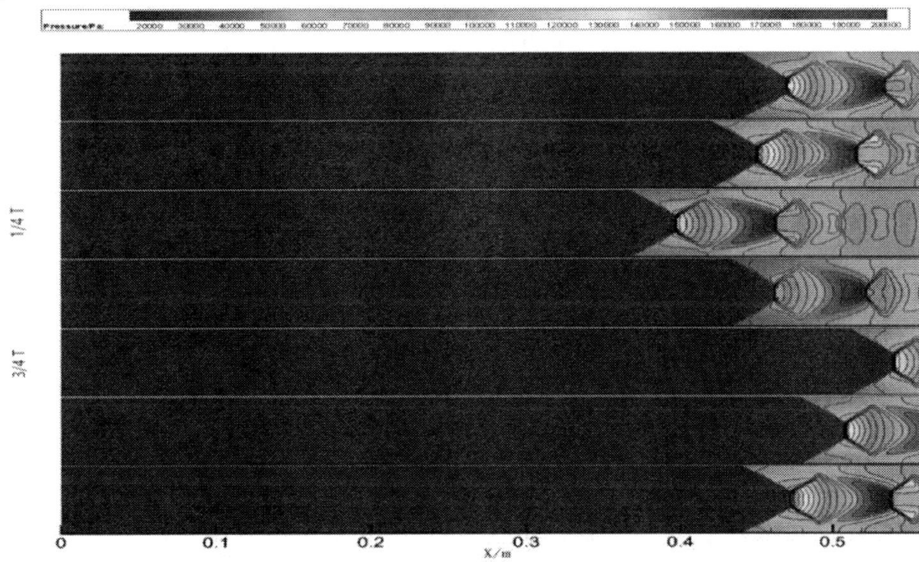

Figure 4. The pressure cloud under a sine wave

4.3 Shock train under outlet pressure of saw tooth wave

In the same way, the backpressure of the outlet is set to the law of saw tooth variation. Figure 5 shows the saw tooth wave change, and this saw tooth wave is a 1000-order Fourier transform of the previous sine wave. Figure 6 shows the change of the start position of the shock train near the wall in the corresponding time interval of Figure 5. Analysis of Figure 5 and Figure 6, in a saw tooth wave cycle, the isolation section of the shock position has also been cyclical changes, but in a cycle there have been three shock changes. In Figure 6, some of the slope of the curve is very large where the shock wave movement speed is fast, and some slope is small where the shock wave movement speed is slow. According to the Fourier transform formula of the saw tooth wave, it is found that the frequency of the shock string position is the same as that of the third order Fourier transform.

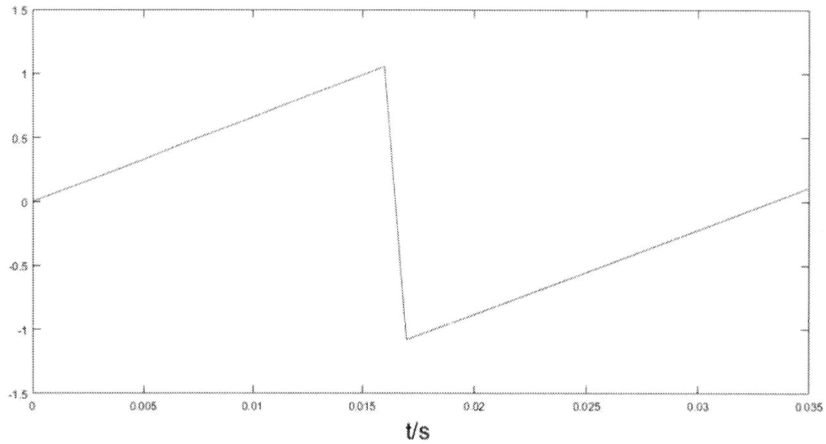

Figure 5. Saw tooth image

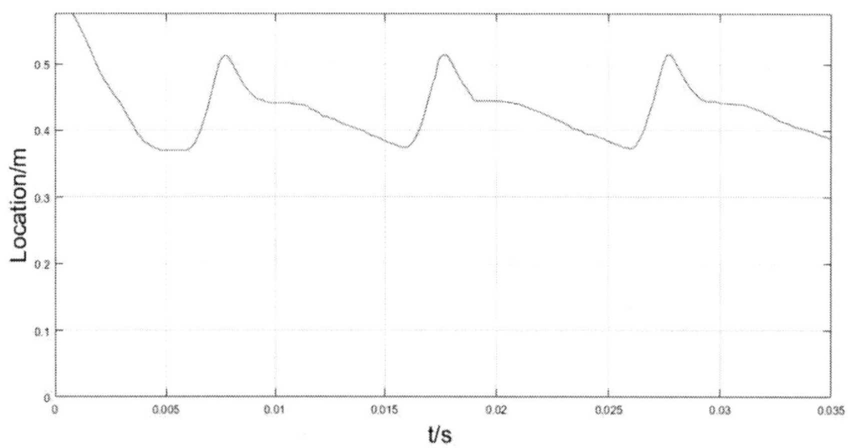

Figure 6. The start position of the shock train

Figure 7 shows the pressure cloud of the minimum and maximum moments of the coordinate position on the axis of the shock train in the isolation section. Compared with Figure 4, it can be seen that the minimum location of the shock wave of the back pressure ratio under the saw tooth wave is smaller than that under the sine wave. It can be seen that when the back pressure ratio changes to the saw tooth wave, the shock string in the isolation section will appear many oscillations. The change position range is almost the same as the position of the shock string when the back pressure ratio is sine wave. The results provide a reference for the design of the combustion chamber.

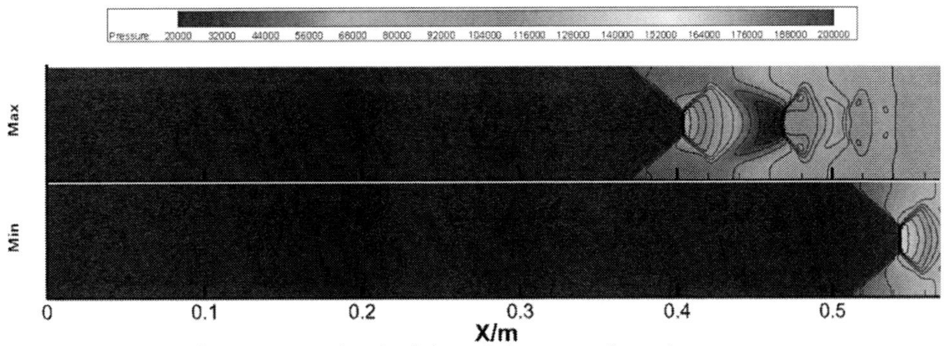

Figure 7. The pressure cloud of the minimum and maximum moments

5. Conclusion

The isolation section can effectively isolate the impact of periodic dynamic backpressure pulsations on the inlet of different waveforms. The increase in the back pressure ratio will cause the shock train to move forward step by step, and the area affected by the back pressure in the isolation section is also enlarged. And for a dual-mode engine, the back pressure lift is usually accompanied by a change in the operating mode. When the back pressure ratio is under the sine wave or saw tooth wave, the length of the shock train in the isolator will change periodically. But there are multiple change cycles of the shock train under the back pressure of the sine wave. The reason needs to further study.

References

[1] HE Can, XING Jian-wen, XIAO Bao-guo and LIU Wei-xiong. Journal of Aerospace Power, 31, 5: 1242-1251 (2016).
[2] Waltrup P J, Billig F S. Structure of Shock Waves in Cylindrical Ducts[J]. Aiaa Journal, 1973, 11(5):1404-1408.
[3] Waltrup P J, Billig F S, Evans M C. Critical considerations in the design of supersonic combustion ramjet /scramjet/ engines[J]. Journal of Spacecraft & Rockets, 2015, 18(4):350.
[4] Hoeger T, King P, Jeffrey D, et al. 2-D Transient CFD Model of an Isolator Shock Train[C]// Aiaa International Space Planes and Hypersonic Systems and Technologies Conference. 2011.
[5] Hoeger T C. Dayton, Ohio: Air Force Institute of Technology (2012).
[6] Peck R I. Dayton, Ohio: Air Force Institute of Technology (2010).
[7] J Chang, B Li, W Bao, D Yu. 2012. Shock train leading-edge detection in an isolator using genetic algorithm. Proceedings of the Institution of Mechanical Engineers, Part G: Journal of Aerospace Engineering 226:11, 1424-1431.
[8] Balu G, Gupta S, Srivastava N, et al. Experimental Investigation of Isolator for Supersonic Combustion[C]// Aiaa/asme/sae/asee Joint Propulsion Conference & Exhibit. 2013.
[9] Concepts R. Fluent user's guide, Version 5.0, [J]. 2014.

The Research on Indirect Modular Adaptable Design Platform on Part Design of Civil Aircraft

Bing Cheng[1]

[1]Shanghai Aircraft Design and Research Institute, Shanghai 201210, China

E-mail: chengbing198805@163.com

Abstract. Based on the traditional modular design and the module division concept of virtual flexible modules in generalized modular design, this paper proposes a design method of indirect modular design. It suggests a way to achieve modular design of products without obvious serialization and hierarchical characteristics integrating modular platform and parameterized platform. The division and application of geometric parameters and numerical parameters of civil aircraft fuselage frame are taken as an example to show how it works. The indirect module is constructed on the basis of the framework model, and the lines and surfaces of the framework model are used as driving parameters. The design idea of adaptable design platform based on indirect modular design is also presented in this paper. The similar modules of the parameterized family modules can be designed quickly by driving the geometric parameters and numerical parameters separately. It makes faster and more accurate response to customer needs.

1. Introduction

The refinement and rapid increase of product requirements require us to have faster design methods to meet customer needs and win market competition. Modular design, as an effective method for serial and rapid design of products, has gradually developed from the early twentieth century and has been applied to many fields. However, the traditional modular design is mainly for products with obvious serialization and hierarchical characteristics. And then do parameterized design for the product or components to achieve modularization of product design. For many non-series products, it is not feasible to simply extract the parameters that can be designed in series at the part level.

However, the similar structure means that the part design steps are similar or there are many identical steps in the part design stage, which can be used as the basis for module division. This article introduces the concept of indirect modular design, expands the application field of modular design theory, and applies the product design platform based on indirect modular design to aircraft component design, further expounds the idea of indirect modular design.

2. Generalized modular design

Professor Y S Xu was the first to propose the concept of generalized modular design [1]. There are three types of the generalized modules, namely parameterized flexible module, virtual flexible module and flexible unit structure [2]. In a complex structural entity, part of its structure may be more standardized in terms of form and parameters, and has a relatively fixed sub-function. It can be further divided into different functional modules, but it cannot be divided in structure. Then comes the virtual flexible module. It only has the meaning of segmentation in CAD software.

Content from this work may be used under the terms of the Creative Commons Attribution 3.0 licence. Any further distribution of this work must maintain attribution to the author(s) and the title of the work, journal citation and DOI.
Published under licence by IOP Publishing Ltd

3. Indirect modular design

Indirect modular design refers to analyzing the design ideas of product parts combining functional analysis, taking an intermediate state in the part design process as the object of modular design, introducing parametric design and variable analysis methods, dividing and constructing the indirect module and product platform with greater adaptability, and then realizing the rapid design of the product through the combination of indirect modules or the derivative of the product platform.

Indirect module is a structure with specific functions and a parameterized structure model. Its interface characteristics depend on the maturity of the indirect module. The maturity of the indirect module is defined as the completion of the indirect module from the final part. If the indirect module is the final module, its maturity is 1 and the interface feature is the interface of the product itself. Otherwise it is a collection of interfaces in the product and indirect module design process.

Indirect modular design is also a method to divide a structure part into different divisions in CAD software. Compared to the virtual flexible module, the module division of indirect module is based on the modeling ideas and steps, and the modules might not have fixed function in the final part. But they have some fixed function in the design process of the part. It doesn't even show up in the final structure. But for mass similar parts, the indirect module and its parameterized family modules, namely flexible indirect module can be the common foundation for design.

The importance of indirect module division is not only to make modular design on some product which is hard to make serial design such as some parts of the civil aircraft, but also to divide the modules at design execution level.

In addition to this, flexible indirect module' parameters mainly come from the points, lines and surfaces of the framework model composed of key design reference information such as points, lines and surfaces, supplemented by a set of values constrained by a certain size range. The main error correction process of the parameter is directed at geometric parameters, which are more convenient and intuitive. Compared to pure numerical parameters, they are more intuitive and easier to adjust.

It is intended to explore the regularity of features and designing ideas in part and product design, and realize the complete combination of regularity and difference. The product design process will go from realizing designing results to exploring rules. Parametric design is intended to emphasize the procedural nature of the design, the logic of the process, and the diversity of results [3].

The basis of the indirect modular design is the framework model, and the lines and surfaces of the framework model are used as driving parameters. We can take the flexible indirect module as the driving object. By changing driving parameters, an instantiated general module could be obtained. Then add optimization features which can be called the special module to the general module to obtain an optimized model. Subsequent optimization of structural static strength, fatigue strength, etc. can be used as a special module.

Take the frame of civil aircraft as an example. A complete frame can be divided into four segments including the upper, the lower, the left, and the right. They all consists of a floating frame and some clips. The designing steps of the floating frame can be divided into about 10 steps as shown in table 1.

Table 1. The design process of the upper frame.

Designing steps	Surfaces used	Geometric parameters	Numerical parameters
Thick surface	Fuselage profile, Frame plane, Level plane, Upper fuselage profile	Fuselage profile, Frame plane, Level plane	Frame plane offset (X_1, X_2), Thick surface offset (A_1, A_2)
Split 1	Upper fuselage profile, Frame plane, Stringer sweeping surface 1, Stringer sweeping surface 2, Upper frame segmented surface (left)	Upper fuselage profile, Frame plane, Stringer sweeping surface 1, Stringer sweeping surface 2	Frame boundary surface (left) offset B_1
Split 2	Upper frame segmented surface (right), Fuselage vertical	Fuselage vertical centerline plane	Frame boundary surface (right) offset B_2

	centerline plane		
Split 3	Frame plane	Frame plane	Frame plane offset X_3
Split 4	Frame plane	Frame plane	Frame plane offset X_4
Split 5	Upper fuselage profile	Upper fuselage profile	Upper fuselage profile offset H_1
Split 6	Frame plane, Upper fuselage profile	Upper fuselage profile	Upper fuselage profile offset H_2
Shell 1	Frame plane	Frame plane	Frame web thickness D
Shell 2	Frame plane	Frame plane	Frame web thickness D
Edge Fillet			R

Fuselage profile denoted by fpr, frame plane denoted by fpl and stringer sweeping surface denoted by sss are selectable geometric parameters. Different parameters can be chosen to design frames at different frame positions, as shown in figure 1 below. The level plane and fuselage vertical centerline plane in table 1 are fixed in a certain type of civil aircraft.

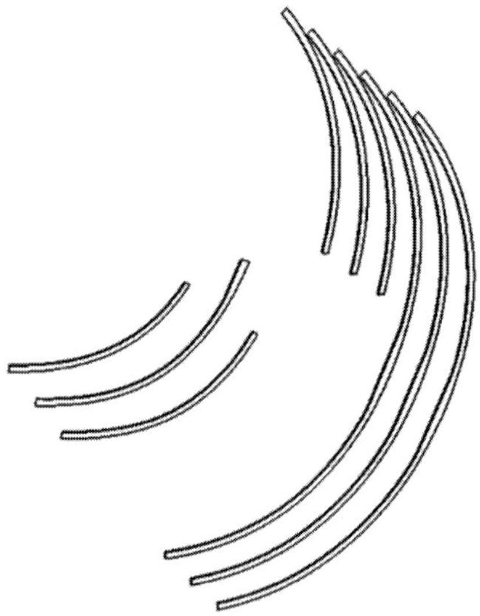

Figure 1. Frames at different frame positions.

$$F_{puf}=f(fpr,fpl,sss_1,sss_2,X_1,X_2,A_1,A_2,B_1,B_2,X_3,X_4,H_1,H_2,D,R) \quad (1)$$

$$F_{puf}=f(G_{puf},N_{puf}) \quad (2)$$

F_{puf} represents all the parameters for designing the upper frame. G_{puf} represents the geometric parameters, and N_{puf} represents the numerical parameters.

Denote these parameters such as *upper frame segmented surface* $(left)$, B_1, *upper frame segmented surface* $(right)$, B_2, X_3, X_4, H_1, H_2, *frame plane*, D and R as I_{puf}. I_{puf} represents interface parameters on the upper frame module. We can make the upper frame cooperate with other modules by changing the parameter I_{puf}.

$$I_{puf}=\left|I_{pup-lf}\,\mathrm{U}\,I_{pup-rf}\,\mathrm{U}\,I_{pup-s}\,\mathrm{U}\,I_{pup-x}\,\mathrm{U}L\right| \quad (3)$$

I_{pup-lf} represents interface parameters with the left frame. I_{pup-rf} represents interface parameters with the right frame. I_{pup-s} represents interface parameters with the aircraft skin. I_{pup-x} represents interface parameters with other structures. As shown in table 2.

Table 2. Interface parameters.

Parameter type	Interface parameters	
I_{pup-lf}	*upper frame segmented surface* (*left*) B_1 H_2 *frame plane* D	
I_{pup-rf}	*upper frame segmented surface* (*right*) B_2 H_2 *frame plane* D	
I_{pup-s}	H_1	
I_{pup-x}	X_3	X_4 H_2 *frame plane*

For the parts and the assembly, we can denote the interface parameters as I_p. The standardization of the interface is the prerequisite for realizing the variant design based on the instance of the modular product [4].

$$I_p = f(I_{pg}, I_{pn}) \tag{4}$$

I_{pg} represents the geometric interface parameters, and I_{pn} represents the numerical interface parameters.

$$M_p = F_P - I_P \tag{5}$$

M_p represents the modeling parameters independent of the interface parameters.

$$F_p = \begin{bmatrix} G_p & N_p \end{bmatrix} = \begin{bmatrix} M_p & I_p \end{bmatrix} \tag{6}$$

F_p represents the parameters that may be used in the frame design process. G_p represents the geometric parameters, and N_p represents the numerical parameters.

Take civil aircraft for an example. Aircraft parts are slowly upgraded, but there are a lot of parts with the same or similar function and similar structure that need to be optimized.

By planning the product family of the parts, the flexible indirect module of the similar structure is divided into general module and special module. The module library creation process of the frame is given as follows.

Firstly, structurally analyze the product on the basis of the design ideas of product parts and functional analysis. The frame's main structure may consist of the upper, the lower, the left, and the right frame, which all consists of a floating frame and some clips.

Then, extract the driving parameters as shown in table 1. They include the geometric parameters and the numerical parameters.

Furthermore, create the flexible indirect model. Create a parametric model of the frame based on the driving parameters.

Finally, design the adaptable module interfaces. Make adaptable design for the interfaces of the other modules as well as the frame module, in order to replace the modules for product upgrading and maintenance.

U_{pi} represents the useful parameters corresponding to a module instance of the flexible indirect module. It contains not only structural design parameters, but also engineering constraints, such as load, stress and strain, and characteristic parameters, such as material.

$$F_p = \left| U_{p1} \, \mathrm{U} U_{p2} \, \mathrm{UL} \;\; \mathrm{U} U_{pi} \, \mathrm{UL} \;\; \mathrm{U} U_{pm} \right|, i = 1, 2, \mathrm{L} \;\; , m \tag{7}$$

In order to make the flexible indirect model more versatile in the instantiation process, it is allowed to have redundant features and redundant parameters in the modeling process. So some parameters

IOP Conf. Series: Materials Science and Engineering **887** (2020) 012032 doi:10.1088/1757-899X/887/1/012032

denoted by F_p may be redundant in some frames. But it makes the parameters in each U_{pi} as uniform as possible and reduce differentiation. When creating a part sketch, the sketch is allowed to contain more information and skeleton structure than what is actually required by any single instanced module. The unified design of multiple similar parts makes a common module applicable to multiple different parts. We call this sketch as flexible sketch. Take the frame as an example. Although most frames can be divided into four parts, some frames will be different with others. Because they will be interrupted by structures such as fuselage door surround structures. But they can be designed with one sketch which contains the information required for all frames, as shown in figure 2. The difference of frames can be achieved by adjusting driving parameters or adding special modules on the adaptable design platform [5], as shown in figure 3.

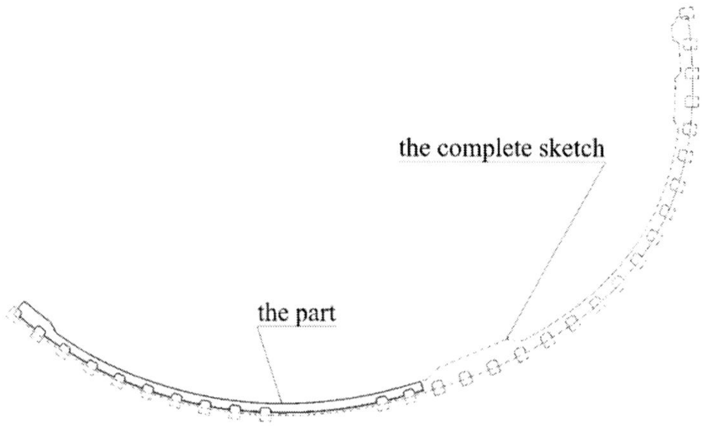

Figure 2. Comparison of the frame and its complete sketch.

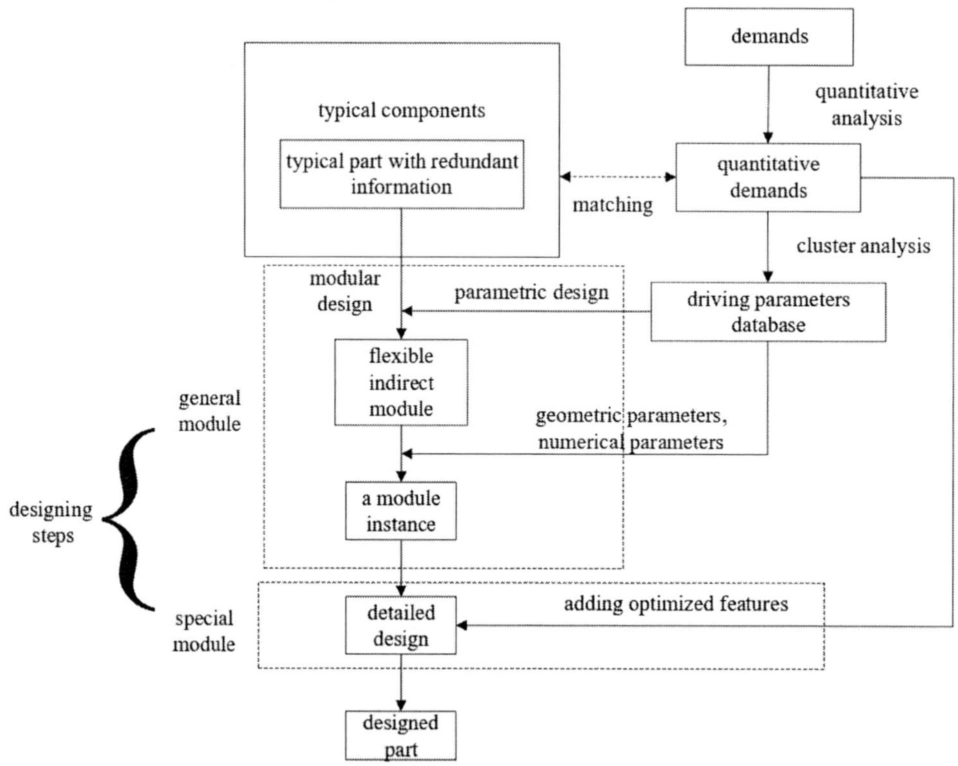

Figure 3. The design process of the part on indirect modular adaptable design platform.

Just as the left frame and the upper frame can be matched by parameter I_{plf}, the frame can be matched with other structures by I_p .When we design the upper frame, the numerical parameters can be discretized, and the geometric parameters can be the only one or discretized. So are other products. For the parameterized product family, parameter matrix is shown below, denoted by $A_{m\times n}$.

$$A_{m\times n}=\begin{bmatrix}F_{puf1} & F_{puf2} & \cdots & F_{pufi} & \cdots & F_{pufn}\end{bmatrix}=\begin{bmatrix}Q_1 & \cdots & Q_i & \cdots & Q_m\end{bmatrix}^T \tag{8}$$

$F_{pufi}=\begin{bmatrix}fpr & fpl_i & sss_{1i} & sss_{2i} & X_{1i} & X_{2i} & A_{1i} & A_{2i} & B_{1i} & B_{2i} & X_{3i} & X_{4i} & H_{1i} & H_{2i} & D_i & R_i\end{bmatrix}$, corresponds to the parameters of the upper frame. $i=1,2,\cdots,n$. n represents the number of the frame that can be modularly designed. Q_i represents a set of discretized values of a driving parameter.

Just as we can establish the upper frame model with F_{pufi} . The module instance M_{ufi} is uniquely determined by F_{pufi} . $M_{ufi}=f\left(F_{pufi}\right)$. The whole product is composed of all the modules.

$$M_{fi}=\left[M_{ufi},M_{lfi},M_{rfi},M_{lfi}\right],\ i=1,2,\cdots,n \tag{9}$$

where, M_{fi}, M_{ufi}, M_{lfi}, M_{rfi} and M_{lfi} represent the module instances of the frame module, the upper frame module, the left frame module, the right frame module and the low frame module.

The discretized values $Q_i(i=1,2,\cdots,m)$ can be inputted into the platform while checking in the module. Then the parameterized family modules are created in the platform.

4. Output the typical module

Indirect modular design not only expands the applicable field of modular design, but also accumulates a large number of typical structures that can be modularized during the design process. Based on this, rapid module construction can be achieved, and more intuitive product types can be provided for the customers. The reference makes the product's response to customer needs faster and more accurate, and enables direct communication between the design and customer needs.

5. Conclusions

The indirect modular technology is a design method to build product parameterized family modules based on the traditional modular design and the generalized modular design. The construction of indirect module provides the approach to modular design for products without obvious serialization and hierarchical characteristics. The design of new products can be quickly realized by reconfiguring, upgrading general modules and replacing special modules on the modular adaptable design platform. This paper tells an example of the frame in civil aircraft design to describe the idea of indirect modular design.

References

[1] Xu Y S, Hou L, Zhang L H and Li S 2001 Theory of the generalized modular design of hydraulic press and it's application *Journal of Machine Design* **7** 1–3,41

[2] Gao W G, Xu Y S, Chen Y L and Zhang Q 2007 Theory and methodology of generalized modular design *Chinese Journal of Mechanical Engineering* **43** 48–54

[3] Cao W Y 2018 Research and application of parametric design in product design [D] North China University of Technology

[4] Xiao X H, Shi M H, Yang X F and Li B 2007 Research on variant design based on modular product case *China Mechanical Engineering* **18** 803–807

[5] Yang Y N, Shi M H and Xiao X H 2008 The research of products family design based on AMPP *Machine Tool & Hydraulics* **36** 260–263

Thermodynamic Investigation of Conventional and Alternative Rocket Fuels for Aerospace Propulsion

Ilyes Ghedjatti, Shiwei Yuan and Haixing Wang

Aerospace Propulsion Laboratory, School of Astronautics, Beihang University, 37 Xueyuan Road, Haidian District, 100191 Beijing, China

Email: ilyes.ghedjatti@buaa.edu.cn

Abstract. The magnitude of the exhaust velocity is dependent on molecular and chemical properties of the propellant, and the expansion ratio of the engine. The main requirement for all propellant combinations is to maximize the energy release per kilogram; in other words, the lower the mass for a given energy release, the higher is the ultimate velocity of the vehicle. This also implies that the mass flow rate depends on the density, while the exhaust velocity depends on the energy contained in the hot gas. The main performance parameters can be defined as the thrust coefficient related to the nozzle performance, and the characteristic velocity related to the propellant and combustion performance. The first parameter reaches its optimum when the exhaust pressure is equal to the ambient pressure while the second parameter is defined by the choice of propellant combination. The objective of this work is to investigate, numerically and thermodynamically, different combinations of oxidizers with conventional and alternative fuels, for different rocket propulsion and energy generation systems. Based on the state-of-the-art and the analysis results on the most promising fuel-oxidizer combinations for the future of aerospace propulsion and space transportation, in a technically and environmentally sound manner, conclusions are made.

1. State of the art

The efficiency of a rocket in terms of spacecraft propulsion depends on the achievable exhaust velocity and mass ratio; thrust is mostly important for the first stage of a launcher. All these parameters are defined by the rocket design and propellants combination. These systems consist of solid, liquid or hybrid rocket motors. For more than five decades, Ammonium Perchlorate (AP) based composite propellants have been the main line of development of solid rocket propulsion [1]. Adding metal-particles to propellants help increase the energy density [2]. Among various metal additives, and for some specific cases, Aluminium (Al) seems to be the most promising one due to its high combustion enthalpy, wide availability and low cost [2]. The burning rate of a double base propellant without aluminium was compared with other double base fuels in which aluminium was added by 2% and 4%. It was found that the burning rates and burning heat of the fuels with aluminium additives improved [3]. Moreover, the decomposition and reaction of AP with HTPB produces gases consisting of Carbon Dioxide (CO_2), Water vapor (H_2O) and Hydrochloric acid (HCl). The added Al particles react with these products to increase the ratio of combustion temperature to gas molecular weight, resulting in higher exhaust velocities [4]. Hydrides can also be added to chemical rocket propellants, as high volumes of Hydrogen (H_2) can be released during combustion. Aluminium hydride is expected to perform effectively; however, its industrial application is still unfeasible because of the lack of its

Content from this work may be used under the terms of the Creative Commons Attribution 3.0 licence. Any further distribution of this work must maintain attribution to the author(s) and the title of the work, journal citation and DOI.
Published under licence by IOP Publishing Ltd

commercial availability [5]. To reduce air pollution during rocket launches, efforts have been done to develop green, or environmentally friendly, propellants with non-toxic combustion products, easier and safer to manufacture and handle than the conventional ones, bringing down the associated transport and storage costs [6]. Several studies on the use of Hydrogen Peroxide/Kerosene (H_2O_2–RP1) as a green propellant combination suited for propulsion applications have been realized [7-12]. Yet, the overall performance can be limited by coking which remains a major problem for current Hydrocarbon cooled rocket engines; but currently and from a different aspect, because of the possibility of harvesting Methane (CH_4) from Mars, the interest of its use for interplanetary missions has increased [13-14]. Another alternative is hybrid systems, as they are known for their simplicity, safety and environmental advantages over conventional ones. However, such systems have not been commercially viable, because they usually use Polymeric fuels with low evaporation rate, and thus low thrust, which limits their use for large launchers. The requirement is to increase the rate at which the fuel evaporates. One approach is to reduce the viscosity in the layer of the melted fuel, by adding a certain amount of Paraffin-wax to the fuel to lower its viscosity [15]. Further studies focused on propellant systems consisting of N_2O/HTPB, N_2O/HDPE and H_2O_2/HDPE [16], bringing also satisfying results. Another approach consists of metals exothermic reaction with air, oxygen, carbon dioxide or even water to generate hydrogen [17-18]. Exploiting metals produced in situ from extra-terrestrial resources as propellants is an alternative which may lower the cost of future space missions by reducing the mass of propellant launched from Earth. Several disadvantages, though, are also inherent in the hybrid metal-oxygen engine concept. A metal fuel grain would act as a high conductivity heat sink, and heat conducted into the fuel could cause more metal to melt than is desired. Nevertheless, some advantages of this concept include the ability to control thrust by controlling the oxygen flow, thus, to throttle, stop, and restart the engine [19].

From the established state of the art, only the most promising alternatives have been selected to be investigated in this work, from a theoretical, thermodynamic and numerical side, by studying the main parameters controlling the rocket design and propellant combination performance; at the same time, providing an overview on the current and future challenges that can encounter the future space exploration and transportation missions. To achieve the current investigation, Python with its package Proptools, as well as the Rocket Propulsion Analysis (RPA) have been used. Proptools helped providing implementation of equations for nozzle flow and rocket structures. It aims to cover most of the commonly used equations in Rocket Propulsion Elements by George Sutton and Oscar Biblarz, and Modern Engineering for Design of Liquid-Propellant Rocket Engines by Dieter Huzel and David Huang. This work is structured as follows: first, the investigated cases are similar to the ones described below in tables 1 and 2; in table 3, concepts of hybrid systems are also presented and examined. Next, results are illustrated and discussed. Finally, conclusions on the most significant results are made.

Table 1. Space Shuttle and Ariane 5 solid rocket boosters

	SRB (Space Shuttle)	MPS (Ariane 5)
Thrust	10.89 MN	5.87 MN
Expansion ratio	11.3	10
Exhaust velocity	2,690 $m.s^{-1}$	2,690 $m.s^{-1}$
Temperature	3,450 K	3,600 K
Pressure	6.5 MPa	6.0 MPa
Propellant	Ammonium perchlorate 69.6% Aluminium powder 16% Polymeric binder 14% Additive iron oxide 0.4%	Ammonium perchlorate 68% Aluminium powder 18% Polymeric binder 14%

Table 2. Liquid-propellants rocket engines

Engine	Raptor	NK-33	RD-0120	RD-701
Propellant	LCH_4/ LOx	RP-1/ LOx	LH_2/ LOx	RP-1&LH_2/ LOx
O/F ratio	3.8	2.7	6.0	6.0
Nozzle ratio	40	27	85.7	133.8
Thrust	2000 kN	1638 kN	1517 kN (SL) 1961 kN (vac.)	4003 kN
Thrust-to-weight ratio	> 170 (goal)	136.7	58	111
Chamber pressure	30 MPa	14.57 MPa	21.8 MPa	29.4
Specific impulse	330s (SL) 380s (vac.)	297s (SL) 331s (vac.)	359s (SL) 455s (vac.)	330s (SL) 415s (vac.)

SL: Sea Level, Vac.: Vacuum

Table 3. Metal-based hybrid systems

Engine concept	Hybrid rocket propellant combinations				
Propellant	Al/ LOx	Al/ H_2O*	Al/H_2O*&LOx	B/ LOx	B/ H_2O*
O/F ratio	1.6	1	1.47 with: H_2O: 80% O_2: 20%	2.5	2.5
Nozzle ratio	15 33	15 33	15 33	15 -	15 -
Chamber pressure	10	10	10	10	10

* Metals mixed with iced water.

2. Analysis results

The specific impulse (Isp) has been calculated for the considered propellant combinations as shown in figure 1, 2 and 3. Isp is first increased, then lowered, and shifted towards lower mixture ratios (O/F) after adding Al, whether to AP-HTPB or AP-Fe_2O_3-HTPB; while it is slightly lowered in the case of LO_2-LH_2. This means that solid fuels in particular demonstrate more efficiency for a certain range of Al additives, outside of this range, Isp decreases. Propellant mixture, optimum O/F ratio, Isp and thrust are significantly influenced by metal addition. The idea behind is that Al addition increases the thrust coefficient (C_F) and combustion temperature (T_C), but also the molecular weight ($\mathfrak{M}$). This is suitable for the first stage, i.e. sea level, where the atmospheric pressure (Pa) is higher, and more thrust is needed to lift off all the payload to upper stages, i.e. higher altitude. Propulsion performance is also considerably influenced by H_2 addition to RP1 and O_2 addition to burn the H_2 generated from Metal-water reaction, as shown in figure 2 and 3 respectively. The reason is that it modifies the bulk density of the considered propellant combination at its current O/F value. Indeed, the mass flow rate depends on density, which is proportional to $(\mathfrak{M}/T_C)^{1/2}$. Figures 4, 5, 6 and 7 show the variation of T_C and $\mathfrak{M}$ with O/F while figures 8, 9, 10 and 11 show the relation between the thermodynamic energy (E_{Th}), defined by: $RT_C/\mathfrak{M}$, the characteristic velocity (C*) and the variation of O/F ratio. C* varies proportionally with E_{Th}, with the variation of O/F ratio, the optimums of both are shifted towards the same O/F stochiometric value. The propellant with high thermodynamic energy is the one for which high Tc and low $\mathfrak{M}$ are achieved.

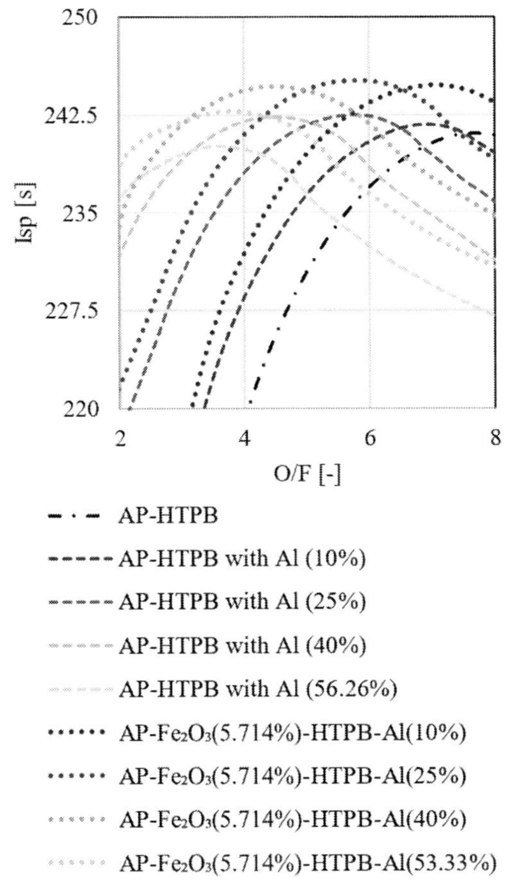

Figure 1. Variation of estimated specific impulse with mixture ratio for solid propellants.

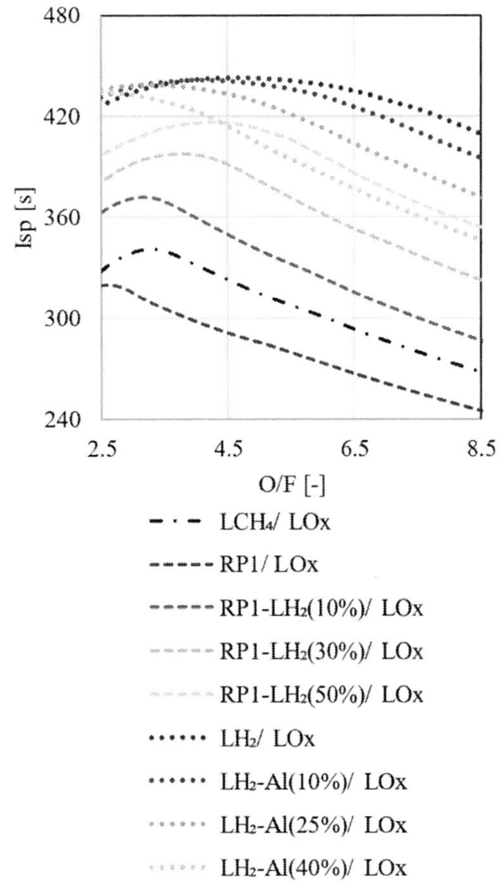

Figure 2. Variation of estimated specific impulse with mixture ratio for liquid propellants.

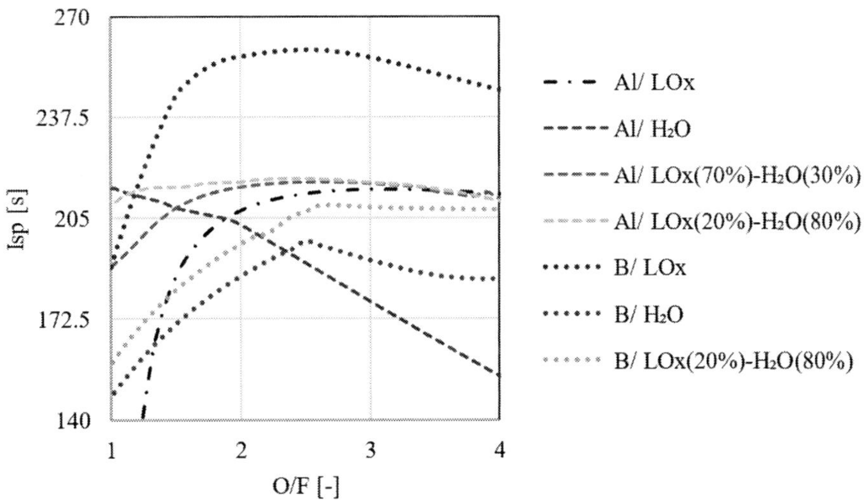

Figure 3. Variation of estimated specific impulse with mixture ratio for metal-based hybrid systems.

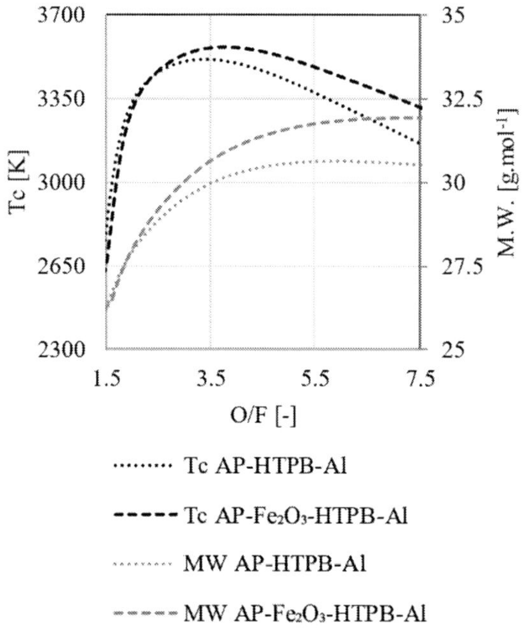

Figure 4. Variation of combustion temperatures and exhaust molecular weight with mixture ratio for solid rocket boosters.

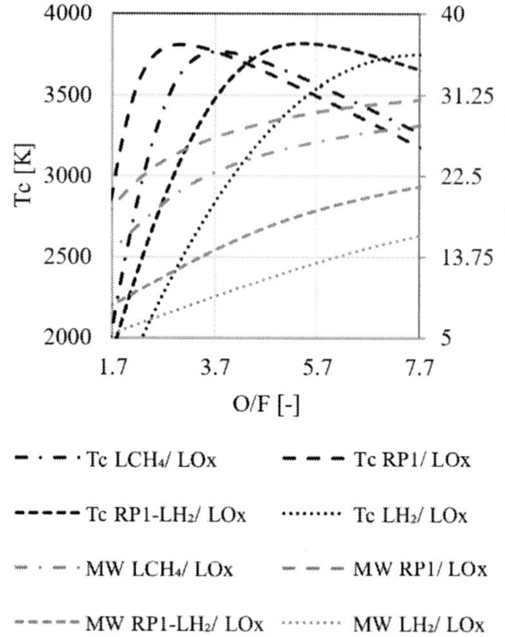

Figure 5. Variation of combustion temperatures and exhaust molecular weight with mixture ratio for liquid propellants.

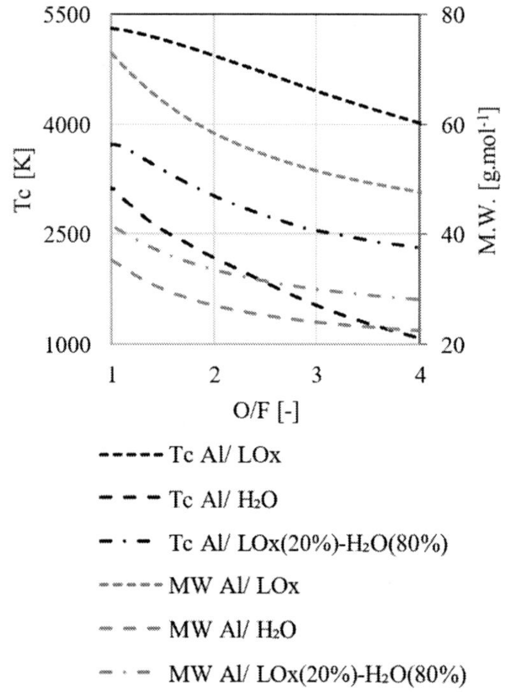

Figure 6. Variation of combustion temperatures and exhaust molecular weight with mixture ratio for Aluminium-based hybrid system.

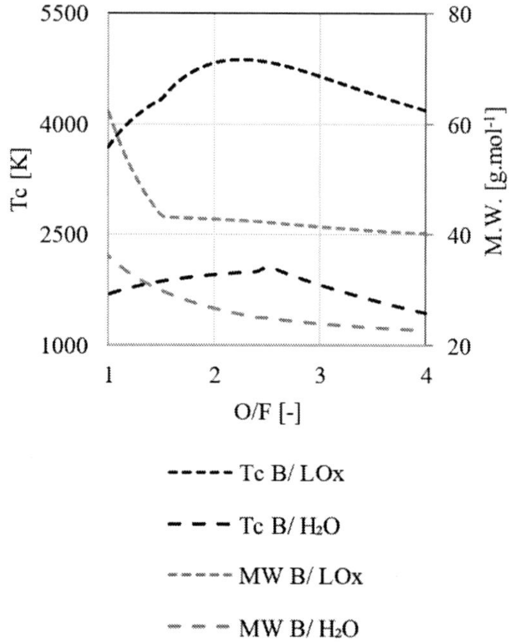

Figure 7. Variation of combustion temperatures and exhaust molecular weight with mixture ratio for boron-based hybrid system.

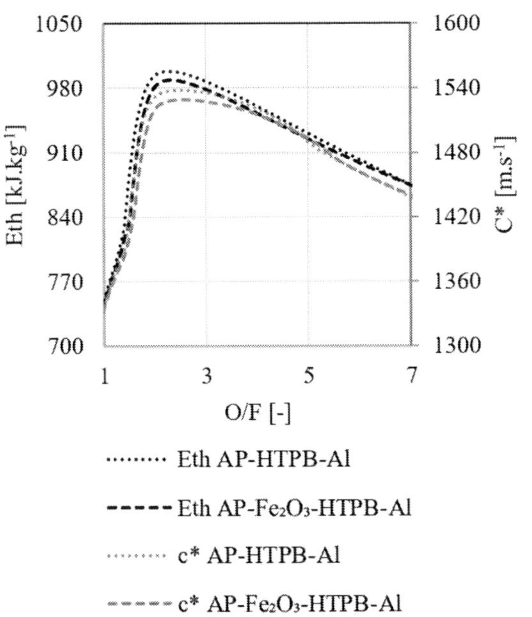

Figure 8. Variation of thermodynamic energy and characteristic velocity with mixture ratio for solid rocket boosters.

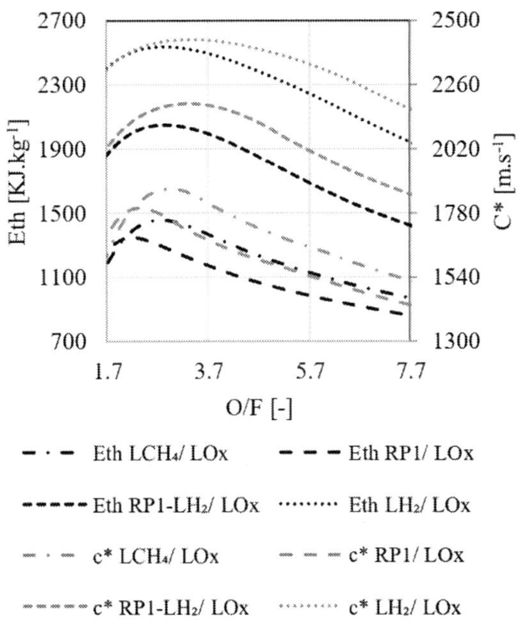

Figure 9. Variation of thermodynamic energy and characteristic velocity with mixture ratio for liquid propellants.

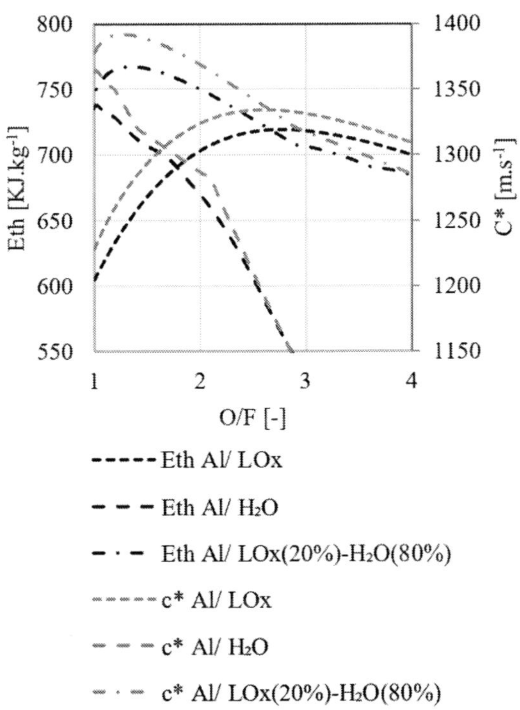

Figure 10. Variation of thermodynamic energy and characteristic velocity with mixture ratio for aluminium-based hybrid system.

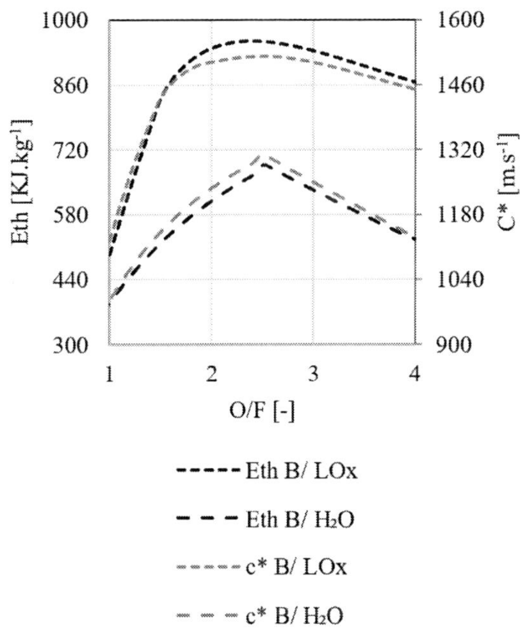

Figure 11. Variation of thermodynamic energy and characteristic velocity with mixture ratio for boron-based hybrid system.

Finally, the calculated results on the main combustion and performance parameters for the considered conditions in tables 1, 2 and 3 are presented in table 4, as shown below:

Table 4. Analysis results on combustion and performance parameters

Propellants Solid = 1 Liquid = 2 Hybrid = 3	Parameters						
	γ	$\left(\frac{T_C}{\mathfrak{M}}\right)^{1/2}$	$E_{Thermodynamic}$	$\dot{m}/A$ at A_t	C*	C_F	Mach number
	-	-	$KJ.kg^{-1}$	$Kg.m^{-2}.s^{-1}$	$m.s^{-1}$	-	-
[1]AP-HTPB-Al	1.1826	10.79	967.81	3934.58	1524.92	1.62	3.23
[1]AP-Fe_2O_3-HTPB-Al	1.1857	10.74	959.51	4284.88	1516.93	1.64	3.32
[2]LCH_4/ LOx	1.1968	12.59	1319.27	16921.63	1772.77	1.80	4.25
[2]RP1/ LOx	1.1908	12.01	1198.93	8605.51	1693.04	1.76	3.86
[2]RP1&LH_2/ LOx	1.1920	13.37	1485.59	15605.09	1883.93	1.92	5.00
[2]LH_2/ LOx	1.2599	15.98	2123.35	9872.37	2208.13	1.82	5.27
[3]Al/ LOx**	1.5995	8.77	639.29	8959.41	1116.12	1.57 [a] 1.62 [b]	5.50 [a] 6.50 [b]
[3]Al/ H_2O(s)**	1.0987	9.35	727.75	7362.07	1358.23	1.72 [a] 1.85 [b]	3.22 [a] 3.62 [b]
[3]Al/H_2O(s)&LOx**	1.1038	9.52	753.31	7248.61	1379.52	1.72 [a] 1.84 [b]	3.23 [a] 3.63 [b]
[3]B/ LOx	1.2092	10.14	854.84	7033.50	1421.74	1.67 -	3.60 -
[3]B/ H_2O(s)	1.1255	9.03	677.77	7696.79	1299.24	1.71 -	3.30 -

** Two cases of expansion ratio are considered for the estimation of C_F and Mach number: $A_e/A_t = 15$ (a) and $A_e/A_t = 33$ (b)

3. Conclusion

Flow density depends positively on the molecular weight; but is inversely proportional to temperatures. This is because at higher temperatures, a high pressure is exerted by the gas; thus, the mass flowing is less. For multistage rockets in which the lower stages have the main task to develop high thrust rather than high velocity, propellants with high molecular weight are preferred. Solid propellant boosters are in this class, and the use of aluminium, to generate aluminium oxide in the exhaust, is advantageous. For a same value of the expansion ratio, propellant combinations lead to different Mach number values, that are proportional to γ. C_F is maximized at the matched expansion condition, where $P_e = P_a$. C_F is thus a measure of the efficiency with which the nozzle extracts energy from the hot stream in the combustion chamber. Therefore, the expansion ratio is selected based on the ambient pressure at which the engine is expected to operate, as previously shown in tables 1 and 2.

On the other hand, methane is keeping sparking interests for space transportation and exploration applications over hydrogen. The reasons are mainly attributed to:

- The benefit of being easily storable than hydrogen and simpler to be operated due to its density.
- The fact that methane does not coke (polymerize) at the operating temperatures of a rocket engine, making it reusable, thus making the implementation of a full-flow stage combustion cycle easier.
- The potential of manufacturing methane in situ from Mars resources.

References

[1] Cai W, Thakre P and Yang V 2008 A model of AP/HTPB composite propellant combustion in rocket-motor environments *J. Combust. Sci. Technol.* **180** 2143–2169
[2] Vummidi S L, Aly Y, Schoenitz M and Dreizin E L 2010 Characterization of fine nickel-coated aluminum powder as potential fuel additive *J. Prop. Power* **26-3** 454-460
[3] Yamana H, Çelika V and Değirmenci E 2014 Experimental investigation of the factors affecting the burning rate of solid rocket propellants *Fuel* **115** 794-803
[4] Yang V, Brill T B and Ren W Z 2000 Solid propellant chemistry, combustion, and motor interior ballistics *AIAA Progress in Aeronautics and Astronautics* (Reston VA: AIAA) 663-687
[5] Maggi F, Gariani G, Galfetti L, Deluca L T 2012 Theoretical analysis of hydrides in solid and hybrid rocket propulsion *Int. J. Hydrogen Energy* **37-2** 1760-1769
[6] Aggarwal R, Patel I K and Sharma P B 2015 Green propellant: A study *Int. J. Latest Trends Eng. Technol.* **6-1**
[7] Sisco J C, Austin B L, Mok J S and Anderson W E 2005 Autoignition of kerosene by decomposed hydrogen peroxide in a dump-combustor configurations *J. Prop. Power* **21-3** 450-459
[8] Cong Y, Zhang T, Li T, Sun J, Wang X, Ma L, Liang D and Lin L 2004 Propulsive performance of a hypergolic H_2O_2/kerosene bipropellant *J. Prop. Power* **20-1** 83-86
[9] Tan J, Liu J, Yang T and Wang Z 2009 On shut-off explosion of high-test hydrogen Peroxide/RP-1 engine *45th AIAA/ASME/SAE/ASEE joint propulsion conf. and exhibit. (Denver, Colorado)* AIAA 2009-4849
[10] Wernimont E and Durant D 2004 Development of a 250 LBFV kerosene – 90% hydrogen peroxide thruster *40th AIAA/ASME/SAE/ASEE Joint propulsion conf. and exhibit. (Fort Lauderdale, Florida)* AIAA 2004-4148
[11] Miller K, Sisco J, Austin B, Martin T and Anderson W 2003 Design and ground testing of a hydrogen peroxide/kerosene combustor for RBCC application *39th AIAA/ASME/SAE/ASEE joint propulsion conf. and exhibit. (Huntsville, Alabama)* AIAA 2003-4477
[12] Ventura M, Wernimont E, Heister S and Yuan S 2007 Rocket grade Hydrogen Peroxide (RGHP) for use in propulsion and power devices – Historical discussion of hazards *43rd AIAA/ASME/SAE/ASEE joint propulsion conf. and exhibit. (Cincinnati, Ohio)* AIAA 2007-5468
[13] Melcher J C and Allred J K 2009 Liquid oxygen/liquid methane test results of the RS-18 lunar ascent engine at simulated altitude conditions at NASA white sands test facility *45th AIAA/ASME/SAE/ASEE joint propulsion conf. and exhibit. (Denver, Colorado)* AIAA 2009-4949
[14] Barsi S, Moder J, and Kassemi M 2008 Numerical investigation of LO_2 and LCH_4 storage tanks on the lunar surface *44th AIAA/ASME/SAE/ASEE joint propulsion conf. and exhibit. (Harford, Connecticut)* AIAA 2008-4749
[15] Cantwell B, Karabeyoglu A and Altman D 2010 Recent advances in hybrid propulsion *Int. J. Energetic Mater. Chem. Prop.* **9-4** 305–326
[16] Chen Y S et al. 2016 Multifunction rocket system development based on advanced hybrid propulsion *14th Int. Conf. Space Operations* 9781624104268
[17] Bergthorson J M, Yavor Y, Palecka J, Georges W, Soo M, Vickery J, Goroshin S, Frost D L and Higgins A J 2017 Metal-water combustion for clean propulsion and power generation *Appl. Energy* **186-1** 13-27
[18] Ghedjatti I, Yuan S and Wang H 2019 Energy generation from metal-water reaction for power systems, underwater and aerospace propulsion applications *16th Int. Bhurban Conf. Appl. Sci. Technol., IBCAST (Islamabad, Pakistan)* 8667177
[19] Meyer M L 1992 Design issues for lunar in situ aluminum/oxygen propellant rocket engines *AIAA Aerospace Design Conf. (Irvine, California)* AIAA-92-1185

Research on Virtual Assembly Simulation System for Aircraft Flight Control System Based on Unity3D

Yan Wang[1], Zhangning Zhou[2,3], Puli Zhou[2,3] and Bo Li[2,3,*]

[1] AVIC Chengfei Commercial Aircraft Co., Ltd, No.358 Wenguang Road, Chengdu, 610091, China

[2] School of Aeronautics and Astronautics, University of Electronic and Technology of China, Chengdu, 611731, China

[3] Aircraft Swarm Intelligent Sensing and Cooperative Control Key Laboratory of Sichuan Province, University of Electronic and Technology of China, Chengdu, 611731, China

[*]Corresponding author: libo@uestc.edu.cn

Abstract. Based on Unity3d's digital assembly simulation technology, this paper designs a virtual assembly simulation system for the flight control system of a certain aircraft nose. Firstly, the structure tree division of the flight control system assembly model based on modular decomposition is studied. Secondly, a virtual assembly simulation system for the assembly operation of the flight control system is designed and implemented by Unity3D. Finally, the system verified the virtual assembly operation of the flight control system of the aircraft nose through the laboratory platform. The assembly guidance and interaction functions of the flight control system are initially realized by this simulation system.

1. Introduction

In order to improve the efficiency and quality of assembly engineers, technical methods such as virtual assembly simulation technology, 3D assembly instructions, and electronical work instructions are gradually applied to the assembly site to provide engineers with a visual and interactive display of the operation process. However, for the assembly process of certain types of aircraft flight control systems, the complicate assembly process brings a series of problems to the actual operation for the engineers. Standard & cumbersome assembly drawings and procedures cannot effectively provide convenient operation guides, and also cannot provide appropriate assembly training and demonstration functions. At the same time, with increase in product development frequency and assembly quality requirements, the contradictions between assembly efficiency, assembly quality and manual assembly operations have become increasingly prominent. Traditional instruction visualization technology has been unable to meet the current urgent requirements[1-2].

Therefore, during the assembly operation phase of the aircraft flight control system, a virtual assembly simulation system capable of visualization and interaction is urgently needed to enable assembly engineers to interact with the assembly guidance system, thereby improving workers' assembly proficiency and quality. With the rapid development of virtual simulation assembly technology, its visualization, interaction and other characteristics have shown great application potential in aircraft assembly research.

Content from this work may be used under the terms of the Creative Commons Attribution 3.0 licence. Any further distribution of this work must maintain attribution to the author(s) and the title of the work, journal citation and DOI.
Published under licence by IOP Publishing Ltd

2. Research Status of Virtual Assembly Simulation Technology

All of the aircraft manufacturers generally adapt the model of "main manufacturer + suppliers". The suppliers are responsible for manufacturing parts and assembly of the airframe structure (excluding the system part), and the main manufacturer is responsible for the entire aircraft butt and system parts manufacturing, installation and commissioning. Because the system parts are installed on the body structure, there are many coordination relationships between them. Therefore, the main manufacturer and suppliers often delay the production progress due to the organization and system coordination problems.

Virtual assembly refers to the use of virtual reality technology, computer tools, artificial intelligence technology and simulation technology to construct a virtual environment, through analysis and prediction of product models, interactive analysis of product data and simulation of the product assembly process and assembly results. Virtual assembly technology enables aircraft assemblers to influence product design, thereby reducing manufacturing costs and avoiding aircraft assembly cycles and costs due to design issues. Simulation technology can find potential problems in the early stages of design maturity and eliminate potential assembly defects. The aircraft assembly process can use virtual simulation technology to simulate the assembly process, thereby evaluating and optimizing the design and process of the aircraft, which can improve production efficiency and reduce product assembly cycle, reducing costs, thus greatly improving the professional level of aircraft assembly [2].

3. Construction of Modular Assembly Process for Flight Control System

The flight control system modular assembly is an assembly technology based on the modular design of components to carry out modular assembly, testing, and delivery. The following analysis will be made from three aspects of modular feature analysis, traditional assembly processes, and modular assembly processes.

3. 1. Modular feature analysis

The flight control system consists of a rudder, an elevator, an aileron, and a central console. The aileron performs roll control of the aircraft, the rudder controls yaw control of the aircraft, the elevator controls pitch of the aircraft, and the central console serves as the control centre, integrating multiple control devices. The rudder, elevator, and ailerons are separated by a central wing. The central control system is located in the middle of the nose compartment. In fact, each module forms a unit structure, which has the basic modular features of the overall assembly. On the rudder, elevator, and aileron, there is an interrelated relationship. Therefore, its modular characteristics include two major characteristics, named structure and system function.

3.2. Analysis of traditional assembly processes

From the perspective of the manufacturing process of a certain aircraft nose structure, in the traditional assembly method, the components of the aircraft control system need to be completed during the component assembly stage, such as the docking assembly, the installation of parts and standard parts and the flight assembly control The testing of the moving state and sensor state of system components and related moving parts is the main task. The flow analysis is performed according to aircraft motion control and structural principles. All control of flight activities is directly through the central wing to the cockpit central console. All operation and control signal data collection and processing are performed in the central console[3]. Flight control of the nose of a certain aircraft The overall process of system traditional assembly is shown in Figure 1.

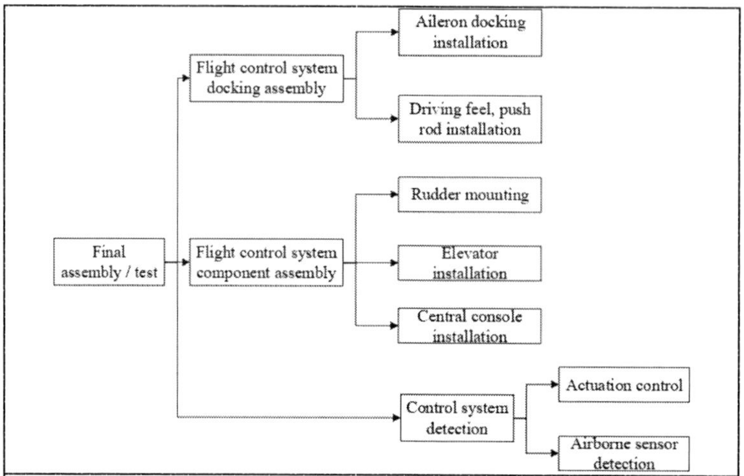

Figure 1. Traditional assembly process of aircraft flight control system.

3.3. Analysis of modular assembly processes

According to the traditional assembly process and characteristic analysis of the flight control of a certain type of aircraft, its structural assembly has been separated from the final assembly. The assembly process is mainly to place the central controller, rudder, and elevator in the local assembly, and the aileron docking and control system inspection to the final assembly phase, the most influential factor for the traditional modular assembly of flight control is the final assembly test. To achieve the traditional modular assembly of flight control, the traditional assembly process of flight control needs to be changed firstly. For assembly integration, docking digitalization, and flexible assembly, the system modification of test unitization and other aspects. The modular assembly process of the aircraft flight control system is shown in Figure 2.

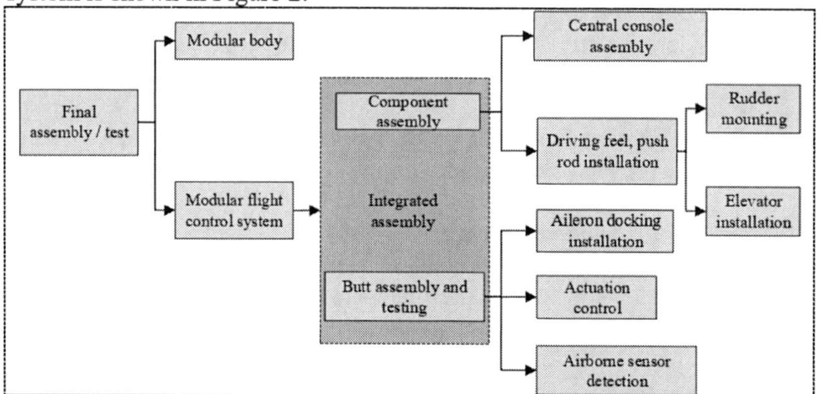

Figure 2. Modular assembly process of aircraft flight control system

4. Design of Virtual Assembly Simulation System for Flight Control System

The process of developing a flight control system virtual simulation system based on Unity3D software includes three main steps. The first step is to package all models according to the divided modules. The second step is to export the .fbx file and import it into the established Unity3D software project. The final step is to add the according components, such as cameras, collision detection, and systems interface, etc. The interface and other components are implemented through C # programming to complete the design of the flight control virtual assembly simulation system, to observe the components, the assembly sequence, and the completion situation, etc., to increase the user's awareness of the flight control system assembly, and to achieve the expected teaching guidance effect[4]. Its system structure diagram is shown in Fig. 3.

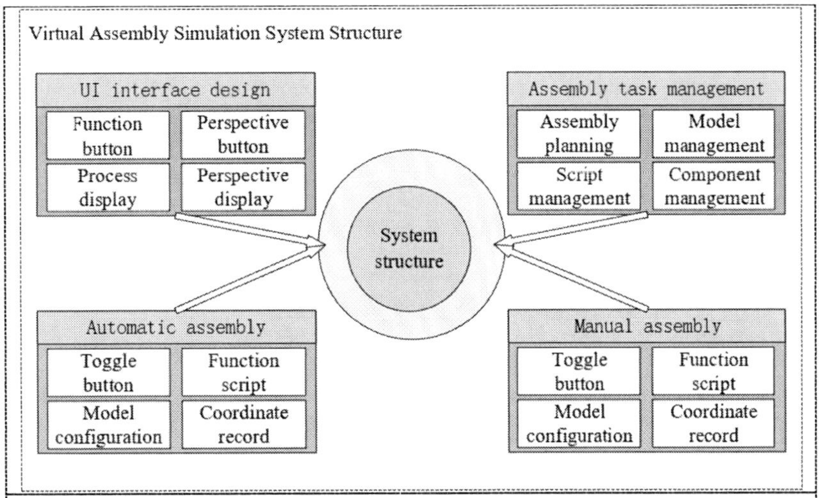

Figure 3. System structure diagram

4.1. Models imported into Unity3D

According to the Catia 3D model provided by an aircraft manufacturer, it is divided based on the modular structure tree; then the CGR format file is exported to 3d Max software, and then the FBX format file is exported, where the model's Y axis is set in the export dialog box , So that it is consistent with the coordinate system in Unity3D software, and then create a new project, import the FBX model exported from 3DMax into the new project[5-7].

4.2. Interface design

To create a new scene, one main camera by default, and two game objects with directional light. The main camera is equivalent to the "eye" that observes the scene. What the eye sees will be displayed in the game view. The directional light is equivalent to the "sun" in the scene, thereby enhancing the brightness of the entire scene. In order to meet the assembly perspective of the front view, left view and side view functions, please add two cameras in the function box and assign IDs. In addition, to add components, such as function buttons, text boxes, and combo boxes, assign IDs accordingly and create script files. The complete interface structure is shown in Figure 4.

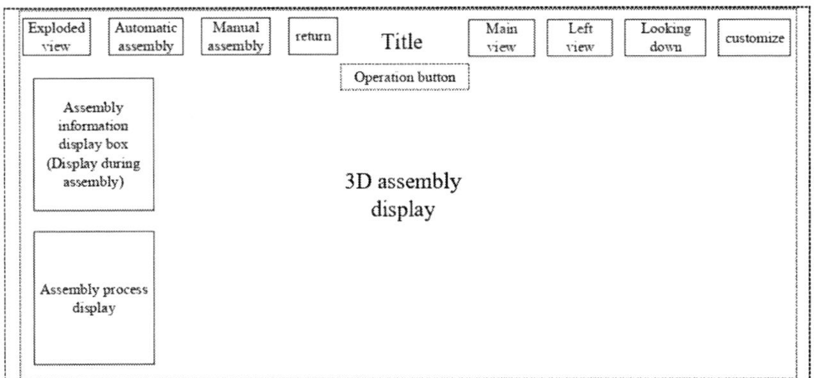

Figure 4. Effect drawing of the flight control system virtual assembly

4.3. Functional design

According to the corresponding function button definition of the interface, to create the corresponding script file, and write the corresponding function code using C# language. The main function scripts include initialization scripts for binding model components, scripts for view display and camera perspective conversion; automatic assembly scripts for automatic display of assembly sequence; manual assembly scripts for manual assembly by users.

To initialize the scripts, bind the model components and interface ID, for subsequent function implementation; camera script, bind three cameras to achieve custom view, left view and side view functions; automatic assembly script to achieve assembly with start-up, The order of the pause and progress display functions is automatically displayed; manual assembly of the script enables the user to manually assemble the next, previous and progress display functions. Finally, you can bind the script to the component and verify the function implementation.

5. Application examples of virtual Assembly simulation systems

5.1. Functions of Flight Control System Virtual Assembly Simulation System

In the actual assembly process of the worker, the layout of the flight control system and the cognition of the assembly process are very important. Therefore, the following system functions have been developed for the flight control system of the nose of a certain aircraft: the switching of the assembly perspective and the animation of the assembly process interface for display, virtual manual assembly, and human-computer interaction[8].

(1)Perspective conversion. The perspective view of the virtual assembly system is divided into custom views, front view, left view and side view. These four views are switched by the interface perspective function buttons.

(2)Animation demonstration of the assembly process. Uses C# script to control component coordinates so that the model can be animated for assembly in a specific order.

(3)Virtual manual assembly function. Uses C# script to control component coordinates to realize manual assembly by the user, with next, previous and progress display functions

(4)User interface. The initial interface of the system and the UI interaction of the information display box are made using the GUI components that come with Unity3D.

5.2. Use case of flight control system virtual assembly simulation system

The virtual assembly simulation system of the flight control system in this paper is exported as a Windows 7/8/10 system executable program file through Unity 3D. The user uses the monitor and mouse control to assemble the scene on the virtual machine head from the first angle of view. The user can observe the assembly process during use, and the corresponding information display box can make the user quickly familiar with the assembly process and details. The simulation system is shown in Figure 5. The assembly progress is shown at the bottom left.

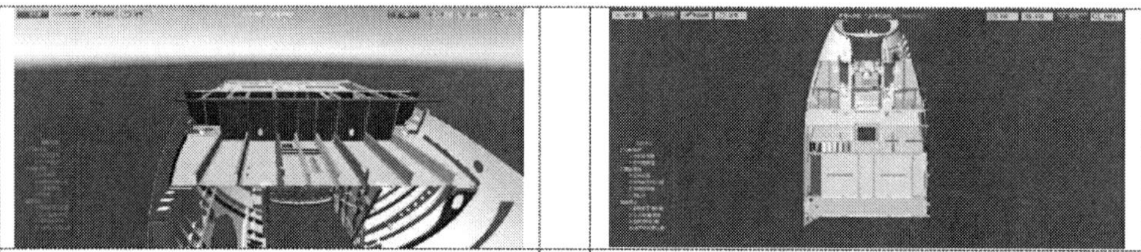

Figure 5(a). Initial interface diagram (front view)

Figure 5(b). Top view

Figure 5(c). Automatic assembly

Figure 5(d). Manual assembly

Figure 5. System example simulation screenshot

In Figure 5, Figure (a) shows the virtual system initialization interface (main interface), and workers can choose the corresponding operation according to their needs. Figure (b) shows the top view of the system, allowing workers to observe from different angles Figure (c) shows the automatic assembly function of the system. There are the start and pause function buttons at the top of the interface, and the assembly progress is displayed in the lower-left corner. Figure (d) shows the manual assembly function of the system. There are previous and next assembly functions at the top of the interface, and an information display box at the top right shows the details of assembly at each step.

6. Conclusion

This article builds a virtual assembly simulation system for an aircraft flight control system based on Unity3D, which has been verified by laboratory environment simulation. The simulation system can guide and train workers to assemble a flight control system in actual aircraft assembly production process. The results show that the system is in use. It can effectively guide assembly and interactive operations. With the development of digital technology and virtual reality technology, the traditional assembly guidance and training methods in the manufacturing industry will be gradually replaced, and the development of virtual reality will also be increasingly perfected to achieve the requirements of improving better production efficiency and quality.

Acknowledgments

This work is supported by Sichuan Science & Technology Program under Grant 2018GZDZX0019 & 2019YFG0383 & 2019YFG0052.

Reference

[1] SUN Shanmin, ZHOU Shuo, GAO Ge, LI Lin. Key Technical Issues in Aero Engine Assembly Simulation[J]. Aeronautical Manufacturing Technology, 2018,61(22):98-103. [In Chinese]

[2] TABRIZIA, SANGUINETTI P, KHOSROWPOUR MEncyclopedia of information science and technology[M]. New Orleans: Information Resources Management Association,2018.

[3] WANG Shouchuan, HAO Ju, LI Xining, WANG Yonghong, WNAG Zhanfeng. Research on Aircraft Wing Modular Assembly Technology[J], Aeronautical Manufacturing Technology, 2018, 61(13): 68-73. [In Chinese]

[4] ZHU Anqing, HU Anchao. Construction of Virtual Simulation Teaching System for Shipbuilding Based on Unity 3D [J] .Laboratory Research and Exploration, 2018,37 (06): 117-120.

[5] ZHU Ran. Research and Design of Intelligent Manufacturing Simulation System Based on Unity3d [D] .Southwest Jiaotong University, 2018. [In Chinese]

[6] ZHANG Qinghui. Application of Virtual Assembly Technology in Aircraft Assembly [J] .Mechanical and Electrical Information, 2018 (36): 97-98. [In Chinese]

[7] TANG Jianjun, YE Bo, GENG Junhao. Exploration and Practice of AR Intelligent Guidance Technology for Aircraft Assembly Operations[J]. Aeronautical Manufacturing Technology, 2019,62(08):22-27. [In Chinese]

[8] Han, P., Zhao, G. Line-based initialization method for mobile augmented reality in aircraft assembly[J]. Vis Computer 33, 1185–1196 (2017).

Dynamic Characteristics Analysis and Flight Test of Free Flight Model with Four Propellers

Yalong Wang*, **Wenqiang Yin and Shenjun Qi**

Chinese Flight Test Establishment, Xi'an, China

*E-mail address: yalong2015@163.com

Abstract. Flight mode characteristics of free flight model with four propellers are analyzed, based on the results and flight test requirement, the control law is designed, then the simulation model is established and analyzed, finally, the control law is applied to free flight model test.The flight test results show: the propeller promoted the velocity stability and Dutch mode characteristic; level and turn flight stall test subjects were effectively accomplished based on the control law; the dynamic model built can simulate the free flight model test processand initially predict the stall characteristics; level stall angel is 10-12°, turn stall angel is 8-10°.

1. Introduction

According to the requirements of the stall demonstration of the CCAR 25 Stall Test Flight Clause 25.201, the stall flight test should be made under zero thrust and thrust required to maintain 1.5VSR1 level flight [1], and the thrust characteristics will affect the stall characteristics, therefore, stall flight test must be conducted under similar dynamic conditions to explore the stall characteristics of a full-size aircraft. However, in the free flight test, the model is dropped from a height of more than 1km. The flight operator has limited field of view when operating on the ground and cannot intuitively feel the motion state of the model. Therefore, there exists error in maneuvering, and at the same time, carrier' airflow and model processing error will also affect the stability of the model, which will have an impact on the test. Therefore, the corresponding control law must be designed before the test to ensure the success of the flight test.

Scholars have carried out a lot of research on flight simulation and test of free flight model. Zheng Hao et al. designed control law of the free flight model and effectively completed the free flight test [2]. Liu Shangmin et al. studied the control law of the fly-by-wire model [3], and gave the flight test method of the static unstable aircraft. Liu Chaojun et al. used a free flight test as an example [4], Through simulation, the deflection law of the control surface during the stall test of the free flight model was studied. Jiao Wencheng et al. studied the propeller's dynamic characteristics of free flight model and simulated the propeller's pull force using similarity criteria. However, the model used did not have a self-driving system and did not simulate the test process [5]. Luo Wumei used the free flight model to study the stability of K8 aircraft, he designed the control stabilization system and carried out simulation experiments, and achieved good results [6]. However, most of these studies are based on flight test data, and there is no calculation and analysis of the dynamic characteristics of the model.

Taking a free flight model with four propellers as an example, the influence of the propeller's pull force on the dynamic characteristics of the free flight model is considered. The control logic and control law suitable for the free flight test are designed. The free flight model test process is carried out through

Content from this work may be used under the terms of the Creative Commons Attribution 3.0 licence. Any further distribution of this work must maintain attribution to the author(s) and the title of the work, journal citation and DOI.
Published under licence by IOP Publishing Ltd

simulation. The simulation and prediction of the stall characteristics were carried out. Finally, the design scheme was applied to the flight test of the free flight model, and the level and turn flight stall subjects were accomplished, and desirable results were obtained.

2. Analysis of propellers' influence on dynamic characteristics

2.1. Theoretical Analysis

In order to ensure the dynamic similarity between the free flight model and the prototype aircraft, it is necessary to simulate the propeller's pull force. In this test, the propeller's pitch is fixed and the forward ratio of the free flight model is the same as the prototype aircraft, which in turn ensure the dynamic similarity of the free flight model to the prototype aircraft at the stall point.

However, during the test, the propeller's pull force can affect the dynamic characteristics of the free flight model due to the fixed pitch and the changing force of the propeller with the incoming flow velocity.

The free flight model has four propellers, according to the formula for calculating the single propeller pull force and forward ratio

$$T = T_c \rho n_s^2 D^4 \tag{1}$$

$$\lambda = \frac{V}{n_s D} \tag{2}$$

Then, the pull force can be obtained as a function of the speed

$$T_V = \frac{4\Delta T}{\Delta V} = \frac{4\Delta T_c \rho n_s^2 D^4}{\Delta\lambda n_s D} = \frac{4\Delta T_c}{\Delta\lambda} \rho n_s D^3 \tag{3}$$

Where,

T_c	Pull force coefficient
T	Pull force
ρ	Air density
n	Speed of the propeller's rotation
V	Air velocity
D	Diameter of the propeller

The wind tunnel test data of the propeller used to obtain the relation between the propeller's pull coefficient and the forward ratio under the thrust required to simulate unpowered and 1.5VSR level flight state, as shown in Figure 1.

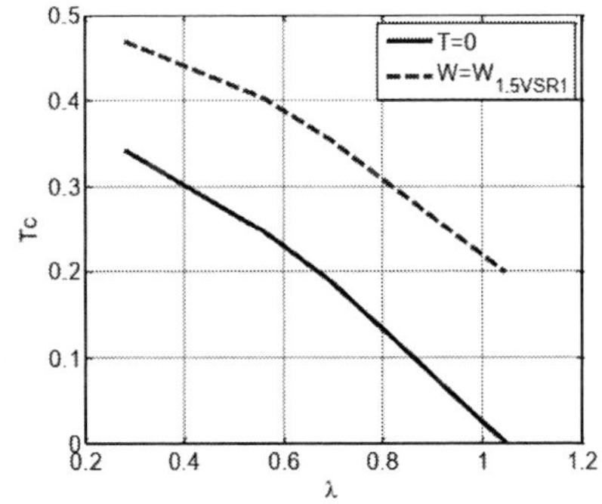

Figure 1. Pull coefficient curve of different state

$\frac{\Delta T_c}{\Delta \lambda}$ can be calculated by linearizing the curves above, and the T_V can be calculated by the equation(3), the results are shown in table1.

Table 1. T_V of different condition

Flight condition	T_V
Tc=0	-4.1
1.5VSR1	-2.67
Without propeller	0

Furthermore, the influence of the distributed propeller's pull force on the directional damping characteristics of the free flight model can be obtained, as shown in Equation 4.

$$\Delta C_{nr} = \frac{\Delta C_n}{\Delta r} = \frac{\frac{1}{2} T_V \Delta r (L_1 + L_2) / (\frac{1}{2} \rho V^2 Sb)}{\Delta r} = \frac{T_V (L_1 + L_2)}{\rho V^2 Sb} \tag{4}$$

Where, L1 and L2 are the distances of one side of the engine from the symmetry plane.

It can be seen that the four-propeller can improve the velocity stability of the free flight model and the directional damp, which in turn affects the dynamic characteristics. Therefore, the influence of the propeller's pull force needs to be considered in the subsequent dynamic analysis, flight simulation, and control law design.

2.2. longitudinal flight mode

When the flight quality is calculated by the small disturbance equation, it is generally considered that $T_V = 0$, in this test $T_V \neq 0$, the longitudinal flight quality calculation results are shown in Table 2. It can be seen that the dynamic characteristics of the propeller significantly improve the long period damping and have no effect on the short period characteristics.

Table 2. longitudinal flight quality

Flight conditon	Phugoid		Short period		CAP
	frequency	damp	frequency	damp	
Tc=0	0.37	0.19	6.4	0.47	5.06
V=1.5VSR	0.33	0.15	6.4	0.47	5.06
Without propeller	0.38	0.05	6.4	0.474	5.06

2.3. lateral and directional flight mode

The results of the lateral and directional flight quality are shown in Table 3. It can be seen that the free flight model Dutch mode and the Roll mode flight quality is improved by the propeller. The spiral mode is unstable, so it need to be stabilized in the free flight test.

Table 3. lateral and diretional flight quality

Flight Condition	Dutch mode		Roll mode		Spiral mode	
	frequency(rad/s)	damp	eigenvalue	T_r(s)	eigenvalue	$T_{1/2}/T_2$(s)
Tc=0	5.39	0.153	-5.43	0.1842	0.0469	14.78
V=1.5VSR	5.84	0.159	-5. 52	0.1810	0.0430	16.12
Without propeller	5.78	0.141	5.72	0.1746	0.0540	12.84

3 Design of control law

3.1 Flight Control Strategy

As shown in Figure 2, Three control modes are designed for flight test.

- Drop control mode
  Drop control mode is designed to achieve the safe separation of the model from the carrier, the lateral trimming after the drop and the desirable attitude maintenance when the umbrella is to be opened. In the test, the pitch angle and the roll angle control law are designed to make the model stable after it's dropped.
- Stall control mode
  The role is to achieve automatic control of the model. In the process of entering the stall, the roll angle control law is used to ensure lateral and directional stability before the model enters the stall.
- Direct control mode
  The main control method is to use the full rod to control the full rudder. The model is rectified after stalling.

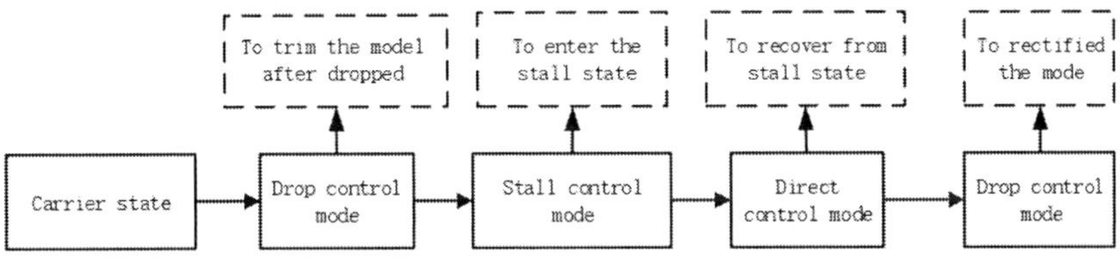

Figure 2. Flight control scheme and process of the free flight test

3.2 Control law

a) Longitudinal control law

The longitudinal control loop circuit input is the pitch angle command, and the feedback is the pitch angle and the pitch rate. The output is the elevator deflection command.

$$\begin{aligned} &\delta_e = k_q q + k_\theta \Delta\theta + k_{\theta i}\int \Delta\theta dt + \delta_{e0} \\ &\Delta\theta = \theta - \theta_c \end{aligned} \tag{5}$$

b) Lateral & directional control law

$$\begin{aligned} &\delta_l = I_p p + I_\phi(\phi - \phi_c) \\ &\delta_r = K_r r \end{aligned} \tag{6}$$

4 Flight test

4.1 Design of the flight process

Step1: The model is taken to 1500m by the carrier aircraft. After the carrier reaches the scheduled drop point, it will be confirmed by the tower, then the model will be dropped by the pilot of the carrier aircraft.

At the same time, the operator manipulates the flight control into the delivery control mode, at which time the longitudinal bar corresponds to the pitch angle control command, and the transverse bar corresponds to the roll angle control command;

Step2: After the model is dropped, the ground operator manually pushes the push rod, the model subducts, and the roll angle remains at zero. At the same time, the carrier pilot pulls the pole and exits the test area.

Step3: The operator switches to the stall control mode, then the model enters the stall. During the flight test of the level flight, the roll angle is controlled to -30° during this process, and the roll angle is controlled to -30° during the test.

Step4: After the model stalls, it enters the direct control mode, and the simulated pilot changes out of stall.

4.2 Analysis of the flight data

The results of the level-flight stall test are shown in Figure.3, and the results of the turn-flight stall test are shown in Figure.4.

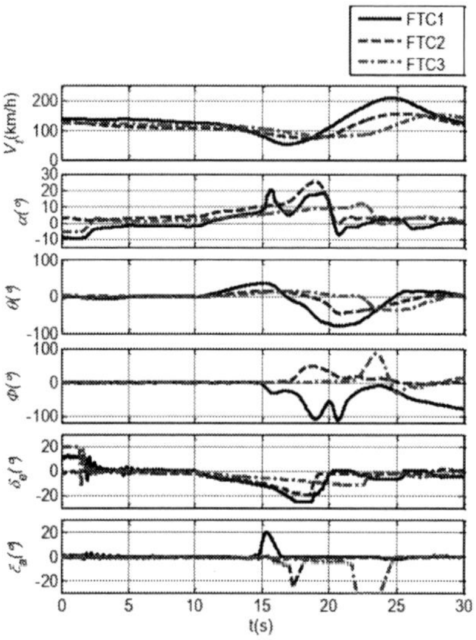

Figure 3. Flight test result of level flight stall

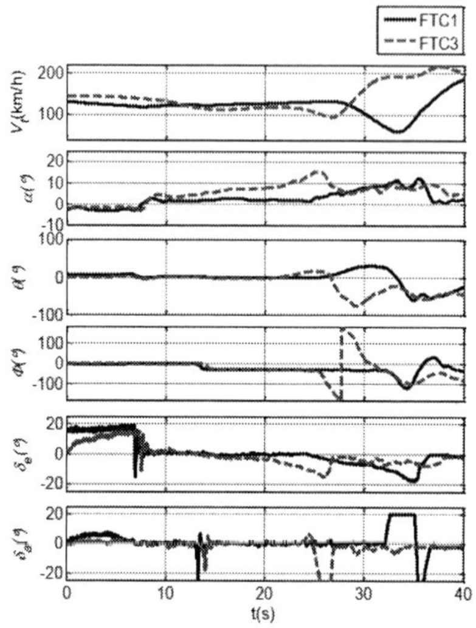

Figure 4. Flight test result of turn flight stall

The following conclusions can be drawn:

1) Under different test conditions, the switch of control mode is stable and there is no control surface jump;

2) The angle of attack of the level flight stall is 10-12°, and the angle of attack of the turn flight stall is 8-10°, which is consistent with the simulation results;

3) Operator evaluation: The propeller improved the stability of the free flight model, which is consistent with the analysis and simulation results.

5 Conclusion

1) The dynamic characteristics of the propeller can improve the speed stability and the directional stability of the free-flying model, which is beneficial to the flight operator's control;

2) The established dynamic model can achieve the simulation of the flight test process of the propeller powered free flight model;

3) The designed control scheme effectively implements the level and turn stall tests;

4) The research procedure is applicable to the stall flight test of the propeller powered free flight model.

References

[1] Chinese Civil Aviation Regulations. *CCAR-25-R4.*

[2] Zheng Hao Zhao Zhijun Zhang Jiaqi. Design of control law for unpowered free flight model [J]. *Flight Mechanics,* 2016.

[3] Liu Shangmin, Zhao Lei. Research on flight control technology of free flight test of teletype aircraft model[J].Flight Mechanics,2012,2.

[4] Liu Chaojun, Sun Yingfei, Wu Xinlong. Simulation Research on Free Flight Test Process of Unpowered Model[J]. China Science and Technology Information, 2016, 9.

[5] Jiao Wencheng. Experimental study on free flight with dynamic model [J]. Flight Mechanics, 1994, 3.

[6] Luo Wumei. Research on Relaxation and Static Stability of K8 Aircraft Free Flight Model[J].Flight Mechanics,1995,9.

Haptic Rendering Algorithm based on Force Feedback Handle in the Teleoperation of the Space Manipulator

X A Qiu, D T Ma, S J Wang, F W Duan and B Zhu

Lanzhou Institute of Physics, Lanzhou 730000, China

qiuxinan502@163.com

Abstract. In this paper, we study the haptic rendering algorithm of the handle for the requirement of designing the force feedback handle which is applied to the teleoperation of the space manipulator, to improve the user experience of the force feedback handle. By the research of arbitrary force and moment at the end of the handle, we propose the operation assisting force based on gravitational potential field method and the precise operation guiding method based on gravitational potential energy model. At the same time, the vibration signal is used in the state information of the slave and state. The experimental results show that the force-aware rendering algorithm applied to the teleoperation of the space manipulator can greatly improve the operation efficiency of the teleoperation task and effectively enhance the telepresence of teleoperation.

Key words: handle of manipulator haptic rendering assisting force

1. Introduction

Force feedback device is a kind of interface that realizes human-computer (Computer or robot) interaction through force. It can collect the movement information of human hands to control the computer or robot. At the same time, it can output regular force feedback to the operator to realize the operator's perception of the physical state of computer or robot. At present, space robotic arm systems that has successfully developed and put into use include SSRMS (Space Station Remote Manipulator)[1] jointly developed by Canada and NASA, ERA (European Robotic Arm)[2] of European Space Agency, and JEMRMS[3] of Japan. For the above-mentioned robotic arm teleoperation system, the master-side handle is the core control component [4], which is operated by professionals and sends control instructions to the slave-side robotic arm. The main control device used in the MSS system of the International Space Station is a pair of 3DOF handles. Professionals can control the movement of the robotic arm outside the cabin by operating the two handles to complete specific tasks. However, the above-mentioned operation handle does not have a force feedback function, cannot provide force feedback to the operator, and cannot realize the operator's perception of the unknown environment at the slave end and the interaction state between the manipulator and the slave end environment through the force channel. Limited by the accuracy and speed of visual information collection, the single visual interaction channel limits the effectiveness, efficiency, and safety of operators in completing professional tasks. Therefore, space robotic arm teleoperation system puts forward a clear and urgent demand for the force sensor interaction channel created by the main-end control handle with force feedback function.

The force rendering of the main-end control handle can effectively enhance the presence of the

Content from this work may be used under the terms of the Creative Commons Attribution 3.0 licence. Any further distribution of this work must maintain attribution to the author(s) and the title of the work, journal citation and DOI.
Published under licence by IOP Publishing Ltd

main-end operator and improve the efficiency of teleoperation. Introducing it into the teleoperation of space robotic arm, it can feedback the interactive information of the unknown environment with uncertainty to the operator, so that professionals can obtain comprehensive information of vision and force perception; It can also provide effective force alert when an emergency or unexpected situation occurs, to avoid damage to the robot arm or the aircraft. The force rendering algorithm of force feedback handle studied in this paper is not only new development of the application field of force feedback device, but also well meets the engineering requirements of China's space robotic arm construction tasks.

The force feedback handle is used as the interactive device and the main-end of the teleoperation of the robotic arm and it is responsible for the input of the robotic arm movement instructions and the output of the robotic arm force and status information in the control loop. The tasks of the control system of the force feedback handle include: (1) Collect the joint angular displacement of the handle in real-time and calculate the attitude of the end of the handle; (2) Transmit motion instructions to the slave-end robot arm control system; (3) Calculate the feedback force based on the interaction status of the robot arm in the slave-end environment Information; (4) Transmit the feedback force command to the control system of the main-end handle.

The force feedback handle studied in this paper is based on a 3-RRR spherical parallel mechanism. Larger components such as motors, encoders, and brakes are installed on the base, and a rope transmission method with no backlash and low friction is used. Due to the small mass of the end, so the impedance control mode is used.

2. Arbitrary force and moment of force calculation at the end

2.1. Moment of force calculation in any direction

Generally, the speed Jacobian matrix is used to describe the speed relationship between the joint space and the operating space of the mechanism. Therefore, the Jacobian matrix is also called the generalized transmission ratio of the mechanism. Reference [5], the following will calculate the speed Jacobian matrix of 3-RRR spherical parallel mechanism.

Set the angular velocity at the end of the mechanism as $\boldsymbol{\omega}$, driving joint angular velocity is $\dot{\boldsymbol{s}}$, the speed relationship of the 3-RRR spherical parallel mechanism can be described as:

$$\boldsymbol{A\omega} = \boldsymbol{B\dot{s}} \tag{1}$$

Among them, A is the first type of Jacobian matrix, and B is the second type of Jacobian matrix.

$$\boldsymbol{A} = \begin{bmatrix} (\boldsymbol{w}_1 \times \boldsymbol{v}_1)^T \\ (\boldsymbol{w}_2 \times \boldsymbol{v}_2)^T \\ (\boldsymbol{w}_3 \times \boldsymbol{v}_3)^T \end{bmatrix} \tag{2}$$

$$\boldsymbol{B} = diag(\boldsymbol{u}_i \times \boldsymbol{w}_i \cdot \boldsymbol{v}_i) \quad i = 1,2,3; \tag{3}$$

$\boldsymbol{u}_i$, $\boldsymbol{w}_i$, $\boldsymbol{v}_i$ $(i = 1,2,3)$ is the direction of the three rotation auxiliary axes of each moving branch chain (the drive shaft, the hinge shaft of the side link and the connecting rod, and the hinge shaft of the connecting rod and the moving platform).

Set J as the speed mapping relationship between the input and output of the mechanism ($\dot{\boldsymbol{s}} = \boldsymbol{J\omega}$), that is, the speed Jacobian matrix,

$$\boldsymbol{J} = \boldsymbol{B}^{-1}\boldsymbol{A} \tag{4}$$

In the control, we need to use the differential of Euler angle to time, so the transformation matrix $\boldsymbol{R}$ is introduced,

$$\omega = R\dot{t} \tag{5}$$

Where $\boldsymbol{t}$ represents the Euler angle vector, $\boldsymbol{t} = [\phi, \theta, \psi]^T$;

$$\boldsymbol{R} = \begin{bmatrix} 0 & \cos\phi & -\cos\theta\sin\phi \\ 0 & \sin\phi & \cos\theta\cos\phi \\ 1 & 0 & \sin\theta \end{bmatrix} \tag{6}$$

Thus, it can be obtained

$$\dot{s} = \boldsymbol{JR}\dot{\boldsymbol{t}} \tag{7}$$

Depending on the task of the operation, it is essential to apply a specific moment of force at the end of the handle. A moment vector can be determined by the axis direction and magnitude of the moment, so the direction and magnitude of the moment need to be given in the force calculation process.

Ignoring the gravity of the mechanism rods and considering the relationship between the moment at the end of the mechanism and the joint driving force. Set the external moment at the end of the handle as $\boldsymbol{M}_t$, and the moment that drives the joint is $\boldsymbol{\tau}$.

According to the principle of virtual work, there are

$$\boldsymbol{M}_t^T \dot{\boldsymbol{t}} = \boldsymbol{\tau}^T \dot{\mathbf{s}} \tag{8}$$

Substituting equation (7) into (8) gives

$$\boldsymbol{\tau} = (\boldsymbol{JR})^{-T} \boldsymbol{M}_t \tag{9}$$

Therefore, the control system of the motor converts the torque signal into a current command, thereby controlling the motor driver to output a corresponding current.

2.2. Force calculation in any direction

In order to simulate the interactive force information of the slave-end robotic arm and the environment during teleoperation, the force feedback handle should be able to simulate forces in any direction and magnitude. As shown in figure 2, the handle simulates a force vector $\boldsymbol{F}$, which is directly acting on a human hand. For a force feedback device, $\boldsymbol{F}$ generates a moment $\boldsymbol{M}$ with respect to the rotation center $\boldsymbol{O}$. The size of $\boldsymbol{M}$ is:

$$|\boldsymbol{M}| = |h \bullet \boldsymbol{F} \bullet \boldsymbol{k}| = h|F|\cos < \boldsymbol{F}, \boldsymbol{k} > \tag{10}$$

In the formula, h represents the distance from the center of rotation of the handle to the grip center of the human hand; k represents the unit vector in the z-axis direction of the dynamic coordinate system O-xyz, as shown in Figure 1.

The axis direction of the moment $\boldsymbol{M}$ is defined as $\boldsymbol{e}_m$, then

$$\boldsymbol{e}_m = \boldsymbol{F} \times \boldsymbol{k} \tag{11}$$

So the moment of force vector

$$\boldsymbol{M} = |\boldsymbol{M}| \boldsymbol{e}_m \tag{12}$$

The calculated torque vector can be controlled according to the calculation method of any torque vector.

IOP Conf. Series: Materials Science and Engineering **887** (2020) 012036 doi:10.1088/1757-899X/887/1/012036

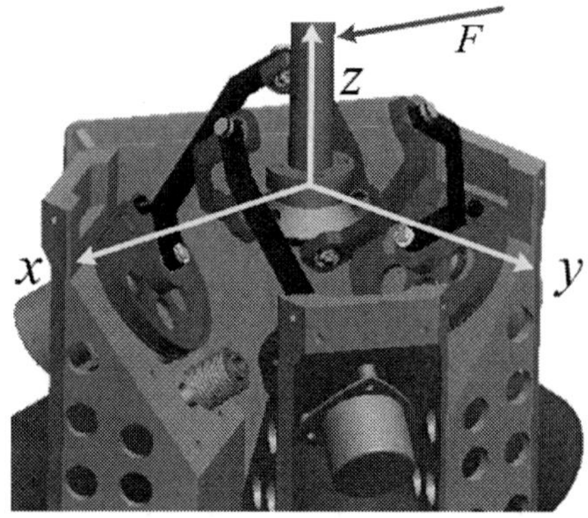

Figure 1. Schematic diagram of arbitrary force vector of the handle

3. Auxiliary Force Study of Operation

In the face of complex space operation tasks, it is very difficult to achieve precise teleoperation tasks only by visual feedback, while ensuring safety and avoiding collisions, and the operator will also face great mental pressure. The introduction of tactile assistance to guide the operator will help reduce the difficulty of operation tasks and greatly improve the efficiency, accuracy and safety of operations.

3.1. Artificial Potential Field

APF (Artificial Potential Field) was proposed by Khatib in 1986[6]. It was originally used in mobile robot obstacle avoidance and path planning to calculate virtual force based on the potential energy function related to the robot position. Subsequently, the artificial potential field method was gradually applied to the fields of robot path planning, bilateral teleoperation, and robot safety warning. The basic theory of artificial potential field is that for a target-oriented robot, an artificial potential field can be defined regardless of the presence of static or moving obstacles in the environment. In this potential field, the target position produces attraction, and the obstacle generates repulsive force [7]. Define the pose at the end of the robot as $\boldsymbol{q}$, the target pose as $\boldsymbol{q_g}$, the obstacle pose as $\boldsymbol{q_w}$, the attractive potential associated with the target pose as $\boldsymbol{U_{att}(q)}$, and the repulsive potential associated with the obstacle pose as $\boldsymbol{U_{rep}(q)}$. When a robot approaching an obstacle in an artificial potential field, it will generate repulsive force, and when approaching a target, it will generate an attractive force. Feedback of these repulsive forces and attractive forces to the operator can guide the operator to make correct decisions. Compared with the traditional model based on collision detection and "spring-mass-damping", this article attempts to provide the auxiliary force for remote operation of the robotic arm with the artificial force field. Compared with the former, the artificial potential field method does not require an accurate dynamic model and does not need to understand the mass and inertial properties of the object, thereby making the auxiliary force perception rendering algorithm simple and robust.

The artificial potential field can use different potential field functions to obtain different virtual potential field forces. The commonly used potential field is the gradient potential field method, and the virtual force is the negative gradient of the potential field, that is:

$$\boldsymbol{F}_{att}(\boldsymbol{q}) = -\nabla \boldsymbol{U}_{att}(\boldsymbol{q}) \tag{13}$$

$$\boldsymbol{F}_{rep}(\boldsymbol{q}) = -\nabla \boldsymbol{U}_{rep}(\boldsymbol{q}) \tag{14}$$

Among them, $\boldsymbol{F}_{att}(\boldsymbol{q})$ represents the attractive force of the potential field, and $\boldsymbol{F}_{rep}(\boldsymbol{q})$ represents the repulsive force of the potential field.

The commonly used potential field model is:

Gravitational field:

$$U_{att}(q)=\frac{1}{2}\xi\rho_g^2(q) \tag{15}$$

Repulsive field:

$$U_{rep}(q)=\begin{cases}\frac{1}{2}\eta\left(\frac{1}{\rho_w(q)}-\frac{1}{\rho_0}\right), & \rho_w(q)\le\rho_0\\ 0, & \rho_w(q)>\rho_0\end{cases} \tag{16}$$

Among them, $\rho_g(\boldsymbol{q})=\|\boldsymbol{q}-\boldsymbol{q}_g\|$ represents the Euclidean distance from the end of the robot to the target pose $\boldsymbol{q}_g$;ρ_0 is a constant, and $\rho_0>0$, which indicates the radius of the range where the obstacle generates repulsive force;$\rho_w(\boldsymbol{q})=min\|\boldsymbol{q}-\boldsymbol{q}_w\|$ represents the minimum distance from the obstacle to the end of the robot;ξ, η represents the proportionality factor;The virtual force generated by the above repulsive and gravitational fields is:

Attractive:

$$F_{att}(q)=\xi(q-q_g) \tag{17}$$

Repulsion:

$$F_{rep}(q)=\begin{cases}\frac{\eta}{\rho^2(q)}\left(\frac{1}{\rho(q)}-\frac{1}{\rho_0}\right)\nabla\rho(q), & \rho(q)\le\rho_0\\ 0, & \rho(q)>\rho_0\end{cases} \tag{18}$$

Where $\nabla\rho(\boldsymbol{q})$ is the unit vector from $\boldsymbol{q}_w$ to $\boldsymbol{q}$, $\nabla\rho(\boldsymbol{q})=\frac{\boldsymbol{q}-\boldsymbol{q}_w}{\|\boldsymbol{q}-\boldsymbol{q}_w\|}$.

3.2. Gravitational potential energy model

The artificial potential field method is of great significance in avoiding obstacles and ensuring the safety of the robot. Using the artificial potential field to guide the feedback force, so that the operator can control the robot arm to move along the safe path. However, the attractive force of the artificial potential field decreases as it approaches the target point. When it is smaller than the human perception threshold, the effect of the guiding force disappears. For precise operation tasks, such as assembly and alignment, when the end of the robot arm is close enough to the target point, small position and attitude deviations are difficult to distinguish through visual feedback. If the guiding force action disappears at this time, it will be difficult for the operator to accurately complete the task. It is of great significance to improve the accuracy of teleoperation through force guidance. Therefore, when approaching the target point, there must be sufficient guidance force, so that the operator can clearly understand the pose error and movement direction of the current end of the robot arm through force perception.

Taking the end of the robotic arm and the target point into two particle gravity models, the attraction of the target point to the end of the robotic arm will increase with the decrease of distance. This law is suitable for precise operation control guidance after the end of the robotic arm approaches the target point. Change the gravitational field in the aforementioned artificial potential field to a gravitational potential energy model:

$$U_{att}(q)=\frac{G}{\rho_g(q)} \tag{19}$$

The attractive force generated by this potential energy field is:

$$F_{att}(q) = -\frac{G}{\rho_g^2(q)} \tag{20}$$

When the end of the robot arm approaches the target point infinitely, the attractive force will tend to infinity. Excessive feedback force will affect the stability of the equipment and the safety of operation. Therefore, the force guidance of the precise alignment operation process is divided into three stages. In stage one, define the domain of attraction A(q) , the end of the robot arm enters the attraction domain, and the attraction of the target point to the robot arm is calculated according to formula (20); In stage two, when the attractive force of the target point increases beyond a certain threshold, the attractive force is changed to a constant force. At this time, only the direction of the force is changed and the magnitude of the force is not changed. In stage three, when the pose error between the end of the robotic arm and the target point is very small, the attraction force will be zero when it is within the allowable error range of the assembly or docking task. Therefore, the formula for attraction is modified as:

$$F_{att}(q) = \begin{cases} -\frac{G}{\rho_g^2(q)}, & \rho_g(q) \geq \rho_a \\ F_{max}, & \varepsilon < \rho_g(q) < \rho_a \\ 0, & \rho_g(q) < \varepsilon \end{cases} \tag{21}$$

Among them, ρa is the radius of attraction domain, ε is allowable pose error, F_{max} is the maximum output force to ensure system stability. The change law of attractive force with distance after correction is shown in Figure 2.

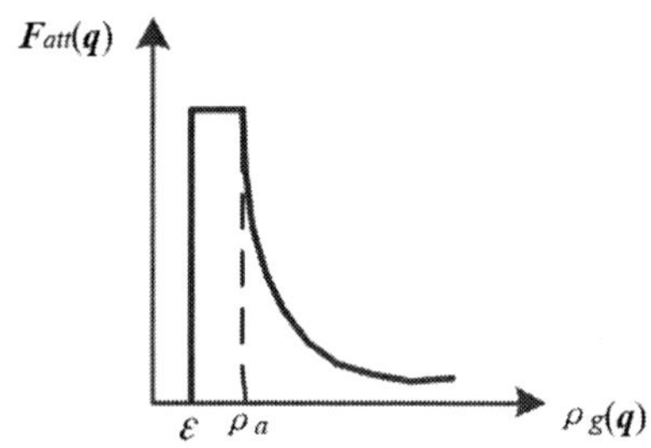

Figure 2. Law of Attraction

3.3. Research on State Warning Signals

The use of haptic vibration information to provide warning signals has been widely studied and applied in many fields. Compared to visual cues and acoustic cues, the haptic signal will not be ignored due to inattention or environmental impact. Among many researches on tactile vibration warning, lane departure warning system (LDWS) is an important application field and a research hotspot in recent years. Generally, the haptic signals of LDWS can be divided into two types: one is the additional torque, and the other is the vibration signal[8]. Both types of haptic signals are applied to the steering wheel of the car, because the driver will always touch the steering wheel, so that he can feel the haptic prompt signal at any time. Compared with the torque signal, the vibration signal can more attract the operator's attention, and the warning effect is stronger. Ziegler [9] and others first studied the application of haptic vibration signals in LDWS, and then more scholars carried out a deep research in this field, such as Rothe [10], Tijerina [11], and so on. The frequency of the vibration prompt signal of LDWS is generally 5 ~ 20Hz, and the amplitude is generally 0.4 ~ 1.5Nm. In addition, Fu et al developed a tactile shoe that uses vibration signals to prompt stock market trading information to help investors understand the stock market in a timely manner [12]. Vibration signals are also widely used in touch screens, mobile phones, and game equipment.

This paper introduces the vibration signal into the force feedback handle. By generating a small high-frequency vibration at the end of the handle, it provides a certain stimulus to the operator and achieves status information such as success, failure, failure, danger, and mode switching of the robotic arm's operating tasks prompt. The factors that determine the properties of the vibration signal are the vibration law, vibration amplitude, and vibration frequency. This article attempts a square wave, triangle, and sine vibration laws. Based on the subjective experience of the participants, the sine vibration signal is considered (Figure 3) to be more conducive to the perception of the operator. By changing the amplitude and frequency of the sinusoidal signal, the operator's subjective perception and evaluation of the vibration prompt signal will also change. Finally, it was determined through experiments that a sinusoidal vibration signal with an amplitude of 40 Nmm and a frequency of 50 Hz was used as a warning signal of the slave-end state. The operator felt that the prompting effect was obvious and the hands felt comfortable.

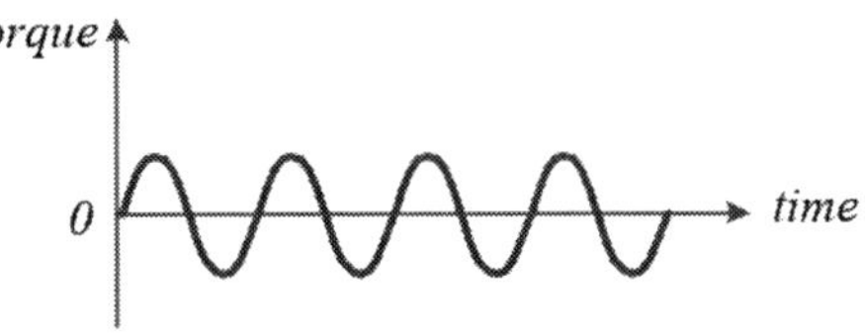

Figure 3. Sine vibration alert signal

4. Experimental verification

The designed simulation task is that the operator completes the capture and auxiliary docking of the spacecraft by remotely operating the space robotic arm with the force feedback handle and a three-degree-of-freedom translation handle developed in this paper, as shown in Figure 4. During the operation, the force feedback needed to be realized includes the auxiliary guidance force of the precision operation of the mechanical arm, the impact force caused by grasping and collision, and the tactile signal to warn the operation status of the task. In order to facilitate control, the entire task process is divided into three stages, and the feedback force is calculated according to the tasks in each stage. The first stage is to capture the spacecraft with a robotic arm, the second stage is to adjust the position and attitude of the spacecraft to approach the docking hatch, and the third stage is to dock the spacecraft.

In the first stage, the operator controls the manipulator to move towards the spacecraft. In this process, the position and attitude of the manipulator must be adjusted continuously. The remote operation adopts the speed mapping mode. The movement speed and direction of the manipulator are controlled by the offset and direction of the control handle offset zero position. When the manipulator is close to the spacecraft, the gravitational potential field is used to guide the operator to accurately adjust the manipulator, so as to quickly capture the spacecraft, as shown in Figure 5. When the spacecraft is captured, the force feedback handle outputs a vibration prompt signal to prompt the operator to capture successfully and enter the next stage of operation.

In the second stage, the robotic arm moves with the spacecraft toward the node module. In this process, the position and attitude of the spacecraft are adjusted continuously, as shown in Figure 6. It is difficult to achieve accurate attitude adjustment through visual feedback, so the artificial potential field model is used to output auxiliary force in the process of attitude adjustment, which greatly reduces the operator's operation difficulty. When the attitude of the spacecraft is adjusted and the direction of the docking hatch axis is the same as that of the docking hatch axis of the node cabin, the handle will output a vibration prompt signal to prompt the operator to enter stage 3 and start the docking of the spacecraft. In the docking stage, the gravitational potential field model is still used to generate the attraction centered on the target position, assisting the operator to complete the precise docking task, as shown in Figure 7. At the same time, the collision force in the docking process is also output to the human hand through the force feedback handle, so that the operator can directly feel the force state of the spacecraft.

Figure 4. Experimental environment for simulated docking tasks

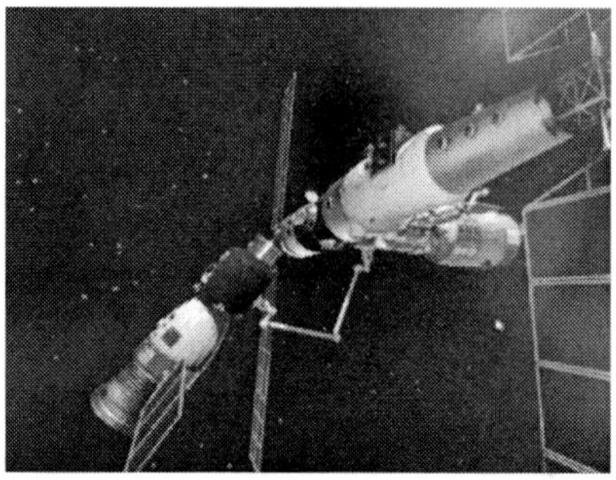

Figure 5. In the first stage, spacecraft capture

Figure 6. In the second stage, adjustment of spacecraft attitude

Figure 7 .The third stage, spacecraft docking

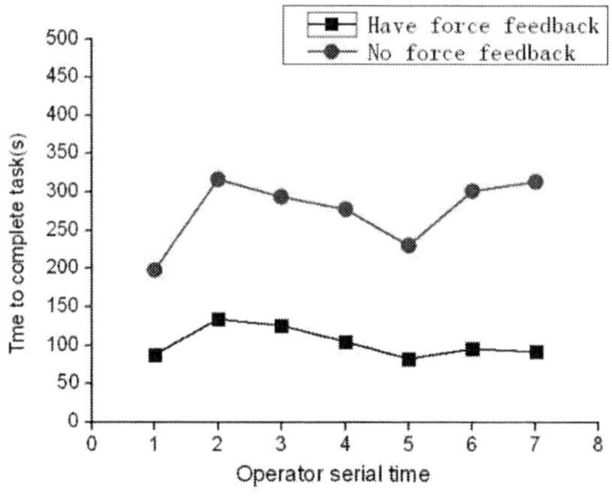

Figure 8 .Time for operator to complete simulation task

Seven operators were invited to complete the simulation tasks in the above three stages. Each operator's experiment was divided into two groups, one was the operation with force perception rendering, and the other was the operation that blocked force perception rendering. The time they took to complete the task was shown in Figure 8. From the experimental results, it can be seen that the time taken for the seven operators to complete the operation tasks after adding auxiliary forces and status prompts is greatly reduced. When rendering incapable, relying only on visual feedback to operate the robotic arm to complete the simulation task, the average time taken by the seven operators is 2.6 times that when operating with force feedback. Through interviews after the experiment, operators generally report that when rendering with force sense, the operation task becomes simpler and the teleoperation process is more realistic. This shows that the robot arm teleoperation force rendering algorithm studied in this paper can greatly improve the operation efficiency of teleoperation tasks and effectively enhance the presence of teleoperation.

5. Conclusion

This paper studies the auxiliary forces in the teleoperation to improve operational efficiency and operational friendliness. The traditional artificial potential field method is introduced into the teleoperation of the space manipulator. For precision operational tasks, a method was proposed that using the gravitational potential field model calculate the attractiveness of the target point. The method can achieve an increase in the guiding force when the mechanical arm is reduced from the target position.It helps the operator to complete tasks such as precision assembly or alignment. At the same time, the amplitude and frequency of the vibration signal suitable for the human hand are proposed when the vibration feedback is used in teleoperation. It can be seen from the experimental results that the haptic rendering algorithm proposed in this paper can greatly improve the operation efficiency of the teleoperation task and effectively enhance the telepresence of teleoperation when the force feedback handle is used for the teleoperation of the space manipulator.

References

[1] Stieber M F, Trudel C P and Hunter D G 1997. *Robotics and Automation*, 3068-3073.

[2] Laryssa P, Lindsay E, Layi O, Marius O, Nara K, Aris L, Ed T 2002 *7th ESA Workshop on Advanced Space Technologies for Robotics and Automation* (Nottingham: ASTRA) p 1-8.

[3] Iwata T 2001 *Proceeding of the 6th International Symposium on Artificial Intelligence and Robotics & Automation in Space*(Canada : Canadian Space Agency, Quebec).

[4] Dengyun Yu, Jing Sun and Xing rui Ma 2008 *Spacecraft Engineering* **16** 1-8.

[5] Birglen L, Gosselin C, Pouliot N, Bruno Monsarrat B 2002 SHaDe, a new 3-dof haptic device *Robotics and Automation* **18** 166-175.

[6] Khatib O 1986 Real-time obstacle avoidance for manipulators and mobile robots *The international journal of robotics research* **5** 90-98.

[7] Rongqi Zhao 2008 Study on Path Planning of Robot Based on Artificial Potential Field *Shandong University*.

[8] Beruscha F, Augsburg K and Manstetten D 2001 Haptic warning signals at the steering wheel: A literature survey regarding lane departure warning systems *Haptics-e, the electronic journal of haptic research* **4** 6.

[9] Ziegler W 1998 Computer vision on the road: A lane departure and drowsy driver warning system *ITS technology collection on CD-ROM: SAE's essential resource for ITS vehicle applications* 1995.

[10] Sato K, Goto T, Kubota Y, Goto T 1998 A study on a lane departure warning system using a steering torque as a warning signal *Tokyo, Society of Automotive Engineers of Japan* p 479-484.

[11] Tijerina L, Jackson J L, Pomerleau D A, Romano R A, Petersen A D 1996 Driving simulator tests of lane departure collision avoidance systems *Proceedings of the 1996 Annual Meeting of ITS America* p 636-648

[12] Fu X, Li D 2005 Haptic shoes: representing information by vibration[C]// *proceedings of the 2005 Asia-Pacific symposium on Information visualisation* vol 45(Australian : Australian Computer Society) p 47-50.

IOP Conf. Series: Materials Science and Engineering **887** (2020) 012037 doi:10.1088/1757-899X/887/1/012037

Evaluation of Dynamic Stability Derivatives for Hypersonic Vehicle

Tong Sheng-Xi[1,2,a], Fan Yan-Ming[1,b], Gao Feng[2,c] and Zhou Guo-Qing[2,d]

[1]Shenyang Aircraft Design and Research Institute, Shenyang 110035, China
[2]Liaoning General Aviation Academy, Shenyang 110136, China

[a] tongshengxi@lgaa.com.cn, [b]michaelfan@yeah.net, [c]gaofeng@lgaa.com.cn, [d]22530447@qq.com

Abstract. The dynamic stability derivative is the key parameter of the aircraft control system and the boundary analysis of the dynamic stability criterion. The accurate evaluation of dynamic derivative is of great importance to the design and flight of the aircraft. A method to predict the derivative of pitch damping is presented. The unsteady aerodynamic force is obtained by solving the unsteady flow field, and the static and dynamic derivatives of the aircraft are obtained by using integral method and least square method. By comparing the aerodynamic derivatives at different equilibrium positions and different centers of gravity, it can be seen that the calculated results are in good agreement with the experiment. It indicates that the present method has satisfactory order and efficiency and be suitable for engineering problems.

1. Introduction

Hypersonic vehicle including space transportation systems, the hypersonic transport, the atmospheric aircraft, and re-entry vehicles have become more and more important. The performance of a vehicle depends not only on steady aerodynamic characteristics, but also its ability to follow a predetermined flight path as a function of stability and control characteristics. During the design and development of the aircraft, it is difficult to obtain the dynamic stability of the flight envelope. The flow separation induced by shock wave and vortex movement brings instability and strong nonlinear characteristics. This usually results in an uncertain aerodynamic effect on the aircraft, even have some devastating consequences.

Dynamic derivatives, also called dynamic stability derivatives, describe the aerodynamic characteristics of aircraft during maneuvers and disturbances. The prediction of aircraft dynamic differentiation will provide key aerodynamic parameters for the design of aircraft control systems. For aircraft open loop systems, dynamic derivative directly determines the convergence characteristics. For aircraft closed loop systems, dynamic derivative determines the gain of the control system. If the feedback gain adjustment is too large, although the flight stability is good, the maneuverability of the aircraft degrades the quality of control and the quality of the flight deteriorates. A robust and reliable approach to dynamic stability analysis can be used to evaluate different aircraft design configurations to minimize design costs and reduce risk.

Since the concept of aerodynamic derivatives was introduced by Bryan[1], it is assumed that the aerodynamic forces and moments are a function of the instantaneous values of disturbing speed, control angle and their rates. Etkin[2] consider the nonlinear effect, the aerodynamic force depends not

Content from this work may be used under the terms of the Creative Commons Attribution 3.0 licence. Any further distribution of this work must maintain attribution to the author(s) and the title of the work, journal citation and DOI.
Published under licence by IOP Publishing Ltd

only on the instantaneous angle of attack, but also on the historical process of angle of attack. The prediction of dynamic aerodynamic derivatives is inherently more difficult than the basic static aerodynamic derivatives.

Traditionally the aerodynamic derivatives have been measured in the wind tunnel tests. The wind tunnel test is greatly influenced by the Reynolds number. This is not only because they are time consuming, but also they have restrictions due to wind tunnel interferences and other factors. The computational fluid dynamics (CFD) represents the most advanced level of nonlinear aerodynamics prediction. It is gradually used to calculate dynamic derivatives. The Unsteady Reynolds average Navier-Stokes equation is solved to identify the dynamic derivative. Many researchers have done a great deal of work on the calculation of dynamic derivatives using CFD. Weinacht[3] applied a steady prediction method, which can reduce the computational time but with accuracy loss. Murman[4] presented a reduced-frequency approach for calculating dynamic derivatives. Ronch[5] evaluate the dynamic derivatives using CFD.

This paper begins with a brief review of dynamic stability derivatives, and then presents the results of rocket and missile. The computed results are compared against wind tunnel data. Finally, the main results of this work are summarized.

2. Numerical Approach

The flow control equation is three-dimensional compressible unsteady Navier-Stokes equation, in general curvilinear coordinate system the non-dimensional government equation is given below

$$\frac{\partial Q}{\partial t}+\frac{\partial F}{\partial \xi}+\frac{\partial G}{\partial \eta}+\frac{\partial H}{\partial \zeta}=\frac{1}{\mathrm{Re}}\left(\frac{\partial F_V}{\partial \xi}+\frac{\partial G_V}{\partial \eta}+\frac{\partial H_V}{\partial \zeta}\right) \tag{1}$$

$$\vec{w}=\begin{bmatrix}\rho\\ \rho u\\ \rho v\\ \rho w\\ \rho E\end{bmatrix},\quad \vec{F}=\begin{bmatrix}\rho(q-q_\omega)\\ \rho u(q-q_\omega)+pi_x\\ \rho v(q-q_\omega)+pi_y\\ \rho w(q-q_\omega)+pi_z\\ \rho h(q-q_\omega)+pq_\omega\end{bmatrix} \tag{2}$$

Where Q is the conserved variable. $\vec{F}$ and $\vec{F}_v$ are the inviscid and viscous fluxes respectively. The mesh movement can be treated as rigid body. The mesh moved once per real time step during the unsteady calculation. Where q_ω is the grid velocity.

The flow field solution adopts the finite volume method based on the structure grid. The convective term is discredited using the Roe scheme[6], to avoid unrealistic solutions like expansion shocks it is necessary to satisfy the entropy correction[7-8] for the Roe scheme. MUSCL interpolation is used provide third-order accuracy and min-mod limiter is employed to prevent numerical oscillations. The implicit Lower-Upper Symmetric Gauss Seidel (LUSGS) is applied to obtain the update solution[9]. The dual time stepping is introduced, and the system can be solved by inner iteration through pseudo time to the steady state. The accelerate convergence methods such as local time step can be used in the inner iteration. The dual time is given by

$$\frac{3V_{i,j,k}^{n+1}w_{i,j,k}^{n+1}-4V_{i,j,k}^{n}w_{i,j,k}^{n}+V_{i,j,k}^{n-1}w_{i,j,k}^{n-1}}{2\Delta t}+R_{i,j,k}(w_{i,j,k}^{n+1})=0 \tag{3}$$

$$V_{i,j,k}^{n+1}\frac{w_{i,j,k}^{n+1,m+1}-w_{i,j,k}^{n+1,m}}{\Delta\tau}+\frac{3V_{i,j,k}^{n+1}w_{i,j,k}^{n+1,m}-4V_{i,j,k}^{n}w_{i,j,k}^{n}+V_{i,j,k}^{n-1}w_{i,j,k}^{n-1}}{2\Delta t}+R_{i,j,k}(w_{i,j,k}^{n+1,m})=0 \tag{4}$$

$$V_{i,j,k}^{n+1}\left(\frac{1}{\Delta\tau}+\frac{3}{2\Delta t}+A+B+C\right)\Delta w_{i,j,k}=-R_{i,j,k}(w_{i,j,k}^{n+1,m})-\frac{3V_{i,j,k}^{n+1}w_{i,j,k}^{n+1,m}-4V_{i,j,k}^{n}w_{i,j,k}^{n}+V_{i,j,k}^{n-1}w_{i,j,k}^{n-1}}{2\Delta t} \tag{5}$$

Where Δt is the physical time step, $\Delta\tau$ is the pseudo time step, A, B, C is the Jacobi metrics of the inviscid flux. The $k-w$ two equation turbulence model[10] is adopted to simulate viscous field. A loosely coupled algorithm is used to integrate the Navier-Stokes and the $k-w$ equations separately. For far-field boundary, the supersonic inflow boundary and supersonic outflow boundary is used. For the wall boundary, the velocity was accounted for nonslip conditions. For the singular axis boundary the value is first-order extrapolation and then averaged.

3 Dynamic Derivative Extraction Methods

Forced oscillation motion is the most commonly used method of estimating dynamic derivatives. This section briefly introduces these methods.

3.1 Integral method

The aircraft performs a single-degree-of-freedom pitch movement. If its center of gravity is constant, the independent state variables of motion are determined only by the angle of attack and speed of the pitch angle. For pitching moment coefficients, according to Etkin assumption, it can be written as

$$C_m = C_m(\alpha(t), \dot{\alpha}(t), \ddot{\alpha}(t), \dots q(t), \dot{q}(t), \ddot{q}(t), \dots) \quad (6)$$

Such expression can reflect the non-stationary and non-linear aerodynamic characteristics, such as wake vortex and separation flow. Given forced movement, $\alpha(t) = \alpha_0 + \alpha_m \sin(kt) = \alpha_0 + \theta$,where α is the instantaneous angle of attack, α_0 is the angle of attack of equilibrium position, α_m is the amplitude, k is reduced frequency, it is defined as $k = \omega d / 2u_\infty$. Expand the C_m with Taylor coefficient

$$C_m = C_{m0} + \left(\frac{\partial C_m}{\partial \theta}\right)_0 \cdot \theta + \left(\frac{\partial C_m}{\partial \dot{\theta}}\right)_0 \cdot \dot{\theta} + \left(\frac{\partial C_m}{\partial \ddot{\theta}}\right)_0 \cdot \ddot{\theta} \quad (7)$$

$$\dot{\theta} = \alpha_m k \cos(kt), \ddot{\theta} = -\alpha_m k^2 \sin(kt) \quad (8)$$

In the case that k is not very large and the influence of higher derivative can be ignored. When the unsteady vibration process reaches the periodic solution, the static derivative and the dynamic derivative can be obtained by numerical integration. The moment coefficient can be written as

$$C_m = C_{m0} + \left[\left(\frac{\partial C_m}{\partial \theta}\right)_0 - k^2 \left(\frac{\partial C_m}{\partial \ddot{\theta}}\right)_0\right] \cdot \theta + \frac{1}{k}\left(\frac{\partial C_m}{\partial \dot{\theta}}\right)_0 \cdot \dot{\theta} \quad (9)$$

$$C_1 = \left[\left(\frac{\partial C_m}{\partial \theta}\right)_0 - k^2 \left(\frac{\partial C_m}{\partial \ddot{\theta}}\right)_0\right], C_2 = \left(\frac{\partial C_m}{\partial \dot{\theta}}\right)_0 \quad (10)$$

Integrate within one period

$$P_1 = \int_t^{t+T} C_m(t) \sin kt dt, P_2 = \int_t^{t+T} C_m(t) \cos kt dt \quad (11)$$

Then

$$C_1 = \frac{P_1 k}{\alpha_m \pi}, C_2 = \frac{P_2}{\alpha_m \pi} \quad (12)$$

3.2 Singular value decomposition method

In the case of the unsteady dynamic pitching moment model, the linear term is retained after Taylor expansion

$$C_m = C_{m0} + \left(\frac{\partial C_m}{\partial \theta}\right)_0 \cdot \theta + \left(\frac{\partial C_m}{\partial \dot{\theta}}\right)_0 \cdot \dot{\theta} + \left(\frac{\partial C_m}{\partial \ddot{\theta}}\right)_0 \cdot \ddot{\theta} \tag{13}$$

The coefficient of pitching moment recorded in a set of experiments or calculations, $x_1 = x_1(t) = \theta$, $x_2 = x_2(t) = \dot{\theta}$, $x_3 = x_3(t) = \ddot{\theta}$ are three basic functions. if $\theta = \alpha_m \sin(kt)$,then $x_1 = \alpha_m \sin(kt)$, $x_2 = \alpha_m k \cos(kt)$, $x_3 = -\alpha_m k^2 \sin(kt)$

Three stability parameters $\left(\frac{\partial C_m}{\partial \theta}\right)_0, \left(\frac{\partial C_m}{\partial \dot{\theta}}\right)_0, \left(\frac{\partial C_m}{\partial \ddot{\theta}}\right)_0$ is recorded as C_1, C_2, C_3, then

$$y = C_0 + C_1 x_1 + C_2 x_2 + C_3 x_3 \tag{14}$$

By experiment or numerical calculation, the value y can be got. The singular value decomposition method can be used to solve the problem. C_1, C_2, C_3 can be identified at one time, that is the stability parameter.

$$y = Ax \tag{15}$$

$$y = \begin{bmatrix} y(t_1) \\ y(t_2) \\ \vdots \\ y(t_n) \end{bmatrix}, A = \begin{bmatrix} 1 & x_1(t_1) & x_2(t_1) & x_3(t_1) \\ 1 & x_1(t_2) & x_2(t_2) & x_3(t_2) \\ \vdots & \vdots & \vdots & \vdots \\ 1 & x_1(t_n) & x_2(t_n) & x_3(t_n) \end{bmatrix}, x = \begin{bmatrix} C_0 \\ C_1 \\ C_2 \\ C_3 \end{bmatrix} \tag{16}$$

Singular value decomposition method can be used to solve the equation

$$A = U \Sigma V^T \ , x = \sum_{i=1}^{i=n} \frac{u_i^T b}{\sigma_i} v_i \tag{17}$$

4 Numerical results

This technique is applied to predict the pitch damping derivative of the projectile model and the hypersonic ballistic model. The effects of gravity position and angle of attack on the equilibrium position are studied.

4.1 Army-Navy spinner rocket

The Army-Navy spinner rocket (ANSR) experiment is performed by the United States Army Ballistic research laboratories. Figure 1 shows the shape of the 5-caliber and 9-caliber ANSR calibration models, where the length of one caliber is 20mm.

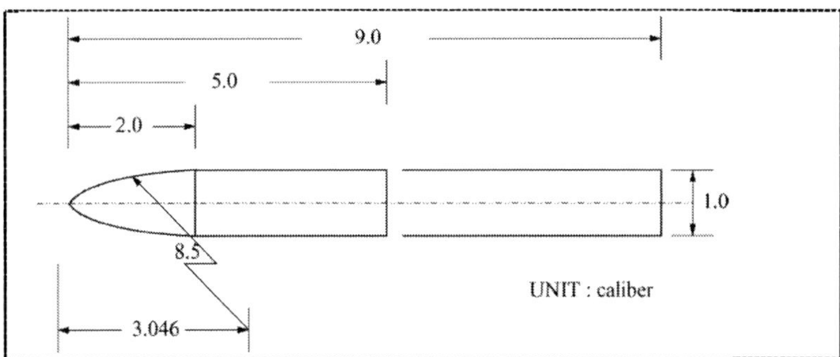

Figure 1. Army-Navy spinner rocket dimensions.

The influence of the center of gravity on the damping derivative is investigated by choosing three different centers of gravity. Table 1 shows the exact location. One is in the front, one in the middle and the other in the back.

Table 1. Position of ANSR center of gravity

Length	Position of center of gravity		
5	2.5	3.0	3.5
9	4.0	5.038	5.885

Because the rocket configuration is axis-symmetry, the three-dimension calculation mesh is generated by rotating the symmetric plane mesh. Figure 2 shows the symmetric plane mesh. the topology is H type. The mesh is fined near the wall and the singular axis boundary. Since the simulation condition is supersonic, the inlet boundary can locate near the lead position of the rocket. The steady flow field is simulated firstly. Figure 3 shows the pressure contour, and the shock wave can be captured clearly.

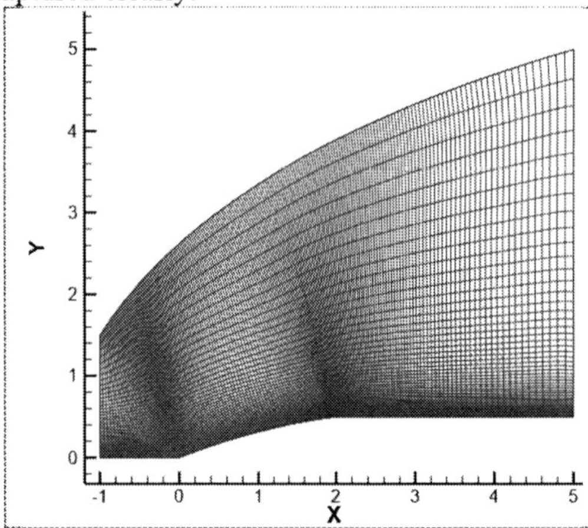

Figure 2. Mesh topology of Army-Navy spinner rocket

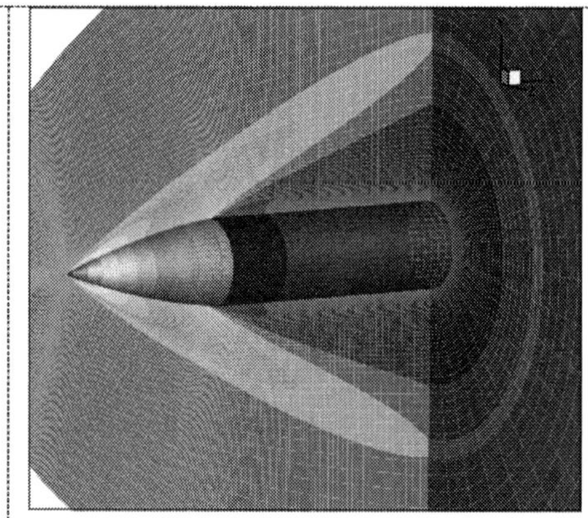

Figure 3. Pressure contour of ANSR flow filed

The grid moves as a rigid body, dual time method is used to solve the unsteady flow field. The forced vibration amplitude a_m is 1° and the non-dimensional reduced frequency is 0.1. The computations are performed using 200 time steps per cycle of motion. The inner iterations number is 50. Figure 4 and Figure 5 present the hysteresis loops for the pitch-damping moments of the unsteady simulations for the flow at zero mean angle of attack of three different center of gravity.

Figure 4. Hysteresis loop of pitch-damping moment.

Figure 5. Hysteresis loop of pitch-damping moment.

The pitch-damping moment coefficients are calculated by integral method. The coefficients are compared with the experimental data. The range data at the forward gravity center in both Figure 6 and Figure 7 deviate from the trend that appears in the experiments. The results are acceptable.

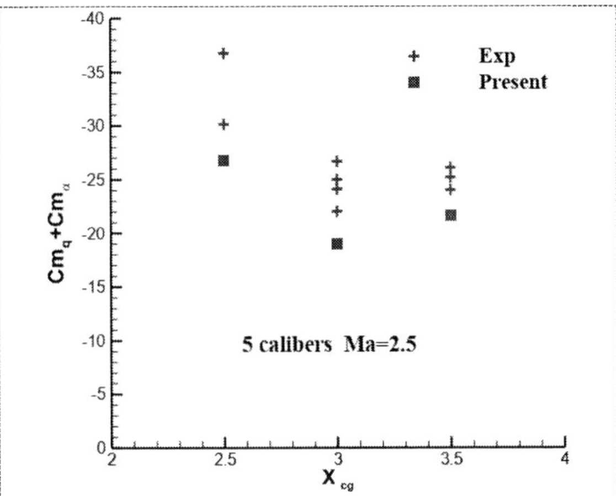

Figure 6. Pitch-damping moment coefficient of 5 calibers ANSR with different gravity center.

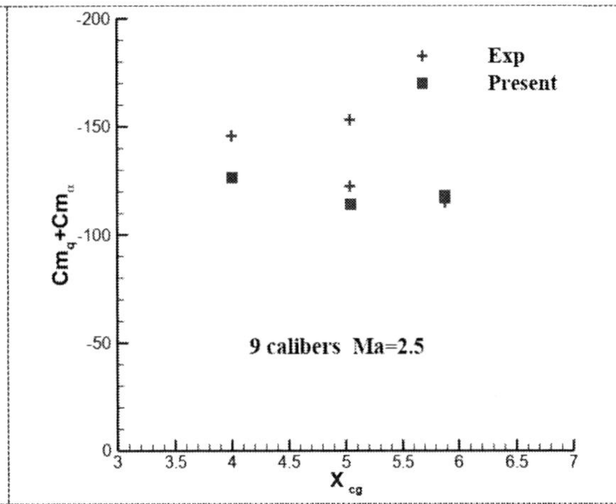

Figure 7. Pitch-damping moment coefficient of 9 calibers ANSR with different gravity center.

4.2 Hypersonic Ballistic Contour Model

Hypersonic ballistic contour model (HBS) is often used to verify the accuracy of the dynamic derivatives prediction. Figure 8 shows the shape of the HBS model. The experiment[11] was carried out in the isentropic light piston tunnel at the University of Southampton. This is an intermittent wind tunnel facility providing an open jet test section of nozzle exit plane diameter 0.21 m with a flow Mach number of M=6.85 for durations of typically 0.5s.

Figure 9 show the mesh of HBS configuration. The mesh topology of the symmetric plane mesh is C type. The three-dimension mesh is generated by rotating the symmetric plane mesh. The mesh is fined near the wall surface. Because the simulation is supersonic, the inlet boundary can locate close to the head of the body.

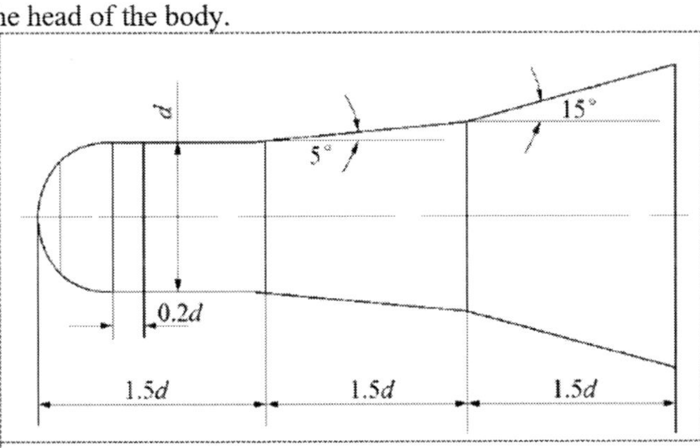

Figure 8. HBS dimensions.

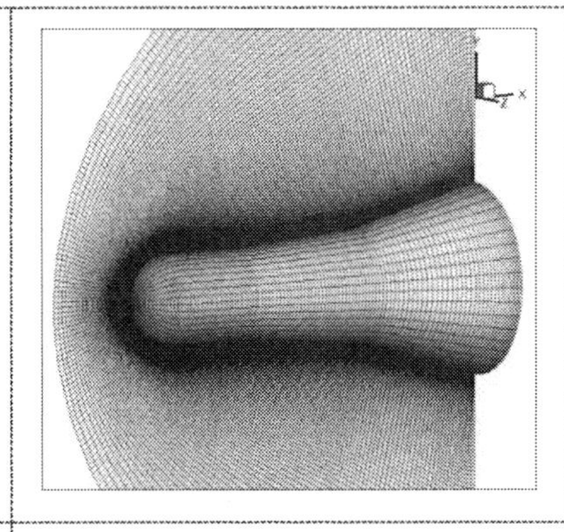

Figure 9. Mesh topology of HBS.

4.2.1 Steady Field Simulation. The steady flow field of different angle of attack is calculated firstly, the initial attack angle is $\alpha_0 = 0^\circ, 5^\circ, 10^\circ, 15^\circ$ respectively. The compute condition is Ma=6.85, and the Reynolds number base on head diameter is 0.72×10^6. Figure 10 shows the Mach number contour, and the shock wave can be captured clearly. With the increase of the angle of attack, the shock wave structure is more obvious.

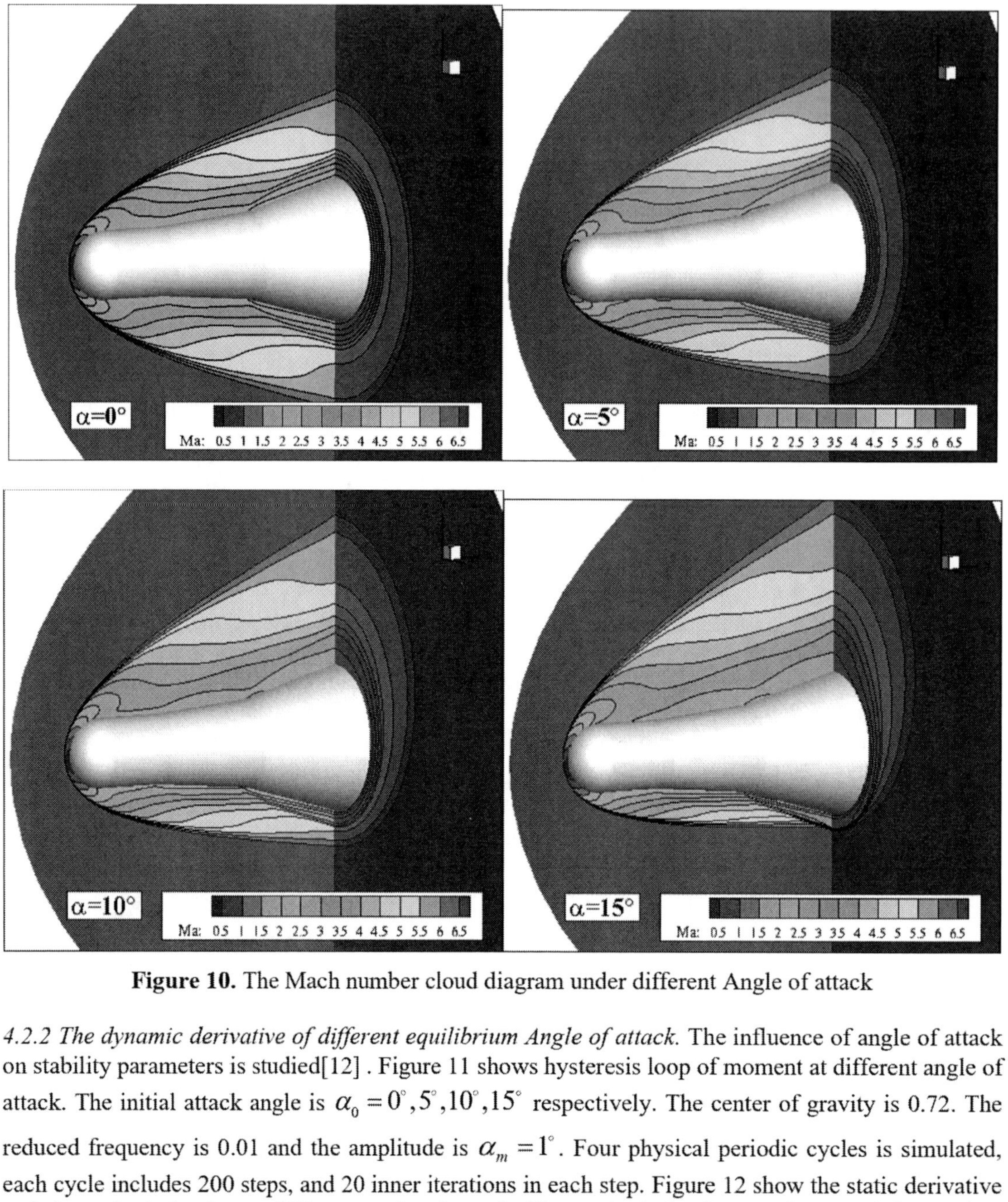

Figure 10. The Mach number cloud diagram under different Angle of attack

4.2.2 The dynamic derivative of different equilibrium Angle of attack. The influence of angle of attack on stability parameters is studied[12] . Figure 11 shows hysteresis loop of moment at different angle of attack. The initial attack angle is $\alpha_0 = 0^{\circ}, 5^{\circ}, 10^{\circ}, 15^{\circ}$ respectively. The center of gravity is 0.72. The reduced frequency is 0.01 and the amplitude is $\alpha_m = 1^{\circ}$. Four physical periodic cycles is simulated, each cycle includes 200 steps, and 20 inner iterations in each step. Figure 12 show the static derivative and dynamic derivative at different angle of attack. The static derivative matches much better than dynamic derivative. The result is acceptable.

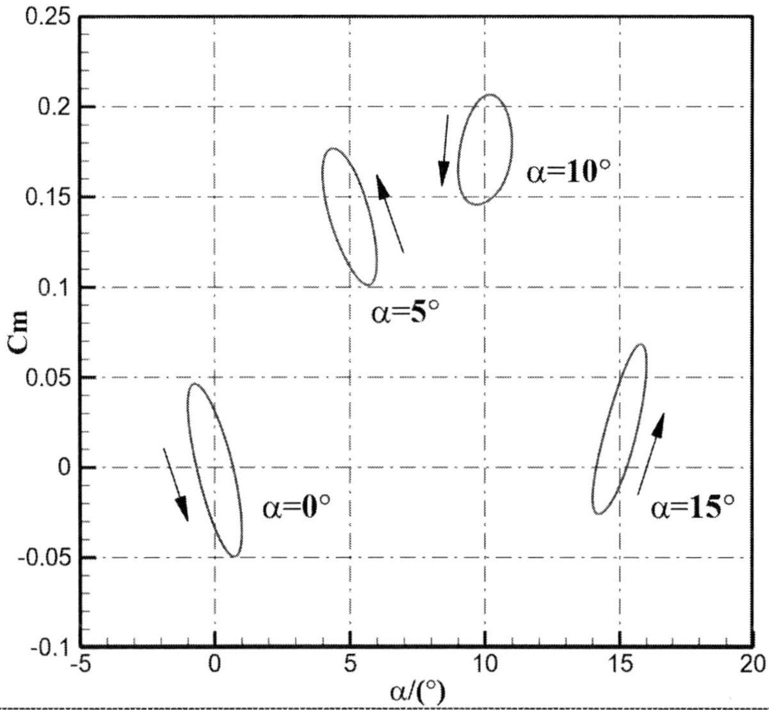

Figure 11. Hysteretic loop of torque curve under different Angle of attack.

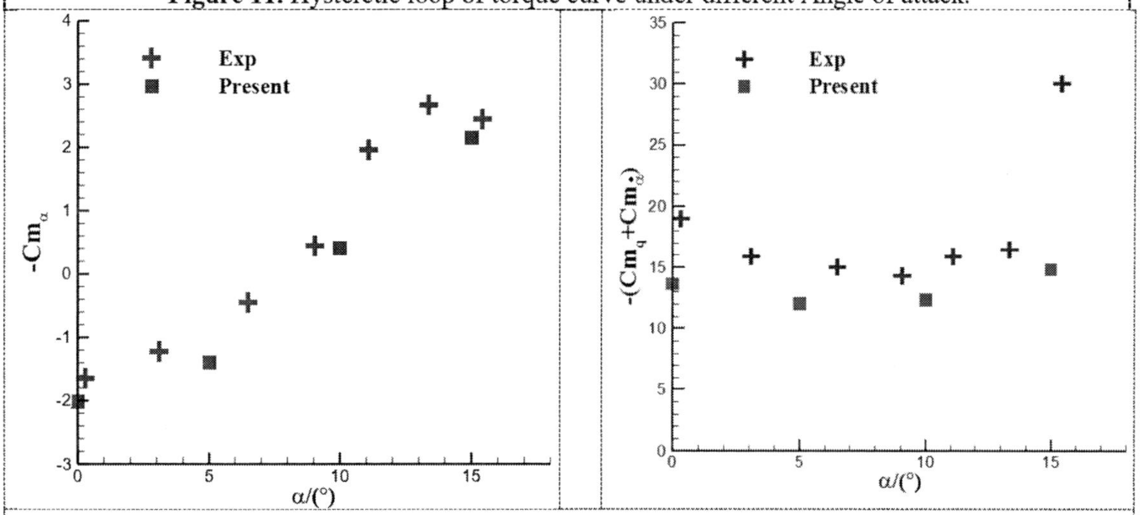

Figure 12. Static derivative and dynamic derivative at different angle of attack.

4.2.3 The Influence of different center of gravity. In order to test the influence of the center of gravity on the stability parameters, six different centers of gravity are calculated. The gravity center is 0.4,0.5,0.6,0.72,0.8,0.9. The initial attack angle is 0°. Figure 13 shows the hysteresis loop of forced pitching motion at different center of gravity. When the center of gravity moves backwards, the hysteresis loop rotates in a clockwise direction. Figure 14 shows the static and dynamic derivatives for different centers of gravity. It can be seen that the numerical method gives a consistent result over a wide range compared to the test results.

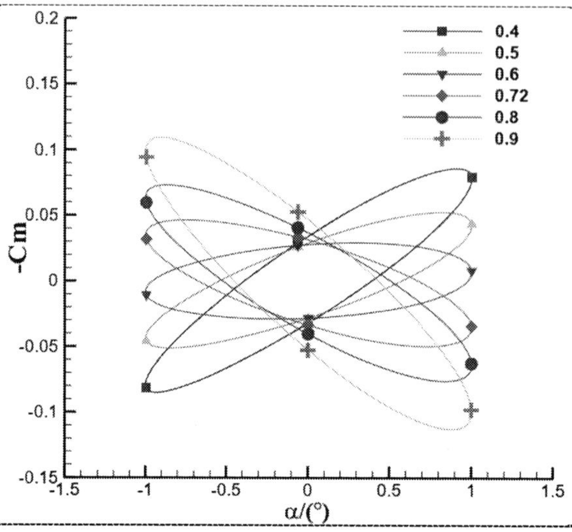

Figure 13. Hysteresis loop of pitch-damping moment.

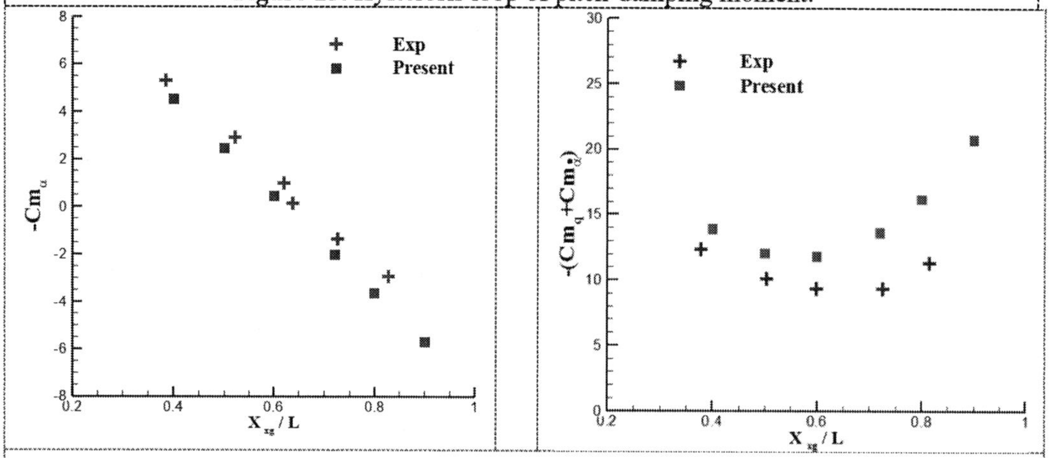

Figure 14. The static derivative and dynamic derivative of different gravity center.

5 Conclusions

A method to predict the derivative of pitch damping is proposed. The unsteady aerodynamic force is obtained numerically and the static and dynamic derivatives of the aircraft are obtained using the integral method and the least square method. First, the steady flow field is simulated, and the shock wave position can be clearly observed. By comparing the aerodynamic derivatives at different equilibrium positions and different centers of gravity, it can be seen that the calculated results are in good agreement with the experiment.

Conflicts of Interest

The author declares that there is no conflict of interest regarding the publication of this paper.

References

[1] Bryan, G.H.,1911. Stability in Aviation, MacMillan, London, 1911, pp 19-37
[2] Etkin,B., 1979. Dynamics of Atmospheric Flight, Courier Dover Publications, New York, NY, USA
[3] Weinacht, P., Sturek, W.B., and Schiff, L.B., 1991. Navier-Stokes Predictions of Pitch-Damping for Axisymmetric Shell Using Steady Coning Motion, AIAA-paper 91-2855

[4] Murman, S.M., 2007. A Reduced-Frequency Approach for Calculating Dynamic Derivatives[J]. AIAA Journal, Vol.47, No.6
[5] Ronch, A.Da., Vallespin, D., Ghoreyshi, M. and Badcock, K.J., 2012. Evaluation of Dynamic Derivatives Using Computational Fluid Dynamics[J]. AIAA Journal, Vol.50, No.2
[6] Roe, P. L., 1981. Approximate Riemann Solvers, Parameter Vector, and Difference Schemes, Journal of Computational Physics, Vol.43, pp357-372
[7] Saxena, S.K. and Ravi, K., 1995. Some Aspects of High-speed Blunt Body Flow Computations with Roe Scheme, AIAA Journal, Vol.33, No.6, June
[8] Kermani, M.J. and Plett, E.G., 2001. Modified Entropy Correction Formula for the Roe Schems, AIAA Paper 2001-0083
[9] Yoon, S. and Jameson, A., 1988. Lower-upper symmetric Gauss-Sediel method for the Euler and Navier-Stokes equations[J]. AIAA Journal, Vol.26, No.9
[10] Wilcox, D.C. and Traci, R.M., 1976. A complete model of turbulence, AIAA Paper 76-351
[11] East, R.A. and Hutt, G.R., 1988. Comparison of Predictions and Experimental Data for Hypersonic Pitching Motion Stability, Journal of Spacecraft, Vol.25, No.3
[12] Bing-Gang Tong and W.H.Hui., 1986. Unsteady Embedded Newton-Busemann Flow Theory, Journal of Spacecraft, Vol.23, No.2, March-April

IOP Conf. Series: Materials Science and Engineering **887** (2020) 012038 doi:10.1088/1757-899X/887/1/012038

Analysis on Aerodynamic Performance of Single-skinned Parawing in the Hypersonic Rarefied flow

Q J Rao[1], S Z Fang[1] and S Zhang[1]

[1] School of Aerospace and Engineering, Beijing Institute of Technology, Beijing 100081, China

2120170090@bit.edu.cn

Abstract. In consideration of the serious aerodynamic heating and overload in the return process of the space vehicles, a single-skinned parawing suitable for the whole return process is proposed, which uses lift deceleration to realize flexible reentry. A three-dimensional steady numerical simulation of the single-skinned parawing with different leading edge sweep angles was conducted by using the direct simulation Monte Carlo method, and the effects of different Mach numbers and altitudes on the aerodynamic performance of the parawing were studied in the hypersonic rarefied atmosphere. The results show that with the increase of the leading edge sweep angle, the aerodynamic performance of the parawing decreases obviously, but the ability of stall resistant is improved. The lift-drag ratio decreases monotonously with the increase of Mach number and altitude. The effect of Mach number on lift-drag ratio is not obvious, but the effect of altitude is more obvious. The maximum lift-drag ratio of parawing is 0.45 at 90 km, but only about 0.1 at 110 km. The peak heat flux always appears near the head, which will decreases obviously by decreasing the flight speed or increasing the flight altitude.

1. Introduction

The technology for the return and reentry of spacecraft has always been an important direction of development in the field of aerospace science, which can solve the problem of high cost of human space activities. There are a variety of spacecraft return schemes, in addition to the traditional circular parachute [1], large ramjet parafoil technology and solid rocket recovery technology also shine internationally. However, serious aerodynamic overload and heating can still be observed during the return process because of all of the schemes are used for the terminal recovery of spacecraft. In recent years, some new concepts for the recovery of spacecraft have been put forward by a few of foreign scholars, such as the dual-use lifting ballute [2] for orbit capture and entry, and the solar kite [3] for atmospheric reentry, which can achieve mild aerodynamic overload and heating in the process of reentry and achieve a safe soft landing with extremely low wing loading. The concept of single-skinned parawing proposed in this paper is almost similar, which has shown good performance of gliding and wind-resistant, as well as the advantages of simple structure, light weight and stable deployment characteristics. In addition, it can be controlled to very high angles of attack and adapted to a wide range of velocity from low speed to subsonic speed. Even at supersonic speed, none of the unfavorable tendencies exhibited by conventional parachutes at the speed have been shown, such as squidding and breathing, which means that a faster speed [4] for the entry or return can be realized. Therefore, the single-skinned parawing has a great prospect in the recovery of rocket and spacecraft.

The concept of single-skinned parawing (flexible wing) was first proposed publicly by Rogallo in

Content from this work may be used under the terms of the Creative Commons Attribution 3.0 licence. Any further distribution of this work must maintain attribution to the author(s) and the title of the work, journal citation and DOI.
Published under licence by IOP Publishing Ltd

1954, and a multitude of experimental investigations have been conducted on a wide range of parawings. A preliminary investigation of the aerodynamic and control characteristics of a flexible glider similar to a parachute in construction have been made to evaluate its capabilities as a reentry glider by Rogallo [5] through a series of wind-tunnel and free-glide tests. An investigation was made to determine the behavior of paragwing models at moderate to high supersonic speeds by Taylor [6]. Pressure distributions on three rigid wings simulating paragwings with varied canopy curvature and leading edge sweep have been obtained for low subsonic [7], transonic [8], and supersonic speeds [9] up to a Mach number of 4.65. An investigation to study the problems of deploying parawings for the recovery of boosters and nonlifting spacecraft have been conducted by Libbey [10]. The studies of Carter [11] show that use of structure could improve the performance of a system, enhancing delivery accuracy and capability, while keeping the overall system complexity low. Computational Fluid Dynamic studies of a rigid parawing at Mach numbers from 0.8 to 4.65 were carried out by Cruz-Ayoroa [12], and the simulation results are in good agreement with the wind tunnel test results [9], which means that modern CFD methods can be used to simulate these parawing concepts and in turn help optimize them to achieve desirable aerodynamic performance.

In contrast, there are few studies on single-skinned parawing in China. It is worth noting that these studies have been limited to the low-level continuous flow region. The separation altitude of the rocket [13] is as high as dozens kilometers, and the separation speed can reach several Mach; the altitude for a reentry of space vehicle is usually lower than 120 km, of which the speed is also close to Mach 30. The flow field at these altitudes has shown an obvious rarefaction effect [14], no longer in line with the continuous medium hypothesis. If the parawing is to be applied to the whole return process of spacecraft, study on its aerodynamic performance in the hypersonic rarefied flow is very important.

The primary motivation for the study of single-skinned parawing, usually flies at a low altitude, is to examine its aerodynamic performance in a high altitude, rarefied region. A three-dimensional steady numerical simulation of the single-skinned parawing with different leading edge sweep angles is conducted by using the direct simulation Monte Carlo method, and the effects of different Mach numbers and altitudes on the aerodynamic performance of the parawing are studied in the hypersonic rarefied atmosphere.

2. Computational method and models

2.1. DSMC method

Over the past few decades, the DSMC method has become the most common computational technique for modeling complex transitional flows of engineering interest. The DSMC method simulates rarefied flow by keeping track of the motion and internal energies of simulated molecule in the flow field, where each simulated molecule represents a fixed number of real gas molecules. The molecular collisions are modeled using the VHS (variable hard sphere) molecular model which treats the molecule as hard sphere as far as the scattering angle distribution is concerned, but the collision cross section depends on the relative speed of colliding molecule. Energy exchange between kinetic and internal modes is controlled by the Larsen-Borgnakke statistical model.

In the present study, the SPARTA code based on the DSMC method is used to estimate the high altitude aerodynamic performance of parawings, which is developed with internal funding at Sandia National Laboratories. It uses a hierarchical Cartesian grid to track and group particles for 3d or 2d or axisymmetric models. Objects emedded in the gas are represented as triangulated surfaces and cut through grid cells.

2.2. Validation Process

The problem of a hypersonic rarefied flow over a flat plate was used to validate the SPARTA code employed in this study. According to Lengrand et al. [15], a flat plate of 100 mm of length and leading-edge thickness of 5 mm was tested. The results are compared with those of wind tunnel test and DAC code based on the DSMC method.

Freestream flow conditions used for the numerical simulations are those given by Lengrand [15] and summarized in table 1.

Table 1. Freestream flow conditions.

Parameter	Value	Unit
Velocity	1504	m/s
Temperature	13.32	K
Number density	3.716×10^{20}	m^{-3}
Wall temperature	290	K
Mean free path	1.6×10^{-3}	m
Gas	N_2	

Figure 1 and figure 2 show that the surface pressure and surface heat flux calculated by SPARTA code are compared with the results of wind tunnel test and DAC code. It can be seen that the results are basically consistent, and the results of the SPARTA code have shown better agreement with that of the wind tunnel test.

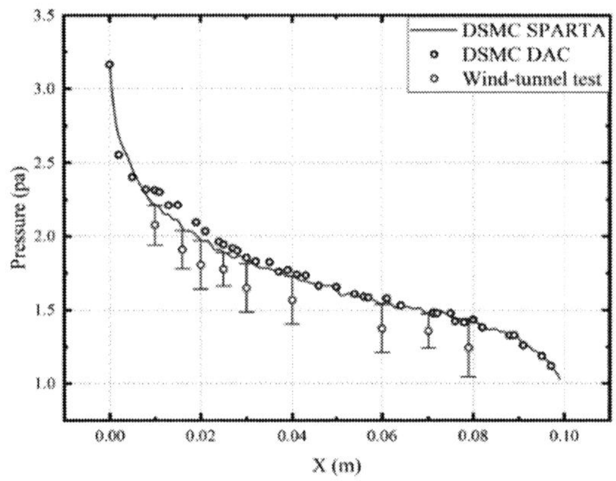

Figure 1. Results on surface pressure of SPARTA against with wind tunnel test and DAC.

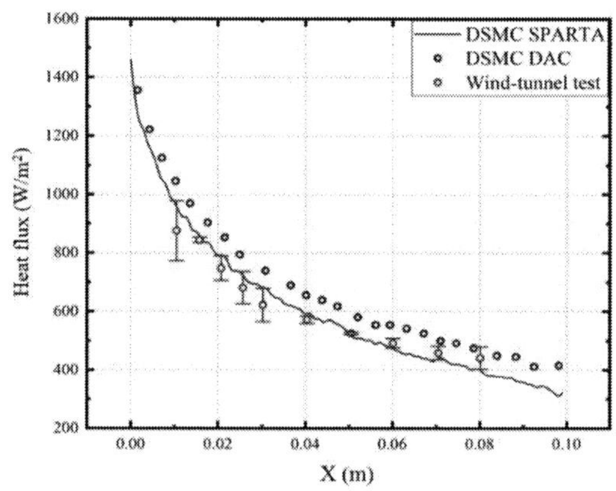

Figure 2. Results on surface heat flux of SPARTA against with wind tunnel test and DAC.

2.3. Model

Parawings are made of very loose or slack cloth whose configuration in flight is maintained by the combination of the aerodynamic forces and the reactions from the load suspension system. The canopy shape of the parawing in flight is very complicated, but the observations of parawing models in previous investigations have shown that the canopy shape could be closely represented by portions of a conical surface. With this assumption, three rigid parawing models of varying leading-edge sweep angles have been constructed.

The geometry of parawing models considered in this work is the same as that presented in Fournier and Bell [9]. Three rigid parawing models simulated a 45° basic flat planform parawing with leading-edge sweep angles of 48.6°, 52.5°, and 61.6°. These configurations resulted in one-half-circle, one-third-circle, and one-quarter-circle semispan trailing-edge curvature when viewed from downstream, which have been defined as Model 1-3, as shown in figure 4. Different curvature represents the different slackness of the canopy (by changing the leading edge sweep angle). As shown in figure 3, the length of the keel and the leading edge is 0.2 m with a radius of about 0.002 m. Considering the actual processing technology and thermal protection design, the passivation with a radius of 0.002 m has been applied to the head of the parawings. The thickness of the parawing is about 0.001 m.

IOP Conf. Series: Materials Science and Engineering **887** (2020) 012038 doi:10.1088/1757-899X/887/1/012038

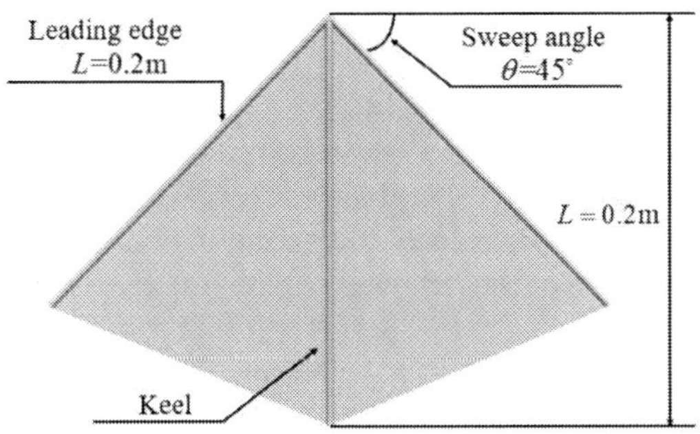

Figure 3. Flat planform parawing structure diagram.

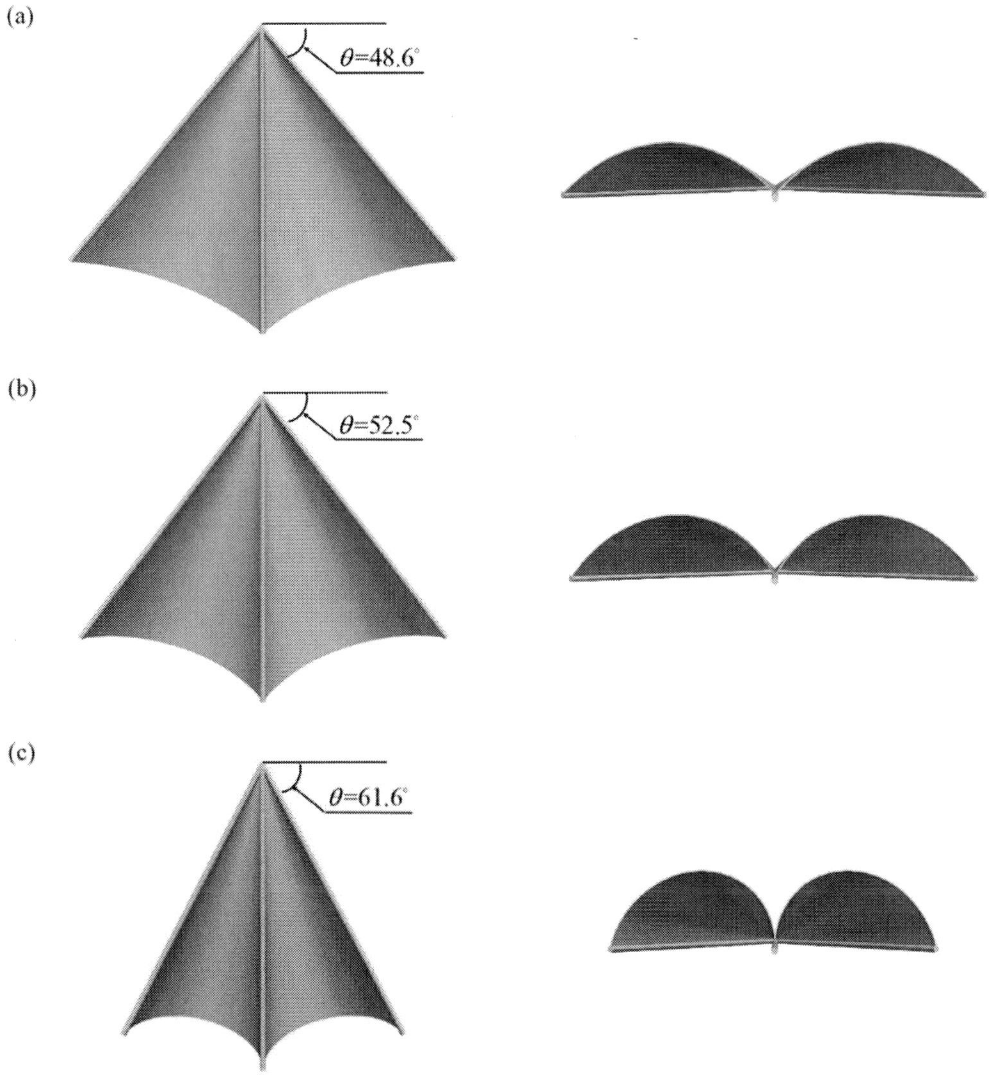

Figure 4. Geometry details of three parawing models: (a) Model 1; (b) Model 2; (c) Model 3.

3. Calculation conditions, results and analysis

Generally speaking, the flow field of the spacecraft has shown a very obvious rarefaction effect at the altitude of 90 km. According to the "reentry corridor" of the non-lift spacecraft [16], it can be determined that the flight velocity of the spacecraft at this altitude is about 28 *Ma*. At this time, the dynamic pressure acting on the parawing surface is only equivalent to that of 9 m/s at the sea level, which is completely bearable for the parawing. By increasing the altitude or decreasing the velocity, the dynamic pressure of freestream will be further reduced. Therefore, the aerodynamic performance of the parawings with different leading edge sweep angles at the height of 90 km and the velocity of 28 *Ma* was first studied; then, the effect of different Mach numbers (8-28 *Ma*) on the aerodynamic performance of the parawing with the best performance (Model 1) was studied at 90 km; Finally, the effect of different flight altitudes (90-110km) on the aerodynamic performance of the parawing with the best performance (Model 1) was explored at a Mach number of 28.

The flow conditions of simulation are as follows: the wall temperature is 300 K, the collision mode between molecules is VHS mode, and the collision model between molecules and object surfaces is complete diffuse reflection model. The other flow conditions are set according to table 2——the atmospheric parameters of different altitudes:

Table 2. Atmospheric parameters at different altitudes.

Altitude/km	Density/(kg·m^3)	nrho	Pressure/Pa	Mean free path /m	Temperature/K
90	3.4163×10^{-6}	7.1162×10^{19}	0.18359	0.0237	186.8
95	1.3934×10^{-6}	2.9202×10^{19}	0.07597	0.0579	188.4
100	5.6041×10^{-6}	1.1885×10^{19}	0.03201	0.1421	195.0
105	2.3248×10^{-7}	5.0211×10^{18}	0.01447	0.3365	208.8
110	9.7077×10^{-8}	2.1440×10^{18}	0.00710	0.7880	240.0

3.1. Effect of leading edge sweep angle on aerodynamic performance

The leading edge sweep angle of Model 1-3 is 48.6°, 52.5°and 61.6°, respectively. The larger the leading edge sweep angle is, the greater the slackness of the canopy is, indicating a larger peak lobe height. Figure 5 to figure 7 show the variation curves of lift coefficient, drag coefficient and lift-drag ratio of three parawing models at the altitude of 90 km and the velocity of 28 *Ma*. It can be seen from figure 5 that the parawing has a negative lift coefficient in the range of small angle of attack. The reason for this phenomenon is that the flow velocity on the upper surface is lower than that on the lower surface due to the large peak lobe height on the trailing edge of the parawing at a small angle of attack, resulting in a negative pressure difference. Considering the flexibility of the parawing surface, for which it is difficult to maintain the shape of canopy under this working condition. A local collapse is inevitable to form, indicating that the flight of the parawing at a small angle of attack is not taken into consideration for the design. Because the slackness of the canopy directly affects the flow separation in the leeward zone, the parawing with the least slackness is the first to reach the stall angle. Increasing the slackness of the canopy (by increasing the leading edge sweep angle) shifts the angle of attack for zero lift increasingly positive and reduces the lift curve slope. which is also observed in the wind tunnel test [17]. To some extent, there is a correlation between rarefied flow and continuous flow in the study of the overall aerodynamic performance of the parawing.

The variation curve of the drag coefficient of Model 1-3 with the angle of attack is basically the same. As shown in figure 6, when the angle of attack is small, the drag coefficient increases slowly, which becomes more fastly when exceeds 30°, and the drag coefficient shows a linear dependence on the angle of attack. The reason for this phenomenon is that the increase of the angle of attack leads to the increase of windward area, resulting in the increase of pressure drag as well as total drag.

It can be seen from figure 7 that the maximum lift-drag ratio of the parawing is about 0.45. When the angle of attack is less than 60°, the smaller the slackness of the canopy is, the greater the lift-drag ratio is, but there is an opposite trend when the angle of attack is greater than 60°. A larger lift-drag ratio can be obtained by increasing the slackness of the canopy, which is directly related to the fact that the parawing with greater slackness of the canopy can obtain a larger lift coefficient at a large angle of attack.

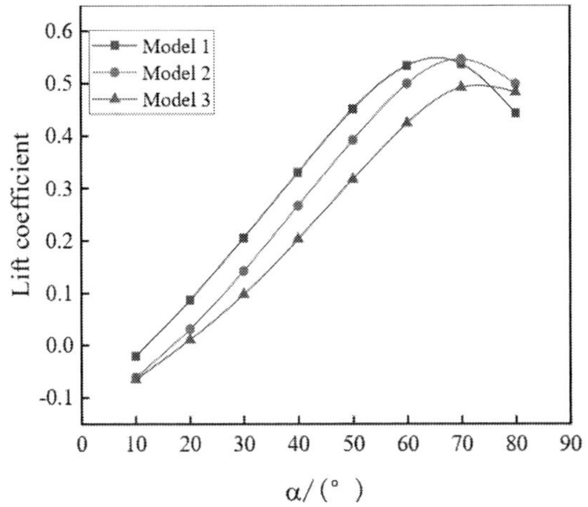

Figure 5. The lift coefficient graph of the parawings.

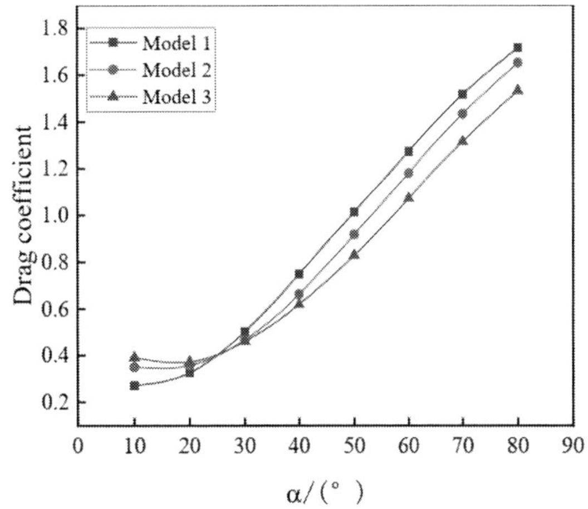

Figure 6. The drag coefficient graph of the parawings.

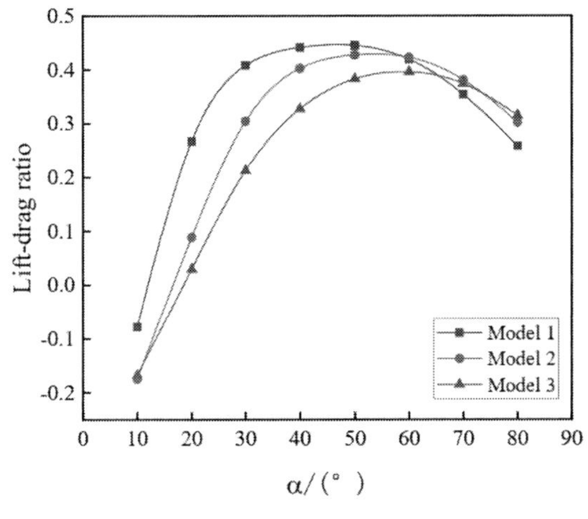

Figure 7. The lift-drag ratio graph of the parawings.

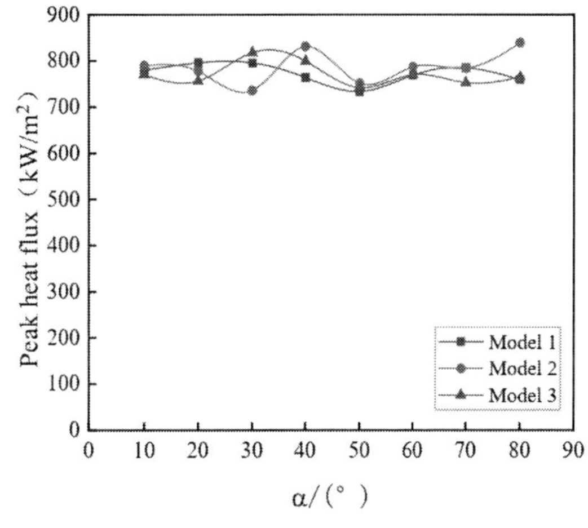

Figure 8. The peak heat flux on surface of the parawings.

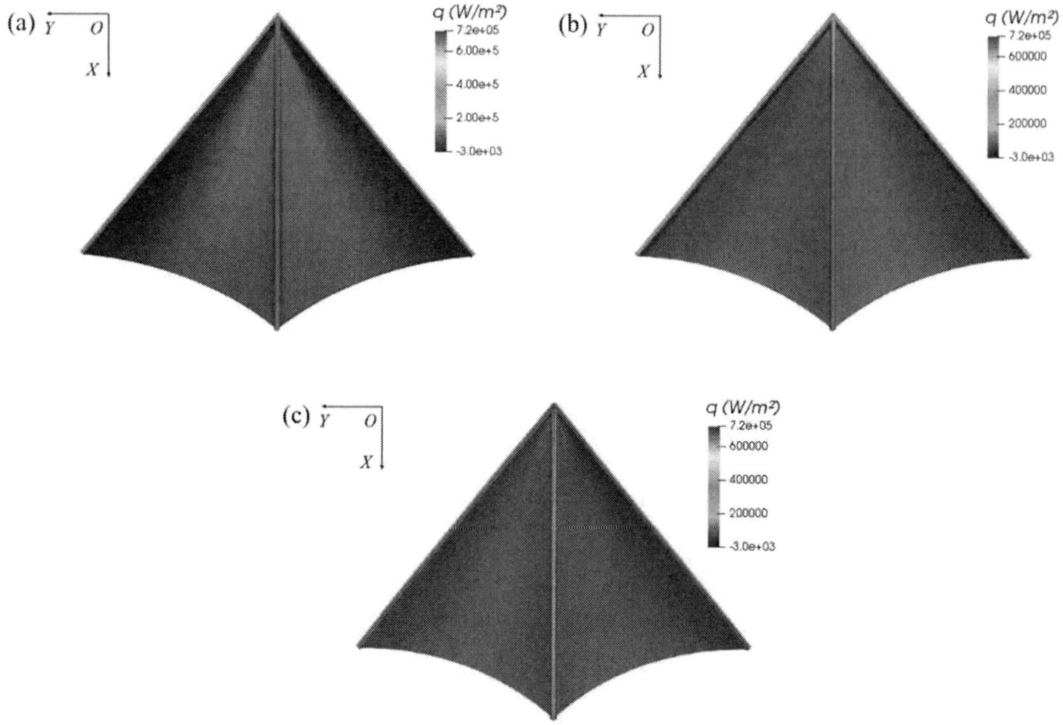

Figure 9. Heat flux distribution on lower surface of Model 1 at different attack angles: (a)α=30°; (b)α=50°; (c)α=70°.

The variation curve of the peak heat flux on surface of the parawings with the angle of attack is shown in figure 8, from which it can be seen that the peak heat flux fluctuates slightly with the increase of the angle of attack. The main reason for this phenomenon is that the peak heat flux (about 720 kW/m^2) mainly occurs at the head of parawings, which have the same passivation treatment. The stationary point is always near the head, which does not change with variation of the angle of attack. And there are also some errors in numerical calculation. The heat flux distribution on the lower surface of Model 1 at the angles of attack of 30°, 50°and 70° is shown in figure 9. It can be seen that the heat flux of the lower surface is much lower than that of the leading edge, but increases obviously with the increase of the angle of attack.

3.2. Effect of velocity on aerodynamic performance

It is necessary to analyze the aerodynamic performance of the parawing in a wide range of velocity, considering the flight velocity of the spacecraft in the whole return process. The previous analysis shows that the aerodynamic performance of Model 1 with the leading edge sweep angle 48.6° is the best. Therefore, the Model 1 is selected for further research to study the effect of different Mach numbers (8 -28 *Ma*) on the aerodynamic performance of parawing.

The variation curves of the lift coefficient, drag coefficient and lift-drag ratio of the parawing at different Mach numbers are shown as in figure 10 to figure 12, respectively, at an altitude of 90 km. It can be seen from figure 10 that at the same angle of attack, the lift coefficient decreases with the increase of Mach number, but this effect weakens obviously and finally tends to a constant value with the increase of Mach number. The stall angle increases slightly with the increase of Mach number. In contrast, the effect of the Mach number on the drag coefficient is very small. It can be seen from figure 11 that the drag coefficient decreases with the increase of the Mach number at a small angle of attack, and the effect begins to weaken. An opposite trend begins to appear with the increase of the angle of attack, which also directly leads to the decrease of the lift-drag ratio of the parawing with the increase of the Mach number. But this effect weakens obviously with the increase of the Mach number, as shown in figure 12.

The previous results show that the peak heat flux on surface of the parawing is little affected by the angle of attack, which mainly occurs near the head of the parawing. Figure 13 shows the variation curve of the peak heat flux on surface of the parawing at $\alpha = 50°$. The red line is the data fitting curve, and the function of fitting curve is y=1.69092 $\times 10^{-9} \cdot x^3$, which is in good agreement with the experimental results and also consistent with Fan Jing's theory that the aerodynamic heat flux at the stationary point is proportional to the cubic power of the freestream velocity. Therefore, the aerodynamic heat flux can be significantly reduced by reducing the flight velocity of the parawing.

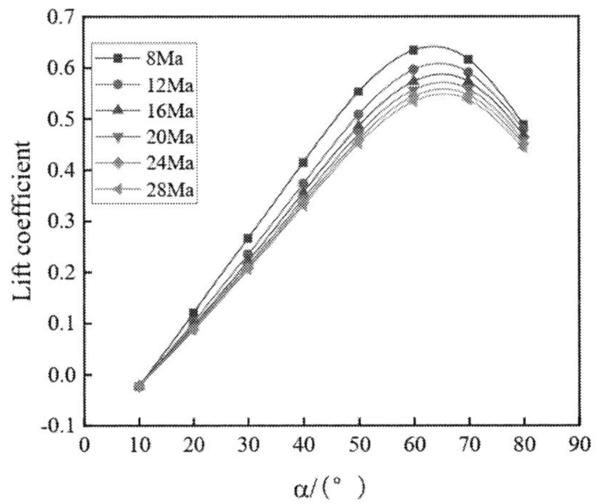

Figure 10. The lift coefficient graph of the parawing at different Mach numbers.

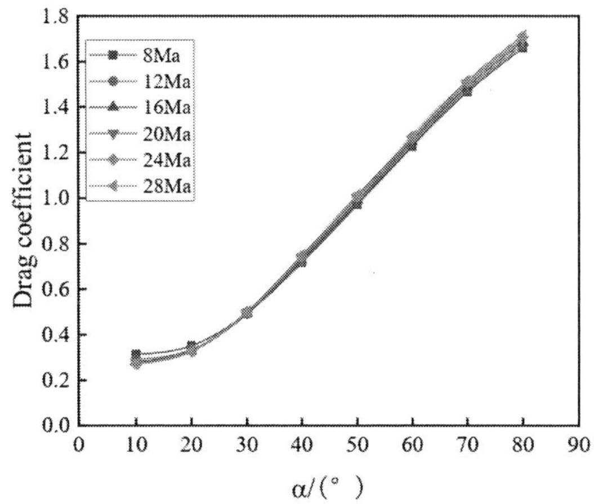

Figure 11. The drag coefficient graph of the parawing at different Mach numbers.

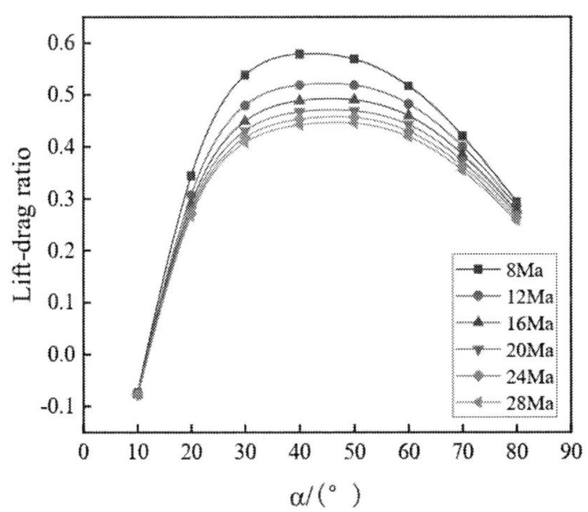

Figure 12. The lift-drag ratio graph of the parawing at different Mach numbers.

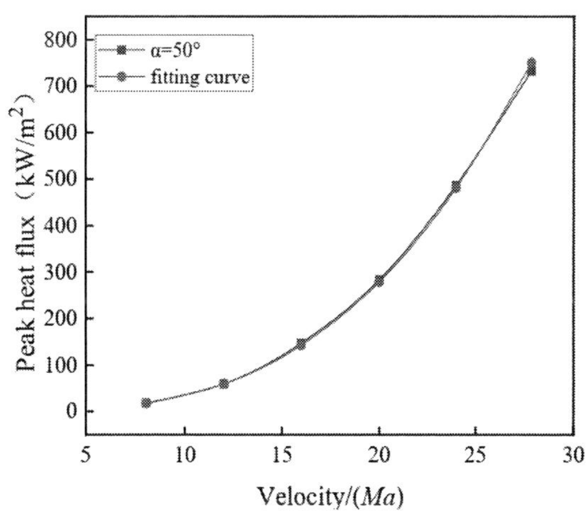

Figure 13. The peak heat flux on surface of the parawing at α=50°.

3.3. Effect of altitude on aerodynamic performance

According to the variation curve of the flight velocity in the whole return process of spacecraft [17], it can be found that the spacecraft basically maintains a constant velocity above 90 km, so the effect of different altitudes on the aerodynamic performance of parawing at a Mach number of 28 is also studied in this paper. The higher the flight altitude is, the lower the gas density is, and the more obvious the rarefaction effect is.

The variation curves of lift coefficient, drag coefficient and lift-drag ratio of the parawing at different altitudes (90-110 km) are shown in figure 14 to figure 16, respectively. At the same Mach number, the lift coefficient decreases obviously with the increase of altitude, and this trend is more and more obvious with the increase of the angle of attack. The maximum lift coefficient of the parawing at 90 km is almost twice that at 100 km. On the contrary, the drag coefficient shows an opposite trend as shown in figure 15, which increases with the increase of altitude, but this trend weakens obviously and finally will tend to a constant value. The similar results had been obtained from the simulation of a flat plate [14]. The lift-drag ratio of parawing decreases with the increase of altitude due to the continuous decrease of lift coefficient and the increase of drag coefficient, but it is worth noting that the angle of attack corresponding to the maximum lift-drag ratio is gradually decreasing, which means that to obtain a better aerodynamic performance, a smaller angle of attack is necessary for the parawing at a higher altitude.

Figure 17 shows the variation curve of peak heat flux on surface of the parawing at $\alpha = 50°$. It can be seen that with the increase of altitude, the aerodynamic heat flux acting on the parawing decreases significantly due to the thinner gas. Therefore, as long as the dynamic pressure is big enough, the parawing needs to open as soon as possible, the higher the altitude, the thinner the air and the smaller the aerodynamic heat flux. At the same time, the lift can also control the increase of the vertical velocity, reduce the dynamic pressure and aerodynamic heating effect, as well as avoid overload and ablation of structure.

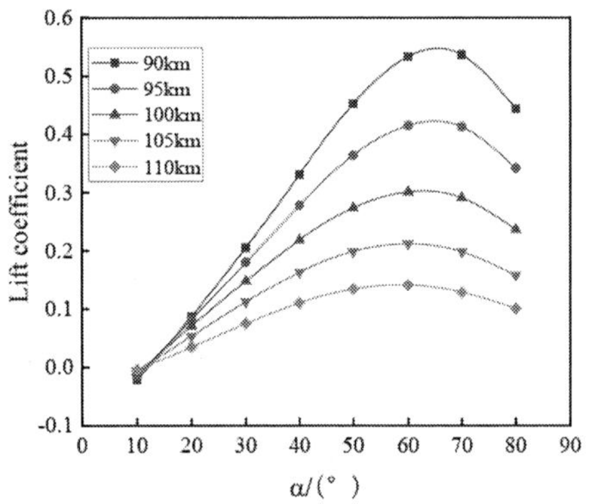

Figure 14. The lift coefficient graph of the parawing at different altitudes.

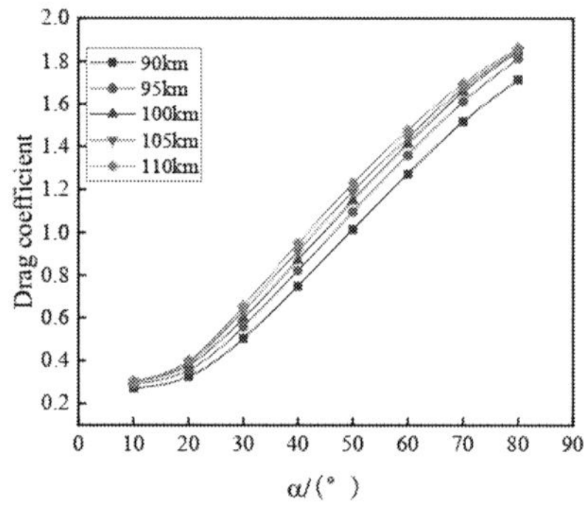

Figure 15. The drag coefficient graph of parawing at different altitudes.

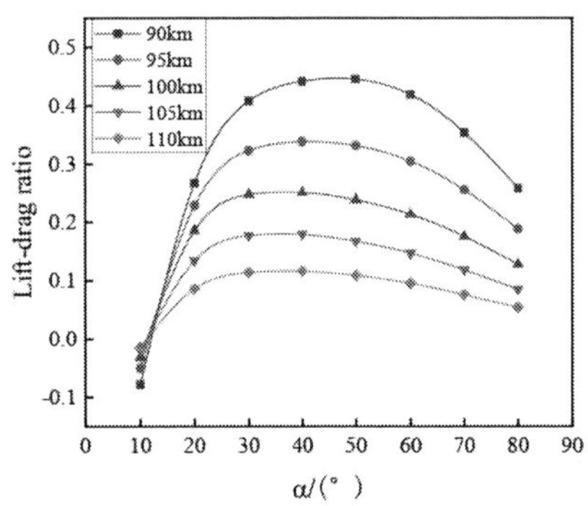

Figure 16. The lift-drag ratio graph of the parawing at different altitudes.

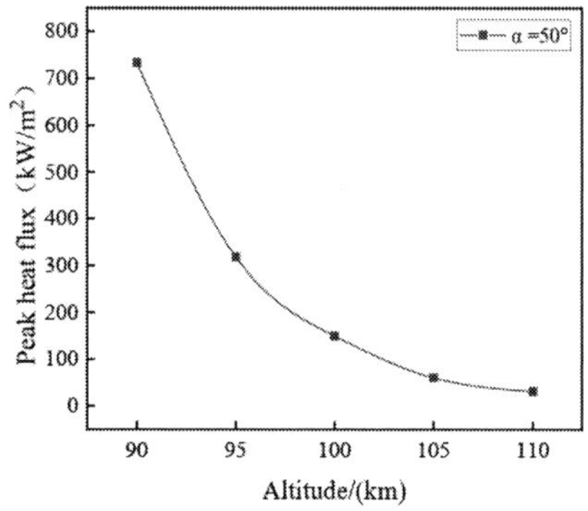

Figure 17. The peak heat flux on surface of the parawing at $\alpha = 50°$.

4. Conclusions

A three-dimensional steady numerical simulation of the single-skinned parawing with different leading edge sweep angles was firstly conducted by using the direct simulation Monte Carlo method, and the effects of different Mach numbers and altitudes on the aerodynamic performance of the parawing were studied in the hypersonic rarefied atmosphere. Increasing the slackness of the canopy (by increasing the leading edge sweep angle) shifts the angle of attack for zero lift increasingly positive and reduces the lift curve slope. The maximum lift-drag ratio of the parawing decreases by about 20%, while the stall angle increases by about 10° when the sweep angle changes from 48.6°to 61.6°. At the same altitude, the lift-drag ratio of parawing decreases with the increase of Mach number, and finally tends to a constant value; at the same Mach number, the lift-drag ratio of parawing decreases with the increase of altitude. Generally speaking, the effect of Mach number on lift-drag ratio is not obvious, while the effect of height is very obvious. The peak heat flux always appears near the head, which will decreases obviously by decreasing the flight velocity or increasing the flight altitude. Therefore, as long as the dynamic pressure is big enough, the parawing will be opened as soon as possible, the higher the altitude, the thinner the air and the smaller the aerodynamic heat flux. At the same time, the lift can also control the increase of the vertical velocity, reduce the dynamic pressure and aerodynamic heating effect, as well as avoid overload and ablation of structure.

References

[1] Li Y L and Qi F R 2011 Optimization and implementation of China SHENZHOU spaceship's system and return technology scheme *Spacecraft Recovery & Remote Sensing* **32** 1–13
[2] Kristin L G, Angus D M, Kerry T N, and Kim M A A dual-use lifting ballute for orbit capture and entry *AIAA/AAS Astrodynamics Specialist Conf.* 2010 Toronto Ontario Canada
[3] Andreas T, Christian M and Chris W Entry of a solar kite into a planetary atmosphere *56th Int. Astronautical Congress of the Int. Astronautical Federation* 2005 Fukuoka Japan
[4] Gao S Y and Huang W 2018 The development and prospect of Chinese spacecraft recovery and landing technology in the last sixty years *Spacecraft Recovery & Remote Sensing* **39** 70–8
[5] Rogallo F M *et al* 1960 Preliminary investigation of a paraglider *NASA TN* D–443
[6] Taylor R T 1961 Wind-tunnel investigation of paraglider models at supersonic speeds *NASA TN* D–985
[7] Fournier P G and Bell B A 1961 Low subsonic pressure distributions on three rigid wings simulating paragliders with varied canopy curvature and leading-edge sweep *NASA TN* D–983
[8] Fournier P G and Bell B A 1962 Transonic pressure distributions on three rigid wings simulating paragliders with varied canopy curvature and leading-edge sweep *NASA TN* D–1009
[9] Fournier P G 1963 Pressure distributions on three rigid wings simulating parawings with varied canopy curvature and leading-edge sweep at mach numbers from 2.29 to 4.65 *NASA TN* D–1618
[10] Libbey C E 1965 Deployment of parawings for use as recovery systems *J. Spacecraft Rockets* **2** 274–5
[11] James C, Peter W and John W Improving the performance of soft canopy deceleration systems with flexible composites structure *17th AIAA Aerodynamic Decelerator Systems Technology Conf. and Seminar* 2003 Monterey California USA
[12] Cruz-Ayoroa J G, Garcia J A and Melton J E 2014 Validation of CFD codes for parawing geometries in subsonic to supersonic flows *NASA TM* 218300
[13] Huang W 2017 The summary of launch vehicle parachute recovery and landing technologies *Spacecraft Recovery & Remote Sensing* **38** 1–12
[14] Huang F, Cheng X L and Shen Q 2009 Numerical investigation of hypersonic aerodynamics of flat plate flying in near space *J. Astronautics* **30** 900–7
[15] Lengrand J C, Allegre J, Chpoun A, and Raffin M Rarefied hypersonic flow over a sharp flat plate: numerical and experimental results *Proceeding of the 18th International Symposium on Rarefied Gas Dynamics* 1992 Vancouver British Columbia Canada
[16] R. A. Harris Aerothermodynamics for space vehicles *Proceedings of the 3rd European Symposium on Aerothermodynamics for Space Vehicles* 1998 ESTEC Noordwijk The Netherlands
[17] Naeseth R L and Gainer T G 1963 Low-speed investigation of the effects of wing sweep on the aerodynamic characteristics of parawings having equal-length leading edges and keel *NASA TN* D–1957

IOP Conf. Series: Materials Science and Engineering **887** (2020) 012039 doi:10.1088/1757-899X/887/1/012039

Research on Combination of Multiple Methods for Spacecraft Propellant Consumption Prediction

Wenjuan Yin[1], Zhiwei Cao[1] and Zhen Lin[1]

[1]Beijing institute of Control Engineering, 100094 Beijing, China

yinwenjuan1221@126.com

Abstract. In order to further improve the accuracy of propellant gauging during the geosynchronous satellite transfer orbit, combined with the existing propellant gauging method, a time series analysis method for satellite propellant consumption fusion algorithm is proposed. The error characteristics of the prediction data are analyzed according to the principle of different gauging methods. On the basis of this, the variance of each gauging method is estimated by time series analysis and the optimal weighted fusion estimation based on least squares method is given. The practical application shows that the fusion algorithm can make full use of the advantages of different gauging methods and provide a more reasonable estimation of the real-time propellant consumption.

1. Introduction

For geosynchronous satellites, the propellant carrying capacity is the main factor limiting the life of the satellite. The propellant prediction will also affect the satellite's de-orbiting operation. Therefore, the high-precision propellant remaining prediction and the propellant depletion time are important tasks of satellite on-orbit management.

For the orbital transfer phase, the commonly used propellant gauging methods include the BK (book-keeping) method, the PVT (pressure, volume, temperature) method, and the ultrasonic flow meter method. The measuring equipment of the BK method and the PVT method is simple, and there is no special requirement for the hardware of the satellite propulsion system. Compared with the PVT method, the BK method and the ultrasonic flowmeter method have higher accuracy, but rely on the integration method, so there is a cumulative error. In practical applications, the method with the highest accuracy is generally used as the main prediction method, and the other methods are used for recheck, and the accuracy of the propellant remaining gauging cannot be further improved. This article is aimed at the characteristics of different gauging methods. Considering the accumulative error effect of propellant gauging methods, a fusion method of propellant consumption prediction based on least squares method and time series analysis is proposed. Practical application shows that this method can effectively utilize multiple gauging methods, provide a new way to reduce the uncertainty of satellite propellant consumption estimation.

2. Propellant consumption gauging method

The main real-time gauging methods during orbital transfer phase are as follows [1~4]:

2.1. Bookkeeping method

The BK method first calculates the flow rate of the propellant based on the telemetry parameters and

Content from this work may be used under the terms of the Creative Commons Attribution 3.0 licence. Any further distribution of this work must maintain attribution to the author(s) and the title of the work, journal citation and DOI.
Published under licence by IOP Publishing Ltd

the ground calibration data, and then accumulates flow rate by working time to obtain the propellant consumption. This method is simple but practical and its uncertainty directly from the uncertainty of the flow rate during LAE and thruster firings. So the BK method has high accuracy especially when the apogee engine is working. The BK method is currently only available for liquid propellant gauging. It requires fully reliable test data of propulsion system performance, but there is accumulative error.

2.2. PVT method

The PVT method is based on the Ideal gas law. The method calculates the pressurizing gas (Helium) volume in the propellant tank according to the telemetry data including pressure and temperature data of pressurizing gas, and then obtains the liquid volume and liquid mass in the propellant tank according to the total volume of the propellant tank and propellant density. The structure of the PVT method is particularly simple, and there is no accumulative error. The PVT accuracy greatly depends on the pressure transducer accuracy and decreases when the amount of propellant in the tank decreases.

2.3. Ultrasonic flowmeter method

The basic principle of the ultrasonic flowmeter method is to calculate the flow velocity of the liquid by measuring the propagation velocity of the ultrasonic pulse in the liquid. The ultrasonic flowmeter is installed at the outlet of the propellant tank, and the real-time flow of the propellant in each tank (including the parallel tanks) can be measured, and the total propellant expelled from the propellant tank can be obtained by integration, thereby calculating the propellant remaining in the tank. Ultrasonic flowmeter has a large uncertainty for small flow, and it relies on the integral method to obtain the propellant remaining amount, so the accuracy of the ultrasonic flowmeter method increases with time.

In general, the ultrasonic flowmeter method has the highest single-measurement accuracy; the BK method accuracy is lower, PVT method has the lowest accuracy compared with the other two gauging methods [5~6]. However, if the accumulative effect is considered, the accuracy of these methods will change. Therefore, the selection of a suitable propellant prediction fusion method can maximize the gauging accuracy.

3. Calculation model

3.1. Error characteristic analysis

For different measurement methods, the measurement data X_i can be regarded as the sum of the true value plus the noise signal, ie

$$X_i = X + e_i \qquad i = 1,2,3 \tag{1}$$

Where, X denotes the true value, e_i denotes the error, the subscript "i"denotes the gauging method. It is assumed that each measurement X_i is independent of each other and is an unbiased estimate of the true value, and the measurement variance is σ_i^2.

For the BK method and the ultrasonic flow meter method, the measurement result is obtained by accumulating the result at the last sampling time, thus, the propellant consumption at the current time can be expressed as

$$X_i(k) = X_i(k-1) + q(k)T + e_i(k) \qquad i = 1,2 \tag{2}$$

Where, q is the instantaneous propellant consumption within one sampling period; T is the sampling period, equation (2) can be expressed as

$$X_i(k)=\sum_{j=1}^{k}q(j)T+\sum_{j=1}^{k}e_i(j) \quad i=1,2 \tag{3}$$

Obviously, the total error characteristic of the BK method and the ultrasonic flow meter method has an accumulative effect. However, the propellant consumption of PVT method is obtained according to the change of the state of the pressurizing gas at the current time and the initial time. The current propellant consumption is independent of the propellant consumption at the previous moment, so the propellant consumption can be expressed as

$$X_3(k)=X(k)+e_3(k) \tag{4}$$

3.2. fusion algorithm

The propellant consumption gauging methods involve many uncertain factors, especially the PVT method and the BK method. Some uncertain factors cannot be accurately defined, such as engine performance degradation, helium leak, and mixture ratio deviation. In order to obtain real-time propellant consumption fusion results, a large amount of real-time measurement data can be utilized to estimate the variance.

For different measurement methods p , q, the measured values are X_p, X_q respectively, and the corresponding measurement error is e_p, e_q, and if the measurement methods are not related to each other, then the autocovariance function R_{pp} of X_p satisfies[7~9]

$$R_{pp}=E(X_p-X)(X_p-X)=E(X_p^2)-E(X^2) \tag{5}$$

The cross-covariance X_p, X_q variance function R_{pq} satisfies

$$R_{pq}=E(X_p-X)(X_q-X)=E(X_pX_q)-E(X^2) \tag{6}$$

The variances X_p are available as follows

$$R_{pp}\text{ - }R_{pq}=E(X_p^2)-E(X_pX_q)=E(e_p^2)=\sigma_p^2 \quad p\neq q \tag{7}$$

Assuming the number of propellant consumption calculation data is k, the time domain estimate of R_{pp} is $R_{pq}(k)$, the time domain estimate of R_{pq} is $R_{pq}(k)$, suppose

$$\tilde{D}_p(k)=X_p(k)X_p(k)-X_p(k)X_q(k) \tag{8}$$

The estimated values of $\tilde{D}_p(k)$ are as follows:

$$D_p(k)=R_{pp}(k)\text{ - }R_{pq}(k)=\frac{1}{k}\left(\sum_{i=1}^{k}X_p(k)X_p(k)-\sum_{i=1}^{k}X_p(k)X_q(k)\right) \tag{9}$$

Further available:

$$D_p(k)=\frac{k-1}{k}D_p(k-1)+\frac{1}{k}\left(X_p(k)X_p(k)-X_p(k)X_q(k)\right) \tag{10}$$

Where, $X_p(k)X_q(k)=\frac{1}{n-1}\sum_{\substack{q=1\\q\neq p}}^{n}X_p(k)X_q(k)$, n denotes the number of gauaging methods.

For the BK method,

$$\begin{aligned}\tilde{D}_1(k)&=\left[\sum_{j=1}^{k}q(j)T+\sum_{j=1}^{k}e_1(j)\right]^2-\left[\sum_{j=1}^{k}q(j)T+\sum_{j=1}^{k}e_1(j)\right]\cdot\left[\sum_{j=1}^{k}q(j)T+\sum_{j=1}^{k}e_2(j)\right]\\&=\left[\sum_{j=1}^{k}e_1(j)\right]^2+\sum_{j=1}^{k}q(j)T\cdot\sum_{j=1}^{k}\left[e_1(j)-e_2(j)\right]-\sum_{j=1}^{k}e_1(j)\cdot\sum_{j=1}^{k}e_2(j)\end{aligned}\tag{11}$$

Assume e_i is independent of the instantaneous flow q, combined with the nature of the time domain estimate, then

$$E\left(\tilde{D}_1(k)\right)=E\left(\sum_{j=1}^{k}e_1^2(j)\right)=k\sigma_1^2\tag{12}$$

It can be noted that for the BK method and the ultrasonic flow meter method, the statistical characteristics of the error signal change with time, and the mathematical expectation of the real-time variance of the propellant consumption increase with time. Therefore, $\tilde{D}_i(k)$ satisfies,

$$\frac{\tilde{D}_i(k)}{k}=o\left(\sigma_i^2\right)\tag{13}$$

Assume $E_p(k)=\dfrac{D_p(k)}{k}$, the iterative relationship can be written as:

$$E_p(k)=\frac{k-1}{k}E_p(k-1)+\frac{1}{k}\frac{\tilde{D}_p(k)}{k}\tag{14}$$

However, it can be noted that when the value k is increased to a certain value, the correction effect is minimal. In order to improve the adaptability of the fusion algorithm, optimize the algorithm by adding data window. It means that only the latest batch of continuous data is processed and calculated, and the previous historical data is discarded. When the data window width is N, the recursion formula is as follows:

$$D_i(k)=\begin{cases}0 & k=0\\ \dfrac{k-1}{k}D_p(k-1)+\dfrac{1}{k}\tilde{D}_p(k) & 1\le k\le N\\ \dfrac{N-1}{N}D_p(k-1)+\dfrac{1}{N}\tilde{D}_p(k) & k>N\end{cases}\tag{15}$$

$$E_i(k)=\begin{cases}0 & k=0\\ \dfrac{k-1}{k}E_p(k-1)+\dfrac{1}{k}\dfrac{\tilde{D}_p(k)}{k} & 1\le k\le N\\ \dfrac{N-1}{N}E_p(k-1)+\dfrac{1}{N}\dfrac{\tilde{D}_p(k)}{k} & k>N\end{cases}\tag{16}$$

Where, equation (16) is suitable for the BK method and the ultrasonic flow meter method, equation (15) is suitable for the PVT method.

According to the constraint that the variance of the fusion result is the smallest, it is easy to find the optimal real-time propellant consumption dynamic weighted fusion result by least square method.

$$\hat{X}(k)=\omega_1(k)X_1(k)+\omega_2(k)X_2(k)+\omega_3(k)X_3(k) \tag{17}$$

Where the weighting factor is

$$\omega_i(k)=\frac{\frac{1}{D_i(k)}}{\frac{1}{D_1(k)}+\frac{1}{D_2(k)}+\frac{1}{D_3(k)}} \tag{18}$$

4. Results analysis

In order to test the actual effect of the fusion algorithm, taking the fusion of the BK method and the PVT method as an example, the sampling period is 1 second and the data window width is 1000. The propulsion system uses a 490N engine, and the orbital firing time is 4280s, 4921s, and 3976s, respectively.

The calculation results of the propellant consumption in the three ignition cycles are showed in figure1~6. Figure 1, 3, 5 are the comparison of the oxidant consumption predictions, Figure 2, 4, 6 are the comparison of the fuel consumption predictions,.

It can be noted that in the initial stage, the uncertainty of the BK method is significantly smaller than that of the PVT method. Therefore, the weighting coefficient of the BK method is larger, and the fusion results are closer to the BK method. As time increases, the error of the BK method accumulates. The proportion of the weighting coefficient gradually becomes smaller. These results are consistent with the theoretical calculation results of the uncertainty of BK method and PVT method.

The comparison of the calculation results of propellant consumption before and after fusion is shown in Table 1. "O" denotes the oxidant consumption; "F" denotes the fuel consumption, the subscript denotes the gauging method.

It can be seen from Table 1 that the accuracy of the BK method is significantly better than the PVT method in the initial stage of measurement. With the accumulation of time, the calculation result of the PVT method is more reliable. The fusion results of the propellant consumption are consistent with the trend of uncertainty, and the data is more reasonable than t only using the BK method or the PVT method.

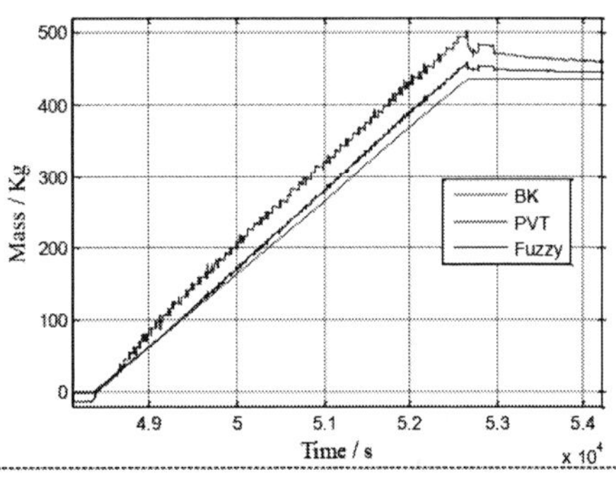

Figure 1. Comparison of oxidant consumption predictions during 1st orbit transfer.

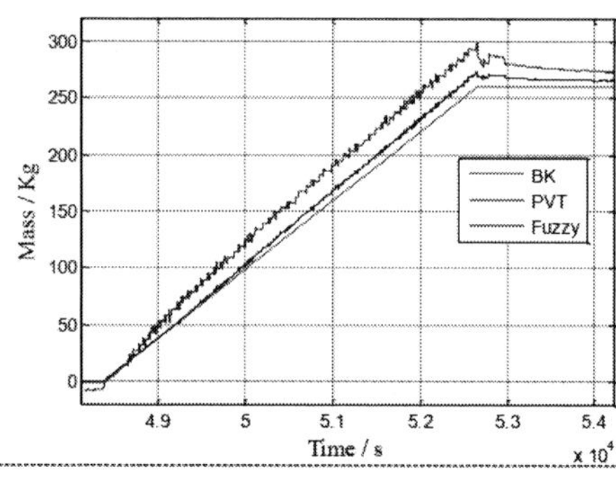

Figure 2 Comparison of fuel consumption predictions during 1st orbit transfer.

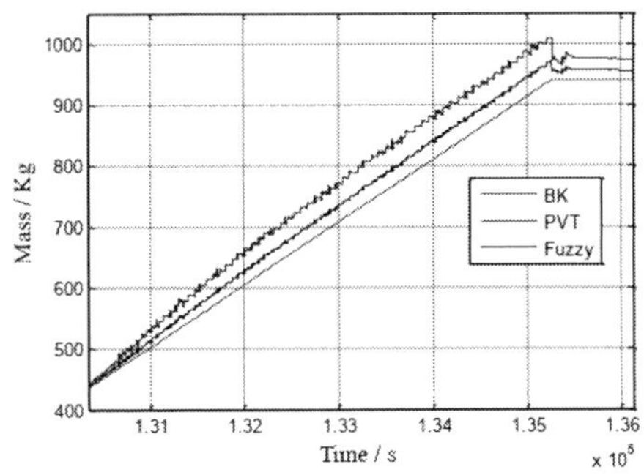

Figure 3. Comparison of oxidant consumption predictions during 2st orbit transfer.

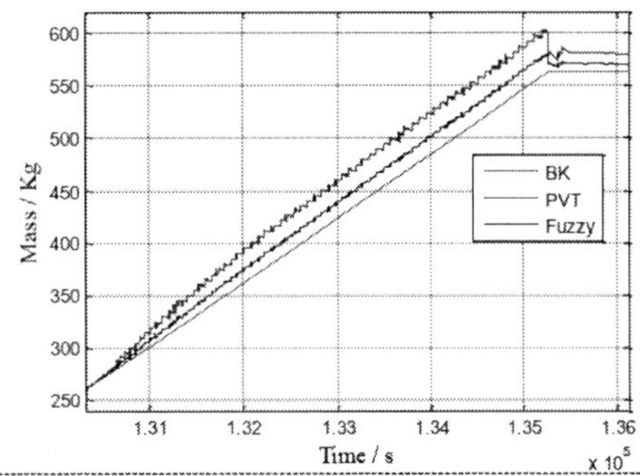

Figure 4. Comparison of fuel consumption predictions during 2st orbit transfer.

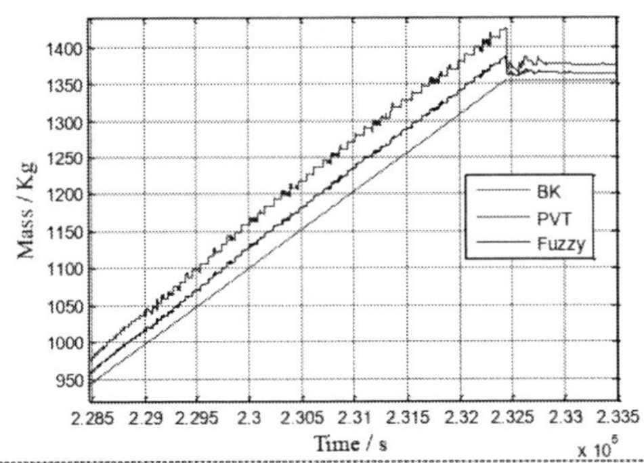

Figure 5. Comparison of oxidant consumption predictions during 3st orbit transfer.

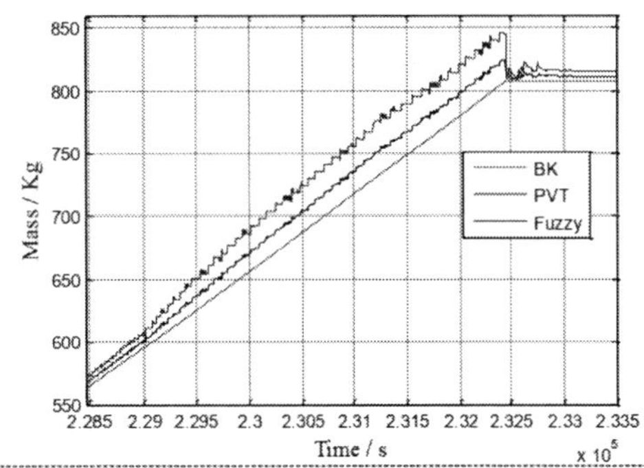

Figure 6. Comparison of fuel consumption predictions during 3st orbit transfer.

Table 1. Comparison of fusion results with theoretical values (units/kg)

	$O_{Theoretical}$	O_{BK}	O_{PVT}	O_{Fussy}	$F_{Theoretical}$	F_{BK}	F_{PVT}	F_{Fussy}
1	439.723	434.458	456.191	443.712	262.783	260.251	271.404	265.563
2	951.097	940.439	960.569	949.567	567.105	562.209	572.309	566.554
3	1369.558	1354.170	1374.849	1363.961	815.362	807.462	815.745	811.311

5. Conclusion

In this paper, the error characteristics of the prediction data are analyzed according to the principle of different gauging methods. Based on this, the variance of each measurement method is estimated by time domain estimation method. The optimal weighted fusion results based on the least squares method of variance are given. The practical application shows that the fusion algorithm can make full use of the advantages of different gauging methods and provide a more reasonable real-time estimation of the propellant consumption, which provides a new way to reduce the uncertainty of satellite propellant remaining prediction.

References

[1] Matthijssen R, Van P P. Ultrasonic flow meter for satellite propellant gauging and ground test facilities[C]. AIAA 2008-4854

[2] Matthijssen R, Van P P. State-of-the art gauging components for improved propellant

IOP Conf. Series: Materials Science and Engineering **887** (2020) 012039 doi:10.1088/1757-899X/887/1/012039

management on 3-axis stabilized spacecraft[C]. AIAA 2006-4714
[3] Yendler B. Review of propellant gauging methods[C]. AIAA 2006-939
[4] Tso-Ping Y. Bipropellant propulsion system performance and propellant remaining prediction[C]. AIAA 89-2512
[5] Chang Y K, Yong M B, Chae Y S, et a1. Error analyses for spacecraft propellant gauging techniques[C]. AIAA 94-2792
[6] Lal A, Raghunandan B N. Uncertainty Analysis of Propellant Gauging System for Spacecraft[J], Journal of Spacecraft and Rockets, Vol. 42, No. 5, September-October 2005, pp 943-946
[7] Luo Ren C et al. Dynamic Multisensor Data Fusion System for Intelligent Robots. IEEE Journal of Robotics and Automation, 1988; 4(4):386～396
[8] Zhai Yili, Dai Yisong. Study of adaptive weighted fusion estimated algorithm of multisensor data [J]. Acta Metrologica Sinica, 1998, 19(1): 69-75
[9] Wang xinfeng, Wang hongzhi. Fusion algorithm for transient loading volume with accumulative error [J]. Journal of Beijing University of aeronautics and astronautics, 2012, 38(5):670-6

On-orbit Engine Thrust Prediction Algorithm for Geosynchronous Satellites based on Neural Network

Wenjuan Yin[1], Tao Song[1] and Zhen Lin[1]

[1]Beijing institute of Control Engineering, 100094 Beijing, China

yinwenjuan1221@126.com

Abstract. Focused on the problem of complex in-orbit engine thrust prediction algorithm the geosynchronous satellite, a thrust prediction method based on the composition of the propulsion system is proposed in the paper. The method integrates theoretical model and neural network model, and the prediction model is simple and practical. In order to further improve the accuracy of the model, the in-orbit data was used to optimize the modified parameters using the multi-objective genetic algorithm NSGA-II. Finally, the optimized model is applied to different cases. The comparison between the simulation results and flight data shows that the accuracy of the optimized model is significantly improved, and the in-orbit thrust prediction method is correct and effective.

1. Introduction

The geosynchronous satellites generally use bi-propulsion systems, which use nitrogen tetroxide as the oxidant, methyl hydrazine as the fuel. The orbit change is performed by a liquid apogee boost motor and the attitude control during the orbit change is performed by the thrusters.

Before the orbit transfer, it is necessary to predict the engine thrust to determine the final orbital strategy [1]. In addition, due to the deviation between the ground test and the in-flight performance, it is often necessary to re-calibrate the engine thrust through a certain orbit maneuver effect, and then correct the engine thrust of the next orbit transfer according to the calibration result to improve the orbit maneuver effect, and avoid waste of propellant. In general, it is necessary to calculate the performance by constructing complex mathematical models of the propulsion system to obtain the predicted thrust. This way is necessary to establish rigorous mathematical models for the products, especially the pressure regulator and the check valves of the pneumatic system, and the versatility is poor. Once the model needs to be corrected, there will be too many parameters involved, and there will be a large randomness.

Focused on the above problems, this paper proposes a neural network-based on-orbit engine thrust prediction algorithm for geosynchronous satellites. This method establishes the basic models of the pneumatic system based on ground data and avoid complex mathematical formulas according to the working principles. The method is simple and has strong versatility. Meanwhile, flight data can be used to modify the model through the multi-objective optimization method [2-5] to improve the engine thrust prediction accuracy. The calculated output of the optimized model is compared with the experimental data to verify the effectiveness of the algorithm.

2. Calculation model

The unified propulsion system mainly comprises the apogee engine, attitude control thrusters

Content from this work may be used under the terms of the Creative Commons Attribution 3.0 licence. Any further distribution of this work must maintain attribution to the author(s) and the title of the work, journal citation and DOI.
Published under licence by IOP Publishing Ltd

IOP Conf. Series: Materials Science and Engineering **887** (2020) 012040 doi:10.1088/1757-899X/887/1/012040

branches, the propellant tanks, helium pressurization system including helium tanks, pneumatic regulator, check valves, other pressurant control and isolation assembles. Figure 1 shows a typical satellite unified propulsion system scheme.

The high-pressure gas flows out of the helium tank is depressurized by the pressure regulator. The gas supplied to the propellant tanks at steady flow rate, and then transfer propellants from the tanks through the pipeline, and finally to the engine and the thrusters. During the satellite orbit transfer, the pressure of the helium tank will decrease with the outflow of helium. The output of the pressure regulator and the check valves may vary due to the output characteristics, which will affect the inlet pressure of the engine. Accordingly, the engine thrust changes.

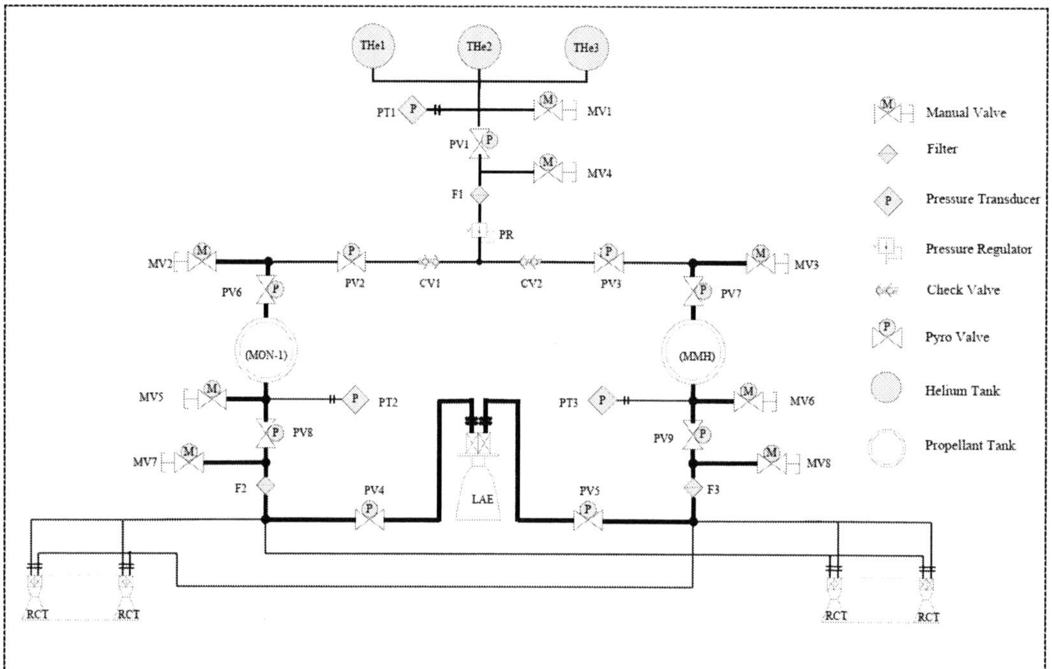

Figure 1. Schematic diagram of propulsion system

2.1. Helium Tank

The high pressure helium tank can be regarded as a combination of an air chamber and a gas outlet port. The dynamic model can be obtained from the mass flow equation of the helium, the continuous equation and the energy equation. Assuming that the gas is ideal, the dynamics equations of helium tank can be can be expressed as

$$\frac{d\rho_g}{dt} = -\frac{Q_g}{V_g} \tag{1}$$

$$\frac{dp_g}{dt} = -\frac{\gamma}{V_g}\frac{p_g}{\rho_g}Q_g - \frac{\gamma - 1}{V_g}\dot{q}_g S_g \tag{2}$$

Where, V_g is the nominal volume of the helium tank, Q_g is the mass flowrate. ρ_g is the density of helium, t is time. $\dot{q}_g$ is the heat flux density per unit area of the gas in the helium tank (usually considered as convective heat transfer), S_g is the contact surface area between the gas and the tube wall in the helium tank, γ is the adiabatic coefficient of helium.

To ensure the in-flight performance of the pneumatic system, multiple pneumatic system tests are often performed on the ground. Through the pneumatic system test, the output pressure of the check valves can be obtained under different inlet pressures of the pressure regulator. The structure of the pressure regulator and the check valves in the pneumatic system is complicated, and the general mathematical model often has the characteristics of slow calculation and poor versatility. The neural network can effectively solve the above problems.

2.2. Pressure regulator

The neural network model of the pressure regulator is as follows:

$$p_{\mathrm{reg}} = net_1(P_{\mathrm{g}}, Q_{\mathrm{g}}) \tag{3}$$

Where, P_g, Q_g are the input vector of the network model. The output vector p_{reg} is the pressure regulator output pressure.

2.3. Check valve

The neural network model of the check valve models are as follows:

$$Q_{\mathrm{go}} = net_2(\mathrm{P}_{\mathrm{reg}} - P_{\mathrm{o}}) \tag{4}$$

$$Q_{\mathrm{gf}} = net_3(\mathrm{P}_{\mathrm{reg}} - P_{\mathrm{f}}) \tag{5}$$

$$Q_{\mathrm{g}} = Q_{\mathrm{go}} + Q_{\mathrm{gf}} \tag{6}$$

The input vector of the network model net_2 is the difference between CV1 inlet pressure $\mathrm{P}_{\mathrm{reg}}$ and the oxidant tank pressure P_{o}, and the output vector is the CV1 output flow Q_{go}.

The input vector of the network model net_3 is the difference between CV2 inlet pressure $\mathrm{P}_{\mathrm{reg}}$ and the fuel tank pressure P_{f}, and the output vector is the CV2 output flow Q_{gf}.

Neural network is suitable for system identification and BP model is one of the most widely used neural network models [6-7]. According to the output characteristics of pressure regulator and check valves, a three-layer BP neural network is selected, the transfer function of the hidden layer and the output layer are recommended to select "purelin". Preliminary training of the network was carried out using ground test data to obtain a preliminary neural network model.

2.4. Tank

As the model shown in figure 2, the tank can be seen as a combination of an air chamber and a gas inlet port, a liquid chamber and a liquid outlet port. The diaphragm between the two chambers is considered to be an ideal geometric isolation surface with equal pressure on both sides. The dynamics equations of helium tank can be can be expressed as follows.

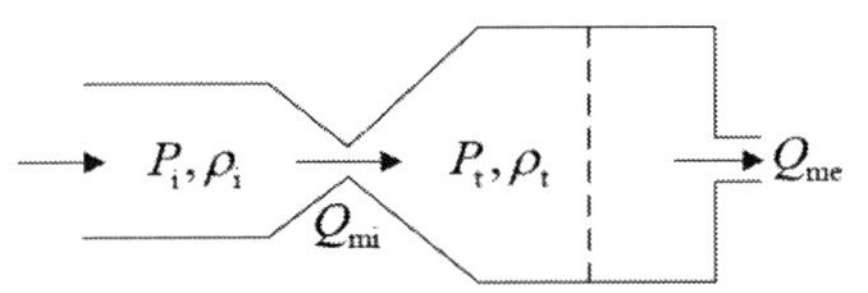

Figure 2. Tank model

$$\frac{\mathrm{d}\rho_{\mathrm{t}}}{\mathrm{d}t} = \frac{1}{V_{\mathrm{t}}}\left(Q_{\mathrm{mi}} - \rho_{\mathrm{t}}\frac{\mathrm{d}V_{\mathrm{t}}}{\mathrm{d}t}\right) \tag{7}$$

$$\rho_{\mathrm{t}}\frac{\mathrm{d}V_{\mathrm{t}}}{\mathrm{d}t} = Q_{me} \tag{8}$$

$$\frac{\mathrm{d}p_t}{\mathrm{d}t} = \frac{\gamma}{V_t}\left(p_i \frac{Q_{mi}}{\rho_i} - p_t \frac{\mathrm{d}V_t}{\mathrm{d}t}\right) \tag{9}$$

Where, Q_{mt} is the gas mass flow into the tank, Q_{me} is the propellant mass flow from the tank. V_t , p_t, ρ_t are the gas volume , pressure and density of the propellant tank, respectively, p_i, ρ_i are the gas pressure and density of the tank inlet chamber.

2.5. Engine

The engine flow is related to the pressure and temperature at the engine inlet. The relationship can be obtained from the ground test and given in the form of a small deviation equation.
The inlet pressure of the engine can be obtained by subtracting the flow resistance of the tank from the pressure of the tank.

$$P_{eo} = P_o - \Delta P_{lo} \tag{10}$$

$$P_{ef} = P_f - \Delta P_{lf} \tag{11}$$

Where, P_{eo}, ΔP_{lo} are the engine inlet pressure and flow resistance of oxidizer branch , P_{ef}, ΔP_{lf} are the engine inlet pressure and flow resistance of fuel branch.

3. Optimization model

Before the satellite orbit transfer, the theoretical thrust value is given according to the ground test results and the current propulsion system state, and then the subsequent orbit maneuver strategy is formulated. Due to the deviation between the ground test and the in-flight performance, it is often necessary to re-calibrate the engine thrust through a certain orbit maneuver effect to improve the orbit maneuver effect.

The neural network correction model of the pressure regulator is as follows:

$$p_{reg} = net_1(P_g, Q_g) + a_1 \tag{12}$$

The neural network correction models of the check valves are as follows:

$$Q_{go} = net_2(\mathrm{P}_{reg} - P_o)\cdot b_1 \tag{13}$$

$$Q_{gf} = net_3(\mathrm{P}_{reg} - P_f)\cdot b_2 \tag{14}$$

Where, a_1, b_1, b_2 are the correction factor.

The flight data available in the existing configuration of the propulsion system is used as a reference value, and the objective function is set as follows:

$$\min f_1 = \overline{P_o} - \overline{p_o} \tag{15}$$

$$\min f_2 = \overline{P_f} - \overline{p_f} \tag{16}$$

$\overline{P_o}$, $\overline{P_f}$ are the average pressure of the oxidant tank and the fuel tank obtained by simulation, $\overline{p_o}$, $\overline{p_f}$ are the average pressure of the oxidant tank and fuel tank measured by the pressure sensors PT2 and PT3, respectively.

Meanwhile, taking into account the maximum deviation of the pneumatic system between in-flight performance and ground test, we can set the constraint function as follows:

$$-0.3 \le a_1 \le 0.3; 0.5 \le b_1 \le 2; 0.5 \le b_1 \le 2 \tag{17}$$

In this paper, the genetic algorithm is used to optimize the model. The specific calculation process is as follows:

1) Try to select the correction factor a_1, b_1, c_1, and if not, a_1 take 0, b_1, c_1 take 1;

2) Calculate the current engine flow rate through the small deviation equation by the tank pressure and temperature data of the propulsion system before the orbit transfer;

3) Calculate the tank pressure(P_o, P_f)by the current engine flow rate;

4) Calculate the flow rate of check valves (Q_{go} , Q_{gf})by the tank pressure and the pressure regulator pressure;

5) Calculate the pressure output of the pressure regulator (p_{reg}) and the helium tank (p_g) according to the the check valve flowrate;

6) Cycle steps 2) to 5) until the end of the orbit transfer;

7) Subtract the o on-orbit average tank pressure of the orbit transfer process from the pipeline flow resistance, obtain the inlet pressure of the engine, and substitute the engine equation to obtain the average predicted thrust of the orbit transfer;

8) Taking a_1, b_1, c_1 as the parameters to be optimized, loop steps 1) to 7) to obtain the fitness function, and solve the optimization model through the multi-objective genetic algorithm NSGA-II, when obtaining a feasible solution that meets the accuracy requirements, the calculation is terminated. Output the optimal solution to the model optimization problem.

9) Substitute the optimal solution into step 1), and calculate the predicted engine thrust of the current orbit transfer by step 2) to step 6).

4. Examples and results analysis

Related parameters settings:

The helium tank volume is 150L and the initial pressure is 23MPa; the nominal volume of the propellant tank is 1412L, the initial pressure of the oxidant tank is 1.505MPa, and the initial pressure of the fuel tank is 1.523MPa. The initial mass of the MON-1 is 1518.808 kg, and the initial mass of the MMH is 916.357 kg. The ignition time is 3208s.During ignition, the on-orbit average pressure of oxidant tank $\overline{p_o}$ is 1.443 MPa and the on-orbit average pressure of fuel tank $\overline{p_f}$ is 1.441 MPa.

The modified parameters a_1, b_1, c_1 are optimized by the NSGA-II algorithm. After the algorithm is finished, the pareto optimal solutions are obtained. A set of optimal correction parameters (a_1=0.0177, b_1=0.9719, b_2=0.9719) are selected in the pareto points, while $f_1 = 4.7306$ Pa , $f_2 = 1.4642$Pa .

Figure 3 and Figure 4 show the pressure variation curves of the oxidant tank and the fuel tank during the ignition process. It can be seen that by optimizing the calculation, the simulation curve is almost coincident with the in-flight performance curve.

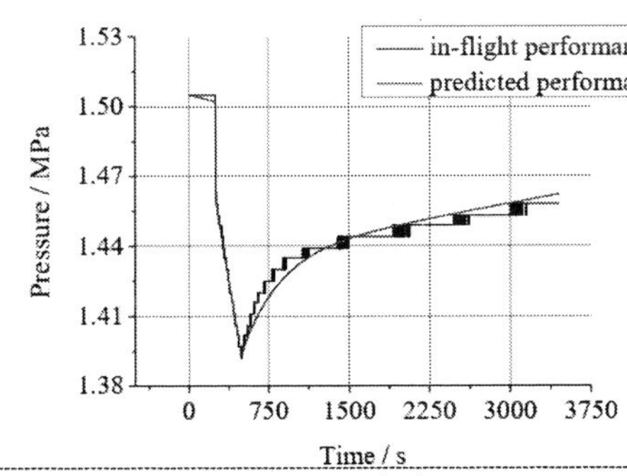

Figure 3. Pressure variation curves of the oxidant tank during the ignition process.

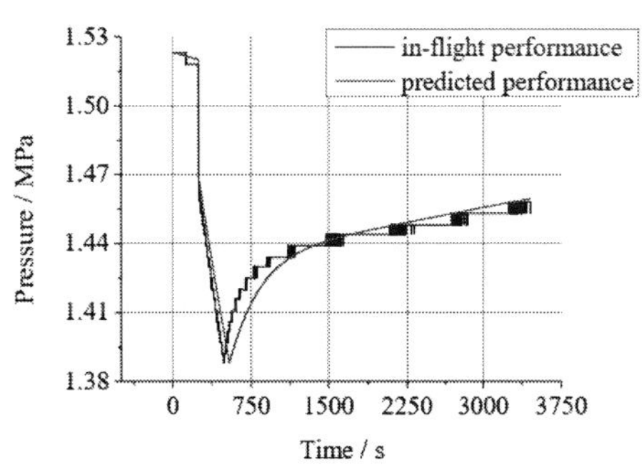

Figure 4. Pressure variation curves of the fuel tank during the ignition process.

In order to further verify the effectiveness of the optimization model, the optimization model is used to predict the engine thrust under different working conditions. The results are given in Table 1.

$\overline{F_c}$ is the calibration thrust thrust that is maintained after the satellite orbit transfer; $\overline{F_p}$ is the prediction thrust value of the maintained model based on the parameters before orbit transfer, T is ignition time.

It can be seen that the average pressure of optimized model calculation results are basically the same as average pressure of the test results. The difference between calibration thrust and prediction thrust are small, the relative error is within 0.2%, which can meet the requirements for on-orbit use.The results show that the optimization model is correct and effective.

Table 1. Optimization results under different cases.

	T / s	$\overline{p_o}$ / MPa	$\overline{p_f}$ / MPa	$\overline{F_c}$ / N	$\overline{P_o}$ / MPa	$\overline{P_f}$ / MPa	$\overline{F_p}$ / N
1	3207.936	1.443	1.441	503.376	1.443	1.441	503.250
2	3917.056	1.463	1.461	508.931	1.463	1.460	508.508
3	2448.384	1.470	1.472	511.711	1.473	1.471	511.271
4	4250.880	1.425	1.423	498.694	1.427	1.427	499.137

5. Conclusion

In this paper, a neural network-based thrust prediction method is proposed for the engine thrust prediction during geostationary satellites.Combined with on- orbit data, the calculation model is studied. The results show that:

1) Combined with ground test data, the prediction model is simpler, more practical and more versatile by selecting a suitable neural network structure to replace the complex valve model.

2) Focused on the difference between the ground test and the in-flight performance of the propulsion system, the multi-objective genetic algorithm NSGA-II is used to optimize the key parameters. The model performance is similar to the in-flight performance, and the model correction method is effective.

3) The optimized model is verified by different cases. The engine thrust prediction error is kept within 0.2%, which satisfies the requirements for on-orbit use. The on-orbit thrust prediction method is correct and effective.

References

[1] Tang Geshi Analysis and Implement of Determining the Orbit Maneuver Effect of CE-1 Satellite. Chinese Space Science and Technology, 2009(6), 1-6

[2] Fonseca C M, Fleming P J. Multiobjective Genetic Algorithms Made Easy: Selection, Sharing and Mating Restriction[R]. Genetic Algorithms in Engineering systems: Innovations and Applications, IEE, 1995.

[3] Horn J, Nafpliotis N, Goldberg D E. A Niched Pareto Genetic Algorithm for Multiobjective Optimization[C]. Proceeding of the 1th IEEE Congress on Evolutionary Computation, IEEE, 1994, 82-87

[4] Zitzler E, Thiele L. Multiobjective Evolutionary Algorithms: A Comparative Case Study and the Strength Pareto Approach [J]. IEEE Transactions on Evolutionary Computation, 1999, 3(4):257-271

[5] Deb K, Agrawal S, Pratap A, et a1. A Fast Elitist Non-Dominated Sorting Genetic Algorithm for Multi-Objective Optimization: NGSA-II[C]. Proceedings of the Parallel Problem Solving from Nature VI Conference, Paris, France: 2000.

[6] Han J, Kamber M. Data mining: concepts and techniques[M]. 2rd ed. San Francisco: Morgan Kaufmann, 2011.

[7] Haykin S. Neural networks and learning machines[M]. 3rd ed. New Jersey: Prentice Hall, 2009.

Research on the Influence and Control of Obstacles Around the DVOR

Jiaquan Ye [1], Yuankai Li [1,*], Fei Liang[1]and Jian Xu[1]

[1.]The Second Research Institute of CAAC, No. 17th, South two of second ring road, Chengdu, Sichuan, 610041, China

* liyuankai@caacsri.com

Abstract. This paper studies the influence and control of obstacles around the DVOR and propose the control methods by combining the standards and theories. First of all, according the relevant technical standards of GB6364-2013 and MH/T4003.1-2014 to control the height of obstacles, then perform modeling and simulation and analyze whether obstacles will affect the signal coverage of DVOR. In order to prevent attenuation, this paper strictly controls the height of obstacles based on the theory of signal propagation, the Fresnel Zone principle and Diffraction theory. The results show that technical standards are the minimum operating requirements when used. When we set the height of obstacles only reply on the technical standards, the signal will not fully meet the signal coverage of requirements. It is necessary to analyze and control the obstacles deeply according to actual condition.

1. Introduction

DVOR is a navigation device that usually works in conjunction with an onboard receiver, which provide the aircraft with comprehensive guidance information and guide the aircraft to fly, enter, depart, and approach along a predetermined route (line) [1] [2]. With the rapid development of the city in recent years, the obstacles around the DVOR are increasing in height above vertical elevation. This seriously affects the signal quality of the navigation station and threatens aviation safety. Therefore, it is urgent to analyze the impact of obstacles and propose the reasonable and effective control methods. The paper analysis how the obstacles with different vertical elevation angle affect the signals on the basis of standards, and propose the control method based on the theory of space signal propagation.

The current control methods for the obstacles around the DVOR currently rely on the provisions of GB6364-2013[3] and MH / T4003.1-2014 [4] for vertical elevation angles. Taking the centre of the DVOR antenna as the reference point and the plane of the antenna counterpoise as the reference plane:(1) There should be no obstacles beyond the height of the reference plane within a radius of 100 m;(2) There should be no obstacles such as roads, buildings, dams, hills, etc. that exceed the height of the reference surface within a radius of 200 m;(3) The vertical elevation angle of trees with a radius of 100 m to 200 m with respect to the reference plane shall not exceed 1.5 °, and the horizontal angle shall not exceed 7 °;(4) The vertical elevation angle of obstacles with a radius of 200 m to 300 m with respect to the reference plane shall not exceed 1.5 °, and the horizontal angle shall not exceed 10 °;(5) There should be no railway beyond the height of the reference plane within a radius of 300 m; the vertical elevation angle of obstacles outside the radius of 300 m with respect to the reference plane should not exceed 2.5 °.

Content from this work may be used under the terms of the Creative Commons Attribution 3.0 licence. Any further distribution of this work must maintain attribution to the author(s) and the title of the work, journal citation and DOI.
Published under licence by IOP Publishing Ltd

IOP Conf. Series: Materials Science and Engineering **887** (2020) 012041 doi:10.1088/1757-899X/887/1/012041

The lowest signal field strength in the VOR signal coverage area is 90 μ V/m(-107dBW/m^2 or -78dBV/m) [5] [6]. If the site is not adequately protected, the signal strength may not be satisfactory.

2. Simulating

Take for example, the obstacle located the south of DVOR, the distance of the nearest point of the obstacle and DVOR is 577 meters,the counterpoise of DVOR is 15.7 meters, the height of obstacle is 54 meters, the horizontal angle is 30 °(164°~194°). The relative position is shown in the Figure 1.

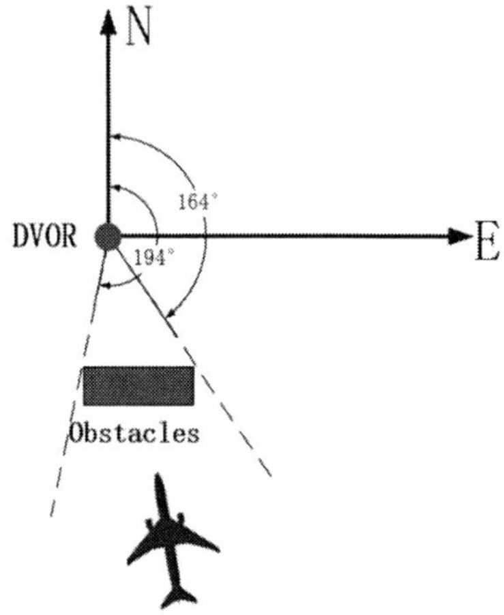

Figure 1. The relative position

Assuming the ground is horizontal, after calculation the vertical elevation angle of the obstacle relative to the DVOR is 3.8 °, which does not meet the standard requirements. The comparison of simulation results of signal coverage is shown in the Figure2.

(a)vertical angle=0° (b)vertical angle=3.8°

(a) the signal coverage when the vertical elevation angle=0°.
(b) the signal coverage when the vertical elevation angle=3.8°.

Figure 2. The comparison result of 0° and 3.8°

In this position (164°~194°), the signal coverage reduced 18km.According to GB6364-2013 and MH / T4003.1-2014, the vertical elevation angle of obstacles outside the radius of 300 m with respect to the reference plane should not exceed 2.5 °. The height of the obstacle was need to reduce to meet the requirements of the standard. We reduce the obstacle height to 40m, and the vertical elevation angle is 2.41°. The signal coverage results are shown in the Figure 3.

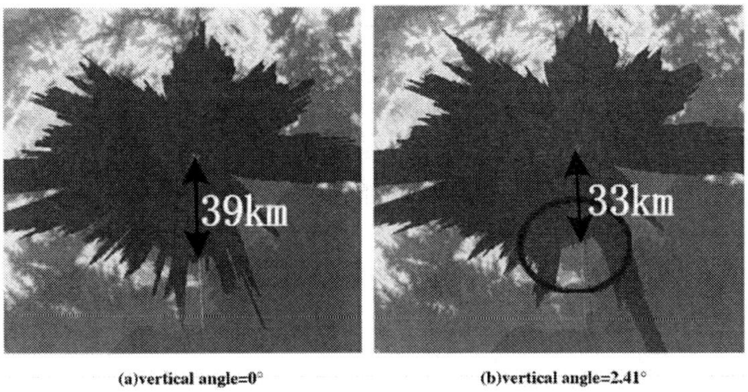

(a) the signal coverage when the vertical elevation angle=0°.
(b) the signal coverage when the vertical elevation angle=2.41°.

Figure 3. The comparison result of 0° and 2.41°

In Figure 3(a), when the vertical elevation angle=0°, the maximum signal coverage distance in the obstacle azimuth is 39km. In Figure 3(b), when the vertical elevation angle=2.41°, the maximum signal coverage distance in the obstacle azimuth is 33km. Compare these two pictures, although the height of obstacle has been reduced according to standards, the max distance of the coverage is 33km, which is also small than 39km.So the signal coverage of the DVOR in the direction of the obstacle is still greatly affected.

3. Attenuation analysis

Taking 170° azimuth as an example, the signal field intensity profile when the vertical elevation angles are 0° and 2.41° shown in Figure4.

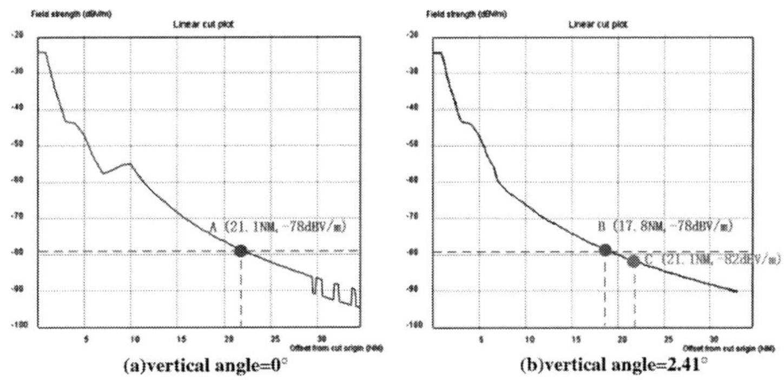

(a) a cross-sectional view of the signal strength on a 170° azimuth when the vertical elevation angle=0° .
(b) a cross-sectional view of the signal strength on a 170° azimuth when the vertical elevation angle=2.41°.

Figure 4. The 2D graphics of 0°and 2.41°

If the field strength is low than -78dBV/m, the aircraft couldn't receive the DVOR signal from the transmitter on the ground. In Figure 4(a), the blue curve is the cross-sectional view of the signal strength on a 170° azimuth when the vertical elevation angle=0°, the maximum signal coverage distance is 21.1NM. In Figure 4(b), the blue curve is the cross-sectional view of the signal strength on a 170° azimuth when the vertical elevation angle=2.41°, the maximum signal coverage distance is 17.8NM.

3.1. The Fresnel ellipsoid

According to the The Fresnel zone [7] [8] principle of telecast transmission, the energy from the wave source T to the receiving point R is basically transmitted by some Fresnel zones with T and R as the focus. In order to obtain the free space transmission better, it is necessary to protect the Fresnel zone from being blocked, show in the Figure 5.

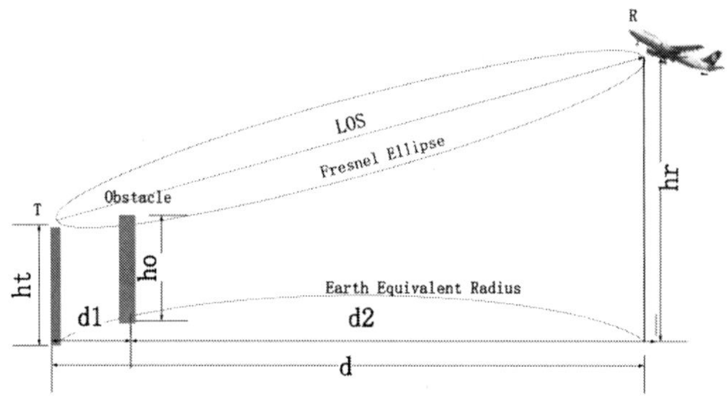

Figure 5. The Fresnel Zone

When the height of transmitter point is 15.7m, the height of receiver point is 900m, the radius of the 1-th Fresnel zones at point M(obstacle) is given by:

$$F1=\sqrt{\frac{300*d1*d2}{f*d}}=38.6\text{m} \tag{1}$$

The radius of the 0-th Fresnel zones at point M(obstacle) is given by:

$$F0=0.577*F1=22.3\text{m} \tag{2}$$

If the obstacle penetrates the 0-th Fresnel zones, the signal strength of free space propagation will be unguaranteed, which greatly affected the signal. Otherwise, the signal strength transmitted in free space is not affected. After calculation, the height of obstacle needs to be reduced to 18.25m.

3.2. Diffraction Theory

Calculate Fresnel parameters v and diffraction loss according to diffraction theory. When the height of transmitter point is 15.7m, the height of receiver point is 900m, the vertical elevation angle of obstacle is 2.41°,the propagation clearance Hc can be calculated, show in Figure 6.

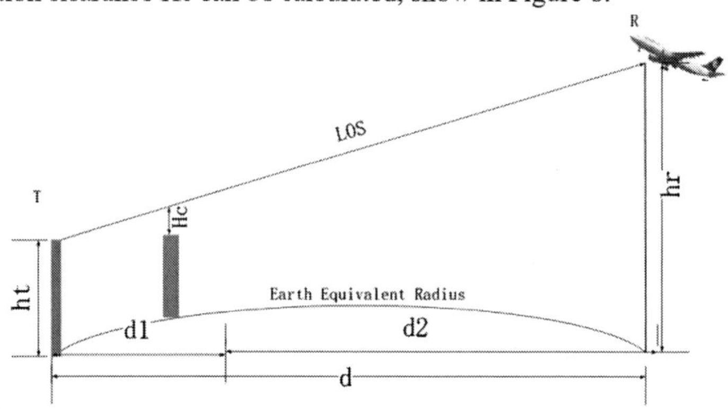

Figure 6. The model of Hc

After calculation, Hc=-0.16m<0(the earth chord height in this same position is 0.68m),and the Fresnel parameter v=0.0057. The curve of diffraction loss as a function of Fresnel parameters [9] is shown in Figure 7. From the curve, if v=0, the diffraction loss is 6dB. And if the Fresnel parameter v<-0.8, the diffraction loss is close to 0.

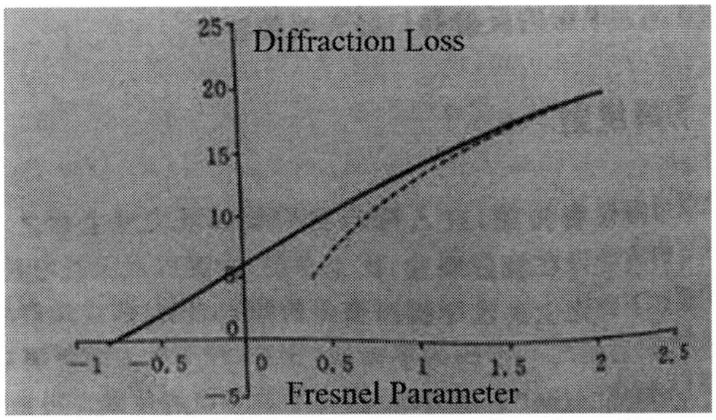

Figure 7. The curve of diffraction loss as a function of Fresnel parameters

As long as the propagation clearance v is smaller than -0.8, which will not affect the propagation of free space signals. After calculation, the height of obstacle needs to be reduced to 18.25m, this requirement is very strict.

3.3. Control method

Comparing the point A in Figure 4 (a) and point C in Figure 4 (b), in the same position from the transmitting antenna 21.1NM, the field strength is -78dBV/m when the vertical elevation angle=0°, and the field strength is -82dBV/m when the vertical elevation angle=2.41°. The attenuation of the obstacle which vertical elevation angle is 2.41°reached 4dB in the same place. Can we continue to reduce the height of the obstacle and reduce the loss by 4dB to improve the signal coverage instead of reduce the height of obstacle to 18.25m.

After calculation, the height of the obstacle need to be reduced to 27m,and the Fresnel parameter v=-0.47. The diffraction loss is 2dBand the vertical elevation angle of obstacle is 1.12°. The simulation results of signal coverage after obstacles are controlled is shown in the Figure 8.

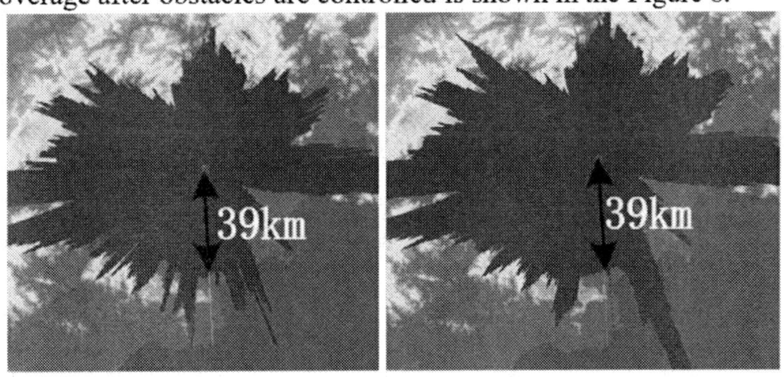

(a)Vertical angle=0° (b)Vertical angle=1.12°

(a) the signal coverage when the vertical elevation angle=0°.
(b) the signal coverage when the vertical elevation angle=1.12°.

Figure 8. The comparison of simulation results

It can be seen from the simulation results that when reduce the height to reduce the loss by 4dB, the coverage distance is the same as when there is no obstacle.

4. Conclusion

The control method of obstacles around the DVOR not only depends on the current existing technical standard specifications, but also combines the actual position of the obstacle and the height of the transmitting and receiving antennas for analysis. If the vertical elevation angle around the DVOR is only controlled according to the standard, the presence of large buildings will cause the signal coverage to be unsatisfactory. But if we control according to Fresnel zone or diffraction theory, the height of obstacles around DVOR will be controlled very strict, which is not possible for most navigation stations. For some obstacles that may affect the critical azimuth signals, they need to be controlled by analysis.

Acknowledgments
This work was supported by China National Key Research and Development Plan Project " The Prevention and Control of Public Safety Risk Prevention and Emergency Technical Equipment" (Fund No. 2018YFC0809505).

References

[1] ICAO Annex 10 Aeronautical Telecommunications, Volume I. Radio Navigation Aids Seventh Edition, July 2018.
[2] Zheng Lianxin 1997 *DVOR VRB-51D Doppler omni-direction Beacon station*, ed Ni Yude (Beiji-ng:The China Civil Aviation Publishing Press):P12-25.
[3] GB6364-2013.Electromagnetic environment requirements for aeronautical radio navigation stati-on.
[4] MH/T4003.1-2014.Specification for aeronautical communication navigation and surveillance station siting criteria—Part 1:Navigation
[5] ICAO DOC 8071. Testing of satellite-based radio navigation systems[Z]. Volume II: Chapters 1.
[6] AC-86-TM-2016-01. Specification for flight calibration of land-based navigation equipment for civil aviation.
[7] Wang Yiping 1987 *Microwave Propagation*, ed Xiao Jingming (People Post Press: Xinhua Bookstore) p 20-31
[8] FAA6820.10 VOR, VOR/DME, and VORTAC SITING CRITERIA (1986)
[9] Christopher Haslett 2012 *Essentials of Radio Wave Propagation* ed Huang Bingke (Xi'an Jiaoton-g University Press) p45-47

IOP Conf. Series: Materials Science and Engineering **887** (2020) 012042 doi:10.1088/1757-899X/887/1/012042

Unmanned Aerial Vehicle Flying Qualities Flight Test Evaluation Technique in Approaching and Landing Phase

Li Jing, Shenjun Qi and Feng Wang

Chinese Flight Test Establishment, Xi'an, China

E-mail address: saintknight372@hotmail.com.

Abstract. Unmanned Aerial Vehicle (UAV) flying qualities in approaching and landing phase relates directly to flight safety. In this paper, autonomous control mode and remote piloted control mode were introduced for analysing approaching and landing flying qualities firstly. Then evaluation criteria about landing flying qualities of these two control modes were proposed according to characteristics and design requirements of UAV and manned aircraft flying qualities specifications. For autonomous control mode, control accuracy and transient response were the main contents of evaluation, but for remote piloted control mode, flight test evaluation method was given in order to lower risk and improve safety, which includes modified Cooper-Harper Rating Scale, closed-loop ground simulation, simulated approaching and landing flight test in low altitude and simulated remote piloted control landing flight test. Finally, flight test was carried out to demonstrate the validity of evaluation criteria and method which is useful and meaningful for UAV landing flying qualities evaluation.

1. Introduction

With the rapid development of flight control and navigation technology, Unmanned Aerial Vehicle (UAV) technology has made a breakthrough. Development and application of UAV has drawn great attentions of all the countries in the world both in military and civilian. Nowadays, US has deployed Global Hawk, Predator and other types to execute Intelligence, surveillance, Reconnaissance (ISR) and strike missions [1], while other countries are also trying to develop their own UAV.

Currently, the accident rate of UAV is so high, which is 10~100 times than manned aircraft [2]. One important reason, which causes high UAV accident rate, is lack of comprehension and inadequate research of UAV flying qualities. Moreover, understanding on concept of UAV flying qualities remain on the understanding of manned aircraft flying qualities and evaluation method and criteria still follow and use manned aircraft evaluation techniques, which does not reflect characteristics of UAV.

In recent years, domestic and foreign experts and scholars conduct study of characteristics of UAV flying qualities, while UAV airworthiness regulations has been issued to promote development of UAV. Mitchell et al [3] redistributed the flight phases of UAV, according to accuracy and rapidity of UAV flight mission. Ding Tuanjie et al [4] analysed influence factors affected UAV remote control mode and proposed preliminarily flying qualities evaluation method using manned aircraft flying qualities specifications. Pan wenjun et al [5] and Wang Feng et al [6] considered characteristics of UAV and manned aircraft flying qualities evaluation methods, analysed fundamental manned aircraft flying qualities requirements including static stability, modes characteristics and time-delay and the applicability of several typical piloted flying qualities requirements like the CAP and the Bandwidth

Content from this work may be used under the terms of the Creative Commons Attribution 3.0 licence. Any further distribution of this work must maintain attribution to the author(s) and the title of the work, journal citation and DOI.
Published under licence by IOP Publishing Ltd

criterion for UAV. Then some effective suggestions were presented. Wang Feng [7] studied landing flying qualities under remote control mode from the perspectives of time-delay and control mode.

Approaching and landing is one of the most accident prone phases of UAV. In this paper, autonomous and remote control mode for UAV approach and landing are introduced firstly which affects flying qualities evaluation. Then flying qualities evaluation criteria are proposed based on characteristics of UAV and UAV flying qualities flight test method is given. After analysing test results, the UAV flying qualities evaluation flight test method is proved effective, which is helpful and meaningful for UAV flying qualities research.

2. Approaching and landing control mode

There are several recovery modes for UAV, including parachute, arresting net, wheel landing and so on. For large and medium sized UAV, wheel landing is the main way for approaching and landing. Nowadays, manned aircraft landing is heavily depended on pilot and auto-landing system is just an auxiliary system, but UAV landing mainly operated by autonomous control system. However, due to complexity of operation, high level of risk and technical maturity, remote piloted control mode would be more useful and helpful for approach and landing in emergency circumstances which need to return to base. Hence, remote piloted control mode would usually be set as standby control mode in order to ensure flight safety. In this section, the theories of autonomous control landing and remote control landing would be introduced briefly.

2.1. Autonomous Control Landing

The most common Auto-control landing systems include Instrument Landing System (ILS), Microwave Landing System (MLS), and Global Position System (GPS). In this paper, the main consideration for auto-control landing is navigation information based on GPS. After finishing missions, flight control system of UAV would control the vehicle on the basis of location information provided by GPS, and approach and land according to pre-planned flight path.

There are six segments for UAV approaching and landing under auto-control landing mode, including initial approach, capturing and gliding, guiding and gliding, level off, heading aligning and landing, shown in figure 1. Auto-control system has two subsystems, longitudinal control system and lateral control system. Control method in every segment is shown as follows:

- Initial approach segment: After finishing tasks, UAV enter landing path. At specified height, landing gear and flap were down, and then UAV maintains this altitude, adjust speed, control lateral direction and align to runway according to pre-planned flight path.
- Capturing and gliding segment: UAV aligns itself to runway in lateral direction, and enter into extension line gliding flight path in longitudinal direction.
- Guiding and gliding segment: UAV flies along pre-planned gliding flight path at steady speed, correct lateral deviation relative to runway centerline.
- Level off segment: when UAV glides to flareout altitude. Then pull up attitude and reduce speed, UAV will touch down in a small sink rate.
- Heading aligning segment: UAV adjusts longitudinal direction attitude and speed in order to land under proper conditions like appropriate pitch attitude and sink rate, aligns heading to runway in lateral direction and level off at roll attitude, which would prevent aircraft from overshooting the runway or prevent wing from scarping the runway.
- Landing and running segment: after touching down, UAV begins to slow down and brake. During touching down, flight control system control attitude in longitudinal direction and deviation in lateral direction.

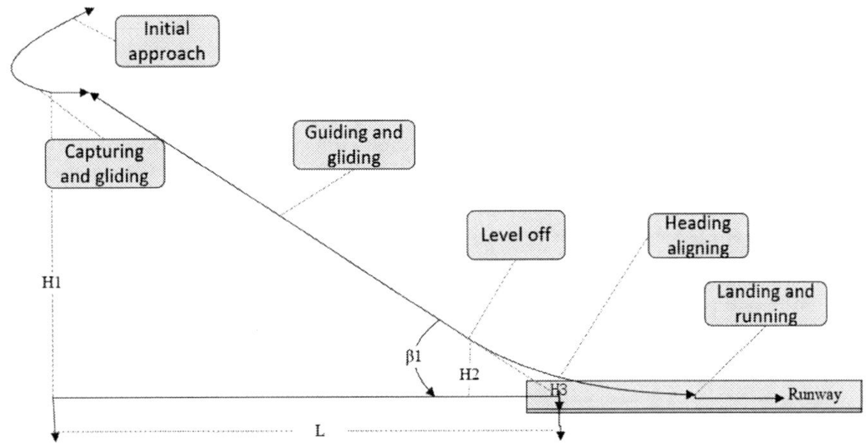

Figure 1. UAV approaching and landing under auto-control landing mode

2.2. *Remote control landing*

UAV remote control landing is similar to manned aircraft landing, which still depend on stick and rudder. The main difference is Situation Awareness. Manned aircraft pilot could get feedback from external environment, but UAV pilot control the vehicle in Ground Control Station (GCS) separated from UAV. Control instructions are transmitted via up-link to flight control system of UAV and all the condition of subsystems and status of UAV are transmitted through down-link to GCS. UAV Remote pilots see the aircraft through on-board sensors and cameras down linking information displayed on the screen of computer in GCS and could not get the same quality of feedback as a manned aircraft pilot feels such as visual weather, smell of smoke, and vibration of fuselage. The figure 2 shows the control flow of manned aircraft and UAV.

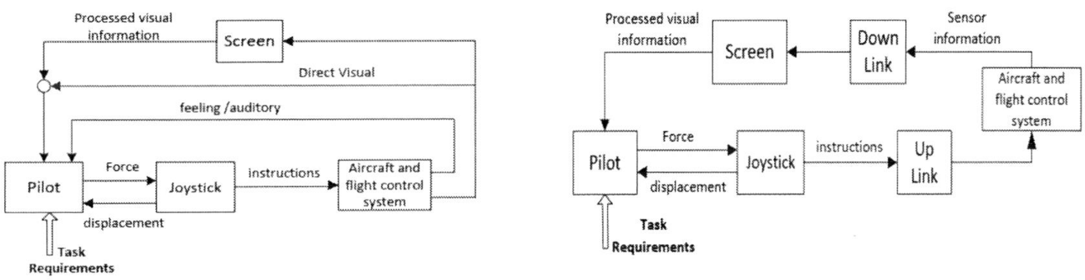

a) Manual control in manned aircraft b) Remote Piloted control in UAV

Figure 2. Control mode in manned aircraft and UAV

3. Flying qualities of autonomous control mode

The idea conditions for UAV autonomous control landing should meet the following requirements. Firstly, the direction of ground speed is the same as runway centerline which means the track angle is equal to runway heading. Secondly, the body axis of aircraft should be in accord with runway centerline that means the course angle should be equal to runway heading. Thirdly, the wing should maintain level and the roll angle is zero. Lastly, UAV should be glide along extension line of runway centerline, and lateral deviation is zero.

Now, there are no requirements and specifications about UAV flying qualities. Hence, the contents of UAV autonomous control landing flying qualities are mainly proposed based on most commonly used UAV design requirements, manned aircraft flying qualities specifications and flight control regulations. The contents of demonstration are as follows: 1) UAV should fly steadily in approaching and landing phase. 2) During transition of configuration, transient response should be smooth and convergent in longitudinal, directional and lateral directions, 3) Control accuracy of UAV should meet requirements of landing safely and precision index should be determined.

For control accuracy, the concerned parameters should be considered from perspective of safe landing as follows, such as initial approaching altitude, longitudinal and lateral deviation in guiding and gliding segment, location of touch down point, sink rate, roll angle, course angle, etc. All accuracies of these parameters should be meet related requirements of control accuracy in approaching and landing phase. Figure 3 is flight test result of UAV autonomous landing. It can be seen that transient response is smooth during transition of configuration and landing phase. From flight test result, UAV landed safely and autonomously and the control accuracy satisfied requirements.

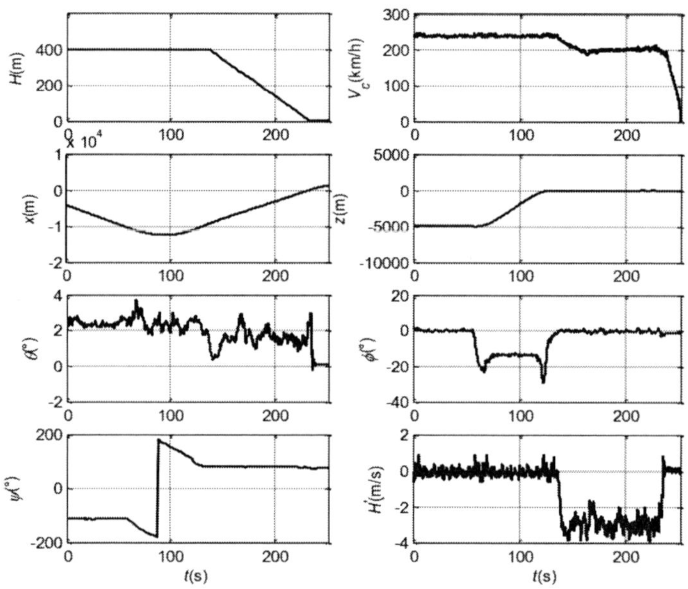

Figure 3. UAV autonomous approaching and landing flight test

4. Flying qualities of remote piloted control mode

4.1. Flight test evaluation method

UAV remote piloted control mode faces great challenge for human factor in human factor. UAV pilot workload is increased due to all operations in GCS, and time-delay increases operating difficulty. So it is needed to fully evaluate pilot workload and system controllability. Then, modified Cooper-Harper Rating Scale was proposed to evaluate flying qualities according to characteristics of UAV and requirements, which is an effective tool for comprehensive evaluation rating by UAV pilot. Modified Cooper-Harper Rating Scale is shown in figure 4.

For UAV landing flying qualities of remote piloted control mode, demonstration flight test should be divided into three steps in order to ensure UAV flight safety. These tests are closed-loop ground simulation, simulated approaching and landing flight test in low altitude and simulated remote control landing flight test.

IOP Conf. Series: Materials Science and Engineering **887** (2020) 012042 doi:10.1088/1757-899X/887/1/012042

Figure 4. Modified Cooper-Harper Rating Scale for UAV

4.1.1. Closed-Loop ground simulation. UAV closed-loop ground simulation usually uses GCS to conduct simulation, while GCS and UAV are connected by data link. So the simulation conditions are almost same as real flight, and UAV closed-loop ground simulation in GCS has high fidelity. The typical closed-loop ground simulation test environment is shown in figure 5. GCS collects operation instructions that are sent to flight control computer in UAV via uplink. After computation of flight control system, the instructions are sent to UAV control surface and engine control system. Then simulation test station of flight management system collects flight state and builds simulation model to calculate dynamic response of aircraft. All of these data are transmitted to GCS through downlink after fulfilment by flight control computer. UAV pilot operate UAV according to information displayed in screens of GCS afterwards. All of these procedures form closed-loop. UAV pilot and operator could know well the flight test procedure and develop risk management measures through closed-loop ground simulation.

4.1.2. Simulated approaching and landing flight test in low altitude. Before simulated remote piloted control landing flight test, simulated approaching and landing flight test in low altitude should be carried out due to safety consideration. Simulated approaching and landing flight test would be conducted in 1000 meters above ground level (AGL) in order to examine UAV time-delay, controllability and stability under remote piloted control mode. Moreover, UAV control mode would be switched to remote piloted control mode before aligning with runway, and then UAV would be controlled remotely by pilot to turn and align with runway. Furthermore, UAV glides to 625 meters AGL with 2.5°glide slope angle. When reaching 625 meters AGL, UAV was controlled to level off, which is the end of simulated approaching and landing flight test in low altitude.

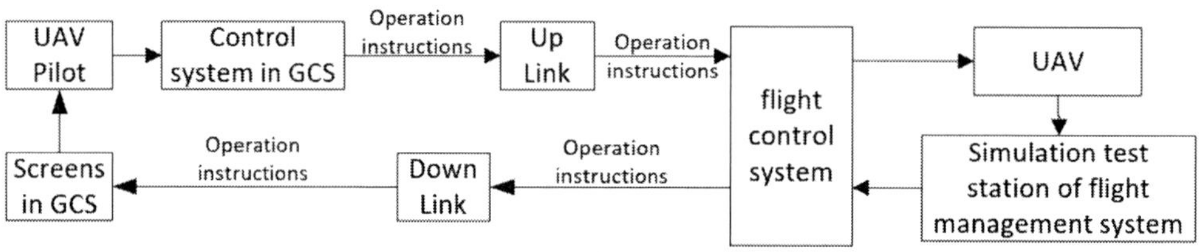

Figure 5. Closed-Loop Ground Simulation Test Environment Diagram

4.1.3. Simulated remote piloted control landing flight test. After analysis, remote piloted control landing test has great risk. For flight test safety, UAV landing flying qualities flight test evaluation would be accomplished through simulated remote piloted control landing. The specific test plan is as follows. Firstly, before aligning runway, UAV pilot sends instructions via GCS to switch control mode from autonomous control mode to remote control landing mode. And then UAV pilot aligns UAV with the designated landing runway and controls UAV gliding along pre-planned gliding flight path to 50 meters AGL. When UAV reaches assigned altitude, UAV pilot controls UAV to go-around.

4.2. Flight test results

Considering significance of flying qualities evaluation to pilot briefing, 5 test pilots who have full of UAV flight experience were chosen to evaluate and rate according to test procedures. UAV test pilots give flying qualities evaluation and rates about approaching and landing under remote control mode according to flight test. Table 1 is the pilot evaluations used by Modified Cooper-Harper scale rating.

Table 1. Modified Cooper-Harper scale rating for test pilots

Pilot	Longitudinal Control	Lateral Control
1	5	4
2	4	4
3	5	4
4	6	6
5	5	5

From table 1, it can be seen that the Cooper-Harper scale locates 4～6 in longitudinal Control, and 4～6 in lateral Control when pilot operate UAV under remote piloted control mode for approaching and landing. UAV pilot could control UAV well in longitudinal and lateral directions in landing phase, but the workload is high.

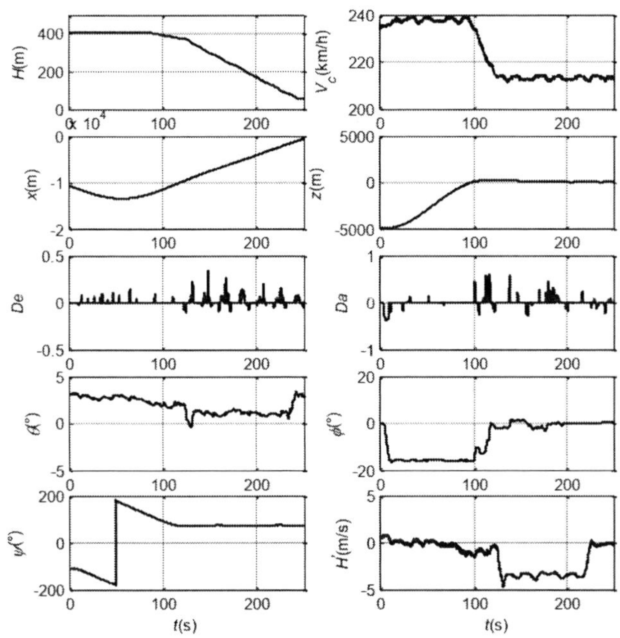

Figure 6. Simulated landing under remote piloted control flight test result

Figure 6 shows that UAV pilot could control UAV flight safety during approaching and landing phase, but frequent operation is needed to align runway and glide along gliding path, which does accord with the results of pilot evaluation on basis of Modified Cooper-Harper Rating Scale.

5. Conclusions

UAV landing flying qualities for autonomous control and remote piloted control were studied theoretically and practically in this paper. Flying qualities of landing in autonomous control mode was evaluated from several aspects, like control accuracy, stability and safety, while for flying qualities of remote piloted control mode, flight test was conducted step by step in order to reduce risk and ensure flight safety, including closed-loop ground simulation, simulated approaching and landing flight test in low altitude and simulated remote piloted control landing flight test. Then modified Cooper-Harper Rating Scale was proposed and used to evaluate landing flying qualities of remote piloted control mode. The results show that the flight test evaluation method given in the study is useful and helpful in evaluation of landing flying qualities of UAV.

Acknowledgments

Authors wishing to acknowledge Mr Xinlong Wu and Mr Yalong Wang of Chinese Flight Test Establishment for their assistance.

References

[1] Paul Gerin F and Thomas James G 2012 *Introduction to UAV Systems* (New York: Wiley) chapter 1 pp 3–15
[2] Yujin T and Jiangpei W 2010 *Flight Dynamics* **28** 13
[3] Mitchell D G, Hoh R H and Aponso B L 1994 *Proposed incorporation of mission-oriented flying qualities into MIL-STD-1797A* (Ohio:Wright-Patterson Air Force Base) p 13
[4] Tuanjie D, Wei F and Feng W 2011 *Flight Dynamics* **29** 17
[5] Wenjun P, Zhanqi F, Minwen W and Xun S 2016 *Flight Dynamics* **34** 6
[6] Feng W, Shengjun Q and Zhe Z 2013 *Flight Dynamics* **31** 389
[7] Feng W 2017 *Flight Dynamics* **35** 73

The Research on Redundant Design in Civil Aircraft Design

Bing Cheng[1]

[1]Shanghai Aircraft Design and Research Institute, Shanghai 201210, China

E-mail: chengbing198805@163.com

Abstract. This article explores the redundant design in the design of civil aircraft. It's based on the meaning of giving redundancy beyond the basic needs to the parts or the products by redundant design. And this article also conducts targeted research on several typical redundant design methods such as redundant design in sketch, redundant design steps in part, integrated design of parts, redundant parts in assembly and redundant assembly of the civil aircraft. The redundant design formulas corresponding to different design methods are obtained by analyzing specific design examples. It is found in the research that redundant designs and components are also effective ways to optimize the design of the structure and to pre-solve a series of problems that may appear in subsequent stages. Moreover, the realizations of the requirements are compared between the step-by-step redundant design and the integrated redundant design. And it is found that the step-by-step redundant design is a better design method.

1. Introduction

The concept of redundant design first appeared in the field of information technology, and its application is mostly used to improve the security of the design, such as improving the reliability of airborne electronic equipment [1] and flight controller [2]. Increasing the redundancy of the product includes two levels of meaning. One is to give redundancy beyond the basic needs, and the other is to provide additional backups in case of need [3].

For the civil aircraft, the pursuit of design optimization is a powerful way to reduce aircraft weight and enhance aircraft competitiveness. However, the optimization of part design does not mean the optimal design of components and products, and the optimal design does not mean the optimal manufacture and assembly. Sometimes in order to achieve the overall optimization of the product, it is necessary to add the redundant part to the component design. That is to say, the design elements that exceed the basic design requirements for components may be given to the components. It eventually produces local redundancy, such as increasing weight, occupying more space, increasing complexity, extending design and manufacturing time, and also increasing costs [4], which is not a local optimal solution. However, for the product as a whole, the redundant design of this part has optimized the design and manufacture for the product. In addition, local redundancy is more reliable than overall redundancy, so the redundant design should start from the most basic units [5].

2. Redundant design

Based on the meaning of the first level of redundant design in other fields, combined with civil aircraft design, the concept of redundant design is given: in the design of parts, in order to carry out further design upgrades in the future, and in product design, to improve the benefits of manufacturing, assembly or operations in subsequent product life cycle, some redundant design elements are added at

Content from this work may be used under the terms of the Creative Commons Attribution 3.0 licence. Any further distribution of this work must maintain attribution to the author(s) and the title of the work, journal citation and DOI.
Published under licence by IOP Publishing Ltd

the beginning of the design. It's beyond the basic design requirements of product design and may cause local redundancy. However, it plays a role in optimizing the design, manufacturing and operational benefits of the overall product.

Redundant models are defined as: product parts are kept in a semi-open or semi-finished state in the design, or design functions or structures that are not basic requirements of product design are attached to provide openings for future design, manufacturing, assembly and operation.

It means that the core of the design is fixed, but the intermediate state of the design can be redefined according to future design needs. We can design the known for the unknown. The part is designed for the product. And the product is designed for its life cycle.

3. Redundant design in sketch

Many parts are designed based on sketches. Sketch information can be selectively expressed in part design. During the sketch design process, design elements can be given to the sketch as many as possible. That is, redundant design is performed in the sketch. It will not only meet the basic design requirements of this part, but also meet the needs of part configuration changes and modifications in the future. For example, the sketch can exceed the design scope of the part instance in the design of the frame, as shown in figure 1. The complete sketch can meet the design requirements of the part in different stations and configurations. The basic design requirements of the part are achieved through feature removal during part design. It facilitates design modifications, improves the convenience of design, simplifies the design process and avoids redundant structures in the physical parts.

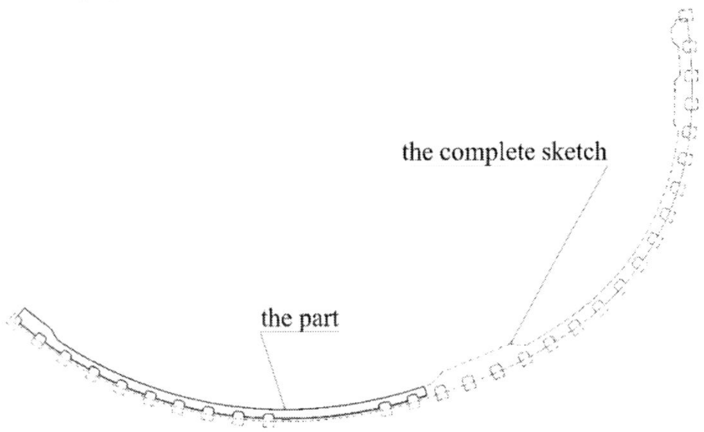

Figure 1. Comparison of the part and its complete sketch.

The corresponding redundant design formula can be summarized as:

$$S = S_0 + S_r^{'} \tag{1}$$

S represents the sketch which contains the sketch meets basic design requirements as S_0 and redundant sketches as S_r.

4. Redundant design steps in part

In the part design process, some redundant design steps that are meaningless to a single part can be added to achieve design optimization. For example, in the longitudinal joint area of the fuselage, the stringer connection is generally achieved by the stringer joints. They are designed by parametric method to control the longitudinal length of stringer joints with uniform numerical parameters. This can already realize the modular design of the stringer joints. But for the stringer joints in the joint area of the contracting skins, the seemingly unified design method will lead to the result that the uniformity of the length of the stringer joints is limited to the equidistant points, due to the contracting skins and the edge fillet treatment of the stringers. And the equidistant point will disappear on the part structure because the stringer joint is rounded to avoid interference with the stringer, as shown in figure 2. This will eventually lead to a result that the stringer joints are not the same distance from the frame plane by the seemingly unified design way. As shown in table 1, the distances of stringer joints from the frame plane at different positions of L1 to 42 are shown. The distance between the equidistant points and the frame plane is 152.5mm. The maximum deviation is about 1.4mm, which brings a certain degree of confusion to subsequent assembly work. In order to solve this problem, a design step is added in the design process of the stringer joint. The tail of the stringer joint is trimmed through the parallel plane of the frame plane, as shown in figure 3. In the figure, line A is the boundary without redundant design, and line B is the boundary with redundant design. And through uniform repair, the consistency of the distances between the tail of the stringer joints and the frame plane is ensured, which facilitates uniform tooling positioning of dozens of stringer joints here during assembly.

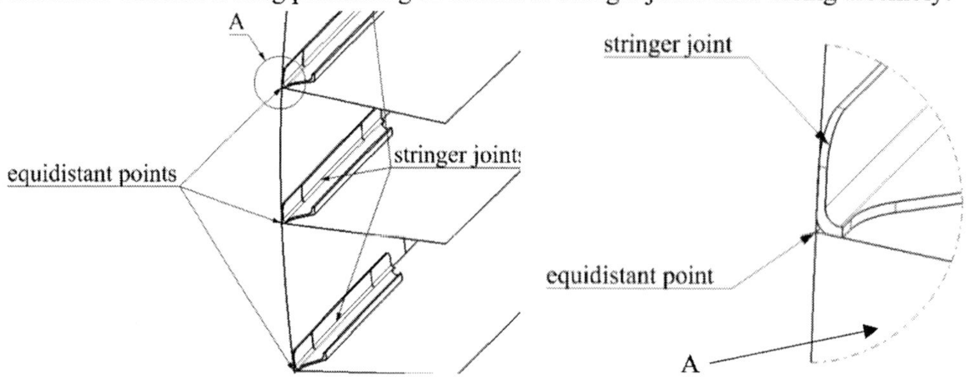

Figure 2. Equidistant points generated during the design of stringer joints.

Table 1. Distances of stringer joints from the frame plane at different positions of the contracting skin.

Stringer	Distance(mm)	Stringer	Distance(mm)	Stringer	Distance(mm)
L1	152.441	L15	N/A	L29	151.703
L2	152.441	L16	152.168	L30	151.676
L3	152.434	L17	152.152	L31	151.723
L4	152.414	L18	152.500	L32	151.129
L5	152.426	L19	152.094	L33	151.621
L6	152.379	L20	152.063	L34	151.605
L7	152.363	L21	N/A	L35	151.578
L8	152.336	L22	151.953	L36	151.547
L9	152.316	L23	151.922	L37	151.539
L10	152.285	L24	151.883	L38	151.535
L11	152.258	L25	151.844	L39	151.535
L12	152.234	L26	151.805	L40	151.539
L13	152.215	L27	151.770	L41	151.547
L14	152.197	L28	151.738	42	151.547

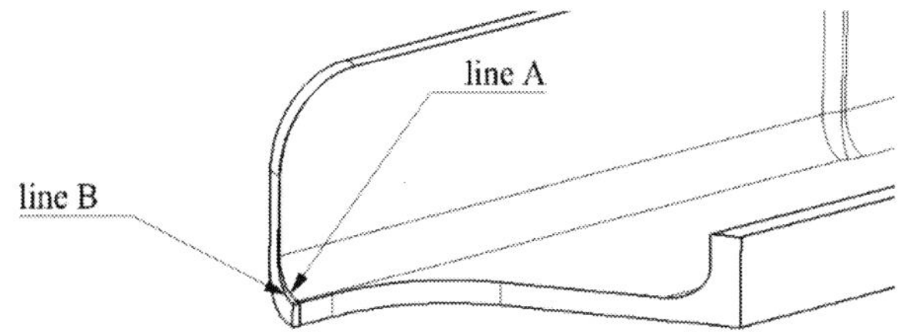

Figure 3. The boundary of the stringer joint.

The corresponding redundant design formula can be summarized as:

$$PS = PS_0 + PS_r \tag{2}$$

PS represents the part design steps which contain the steps meet basic design requirements as PS_0 and redundant steps as PS_r .

5. Integrated design of parts

A certain part needs to be designed due to specific function and structure. When designing, it's tried to avoid additional structures that exceed the basic function. However, when two or more parts are in conflict or close to each other during layout, combine multiple parts into one by integrated design. Although redundant structures may be added to a single part, the layout and design of parts in a local area have achieved an optimized effect.

For example, the trunk bracket and the stringer joint can be integrated to design, so that a single part could have the functions and necessary structures of both parts, as shown in figure 4. For each of the original parts, some structures of the new design are redundant. But it is an optimized design that not only reduces weight, but also guarantees the necessary functions and structures for the both.

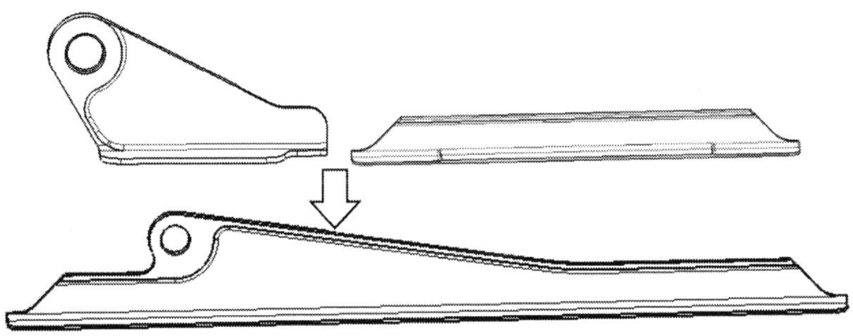

Figure 4. Integrated design of the trunk bracket and the stringer joint.

The corresponding redundant design formula can be summarized as:

$$P = P_1 + P_2 = P_0 + P_r \tag{3}$$

P represents the part which contains key structures of different parts as P_1 and P_2 . One part, denoted by P_r , is redundant for the other, denoted by P_0 .

6. Redundant parts in assembly

In some areas with complex assembly conditions, such as those involving multi-layer connections, the length of the fasteners varies due to the thickness of the connected structure, which can easily lead to the wrong selection of fasteners during subsequent assembly. The length of the fasteners can be unified by adding some parts, such as redundant shim, as shown in figure 5.

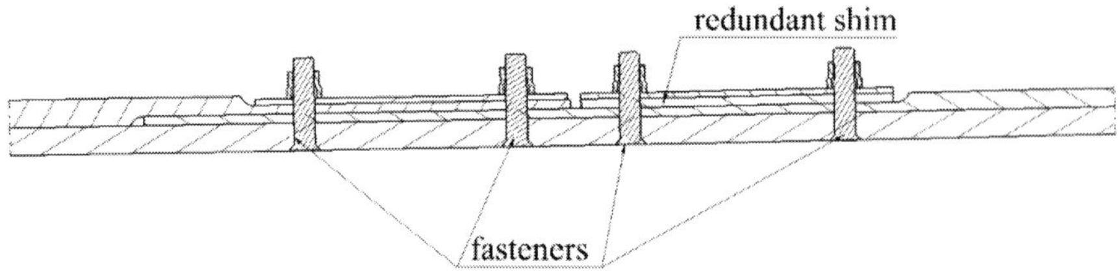

Figure 5. Redundant shim.

These parts are not basic design requirements and can be eliminated. However, the length of the fasteners installed here can be unified by adding the shim. It simplifies the assembly and avoids the wrong arrangement of the fasteners during assembly.

The corresponding redundant design formula can be summarized as:

$$A = A_0 + P_r \tag{4}$$

A represents the assembly which contains the assembly meets basic design requirements as A_0 and redundant part as P_r.

7. Redundant assembly

The realization of the product has captured the stages of design, manufacturing, assembly and operation through requirements. The design cycle may take months or even more than a year. In fact, some of the customer's needs, such as the selection requirements of different internal structures, have little effect on the product's functional structure. Therefore, it is possible to provide diversified options in product design to simultaneously deliver to customers to choose from. For example, in the design of the cabin floor, a diversified design configuration of the seat track is added for customers to use in the selection of various seat configurations in subsequent operations, as shown in figure 6.

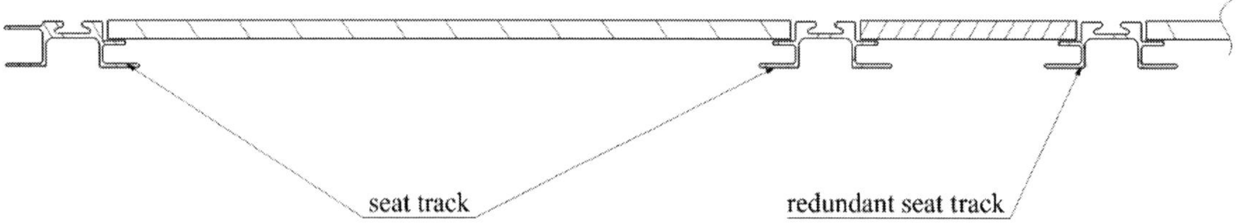

Figure 6. Redundant seat track.

The corresponding redundant design formula can be summarized as:

$$A = A_1 + A_2 = A_0 + A_r \tag{5}$$

A represents the assembly which contains key structures of different assemblies as A_1 and A_2. One assembly, denoted by A_r, is redundant for the other, denoted by A_0. A can be chosen as A_1 or A_2.

For redundant design, redundant sketches, part design steps, parts and assemblies can be combined when necessary. The parts can be described as special assembly, denoted by $P_0 \subset A_0$ and $P_r \subset A_r$. Therefore, the redundant design of the component can be expressed as follows.

$$A = f(S_0, PS_0, A_0, S_r, PS_r, A_r) \tag{6}$$

The logic diagram is shown in figure 7, and the degree of realization of the design process to requirements is recorded as R.

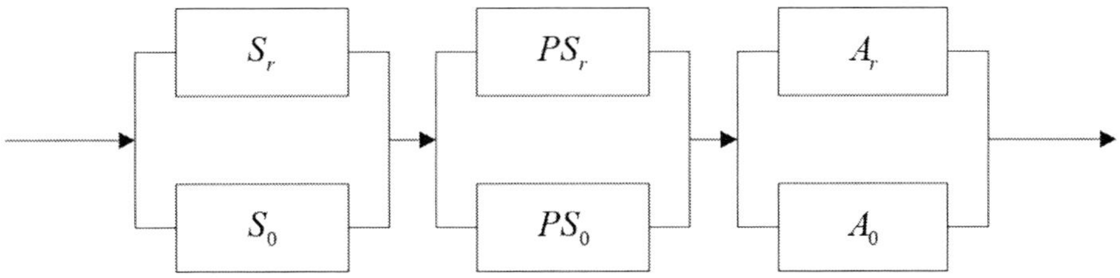

Figure 7. Logic diagram of step-by-step redundant design.

$$R_s = 1-(1-R_{S_0})(1-R_{S_r}) = R_{S_0} + R_{S_r} - R_{S_0}R_{S_r} \tag{7}$$

$$R_{PS} = 1-(1-R_{PS_0})(1-R_{PS_r}) = R_{PS_0} + R_{PS_r} - R_{PS_0}R_{PS_r} \tag{8}$$

$$R_A = 1-(1-R_{A_0})(1-R_{A_r}) = R_{A_0} + R_{A_r} - R_{A_0}R_{A_r} \tag{9}$$

$$R_1 = R_S R_{PS} R_A \tag{10}$$

If we first design for the basic design requirements, and then optimize the design for manufacturing, operations, etc., the logic diagram is shown in figure 8.

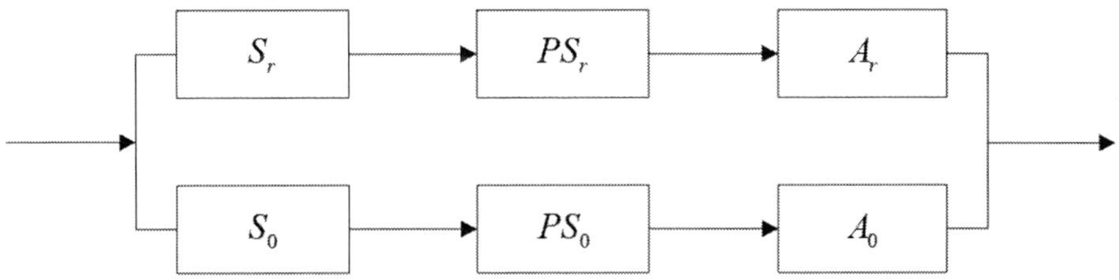

Figure 8. Logic diagram of integrated redundant design.

$$R_2 = 1-(1-R_{S_0}R_{PS_0}R_{A_0})(1-R_{S_r}R_{PS_r}R_{A_r}) \tag{11}$$

We will explore which design method is better.

$$R_1 = R_S R_{PS} R_A = (R_{S_0} + R_{S_r} - R_{S_0}R_{S_r})(R_{PS_0} + R_{PS_r} - R_{PS_0}R_{PS_r})(R_{A_0} + R_{A_r} - R_{A_0}R_{A_r}) \tag{12}$$

$$R_2 = 1-(1-R_{S_0}R_{PS_0}R_{A_0})(1-R_{S_r}R_{PS_r}R_{A_r}) = R_{S_0}R_{PS_0}R_{A_0} + R_{S_r}R_{PS_r}R_{A_r} - R_{S_0}R_{PS_0}R_{A_0}R_{S_r}R_{PS_r}R_{A_r} \tag{13}$$

$$\begin{aligned} R_1 - R_2 &= (R_{S_0} + R_{S_r} - R_{S_0}R_{S_r})(R_{PS_0} + R_{PS_r} - R_{PS_0}R_{PS_r})(R_{A_0} + R_{A_r} - R_{A_0}R_{A_r}) \\ &\quad -(R_{S_0}R_{PS_0}R_{A_0} + R_{S_r}R_{PS_r}R_{A_r} - R_{S_0}R_{PS_0}R_{A_0}R_{S_r}R_{PS_r}R_{A_r}) \\ &= R_{S_r}R_{A_0}(1-R_{S_0})[R_{PS_0}(1-R_{PS_r}) + R_{PS_r}(1-R_{A_r})] \\ &\quad + R_{S_0}R_{PS_r}(1-R_{PS_0})[R_{A_0}(1-R_{A_r}) + R_{A_r}(1-R_{S_r})] \\ &\quad + R_{PS_0}R_{A_r}(1-R_{A_0})[R_{S_0}(1-R_{S_r}) + R_{S_r}(1-R_{PS_r})] \end{aligned} \tag{14}$$

Because $0 < R_{S_0} < 1$, $0 < R_{S_r} < 1$, $0 < R_{PS_0} < 1$, $0 < R_{PS_r} < 1$, $0 < R_{A_0} < 1$ and $0 < R_{A_r} < 1$, we can get $R_1 - R_2 > 0$. It means that the design method of step-by-step redundant design is better.

8. Conclusions

The purpose of redundant design is to pre-solve a series of problems that may appear in subsequent design upgrading, manufacturing, assembly and operation by adding some redundant information and structures during the design phase to give components some functional structures that exceed their basic design requirements. It could achieve design optimization at the product level or during the product life cycle to increase product competitiveness.

References

[1] Zhang Z W 2017 Redundancy design and reliability analysis of airborne electronic equipment *Electro-optic Technology Application* **32** 66–69
[2] Wang Z Q and Gu X S Redundancy design on flight controller *Journal of East China University of Science and Technology (Natural Science Edition)* **35** 158–162
[3] Jiao Z 2018 Theoretical and experimental research on the seismic redundancy design of reinforced concrete frame structure building [D] China Academy of Building Research
[4] Sun W W and Zhou H 2008 Effective redundant design *Electronic Product Reliability and Environmental Testing* **26** 47–50
[5] Sun H Y 2007 Study on the relationship between redundancy design technique and reliability *Chinese Journal of Scientific Instrument* **28** 2089–92

IOP Conf. Series: Materials Science and Engineering **887** (2020) 012044 doi:10.1088/1757-899X/887/1/012044

Numerical and Experimental Research on the Anti-sloshing Characteristic of Swash Bulkhead in LNG Independent Type C Tank

Ge Liu[1], Qimu Wen[1], Yi Yang[1] and Chong Wang[2]

[1] System engineering research institute, Beijing 100094, China.

[2] College of Computer Science and Technology, Jilin University, Changchun 130012, China.

E-mail: SERI_Jiasong@163.com

Abstract. The sloshing induced by partial loaded in LNG (Liquified Natural Gas) independent type C tank will threat the structure of tank. For sloshing suppression, installation of swash bulkhead is a good choice. With model experiment, the sloshing suppression mechanism of perforated swash bulkhead is investigated. Comparing to the tank without swash bulkhead, the suppression efficiency is obtained. Base on the CFD (Computational fluid dynamic) theory, a numerical method is used to sloshing simulation of tank with perforated swash bulkhead, which is verified by the model experiment. Thus, with the same permeation ratio, the effect of anti-sloshing is analysed numerically with comparison of ring baffle and perforated bulkhead. It is found that perforated swash bulkhead has a better performance in sloshing suppression at higher filling level than the ring baffle. And there is a small variation in anti-sloshing efficiency of perforated swash bulkhead

1. Introduction

Liquid sloshing poses a great threat to the stability and structural strength of LNG vessel and tank. The loading limit and installation of swash bulkhead are employed to sloshing suppression in engineering. However, for making LNG tank in good working conditions, it has become a common method to use sloshing suppression device for restraining the motion of liquid and reducing the hazard caused by sloshing[1]. The perforated swash bulkhead is widely used in LNG independent type C tank due to its excellent performance in anti-sloshing[2]. The ring baffle is another choice of sloshing suppression device, which has advantage of simplicity of installation[3] and structure. Usually, the additional strengthening of independent type C tank, like T-bulb can act as the ring baffle.

There are several investigations, which is performed on anti-sloshing devices. Cho and Kim[4] investigated theoretically and experimentally the porous baffle in rectangular tank. It was found that the dual vertical porous baffles can significantly suppress sloshing motions when properly designed by selecting optimal porosity, submergence depth, and installation position. The optimal solidity ratio was given by Wei et al. with investigating the porous baffle in 2D rectangular tank under horizontal excitation. Liu et al.[5] had researched the parameters sensitivity of ring baffle in independent type C tank, including height, position installation, inclined angle, and thickness. It was found that in comparison to other parameters of ring baffle, increasing the height of ring baffle has the greatest influence on anti-sloshing.

Content from this work may be used under the terms of the Creative Commons Attribution 3.0 licence. Any further distribution of this work must maintain attribution to the author(s) and the title of the work, journal citation and DOI.
Published under licence by IOP Publishing Ltd

Due to producing complex flow with installation of perforated swash bulkhead, the experimental study could be helpful to analyse the characteristic of sloshing suppression and validate the numerical method. In this paper, the model experiment is adopted to investigate the sloshing suppression mechanism of perforated swash bulkhead. The numerical method, which is based on CFD theory is employed to compute the sloshing pressure in the tank with perforated swash bulkhead, which is verified by model experiment. Then the ring baffle and perforated swash bulkhead is compared numerically by efficiency of sloshing suppression on the basis of same permeation ratio.

2. Experimental setup

2.1. Experimental Device

The independent type C tank is employed to investigated here. The volume of the tank is 1000 m^3[6]. The ratio of the model tank to the full-scale tank is 1/17.8. The Fig.1 shows main parameters of the model tank: the total length (L_t) is1.910m and inner diameter (D_i) is 0.380m. The test model tank includes main tank and head (see Figure 1). Thickness of the tank is 10 mm. The material of test model tank is plexiglas pipeline and PVC (polyvinyl chloride). The model tank has several holes for measuring sloshing pressure (shown as Figure 2). The experimental fluid is the water, which is dyed by red water-solubility.

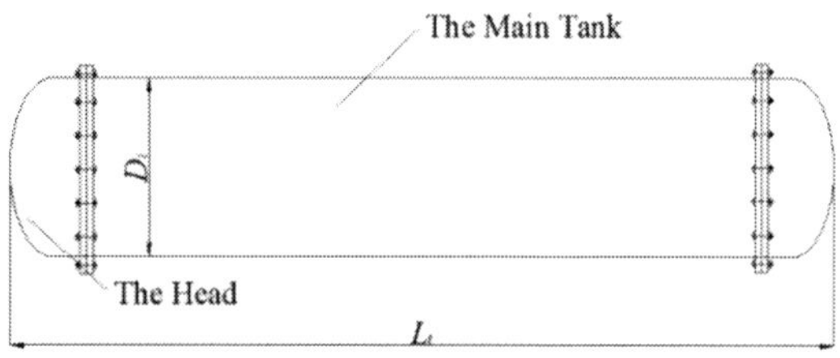

Figure 1. Numerical model for sloshing simulation

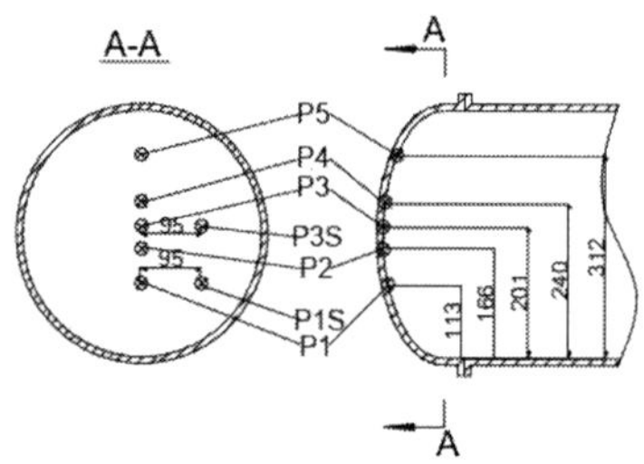

Figure 2. Arrangement of pressure senor (unit: mm)

Perforated swash bulkhead (shown as Figure 3) can be set into the test model tank. The perforated swash bulkhead of this experiment has 21 swash holes which is uniform distributed. The diameter of swash holes is 30 mm. Structural parameters of the perforated swash bulkhead is shown as Table 1, where permeation ratio is the ratio of area of swash hole to the whole area of bulkhead.

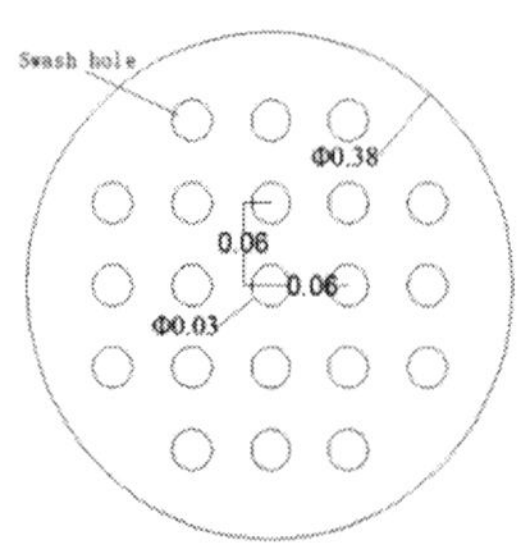

Figure 3. Model of perforated swash bulkhead (unit: m)

Table 1. Structural parameters of perforated swash bulkhead

Parameters	value
Outer diameter of bulkhead/m	0.38
Diameter of swash hole/m	0.03
Quantity of swash holes	21
Area of swash hole/m^2	1.58×10^{-2}
Permeation ratio/%	13.09

The excitation motion for sloshing simulation is driven by a 6-DOFs (degrees of freedom) motion platform. Flush-mounted pressure sensors are employed to achieve loads induced by sloshing. Its

IOP Conf. Series: Materials Science and Engineering **887** (2020) 012044 doi:10.1088/1757-899X/887/1/012044

measuring ranges are 0–10 kPa. The accuracy of them 0.2% FS (full scale). The free-surface shape is traced by high-speed camera, which can record at 50 pictures a second.

2.2. Test case

It can be achieved that for this kind of shape of model tank longitudinal excitation induces sloshing easily. And for a ship, pitch is one of the most common situations[7]. In this study, pitch is selected to excite the test model tank. Three test cases are chosen to analyse the suppression sloshing mechanism of perforated swash bulkhead. Filling level, excitation frequency and excitation amplitude constitute the test case. Test cases are listed in Table 2, where f_n is the natural frequency of liquid on each filling level. The natural frequency is obtained by analysed the free oscillation of experimental fluid with the Fast Fourier transform.

Table 2. The frequency domain of dangerous sloshing condition

Case	Filling level	Excitation frequency	Excitation amplitude /rad
No. 1	$0.40D_i$	f_n (0.28 Hz)	0.052
No. 2	$0.60D_i$	f_n (0.36 Hz)	0.052
No. 3	$0.80D_i$	f_n (0.48 Hz)	0.087

3. Analysis of experimental result

3.1. Free surface shape

The free surface shape of liquid in test model tank are shown as Figure 4~6. It can be found that the perforated swash bulkhead could suppress the motion of liquid significantly. The motion of test model tank is in a quarter period. The excitation angular velocity reaches its maximum. The liquid in the tank, in which the swash bulkhead isn't set, flows with in the form of a bore[8]. At the filling level of $0.8D_i$, the liquid flows to the top of head and break into droplets after swirling. After installing the perforated swash bulkhead, the liquid will flow mildly. The wave transformed into the standing wave. The fluctuation of free-surface is small. And there will not form the height difference between liquid in two sides of tank.

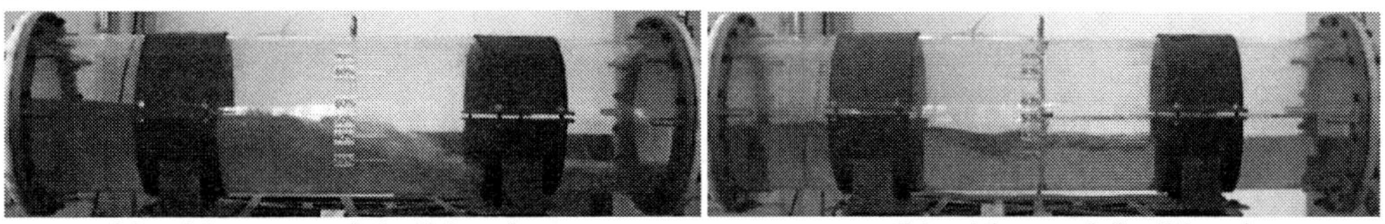

(a) without swash bulkhead (b) with swash bulkhead

Figure 4. Comparison of free surface shape (case No.1)

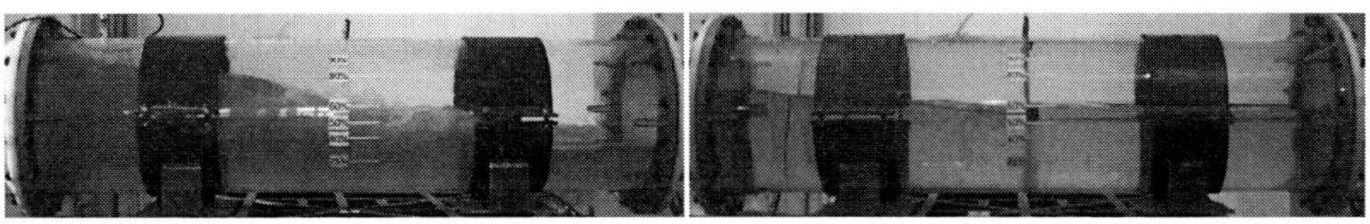

(a) without swash bulkhead (b) with swash bulkhead

Figure 5. Comparison of free surface shape (case No.2)

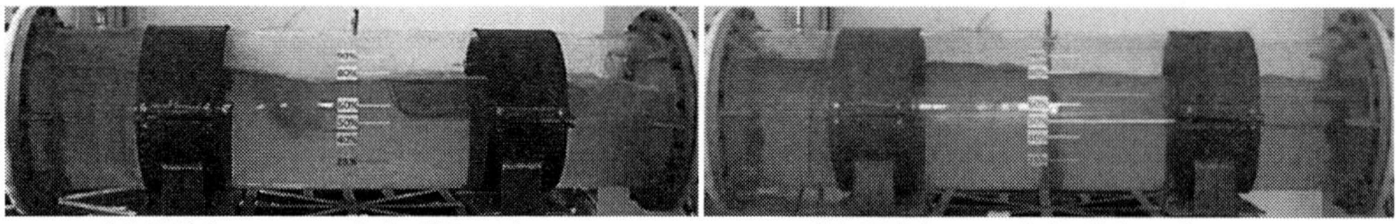

(a) Without swash bulkhead (b) With swash bulkhead

Figure 6. Comparison of free surface shape (case No.3)

3.2. Sloshing pressure

The monitoring points P1S, P2, P3, P4, P5 are selected to compare the sloshing pressure in the tank with or without swash bulkhead. These points are located near the free-surface of each filling level. The sloshing pressure of each points are drawn in Figure 7~9, where P_{WBH} and P_{BH} are the sloshing pressure in normal tank and in tank installed swash bulkhead, respectively. It can be seen that the curve of sloshing pressure in tank with swash bulkhead has a single peak. While in normal tank, there is a double peak in every period, especially, at the higher filling level. As the excitation frequency is close to the natural frequency, the impact pressure due to the violent motion of the free surface will be dominant and the first peak will be formed. Then, when the tank moves to the horizontal position, the liquid will push the tank head due to inertial force and gravity. Thus, it will produce the other peak[5]. Therefore, due to the installation of swash bulkhead, the motion of the free surface is suppressed significantly.

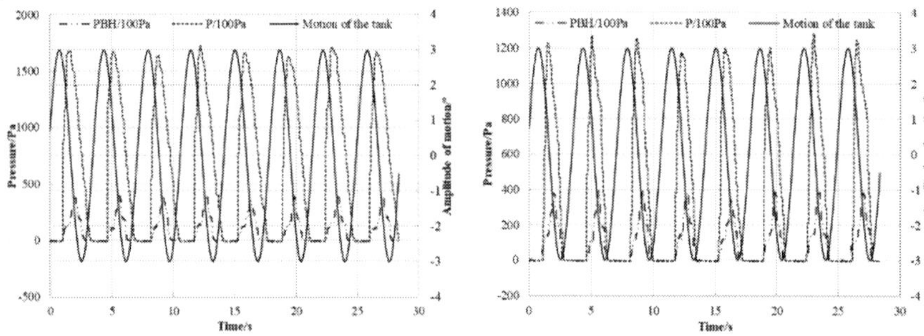

(a) P1S (b) P2

Figure 7. Comparison of anti-sloshing efficiency (case No.1)

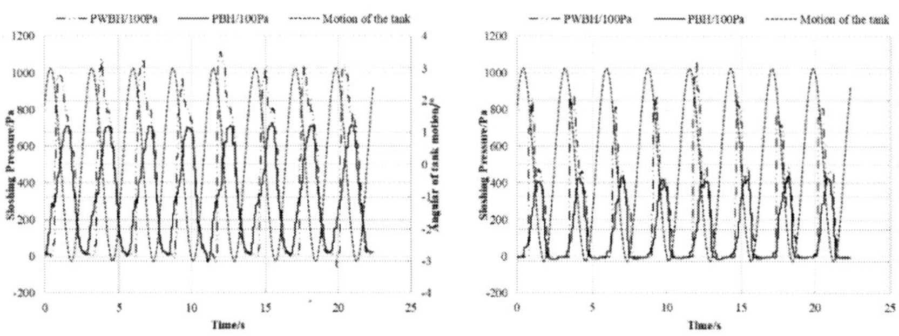

(a) P3 (b) P4

Figure 8. Comparison of anti-sloshing efficiency (case No.2)

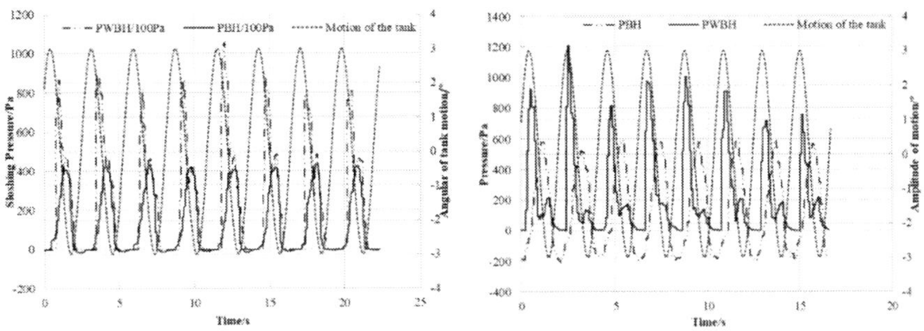

(a) P4 (b) P5

Figure 9. Comparison of anti-sloshing efficiency (case No.3)

The mean and maximum pressure in sloshing pressure are calculated (shown as Table 3). It can be concluded that the perforated swash bulkhead performed the best at the filling level of $0.4D_i$. Compared to the sloshing pressure in the normal tank, the maximum value and mean value of sloshing pressure decrease at 69.99% and 76.45%, respectively. Meanwhile, the pressure of monitoring points at free-surface reduce most compared with other monitoring points. The bulkhead increases the damping of liquid flow, which restricts the motion of the free surface more than other zones in tank.

Table 3. Quantitative analysis of anti-sloshing performance

Case No.1	Sloshing pressure/Pa			
	P_{max}		P_{mean}	
	P1S	P2	P1S	P2
Tank without swash bulkhead	1749.94	1319.17	643.36	362.62
Tank with swash bulkhead	871.49	395.83	460.34	85.39
Efficiency of suppression/%	50.20	69.99	28.45	76.45
Case No.2	P3	P4	P3	P4
Tank without swash bulkhead	1176.14	1033.33	411.68	241.23
Tank with swash bulkhead	735.16	453.33	332.41	123.19
Efficiency of suppression/%	37.49	56.13	19.25	48.93
Case No.3	P4	P5	P4	P5
Tank without swash bulkhead	2499.01	1801.67	848.50	232.60
Tank with swash bulkhead	1343.19	628.41	695.22	135.64
Efficiency of suppression/%	46.25	65.12	18.06	41.68

4. Numerical analysis of perforated swash bulkhead and ring baffle

4.1. Numerical method and model

The lengthwise direction of tank and direction of gravity is set as x and y coordinate, respectively. Thus, the tank rotates about the z axis. The numerical model can be described as Figure 10.

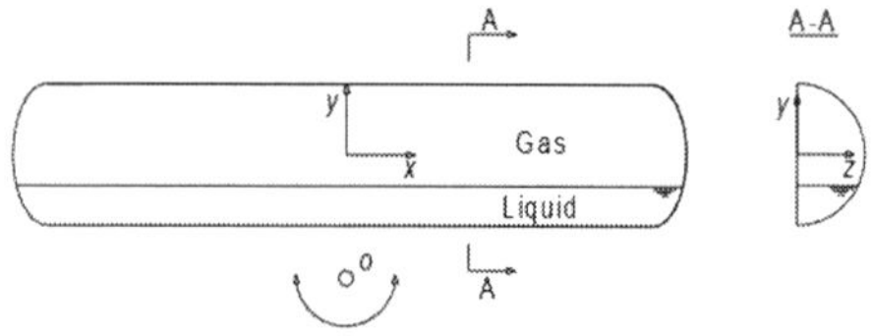

Figure 10. Numerical model for sloshing simulation

The viscous fluid sloshing is subject to a continuity equation and the Navier-Strokes equation. The ideal gas is employed to simulate the air in the tank. Hence, the governing equation also includes the conservation of energy equation. So the governing equations can be listed as follows.

$$\frac{\partial \rho}{\partial t}+\frac{\partial \rho u_i}{\partial x_i}=0, \tag{1}$$

$$\frac{\partial}{\partial t}(\rho u_i)+\frac{\partial}{\partial x_j}(\rho u_i u_j)=-\frac{\partial p}{\partial x_j}+\frac{\partial}{\partial x_j}\left[\mu\left(\frac{\partial u_i}{\partial x_j}+\frac{\partial u_j}{\partial x_i}-\frac{2}{3}\delta_{ij}\frac{\partial u_l}{\partial x_l}\right)\right]+\frac{\partial}{\partial x_j}\left(-\rho\overline{u_i'u_j'}\right), \tag{2}$$

$$\frac{\partial}{\partial t}\rho\left(e+\frac{u_i^2}{2}\right)+\nabla\cdot\rho u_i\left(e+\frac{u_i^2}{2}\right)=\nabla\cdot\left(k_{\mathrm{eff}}\nabla T\right)+\nabla\cdot\left(\mu\cdot u_i\right)+\rho f_i u_i+S_{\mathrm{E}}, \tag{3}$$

Where p is the time averaged pressure; μ is the dynamic viscosity of water; δ_{ij} is the Kronecker delta; $-\rho\overline{u_i'u_j'}$ is the Reynolds stress term; subscripts i, j, and l are the coordinate directions. e is the internal energy, k_{eff} is the effective thermal conductivity, ∇T is the temperature gradient, f_i is the body force vector, and S_{E} is the source term.

For reducing the computational expense, the realizable k-ε turbulence model is employed. The turbulent kinetic energy k and its rate of dissipation ε are achieved by the following equation.

$$\frac{\partial}{\partial t}(\rho k)+\frac{\partial}{\partial x_j}(\rho k u_j)=\frac{\partial}{\partial x_j}\left[\left(\mu+\frac{\mu_t}{\sigma_k}\right)\frac{\partial k}{\partial x_j}\right]+f_c G_k+G_b-\rho(\varepsilon-\varepsilon_0)-\gamma_M+S_k, \tag{4}$$

$$\frac{\partial}{\partial t}(\rho\varepsilon)+\frac{\partial}{\partial x_j}(\rho\varepsilon u_j)=\frac{\partial}{\partial x_j}\left[\left(\mu+\frac{\mu_t}{\sigma_\varepsilon}\right)\frac{\partial\varepsilon}{\partial x_j}\right]+f_c C_{\varepsilon 1}S\varepsilon+\frac{\varepsilon}{k}\left(C_{\varepsilon 1}C_{\varepsilon 3}G_b\right)-\frac{\varepsilon}{k+\sqrt{v\varepsilon}}C_{\varepsilon 2}\rho(\varepsilon-\varepsilon_0)+S_\varepsilon, \tag{5}$$

where μ_t is the eddy viscosity coefficient; G_k is the turbulent production term; f_c is the curvature correction factor; G_b is the buoyancy production term; ε_0 is the ambient turbulence value in the source terms that counteracts turbulence decay; γ_M is the compressibility modification term; S_k and S_ε are user-specified source terms. $C_{\varepsilon 1}$, $C_{\varepsilon 2}$, and $C_{\varepsilon 3}$ are the constant coefficients. σ_k and σ_ε are the turbulent Prandtl numbers for k and ε, respectively. S is the modulus of the mean strain rate tensor. v is the kinematic viscosity.

The volume of fraction (VOF) method is employed to capture the free surface. According to the numerical model, the fluid is divided into two phases, each of which can be written as the following equation.

$$\frac{\partial\alpha_q}{\partial t}+\nabla\cdot\left(\alpha_q u_i\right)=0 \tag{6}$$

where α_q is the qth phase, and all phases are based on the following constraint.

$$\sum_{q=1}^{2}\alpha_q=1 \tag{7}$$

4.2. Verification

The numerical method is verified by comparing the free-surface shape and pressure time history of sloshing in experiment. The condition is selected, in which the test model tank is installed with perforated swash bulkhead. The verification case is No.2, the parameters of which is listed in Table 2. The Cartesian grid is used to mesh the model (see Figure 11). The refinement process is applied to the boundary, free-surface and perforated swash bulkhead. The half model is established and the vertical section is set as symmetrical. After the analysis of grid and time step independence, the quantity of grid and time step is selected as 637 thousand and 0.002s, respectively.

Figure 11. Gird model of the tank

The results of verification are shown in Figure 12 and Table 4. It is found that the numerical profile of the free surface generally coincided well with the experimental profile. Likewise, the curves of pressure time history show good agreement in the case validated.

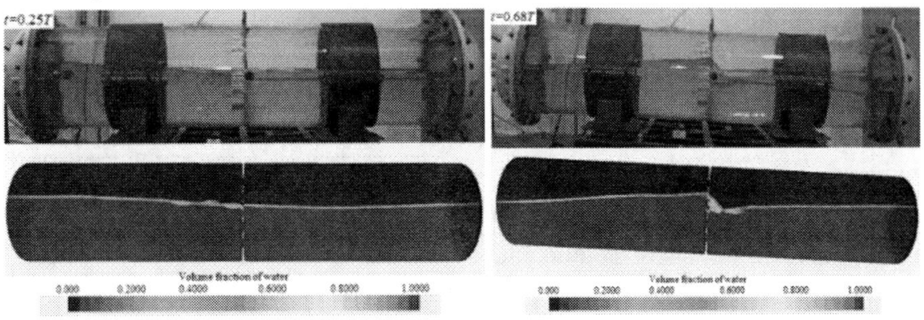

Figure 12. Comparison of free surface shape

4.3. Comparison of anti-sloshing performance

Based on the same permeation area (Table 2), the structural parameters of ring baffle is determined. The comparison of structural parameters is listed in Table 4, where the permeation ratio is the ratio of permeation area and the whole area of swash bulkhead. The ring baffle is also installed in the middle of the tank.

Table 4. Comparison of structural parameters

Form of swash bulkhead	Diameter of swash hole/m	Height of ring baffle/D_i	Permeation area/m^2	Permeation ratio/%
Perforated	0.4	-	2.64×10^{-2}	23.27
Ring baffle	-	0.31	2.66×10^{-2}	23.45

The efficiency of swash bulkhead is taken from the definition of the reference[3]. However, in this paper the water height variation in the equation is replaced by the horizontal force on the head due to liquid sloshing. Thus, the efficiency of the ring baffle is defined as

$$\eta_{\text{anti-sloshing}} = \frac{\sigma_{\text{unbaffled}} - \sigma_{\text{baffled}}}{\sigma_{\text{unbaffled}}}, \tag{8}$$

where σ is the square of deviation of force in x direction at the left head, and the subscript of σ specifies sloshing condition of tank with or without swash bulkhead.

The force time history on the head of the tank in the x direction is achieved in numerical simulation. Finally, the anti-sloshing efficiency of two kinds of swash bulkhead is listed in Table 5, and the corresponding curve graph is shown as Figure 13. It can be found that the sloshing suppression performance of ring baffle is better than perforated swash bulkhead, when the filling level is lower than $0.5D_i$. However, the anti-sloshing efficiency of the two swash bulkheads is almost the same under the filling level of $0.25D_i$. When the filling level exceed the half loaded, the sloshing suppression efficiency of ring baffle will decrease. And the anti-sloshing efficiency of ring baffle has a larger variation on each filling level. The lack of fluid motion suppression in middle of ring baffle is the main reason.

Table 5. Comparison of anti-sloshing efficiency

Filling level/D_i	σ_{WBH}	Perforated		Ring baffle	
		σ_{BH}	$\eta_{anti-slo}$/%	σ_{RB}	$\eta_{anti-slo}$/%
0.25	7.80	3.28	57.95	2.59	59.09
0.40	17.67	8.79	50.24	3.70	53.72
0.60	27.21	16.27	40.23	12.58	36.64
0.80	36.97	19.69	46.73	27.34	26.93

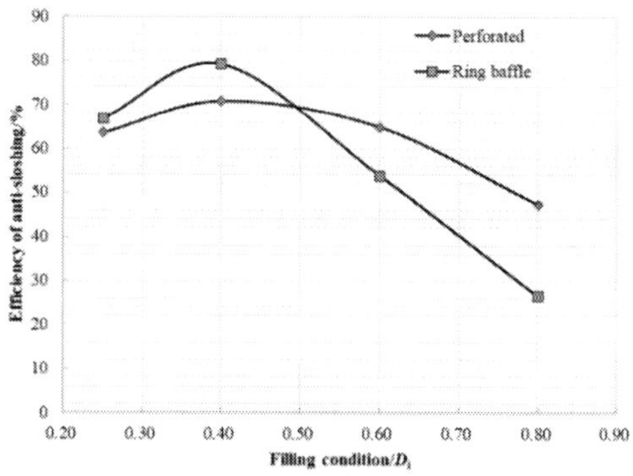

Figure 13. Graph of anti-sloshing efficiency

5. Conclusion

With model experiment, the sloshing suppression mechanism of perforated swash bulkhead is investigated. It can be proved that the perforated swash bulkhead could well restrain the motion free-surface and the impact force of fluid is decrease significantly. The bulkhead performed best under the filling level of $0.4D_i$. The sloshing pressure near the free-surface is reduced by an average of 76.45%. Based on the same permeation ratio, the effect of anti-sloshing is analysed numerically with comparison of ring baffle and perforated bulkhead. With comparison of free-surface shape and sloshing pressure in the experiment, the numerical method is verified to have a good accuracy in anti-sloshing simulation.

The characteristic of anti-sloshing of ring baffle and perforated swash bulkhead is analysed. It can be concluded that ring baffle could suppress fluid sloshing better in the lower filling level. However, the perforated swash bulkhead has a good performance in sloshing suppression at all filling level. Thus, it is suggested that perforated swash bulkhead could be installed where the tank has a great change in filling ratio, like fuel bunker vessel.

Reference

[1] Zhijun, W., 2015 Study on Sloshing and Slamming Load in the Tank Design of the FLNG System Study on sloshing and slamming load in the tank design of the FLNG system, (*Dalian university of technology*)

[2] Lloyd, N., Vaiciurgis, E., Langrish, T.A.G. 2002 The Effect of Baffle Design on Longitudinal Liquid Movement in Road Tankers: An Experimental Investigation (*Process Safety and Environmental Protection* vol 80) Chapter 4 pp 181-185.

[3] Kim, H., Lee, Y. 2008 Optimization Design Technique for Reduction of Sloshing by Evolutionary Methods *Journal of Mechanical Science and Technology* vol 22) Chapter 1 pp 25-33.

[4] LIU, G., LIN, Y., GUAN, G., YU, Y. 2018 Numerical Research On the Anti-Sloshing Effect of

a Ring Baffle in an Independent Type C LNG Tank Numerical research on the anti-sloshing effect of a ring baffle in an independent type C LNG tank (*Journal of Zhejiang University-Science A(Applied Physics & Engineering)* vol 19)Chapter 10 pp 758-773.

[5] LIU, G., LIN, Y., GUAN, G., LI, B. 2017 Experimental Study of Sloshing Pattern On LNG Independent C Type Tank Experimental study of sloshing pattern on LNG independent C type tank (*Journal of Dalian University of Technology* vol 57) Chapter 05 pp 467-475.

[6] Jin, X., Li, L., Wang, D. Jin, X., Li, L., Wang, D. 1998 On Model Experiment of Sloshing in Tanks On Model Experiment of Sloshing in Tanks (*Journal of Shanghai Jiaotong University Journal of Shanghai Jiaotong University*) Chapter 11 pp 116-119.

[7] Olsen, H.A., Johnsen, K.R., Norske, V.O.R.D., 1975. Nonlinear Sloshing in Rectangular Tanks: A Pilot Study on the Applicability of Analytical Models. Det Norske Veritas, Research Department.

[8] Liu, G., Lin, Y., Guan, G., Wang, C. 2017 Experimental Study On Frequency Domain Resonant Characteristic of Sloshing in LNG Independent Type C Tank Experimental study on frequency domain resonant characteristic of sloshing in LNG independent type C tank (*Journal of Zhejiang University (Engineering Science)* vol 51) Chapter 12 pp 2392-2398.

Bearing-only Trajectory Estimation for Ballistic Target Using Sparse Representation

Pu Wang[1,4] Yun Cheng[2] and Qinghong Yin[3]

[1] Colleges of Electronic Science and Technology, National University of Defense Technology , Changsha city, Hunan province, 410073,China
[2] Xichang Satellite Launch Center, Xichang city, Sichuan province, 615000, China
[3] Chengdu Water Resources Bureau Water Conservation Supervision Center, Chengdu city, Sichuan province, 610000, China
[4] Author to whom any correspondence should be addressed

Email:wp421189@nudt.edu.cn

Abstract. This paper discusses the problem of estimating the trajectory of a ballistic target using bearing measurements from one passive sensor. The major challenge of this problem is the ill-conditioning of the estimation problem due to poor observability of the target motion via bearing measurements. We present an estimator based on the sparse representation algorithm for one sensor scenario. The sparse representation can dramatically reduce the amount of data needed to be estimated. With sparse representation, the trajectory of the ballistic target can be estimated in poor observability conditions. Besides, we introduce some complementary constraints to improve the observability of target motion. Simulation results indicate that the proposed estimator is effective.

1. Introduction

The plume generated by a ballistic target makes it quite visible to most optical sensor. With the measurements from optical sensor, we can estimate the flight of ballistic target.

At present, accurate estimation of the trajectory of ballistic target needs bearing measurements from multiple observers. However, in many cases, the ballistic target is observed by only one sensor. Therefore, the study of trajectory estimation based on bearing measurements from one sensor has important value[1].

The bearings-only methods used to estimate the trajectory of a target can be broadly classified as being either profile-free or profile-based models. In profile-free methods, the state vector is estimated using a polynomial motion model. In profile-based methods, it is assumed that the magnitude and orientation of the target thrust vector, thus the acceleration, is known. However, the accurate profile of ballistic target is very difficult to obtain. In many cases, we have to depend on profile-free methods which are hard to obtain accurate estimation[1,2,3].

To solve this problem, we present a trajectory estimation method for ballistic target based on bearing measurements and there is no knowledge of the profile of target. Besides, we assume that the measurements from only one sensor , which means there are only angle measurements (or bearing-only measurements). We introduce sparse-representation into our estimator. The most common method focuses on estimating the status of ballistic target on every time point. In view of sparse-representation, it is unnecessary. Based on the sparse representation of trajectory, we only need to

Content from this work may be used under the terms of the Creative Commons Attribution 3.0 licence. Any further distribution of this work must maintain attribution to the author(s) and the title of the work, journal citation and DOI.
Published under licence by IOP Publishing Ltd

estimate the sparse representation coefficient. With these coefficients, it is easy to recovery trajectory. This enables our estimator with bearing-only measurements to achieve acceptable accuracy.

2. Measurements and sparse representation

2.1. Angle measurements

The geometry relation between target and sensor is showed in figure 1.

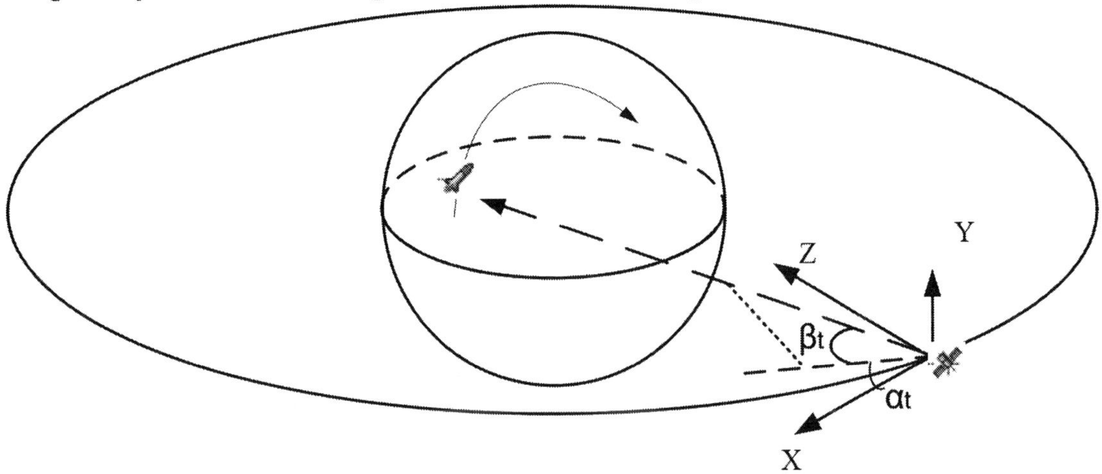

Figure 1. Target, observer geometry

In practice, the measurements of targets contain azimuth bearing α_t and elevation angle β_t measurements at moment *t*.

Let $[x, y, z]^T_{body}$ represent the coordinate of target in satellite coordinate system. $[x, y, z]^T_{target}$ represent coordinate of target in Earth-Centred Inertial (ECI) coordinate system and $[x, y, z]^T_{satellite}$ represent the position of satellite in ECI coordinate system. The relationship among them is listed below:

$$\begin{bmatrix} x \\ y \\ z \end{bmatrix}_{body} = \mathbf{R}^{Body}_{ECI} \bullet \left(\begin{bmatrix} x \\ y \\ z \end{bmatrix}_{target} - \begin{bmatrix} x \\ y \\ z \end{bmatrix}_{satellite} \right) \quad (1)$$

$\mathbf{R}^{Body}_{ECI}$ represent the transformation matrix which convert the position of target in ECI coordinate system to satellite coordinate system. Given the method to obtain $\mathbf{R}^{Body}_{ECI}$ isn't the core of this paper, we don't discuss it in detail. The detail of it is introduced in ref[4].

These position are related to the angle measurements by

$$\begin{bmatrix} 0 \\ 0 \end{bmatrix} = \begin{bmatrix} 1 & 0 & -\sin\beta_t \cot\alpha_t \\ 0 & 1 & -\cos\beta_t \cot\alpha_t \end{bmatrix} \bullet \begin{bmatrix} x \\ y \\ z \end{bmatrix}_{body} \tag{2}$$

The equation (2) is proved in article[3] .

2.2. Sparse representation of trajectory

In practical applications, most of the natural signals can be sparse represented. The sparse representation can expressed as below:

$$\mathbf{f} = \mathbf{\Psi w}$$

The coefficient $\mathbf{w}$ is expressed as sparseness of signal $\mathbf{f}$ in the sparse basis $\mathbf{\Psi}$. The relationship of them is shown in figure 2.

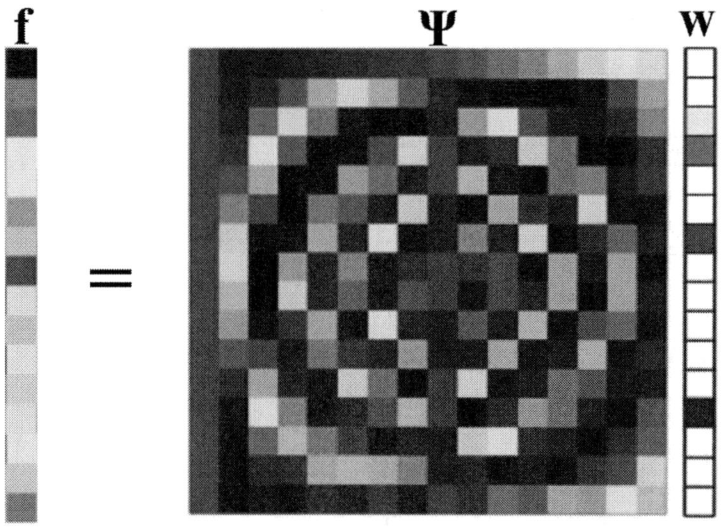

Figure 2. the sparse representation of signal

Most elements of $\mathbf{w}$ are zeros or minor. To estimate signal $\mathbf{f}$, we only needed to estimate the significant elements in $\mathbf{w}$. Sparse representation enable us to reconstruct the original signal with a few observation[5,6].

The trajectory of ballistic target can be represented by a sparse basis. In this paper, we construct a sparse basis of trajectory based on polynomials model. Take the rocket trajectory as an example. In an ideal situation, the space coordinates of the target at time t are $P_t:(x_t, y_t, z_t)$. The track can be expressed as:

$$\begin{aligned} \mathrm{x}_t &= a_0 + a_1 t + a_2 t^2 + a_3 t^3 + a_4 t^4 + \cdots \\ \mathrm{y}_t &= b_0 + b_1 t + b_2 t^2 + b_3 t^3 + b_4 t^4 + \cdots \\ \mathrm{z}_t &= c_0 + c_1 t + c_2 t^2 + c_3 t^3 + c_4 t^4 + \cdots \end{aligned} \tag{3}$$

It should be noted that there are many models describing the target trajectory. The polynomial is just one among them. However, the ballistic target model is not the major of this paper. For the convenience of discussion, this paper chooses a simple polynomial model. In fact, there are other better and more complex models in the sea, which will be introduced in subsequent articles.

The observation of the entire target motion trajectory can be turned to an estimate of the trajectory coefficient. The sparse parameters of trajectory are considered to be fixed during the flight of target. According to this, it can be expressed that the matrix in the middle is a sparse representation matrix, which is completely determined by time and is completely known.

$$\begin{bmatrix} x_t \\ y_t \\ z_t \end{bmatrix} = \begin{bmatrix} 1 & t & t^2 & t^3 & t^4 & \cdots & 0 & 0 & 0 & 0 & 0 & \cdots & 0 & 0 & 0 & 0 & 0 & \cdots \\ 0 & 0 & 0 & 0 & 0 & \cdots & 1 & t & t^2 & t^3 & t^4 & \cdots & 0 & 0 & 0 & 0 & 0 & \cdots \\ 0 & 0 & 0 & 0 & 0 & \cdots & 0 & 0 & 0 & 0 & 0 & \cdots & 1 & t & t^2 & t^3 & t^4 & \cdots \end{bmatrix} \bullet \begin{bmatrix} a_0 \\ a_1 \\ \cdots \\ c_n \end{bmatrix} = \mathbf{\Phi}_t \bullet \begin{bmatrix} a_0 \\ a_1 \\ \cdots \\ c_n \end{bmatrix} = \mathbf{\Phi}_t \bullet \mathbf{w} \tag{4}$$

Where $\mathbf{\Phi}_t$ is the sparse representation matrix of time t, $\mathbf{w}$ representing the trajectory parameters to be estimated.

3. Estimation method for trajectory

The Scheme of proposed method is showed in figure 3.

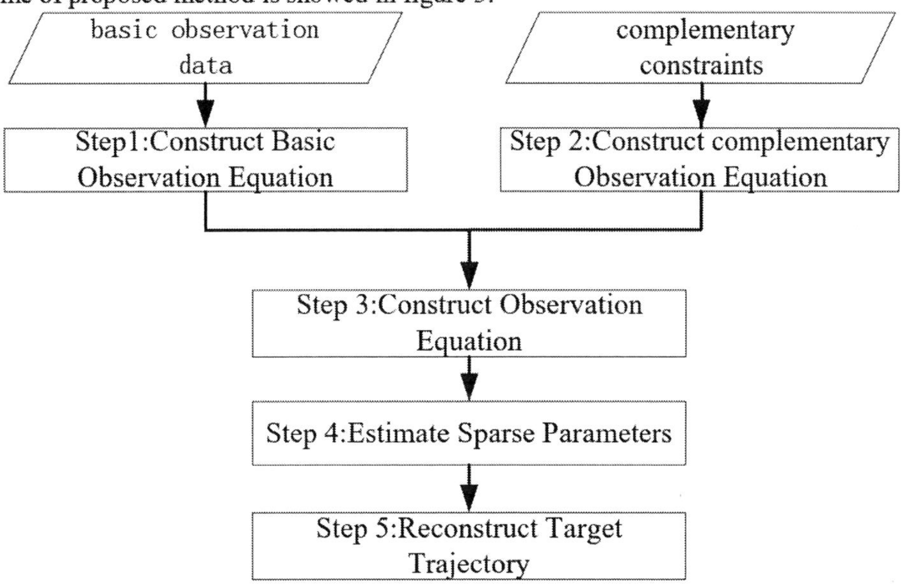

Figure 3. scheme of proposed method

First, we construct the basic observation equation from basic observation data, such as time, the angles of bearing, the positions of the satellite , and the orbit element of satellites. But the basic observation equation usually not enough to estimate the sparse coefficient. To solve this problem, we introduce some complementary constraints to construct a complementary constraint equation. By combining the two types of equations together, we can construct the observation equation. Then, based on the observation equations, we could estimate the sparse parameters. After obtaining the sparse parameters, we could reconstruct the target trajectory. Because steps 4 and 5 are based on classical methods and they are not the focuses of this paper, we don't present an explicit introduction of them in this paper. Only the steps 1-3 are discussed in detail.

3.1. Construct the basic observation equation

The observation angle is β_t , α_t , and represents the azimuth and elevation angles, respectively. Using these two observation angles, an observation equation can be constructed

$$\begin{bmatrix}0\\0\end{bmatrix}=\begin{bmatrix}1 & 0 & -\sin\beta_t\cot\alpha_t\\0 & 1 & -\cos\beta_t\cot\alpha_t\end{bmatrix}\bullet\mathbf{R}_{ECI}^{Body}\bullet\left(\mathbf{\Phi}_t\bullet\begin{bmatrix}a_x\\b_x\\\cdots\\e_z\end{bmatrix}-\begin{bmatrix}x_O(t)\\y_O(t)\\z_O(t)\end{bmatrix}\right) \tag{5}$$

Where ($x_O(t)$, $y_O(t)$, $z_O(t)$) represents the spatial coordinates of satellite at time t. To simplify equation (5), we introduce **b(t)** and **h(t)**, where:

$$\mathbf{b}(t)=\begin{bmatrix}1 & 0 & -\sin\beta_t\cot\alpha_t\\0 & 1 & -\cos\beta_t\cot\alpha_t\end{bmatrix}\bullet\mathbf{R}_{ECI}^{Body}\bullet\begin{bmatrix}x_O(t)\\y_O(t)\\z_O(t)\end{bmatrix} \tag{6}$$

$$\mathbf{h}(t)=\begin{bmatrix}1 & 0 & -\sin\beta_t\cot\alpha_t\\0 & 1 & -\cos\beta_t\cot\alpha_t\end{bmatrix}\bullet\mathbf{R}_{ECI}^{Body}\bullet\mathbf{\Phi}_t \tag{7}$$

With N observations, the observation equation can be constructed

$$\mathbf{B}=\begin{bmatrix}\mathbf{b}(t_1)\\\mathbf{b}(t_2)\\\cdots\\\mathbf{b}(t_n)\end{bmatrix}=\begin{bmatrix}\mathbf{h}(t_1)\\\mathbf{h}(t_2)\\\cdots\\\mathbf{h}(t_n)\end{bmatrix}\bullet\begin{matrix}a_0\\a_1\\\cdots\\c_n\end{matrix}=\mathbf{H}\bullet\mathbf{w} \tag{8}$$

The problem of ballistic target trajectory estimation is converted into a typical problem of estimating the sparse **representation coefficients** of the trajectory. The Basis Pursuit(BP) algorithm is adapted to recovery $\mathbf{w}$. The detail of BP is described in[7].

It had been proved that estimation of $\mathbf{w}$ is feasible when it's sparse and the $\mathbf{H}$ satisfy the Restricted Isometry Property(RIP)[5,6]. However, in the actual situation, the matrix is usually a singular matrix, and does not satisfy the RIP. Therefore, some auxiliary conditions is need to ensure the solvable row.

3.2. Construction complementary constrain equation

3.2.1 Known point constraints. The launch point and some intermediate point can obtained through bearing-only measurements.

1) Launch point constraint

There are plenty of papers devoted to estimate the launch point $\mathbf{P}_0$ of target through bearing-only measurements. However, it is not the major issue of paper. Therefore, we don't introduce the detail of it here. The method to obtain the Launch point is introduced in ref[1].

2) Intermediate point constraint

The tail-fire emission intensity of the ballistic target in the active section is stable. The observed fluctuation of the tail flame radiation is mainly determined by the atmospheric absorption. Using the atmospheric absorption model and the observed target tail flame intensity, the height of target can be estimated when it is passing atmosphere. Under the premise of knowing the height of the target, the positions($\mathbf{P}_L$,..., $\mathbf{P}_M$) of target can obtained. L , ... ,M represent the time point that target position is obtained. The detail of method to obtain $\mathbf{P}_L$,..., $\mathbf{P}_M$ is discussed in ref [8,9].

3.2.2 Coplanar constraints. In the course of flight, ballistic targets and launch vehicles are generally not orbiting in order to save fuel and ensure range. The trajectory of the ballistic target is usually in a plane. Therefore, the target trajectory coplanar constraint can be introduced to further reduce the singularity of the solution matrix. To describe a plane, we need a point $\mathbf{P}_c$ in the plane, a normal vector $\mathbf{V}_N$ which is perpendicular to the plane. It is assumed that $\mathbf{V}_N = \begin{bmatrix} x_{Vn} & y_{Vn} & z_{Vn} \end{bmatrix}^T$. The method to obtain $\mathbf{P}_c$ and $\mathbf{V}_N$ is presented below.

There are K known points and all belong to set C. $\mathbf{P}_c$ can be obtained by the following formula:

$$\mathbf{P}_c = \frac{1}{K} \sum_{P_i \in C} \mathbf{P}_i \tag{9}$$

The normal vector satisfies the following formula:

$$\begin{aligned} \mathbf{0} &= \begin{bmatrix} \mathbf{P}_0 - \mathbf{P}_C & \mathbf{P}_L - \mathbf{P}_C & \cdots & \mathbf{P}_M - \mathbf{P}_C \end{bmatrix}^T \bullet \mathbf{V}_N \\ &= \begin{bmatrix} \tilde{\mathbf{P}}_0 & \tilde{\mathbf{P}}_L & \cdots & \tilde{\mathbf{P}}_M \end{bmatrix}^T \bullet \mathbf{V}_N \end{aligned} \tag{10}$$

Where $\tilde{\mathbf{P}}_0$, $\tilde{\mathbf{P}}_L$, ..., $\tilde{\mathbf{P}}_M$ equals $\mathbf{P}_0$, $\mathbf{P}_L$, $\mathbf{P}_M$ minus the value $\mathbf{P}_c$.

While only the orientation of $\mathbf{V}_N$ is needed, the magnitude can be any value. For the convenience of discussion , the z_{Vn} last variable value is set to 1. The above formula can be rewritten as:

$$\mathbf{0} = \begin{bmatrix} x_{\tilde{\mathbf{P}}_0} & y_{\tilde{\mathbf{P}}_0} & z_{\tilde{\mathbf{P}}_0} \\ x_{\tilde{\mathbf{P}}_L} & y_{\tilde{\mathbf{P}}_L} & z_{\tilde{\mathbf{P}}_L} \\ & \cdots & \\ x_{\tilde{\mathbf{P}}_M} & y_{\tilde{\mathbf{P}}_M} & z_{\tilde{\mathbf{P}}_M} \end{bmatrix} \bullet \begin{bmatrix} x_{Vn} \\ y_{Vn} \\ 1 \end{bmatrix} \tag{11}$$

It is be equivalent to:

$$\mathbf{V}_z = \begin{bmatrix} -z_{\tilde{\mathbf{P}}_0} \\ -z_{\tilde{\mathbf{P}}_L} \\ \cdots \\ -z_{\tilde{\mathbf{P}}_M} \end{bmatrix} = \begin{bmatrix} x_{\tilde{\mathbf{P}}_0} & y_{\tilde{\mathbf{P}}_0} \\ x_{\tilde{\mathbf{P}}_L} & y_{\tilde{\mathbf{P}}_L} \\ & \cdots \\ x_{\tilde{\mathbf{P}}_M} & y_{\tilde{\mathbf{P}}_M} \end{bmatrix} \bullet \begin{bmatrix} x_{Vn} \\ y_{Vn} \end{bmatrix} = \mathbf{H}_{VN} \bullet \begin{bmatrix} x_{Vn} \\ y_{Vn} \end{bmatrix} \tag{12}$$

According to this, x_{Vn} , y_{Vn} can be determined by:

$$\begin{bmatrix} x_{Vn} \\ y_{Vn} \end{bmatrix} = (\mathbf{H}_{VN}^T * \mathbf{H}_{VN})^{-1} \mathbf{H}_{VN}^T * \mathbf{V}_z \tag{13}$$

At this point, the normal vector has been obtained $\mathbf{V}_N$.

3.2.3 Constrained equation. After obtaining the information of known point constraints and coplanar constraints, the complementary constrain equations can be constructed. The method of integration is very simple, and the constrain equations can be directly added behind the original observation matrix. The following equation can be constructed from known points

$$\begin{bmatrix} \mathbf{P}_0 \\ \mathbf{P}_L \\ \cdots \\ \mathbf{P}_M \end{bmatrix} = \begin{bmatrix} \mathbf{\Phi}_0 \\ \mathbf{\Phi}_L \\ \cdots \\ \mathbf{\Phi}_M \end{bmatrix} \bullet \mathbf{w} \tag{14}$$

Le $\mathbf{B}_1 = \begin{bmatrix} \mathbf{P}_0^T & \mathbf{P}_L^T & \cdots\cdots & \mathbf{P}_M^T \end{bmatrix}^T$ and $\mathbf{H}_1 = \begin{bmatrix} \mathbf{\Phi}_0^T & \mathbf{\Phi}_L^T & \cdots\cdots & \mathbf{\Phi}_M^T \end{bmatrix}^T$. Then the above formula can be expressed as:

$$\mathrm{B}_1 = \mathbf{H}_1 \bullet \mathbf{w} \tag{15}$$

It can be seen from the coplanarity that the position $\mathbf{P}_t$ of the track point at any time minus $\mathbf{P}_C$ the coplanar point is perpendicular to the normal vector $\mathbf{V}_N$, and the following equation can be obtained as follows:

$$0 = \mathbf{V}_N^T \bullet (\mathbf{P}_t - \mathbf{P}_C) = \mathbf{V}_N^T \bullet (\mathbf{\Phi}_t \bullet \mathbf{w} - \mathbf{P}_C) \tag{16}$$

It can rewritten in the following equation form:

$$\mathbf{V}_N^T \bullet \mathbf{P}_C = \mathbf{V}_N^T \bullet \mathbf{\Phi}_t \bullet \mathbf{w} \tag{17}$$

In the above formula, $\mathbf{V}_N^T \bullet \mathbf{P}_C$ is a constant that does not change with time. We use **D** to express it. This makes it possible to construct an equation of observation:

$$\mathbf{B}_2 = \begin{bmatrix} \mathrm{D} \\ \cdots \\ \mathrm{D} \end{bmatrix} = \begin{bmatrix} \mathbf{V}_N^T \bullet \mathbf{\Phi}_0 \\ \cdots \\ \mathbf{V}_N^T \bullet \mathbf{\Phi}_N \end{bmatrix} \bullet \mathbf{w} = \mathbf{H}_2 \bullet \mathbf{w} \tag{18}$$

3.3. Construct observation equation and estimate

Integrating the above equations into the original observation equation yields:

$$\begin{bmatrix} \mathbf{B} \\ \mathbf{B}_1 \\ \mathbf{B}_2 \end{bmatrix} = \begin{bmatrix} \mathbf{H} \\ \mathbf{H}_1 \\ \mathbf{H}_2 \end{bmatrix} \bullet \mathbf{w} \tag{19}$$

So far, we have obtained an observation equation with additional constraints. By adding additional constraints, the singularity of the observation matrix can be effectively reduced and the solvability of the equation can be improved.

As mentioned before, we could use BP method to obtain $\mathbf{w}$. By multiplying $\mathbf{w}$ with the sparse representation matrix $\mathbf{\Phi}_t$, the trajectory of target at time t can be obtained.

4. Simulation

In order to verify the feasibility of the above algorithm, we use the software Satellite Tool Kit(STK) to simulate a 3000-km range trajectory of target and the orbit of satellite. Then, based on the trajectory and orbit data, we evaluate the corresponding bearing measurements.

After adding gauss noise which mean is 0 and standard deviation is 10Km, the launch point of the target and the trajectory point of the boost phase (altitude between 20-40km) are used as complementary constraints. The simulation condition is shown in figure 4.

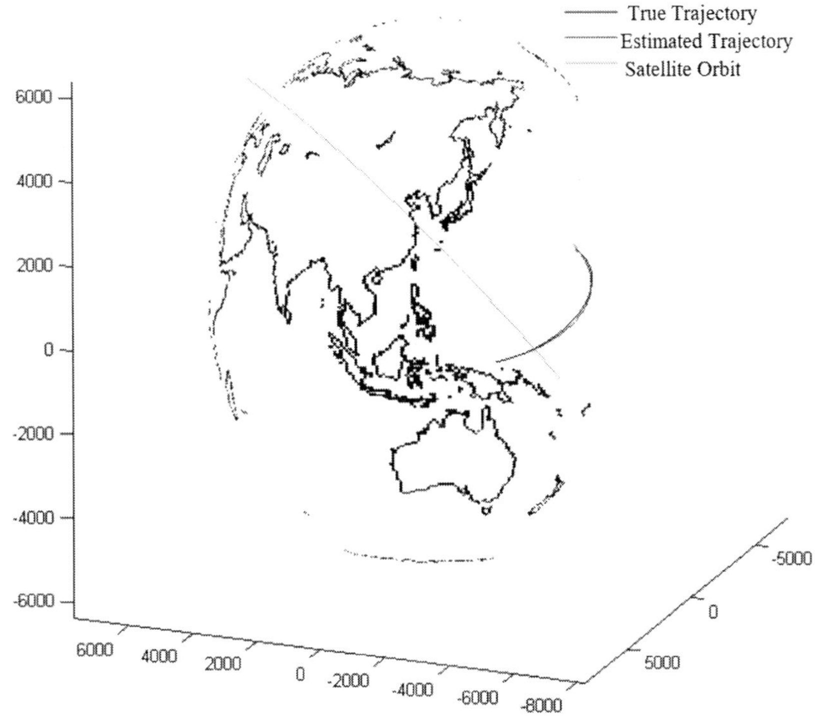

Figure 4. The Estimated trajectory of target

The estimation error of the entire trajectory is shown in the figure below:

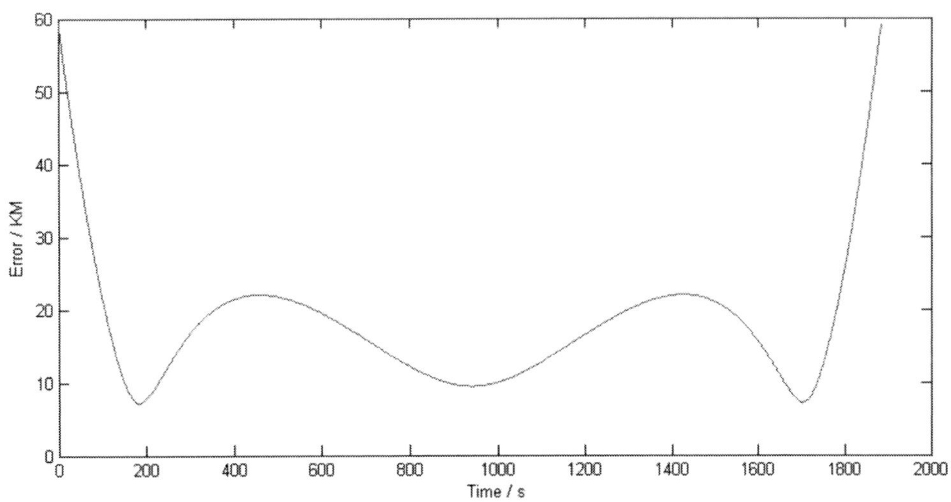

Figure 5. the estimation error of estimated trajectory

Figure 5 shows that the estimation error is less than 2% of the range of target. In ref[10], it is reported that, with bearing-only measurements from only one sensor, the estimation error is 5%-10% of the range. It can be seen that the proposed method can effectively track and locate the ballistic target trajectory under the condition of bearing-only measurement.

5. Conclusion

This paper proposes a new ballistic target trajectory estimation method based on sparse representation which can obtain the position of target with bearing-only measurements from one sensor. What's more, the method doesn't depend on the profile of target. The sparse representation algorithm for ballistic target trajectory is used to convert the trajectory tracking of the ballistic target into an estimation of the sparse representation coefficient. To evaluate these coefficients, RIP should be satisfied. However, RIP is usually unsatisfied when there is only angle measurement from one sensor. To solve this problem, we discussed how to obtain complimentary constraints and construct constraint equations. With constraint equations, the sparse representation coefficient can be estimated. Using these coefficients, the position of target can be easily obtained. An experiment is carried out in this paper and the result shows that our method is effective.

References

[1] Yicong Li, Murali Yeddanapud, (1999)Trajectory and Launch Point Estimation for Ballistic Missiles from Boost Phase LOS Measurements'[C]. 1999 IEEE Aerospace Conference. Proceedings. (Mar. 1999)

[2] Foy, W.H. (1976) Position-location solutions by Taylor-series estimation[J]. IEEE Transactions on Aerospace and Electronic Systems, AES-12. 2 (Mar. 1976), 187-194.

[3] Sherry,E.H., and Vincent J. AIDALA (1985) Observability requirements for three-dimensional racking via angle measurements[J]. IEEE Transactions on Aerospace and Electronic Systems,AES-21, 2 (Mar. 1985), 200-207.

[4] Liu L, (2000),Orbit theory of space craft. (BeiJing National Defence Industry Press.)

[5] Candès, E.J.; Romberg, J. K.; Tao, T.; Stable Signal Recovery From Incomplete and Inaccurate Measurements, *Commun. Pure Appl. Math.*, 2006, 59, 1207–1223.

[6] Candès, E.J.; Wakin, M.B. An Introduction to Compressive Sampling, *IEEE Signal Process. Mag.*, 2008, 25, 21–30.

[7] Chen, S.S.; Donoho, D.L.; Saunders, M.A. Atomic Decomposition by Basis Pursuit, *SIAM J. Sci. Comput.*, 1998, 1, 33–61.

[8] AN Yongquan, LI Jinhua, WANG Zhibin, et al. Mono-station and single-band passive ranging based on oxygen spectrum[J]. Acta Phys. Sin., 2013, 62(14): 1442101-1442107.

[9] Dowski E, Cathey T. Single-lens single-image incoherent passive-ranging systems[J]. Applied Optics, 1994(33): 6762-6773.

[10] A. J. Perella , W. W. Kuhn, "Cueing Performance Estimation Using Space Based Observations During Boost Phase", Proc. 1996 Summer Comp. Sim. Conf., Portland, OR, July 1996.

IOP Conf. Series: Materials Science and Engineering **887** (2020) 012046 doi:10.1088/1757-899X/887/1/012046

Optimization of UAV Flight Control Algorithm and Flight Simulation in Two-dimension

Xiao Xu, Jie Li, Chengwei Yang*, Ziquan Wang, Jin Xiong and Chang Liu

School of Mechanical and Electrical Engineering, Beijing Institute of Technology, Beijing 100081, China

*Corresponding author e-mail: yangchengwei@bit.edu.cn

Abstract. Large-scale Unmanned Aerial Vehicle (UAV) groups flight simulation in two-dimensional planes is commonly applied to UAV cluster mission planning algorithm design. In this paper, an UAV control algorithm optimized for two-dimensional planes is designed. The tandem structure of closed loops, as well as the control laws of L1 and Total Energy Control System (TECS), are transplanted to an UAV model that is simplified to a moving particle of a two-dimensional space, and UAV flight simulation is performed based on the flight control algorithm. Compared with the traditional 3D space flight simulation, it can save hardware resources and improve the simulation efficiency. Compare with the flight simulation based on the geometric method, on the premise of maintaining the dynamics basis, the trajectory and dynamics curves are closer to the actual flight results in 3D space, and the dynamics data onto the entire flight can be recorded. It has been verified that 50 UAVs need only 0.8s to perform 100 square-area snake-like search simulation experiments, and the fitting degree of the UAV's flight curve to the three-dimensional space is greater than 80%. The two-dimensional plane flight control algorithm and flight simulation proposed in this paper provide a new simulation method for the design of UAV cluster mission planning algorithms, therefore can help promote the development of the subject.

1. Introduction

In modern wars, UAVs, as an important reconnaissance force, have been increasingly used by various countries. The UAV flight process is mainly divided into take-off, cruise and descent recovery (final guidance) stages.[1] The main flight process is cruise at fixed altitude. Therefore, in the simulation experiments of the UAV task planning algorithm, the planned waypoints are often set on a two-dimensional plane. For many long-time flight simulations involving large number of UAVs, such as machine learning-based UAV mission planning algorithms, traditional 3D dynamic simulation is inefficient due to excessive computing power consumption, and the number of simulation nodes is limited by hardware resources ; The path simulation based on geometric method lacks a dynamic basis, which results in a large difference between the simulation result and the actual trajectory, low reliability, and unable to perform dynamic data recording and analysis.[2] Therefore, it is necessary to design UAV trajectory simulation method of a dynamic basis that can simulate the 3D space UAV dynamic trajectory on a 2D plane, can track and record the dynamic data, and only consume small amount of calculation and hardware resources.

This paper designs a fixed-wing UAV flight control algorithm optimized for 2D planes for flight simulation in mission planning. The track points are input in a two-dimensional plane with a certain

Content from this work may be used under the terms of the Creative Commons Attribution 3.0 licence. Any further distribution of this work must maintain attribution to the author(s) and the title of the work, journal citation and DOI.
Published under licence by IOP Publishing Ltd

height. Based on the UAV adaptive control algorithm in three-dimensional space, the UAV flight control algorithm in two-dimensional space is designed to quickly and stably generating a flight trajectory that simulates the real environment, and record the kinetic data in flight process.

The structure of this paper is as follows: Section 2 introduces the algorithm background, that is, the problem of transplanting UAV flight control algorithm in three-dimensional space, normal L1 and tangential TECS (Total Energy Control System) flight control algorithm to two-dimensional environment.[3] The third section introduce the cascade closed-loop flight control algorithm of UAV in two-dimensional plane separately from the outer and inner loops. Section 4 uses the personal PC as the hardware foundation, and simulates the algorithm on the Pyhon3.5.2 platform, and compares it with the geometric method and three-dimensional space flight dynamics simulation curves. Section 5 summarizes the advantages of the algorithm and its value of the development of UAV cluster task planning algorithms.

2. Background Description

The solution proposed in this paper is to design a two-dimensional plane UAV flight control algorithm. The L1 and TECS flight control algorithms in three-dimensional space are transplanted to a two-degree-of-freedom UAV model that is simplified to two-dimensional moving particles.

This section briefly analyses the UAV motions model in the two-dimensional space, then proposes the main problems and solutions.

2.1. Two-dimensional plane UAV motion model

In order to improve the efficiency of the UAV control in the two-dimensional plane and reduce the complexity of the control algorithm as much as possible, this paper first discusses the simplified model of the UAV. The simplified UAV dynamics model is the basis of the two-dimensional planar UAV control algorithm. It should meet the following three requirements[4]:

- Able to do two-degree-of-freedom motion in a two-dimensional plane;
- Able to transparent L1 and TECS adaptive control laws in a two-dimensional plane;
- Under the first two conditions, the model dynamic parameters are minimized.

In a two-dimensional coordinate plane, simplifying an UAV into a moving particle of a velocity direction is the limit of the simplification of a motion model that can be achieved. So the UAV body model assumes the following:

- The UAV is a particle, and its speed direction coincides with the direction of the UAV axis;
- UAV propulsion controls the thrust by adjusting the throttle;
- The resistance to the UAV is constant;
- UAV cruise altitude and desired airspeed are constant.

This model involves coordinate systems including cruise plane coordinate system $S_e(o_e x_e y_e)$ and airflow coordinate system $S_w(o_w x_w y_w)$, which are defined as follows:

- Cruise plane coordinate system $S_e(o_e x_e y_e)$: The origin is selected at the position of the center of mass when the UAV starts cruising. In the horizontal plane, $O_e Y_e$ points to the north direction and $O_e X_e$ points to the east direction;
- Airflow coordinates system $S_w(o_w x_w y_w)$: The origin is at the center of mass of the UAV, the $o_w x_w$ direction is consistent with the speed direction, and the $o_w y_w$ direction and $o_w x_w$ direction are at an angle of 90 ° clockwise. The general diagram are shown as figure 1.

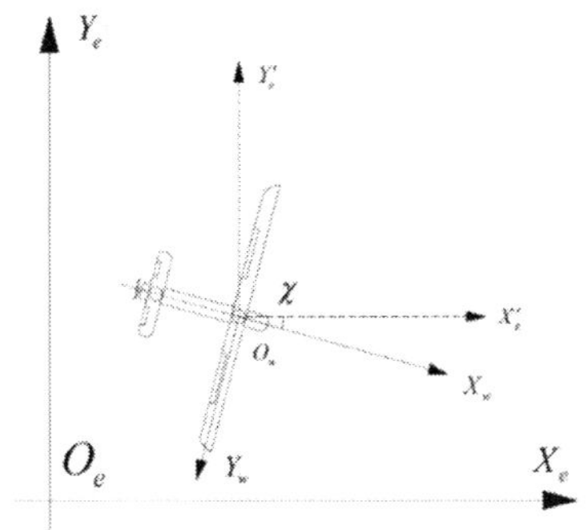

Figure 1. Schematic diagram of cruise plane coordinate system and airflow coordinate system

Among them, the angle between the $o_w x_w$ -axis of the air current coordinates system and the $o_e x_e$ -axis of the cruise plane coordinate system is the azimuth of the track in a two-dimensional plane, which is denoted as χ . The UAV motion model can be described by the following equation:

$$\begin{cases} \dot{x}_w = \dot{x}_{w\_obs} + a_{wl} \times dt \\ \dot{y}_w = \dot{y}_{w\_obs} + a_{wt} \times dt \end{cases} \quad (1)$$

Where $\dot{x}_w, \dot{y}_w$ is the tangential and normal speed of the UAV at time $t+1$; $\dot{x}_{w\_obs}, \dot{y}_{w\_obs}$ is the actual tangential and normal speed of the UAV at time t ; a_{wl}, a_{wt} is the tangential and normal acceleration of the UAV at time t ; dt is integral The step size, the minimum time interval, depends on the frequency of the UAV status sensor.

By projecting the UAV speed in the air current coordinate system to the cruise plane coordinate system, we get the speed $\dot{x}_w, \dot{y}_w$ of the UAV in the direction of cruise plane x and y :

$$\begin{cases} \dot{x}_e = \dot{x}_w \times \cos\chi - \dot{y}_w \times \sin\chi \\ \dot{y}_e = -\dot{y}_w \times \sin\chi - \dot{y}_w \times \cos\chi \end{cases} \quad (2)$$

2.2. Improvement plan

According to the definition of the UAV motion model in Section 2.2, the default UAV height will not change in the two-dimensional plane, which is an environmental constraint and therefore cannot be controlled. Think of the UAV as a moving mass, and it does not involve attitude angle and steering gear control. Based on the establishment of a two-dimensional plane UAV motion model, the research problem of this paper is transformed into a three-dimensional space UAV control algorithm being transplanted to a two-dimensional space simplified motion model. The control objects are the normal and tangential acceleration of the UAV, and the purpose is to achieve rapid tracking of the mission planning trajectory and airspeed.[5]

Similar to the effect of the lateral acceleration of the UAV in three-dimensional space[1], the normal acceleration of the UAV controls the yaw distance of the UAV so that the UAV approaches the desired track and is controlled by the L1 algorithm; the tangential acceleration controls the UAV Airspeed, controlled by TECS algorithm.

In summary, the design steps of the two-dimensional planar UAV control algorithm can be summarized as:

- Decompose the UAV control algorithm into outer and inner loop controllers, and decompose the outer loop controller into L1 controller with normal acceleration control and TECS controller with tangential acceleration control;
- Design TECS and L1 control algorithms for two-dimensional planes;
- Design the inner loop PID control algorithm;
- Debug the outer loop and inner loop control parameters to optimize the control effect;

- Simulate outer loop and inner loop control effects.

3. UAV flight control algorithm based on L1 and TECS

In UAV mission planning algorithms, UAVs are often reduced to two-degree-of-freedom aircraft on a two-dimensional plane. For many long-time flight simulations involving a large number of UAVs, such as machine learning-based UAV mission planning algorithms, traditional 3D dynamic simulation is inefficient due to excessive computing power consumption, and the number of simulation nodes is limited by hardware resources ; The path simulation based on geometric method lacks a dynamic basis, which results in a large difference between the simulation result and the actual trajectory, low reliability, and unable to perform dynamic data recording and analysis.

3.1. Cascade closed-loop control strategy

The structure of the UAV control system is shown as figure 2.

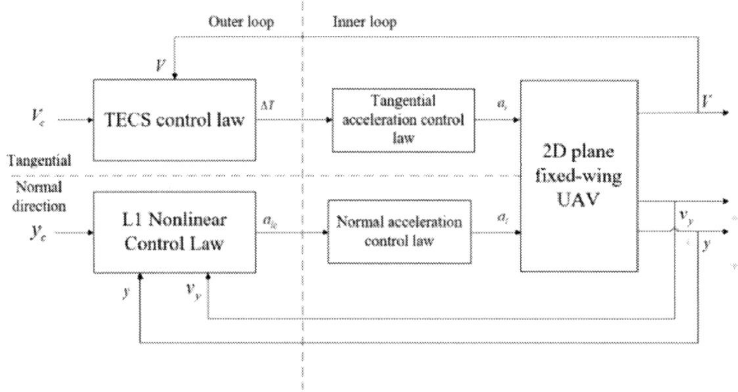

Figure 2. Schematic diagram of two-dimensional planar UAV cascade control

According to the function of the control loop, the inner loop is the acceleration control system of the UAV, which is used to control the tracking of the UAV's acceleration to the expected acceleration. The outer loop is a trajectory control loop that calculates the expected acceleration, controls the speed and yaw distance of the UAV, and enables the UAV to fly along the predetermined route and airspeed at the desired speed.

According to the control channel decomposition, the control system can be divided into tangential control system and normal control system. The tangential control system controls the acceleration of the UAV along the speed direction, using the TECS control algorithm; the normal control system controls the acceleration of the UAV along the vertical speed direction, and uses the L1 control algorithm.

3.2. Lateral acceleration algorithm based on L1 adaptive control

The L1 control algorithm of the UAV in the two-dimensional plane also makes the motion of the UAV close to the desired trajectory in the two-dimensional plane. The application scenario is as follows:

- In a two-dimensional plane, the position $P(x_p, y_p)$ and speed $V(v_x, v_y)$ of the UAV at a certain moment are known;
- The desired track segment i is a straight line segment from the initial waypoint $A_i(x_a^{\,i}, y_a^{\,i})$ to the end waypoint $B_i(x_b^{\,i}, y_b^{\,i})$. If the UAV reaches the ending waypoint at this time, and the ending waypoint is not at the end of the waypoint list, then the ending waypoint is set as the initial waypoint at the next time, and the next waypoint in the waypoint list is the ending waypoint . If the end waypoint is not reached at that time, the initial waypoint and the end waypoint at the next time are unchanged;

- The L1 control law calculates the expected normal acceleration a_l^{exp} pointing to the desired track perpendicular to the speed direction.

According to the position relationship between the position of the UAV and the desired track, the angle η between the speed direction and the virtual waypoint is calculated in three cases, so that the UAV can approach the endpoint of the desired track segment as soon as possible when it is away from the desired track segment . In either case, the L1 length is always determined by the product of speed and natural frequency:

$$L_1 = V \cdot L_1^{ratio} \tag{3}$$

Among them, L_1 is the distance from the current position to the virtual waypoint, and L_1^{ratio} is the natural frequency, which represents the ratio of the expected airspeed to L_1.

Scenario 1. From the UAV position to the desired flight path segment, foot $C(x_c, y_c)$ is located outside the desired flight path segment and is close to the initial waypoint $A(x_a, y_a)$, as shown in the figure 3. $O(x_o, y_o)$ is the virtual turning circle centre.

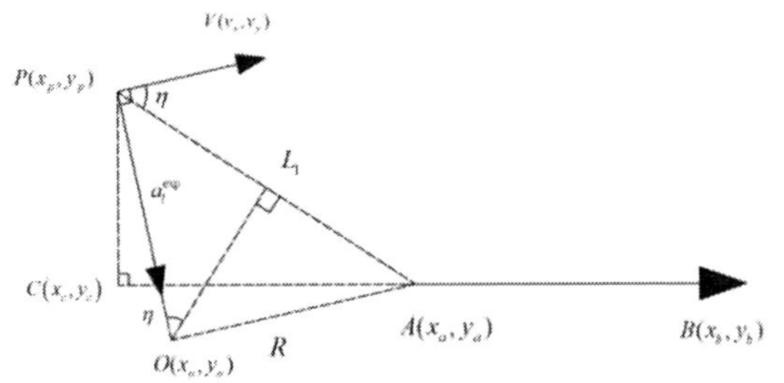

Figure 3. Schematic diagram of scenario 1

That is, when $\cos\angle PAB < 0$, the virtual waypoint is set as the initial waypoint. The size of L1 is independent of the relative initial waypoint position, and η is the angle from the UAV speed to PA. The calculation formula for η and virtual turning radius R is as follows:

$$\cos\eta = \frac{\vec{V} \cdot \overrightarrow{PA}}{|\vec{V}| \cdot |\overrightarrow{PA}|} \tag{4}$$

$$R = \frac{L_1}{2\sin\eta} = \frac{L_1}{2\sqrt{1-\cos^2\eta}} \tag{5}$$

Therefore, the expected lateral acceleration a_l^{exp} is calculated as:

$$a_l^{exp} = \frac{V^2}{R} = \frac{V^2}{L_1} 2\sqrt{1-\cos^2\eta} \tag{6}$$

With equations 5 and 6, the expected lateral acceleration is expressed as:

$$a_l^{exp} = \frac{2V}{L_1^{ratio}} \sqrt{1 - \left(\frac{\vec{V} \cdot \overrightarrow{PA}}{|\vec{V}| \cdot |\overrightarrow{PA}|} \right)^2} \tag{7}$$

Scenario 2. UAV positions to the desired track segment, foot $C(x_c, y_c)$ is located outside the desired track segment, and is close to the ending waypoint $B(x_b, y_b)$, as shown in the figure 4:

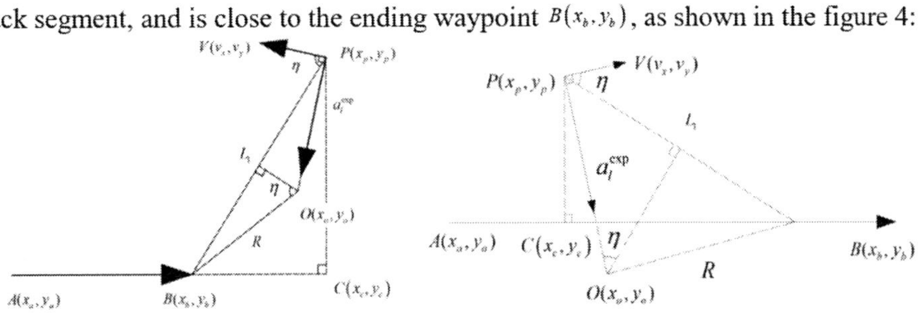

Figure 4. Schematic diagram of scenario 2 (left) and Schematic diagram of scenario 3 (right)

When $\cos\angle PBA < 0$, set the ending waypoint as a virtual waypoint. As in the first case, although the length of L1 is proportional to the speed of the UAV, the selection of the virtual waypoint has nothing to do with the size of L1 until the end waypoint is reached.

$$\cos\eta = \frac{\vec{V}\cdot\overrightarrow{PB}}{|\vec{V}|\cdot|\overrightarrow{PB}|} \tag{8}$$

$$a_l^{\exp} = \frac{2V}{L_1^{ratio}}\sqrt{1-\left(\frac{\vec{V}\cdot\overrightarrow{PB}}{|\vec{V}|\cdot|\overrightarrow{PB}|}\right)^2} \tag{9}$$

Scenario 3. From UAV position to desired track segment, foot $C(x_c, y_c)$ is located in the desired track segment, as shown in the figure 4:

At this time, $\cos\angle PAB > 0$ and $\cos\angle PBA > 0$ are standard L1 algorithm scenarios, and the virtual waypoint is located on track segment AB, and the distance to P_t is L1.

Let η_1 be the angle between $\vec{V}$ and $\overrightarrow{AB}$, and η_2 be the angle between $\overrightarrow{PC}$ and L_1, then the angle η can be expressed as:

$$\eta_1 = \arccos\left(\frac{\vec{V}\cdot\overrightarrow{AB}}{|\vec{V}|\cdot|\overrightarrow{AB}|}\right),\ \eta_2 = \arcsin\left(\frac{|\overrightarrow{PC}|}{L_1}\right) \tag{10}$$

$$\eta = \frac{\pi}{2} + \eta_1 - \eta_2 \tag{11}$$

$$\sin\eta = -\cos(\eta_1 - \eta_2) \tag{12}$$

Therefore, the virtual turning radius formula is:

$$R = \frac{L_1}{2\sin\eta} = \frac{V\cdot L_1^{ratio}}{2\sin\eta} = -\frac{V\cdot L_1^{ratio}}{2\cos(\eta_1-\eta_2)} \tag{13}$$

The formula of the normal acceleration of the UAV is:

$$a_l^{\exp} = \frac{V^2}{R} = \frac{V^2}{L_1}2\sin\eta = \frac{-2V}{L_1^{ratio}}\cos(\eta_1-\eta_2) \tag{14}$$

The empirical parameter K_{L1} is introduced here to adjust the problem that the normal acceleration of the UAV in the two-dimensional plane is too large and serious overshoot. Therefore, the expected normal acceleration $a_l^{\exp}$ after optimization is:

$$a_l^{\exp} = a_l^{\exp}\cdot K_{L1} \tag{15}$$

The normal acceleration direction is perpendicular to the speed and points to the track segment. In summary, the expected normal acceleration of the UAV in the two-dimensional plane can be obtained.

3.3. Algorithm for Tangential Acceleration of Outer Ring Based on TECS Control

TECS control the size of the UAV's throttle to change the output power of the UAV's engine, and controls the UAV's elevator to change the power ratio of the UAV's potential energy and kinetic energy to maintain or adjust the UAV's flying height[6]. In a two-dimensional plane, the height of the UAV does not need to be adjusted, so the potential energy of the UAV is constant. Based on the most simplified principle, the UAV altitude potential energy is set to zero. The TECS algorithm for a two-dimensional plane can change the output power of the UAV by controlling the throttle valve without changing the potential energy, so as to track the desired speed. The application scenario is described as follows:

- The UAV's flight speed is V_{state} at a certain moment, and the expected airspeed given by the mission plan is V_{set};
- Limited by the performance of the UAV engine, the change rate of the total specific energy of the UAV is constrained within the interval $[STErate_{\min}, STErate_{\max}]$, where $STErate_{\min} < 0 < STErate_{\max}$;

- Integrate UAV airspeed and desired airspeed, and calculate the expected throttle opening rate $throtle_p$ of the UAV under the constraints of the total specific energy change rate and the controllable change interval of the throttle valve.

First, the expected airspeed change rate $TASrate_{set}$ is the airspeed error at that moment, that is, the ratio of the expected airspeed to the actual airspeed difference and the time interval ΔT. Introduce the empirical proportional parameter K_V to adjust the change rate to avoid the failure of the speed control due to the expected change rate.

$$TASrate_{set} = \left(TAS_{set} - TAS_{state}\right) \cdot \Delta T^{-1} \cdot K_V \tag{16}$$

From the specific energy calculation formula, the actual total specific energy STE_{state} and the expected total specific energy STE_{set}, and the total specific energy error STE_{error} can be obtained:

$$STE_{error} = STE_{set} - STE_{state} = \frac{1}{2}\left(TAS_{set}^2 - TAS_{state}^2\right) \tag{17}$$

The expected total specific energy change rate $STErate_{set}$ is the expected change rate of the total specific energy obtained by multiplying the specific energy magnitude and the desired speed change rate at that moment.

$$STErate_{set} = TAS_{state} \times TASrate_{set} \tag{18}$$

When the airspeed error rises, the expected airspeed change rate rises accordingly, which may exceed the UAV's engine controls range. Therefore, according to the actual control ability of the UAV, the calculation formula of $STErate_{set}$ is as follows:

$$STErate_{set} = \begin{cases} STErate_{set},\ STErate_{\max} > STErate_{set} > STErate_{\min} \\ STErate_{\max},\ STErate_{set} > STErate_{\max} \\ STErate_{\min},\ STErate_{\min} > STErate_{set} \end{cases} \tag{19}$$

From the principle of UAV dynamics, it is expected that the total specific energy change rate is linearly related to the throttle opening rate, so the theoretical throttle opening rate $throtle_p$ can be calculated. The reference throttle value of the UAV is $throtle_{const}$. When the expected total specific energy change rate is greater than 0, the total specific energy of the UAV increases, and the throttle opening rate are directly proportional to the maximum opening rate $throtle_{\max}$. In contrast, when the expected total specific energy change When the rate is less than 0, the total specific energy of the UAV decreases, and the throttle opening rate are proportional to the minimum opening rate $throtle_{\min}$. Calculate as follows:

$$throtle_p = \begin{cases} throtle_{const} + \dfrac{STErate_{set}}{STErate_{\max}} throtle_{\max}\ , STErate_{set} > 0 \\ throtle_{const} + \dfrac{STErate_{set}}{STErate_{\min}} throtle_{\min}\ , STErate_{set} < 0 \end{cases} \tag{20}$$

In this case, the inner loop PID control algorithm for the tangential acceleration of the UAV can be designed by the specific energy error STE_{error}, the throttle opening rate $throtle_p$, and the expected total energy change rate $STErate_{set}$ of the UAV.

3.4. Inner loop PID control algorithm

The UAV in the two-dimensional plane does not have the rudder surface control structure nor the attitude angles control capability. Therefore, the desired acceleration command a_l^{exp} and throttle opening rate $throtle_p$ given by the outer ring controller of the inner ring controller are the control targets. Tangent two parts, through the PID controller to simulate the operation process of the inner loop controller in three-dimensional space.

The normal control law of the inner loop of the UAV is:

$$a_l^{true} = K_l^P \cdot \Delta a_l + K_l^I \cdot \int \Delta a_l dt + K_l^D .a_l' \tag{21}$$

Among them, Δa_t is the difference between the expected tangential acceleration and the actual tangential acceleration of the UAV at this moment, and K_t^P, K_t^I, and K_t^D are the coefficients of the proportional link, the integral link, and the differential link.

The tangential control law of the inner loop of the UAV is:

$$throtle_{True} = throtle_P + \left(STE_{error} + STErate_{error} \times K_t^D\right) \times K_t^P + K_t^I \times \int STE_{error} dt \tag{22}$$

Among them, $TASrate_{error}$ is the difference between the expected airspeed change rate and the actual airspeed change rate at this moment, and K_t^P, K_t^I, and K_t^D are the coefficients of proportional link, integral link, and differential link. Because the tangential acceleration of the UAV is proportional to the throttle opening rate, the actual tangential acceleration of the UAV at this time is:

$$a_t^{true} = K_{acct} \times throtle_{True} \tag{23}$$

K_{acct} is the tangential acceleration proportionality factor.

4. Generate trajectory and simulation calculation of dynamic parameters

In this paper, python3.5 is selected as the simulation platform. Using a two-dimensional plane flight controls algorithm, given a serpentine search waypoint sequence, patrol at a constant speed of 30m / s and a fixed height of 50m relative to the ground. The two-dimensional simulation trajectory, traditional calculation path based on the Dobbins curve and the flight path of the UAV dynamics simulation with the same waypoint in 3D space obtained is as figure 5:

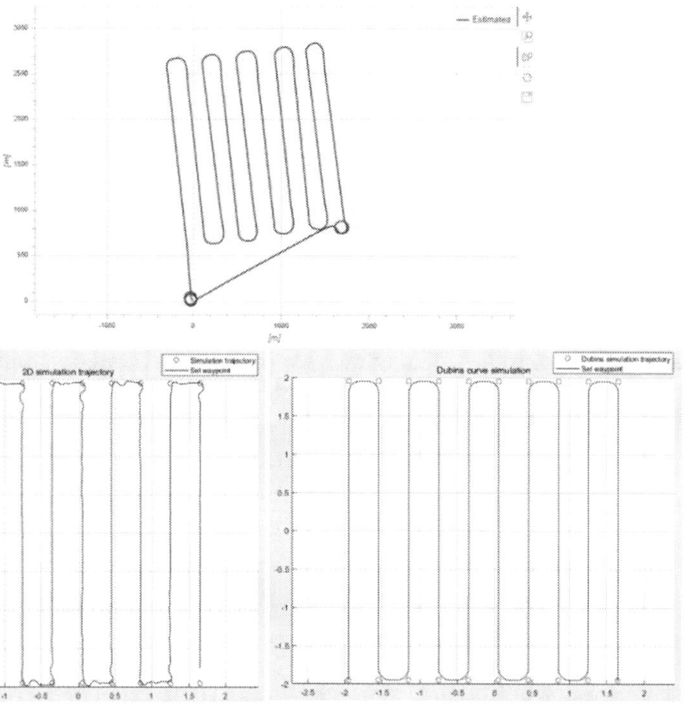

Figure 5. Three-dimensional flight simulation trajectory of a fixed-wing UAV (previous page);

Trajectory simulation based on two-dimensional plane control algorithm (left) and Trajectory Simulation Based on Geometric Algorithm (right)

It can be seen that the two-dimensional simulation trajectory can simulate the track characteristics of the UAV slightly overshooting after reaching the waypoint, which is better than the Dubins curve track. The clue-aware trajectory similarity (CATS) algorithm is used to calculate the similarity with the 3D simulation trajectory[7]:

$$score_{\varepsilon,\tau}\left(p_{i,l},T_j\right)=\max\left\{f_\varepsilon\left(p_{i,l},p_{j,k}\right)\mid p_{j,k}\in T_j \text{ and } t_{j,k}\in\left[t_{i,l}-\tau,t_{i,l}+\tau\right]\right\} \tag{24}$$

$$CATS_{\varepsilon,\tau}\left(T_i,T_j\right)=\frac{1}{|T_i|}\times\sum_{p_{i,l}\in T_i} score_{\varepsilon,\tau}\left(p_{i,l},T_j\right) \tag{25}$$

The approximation of the two-dimensional trajectory simulation curve is 0.83. The similarity is greater than 80%, indicating that the two-dimensional trajectory simulation is basically consistent with the three-dimensional trajectory simulation.

The normal and tangential speed curves of the two-dimensional plane UAV flight control algorithm, and the comparison diagram of the UAV speed curve in three-dimensional space are as figure 6 and figure 7.

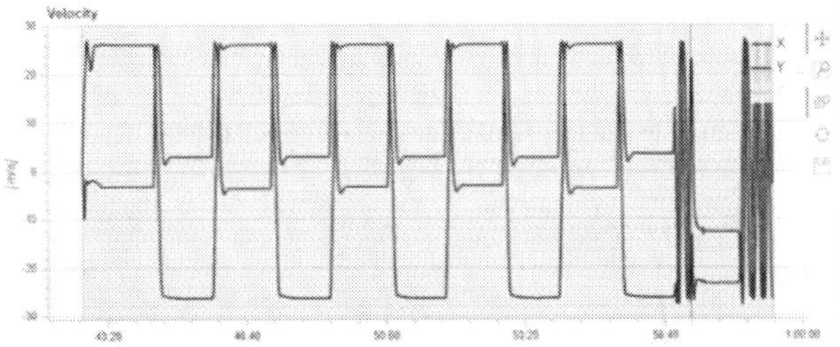

Figure 6. Three-dimensional space velocity curve

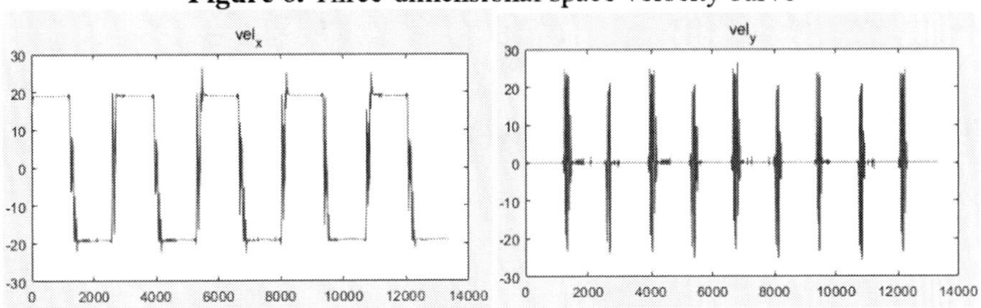

Figure 7. Two-dimensional space tangential velocity curve (left) and normal velocity curve (right)

The tangential velocity in 2D simulation is basically consistent with the tangential velocity curve in 3D space simulation. When changing the direction of flight speed, the UAV's normal speed changes drastically, but it can converge to the 3D simulation value.

The tangential acceleration and normal acceleration curves of the two-dimensional simulation UAV are as figure 8:

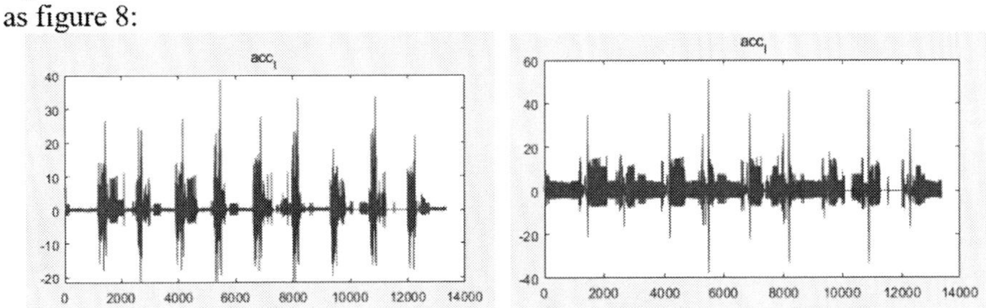

Figure 8. Two-dimensional space tangential acceleration curve (left) and normal acceleration curve (right)

Fifty UAV serpentine formation searches flights took a total of 0.8ses to calculate on a personal computer. The computing power of a three-dimensional dynamic simulation personal computer for the same scenario cannot be supported. Compared with the simulation trajectory obtained by the Dubins curve, it improves the approximation of the flight trajectory of a real UAV, and can output the dynamic parameters of a single UAV at any time.

IOP Conf. Series: Materials Science and Engineering **887** (2020) 012046 doi:10.1088/1757-899X/887/1/012046

5. Conclusion

This paper studies an UAV control algorithm optimized for two-dimensional planes. The core idea is to borrow the mature outer-loop control algorithm in three-dimensional space, that is, the TECS control law of the pitch channel and the L1 control law of the yaw and roll channels, and transplant it to a dynamic model that is simplified to a two-degree-of-freedom particle in the two-dimensional plane.

In this way, it can simulate similar motion trajectories and changes in dynamic parameters in three-dimensional space, and greatly reduce the computing power requirements of three-dimensional space dynamics simulation. Especially in the simulation of mission planning at the UAV cluster level, it can greatly save hardware resources and time costs. For simulation experiments that require a large number of UAVs for a long time and multiple flights, such as machine learning UAV mission planning algorithm design a large amount of simulation training is required. This algorithm can solve the problems of traditional path planning simulations such as the lack of dynamic basis in the Dubins curve and the low credibility of the planned path, making dynamic simulation under this kind of problem possible.

References

[1] Saifei W 2016 Research on Precise Landing Technology of Shipborne UAV Based on Visual Information Guidance (Master Thesis) Nanjing University of Aeronautics and Astronautics
[2] Jian Z and Xiaoming N, Jun L 2014 *Opt. Laser. Technol.* **64** 319-323
[3] Chad G, Zhaodan K, Berenice M 2010 *J. Intell. Robot. Syst.* **57** 65-100
[4] Chih-Chieh H, Wen-Chih P, Wang-Chien L 2015 *VLDB. J.* **24** 169-192
[5] Andrea C, Emanuele F, Adriano M, Primo Z 2010 *J. Intell. Robot. Syst.* **57** 233-257
[6] Alexander F, Boris A 2005 *Eur. J. Control.* **11** 71-79
[7] Hyunjin C, Youdan K 2014 *Control. Eng. Pract* **22** 10-19

Aerodynamic Performance Analysis of a Curved-channel ABLE Airfoil in Supersonic Continuous and Hypersonic Rarefied Flows

S Zhang[1,*], S Z Fang[1] and J Guo[1]

[1] School of Aerospace and Engineering, Beijing Institute of Technology, Beijing 10081, China

[*] 317898841@qq.com;

Abstract. An airfoil, equipped with a curved-channel ABLE, is a good alternative to increase the lift-to-drag ratio. For this purpose, a curved-channel artificial blunted leading edge (ABLE) was applied to several supersonic/hypersonic symmetrical airfoils to analyse the aerodynamic performance in low-altitude supersonic continuous flow and high-altitude hypersonic rarefied flow during the process of re-entry into the atmosphere. The compressible Navier–Stokes equations were solved using the finite volume method to analyse the aerodynamic performance and determine the optimal curved-channel ABLE airfoil for supersonic continuous flow. Thereafter, the drag reduction and lift increment efficiency of the optimal ABLE airfoil were analysed in the hypersonic rarefied flow. The variation in the aerodynamic performance of the ABLE airfoil with the Knudsen number and freestream Mach number was determined using the direct simulation Monte Carlo (DSMC) method. The results showed that with suitable ABLE configuration and parameter values, the concept of ABLE can help reduce the drag coefficient, improve the lift coefficient, and obtain a better lift-to-drag ratio in both continuous and rarefied flows without considerable aerothermal penalty. However, the drag reduction effect of the curved-channel airfoil in the rarefied flow is much lower than that in the continuous flow, because of the lower proportion of the wave drag in the total drag, and the overall aerodynamic performance is significantly deteriorated because of the rarefaction effect of the atmosphere.

1. Introduction

The lift-to-drag ratio (*L/D*) of supersonic and hypersonic vehicles, such as space shuttles and re-entry vehicles, significantly affects performance criteria such as the cross range, manoeuvring requirements, and payload mass fractions, as well as their economic viability. The drag acting on supersonic/hypersonic vehicles can be classified into three categories: 1) skin friction drag, 2) drag due to lift, and 3) zero-lift bluntness (wave) drag. The skin friction drag is due to the fluid viscosity and is a function of the total wetted surface area of the vehicle, and the drag due to lift includes the induced drag and the wave drag component[1]. During re-entry, the vehicles first fly in a hypersonic rarefied flow and then in a hypersonic or supersonic continuous flow. This requires the vehicles to have an optimal aerodynamic performance under both the flow conditions.

For supersonic or hypersonic continuous flows, Gupta and Ruffin introduced the concept of a curved-channel artificial blunted leading edge (ABLE) to reduce wave drag under the condition that the channel is sized such that a choked flow condition exists. This concept allows for a hollow channel

Content from this work may be used under the terms of the Creative Commons Attribution 3.0 licence. Any further distribution of this work must maintain attribution to the author(s) and the title of the work, journal citation and DOI.
Published under licence by IOP Publishing Ltd

IOP Conf. Series: Materials Science and Engineering **887** (2020) 012047 doi:10.1088/1757-899X/887/1/012047

to be opened in the airfoil while cruising at supersonic/hypersonic speeds. The channel thickness should be sufficiently small, a normal shock should exist in front of the channel, and the flow should enter the channel subsonically. They studied the aerodynamic characteristics of the ABLE airfoil, such as the force coefficient, lift-to-drag ratio, and maximum heat transfer rate by conducting Navier–Stokes simulations. The ABLE concept helped reduce wave drag, slightly increase the skin friction drag, reduce the total drag, and increase the lift-to-drag ratio while keeping the same aerothermal load level as that of a no-channel airfoil[1-4].

Two types of ABLE concepts have been proposed: a flow-through channel ABLE and a curved-channel ABLE. The latter has a shorter channel, opened toward the nose of the vehicle or the leading edge of the airfoil and exhausted near the leading edge. With the channel opened, a part of the high-pressure surface of the nose or the leading edge of the airfoil is removed, thus reducing the wave drag. In addition, the jet-flow acting on the upper part wall results in a significant lift increment. Thus, the lift-to-drag ratio is improved. The flow-through channel ABLE airfoil has a lower total drag, and the flow expanding over the channel lip significantly increases the lift relative to the baseline geometry; the curved-channel ABLE has more advantages compared with the flow-through channel ABLE, such as better volumetric efficiency, lower internal structural/thermal load penalties, and lower viscous drag penalties.

It is reasonable to apply the curved-channel ABLE to re-entry vehicles as the leading edge of the airfoil for high-altitude hypersonic rarefied flows. Figure 1 shows the typical parameter–time maps of an unmanned space vehicle (USV)[5], from which the heat flux occurring in the stagnation point region when re-entering the earth's atmosphere can be read. During re-entry, a space shuttle undergoes a relatively significant aerothermal heating at altitudes in the range of 60–110 km, where the flow regime includes free molecular and transitional regions.

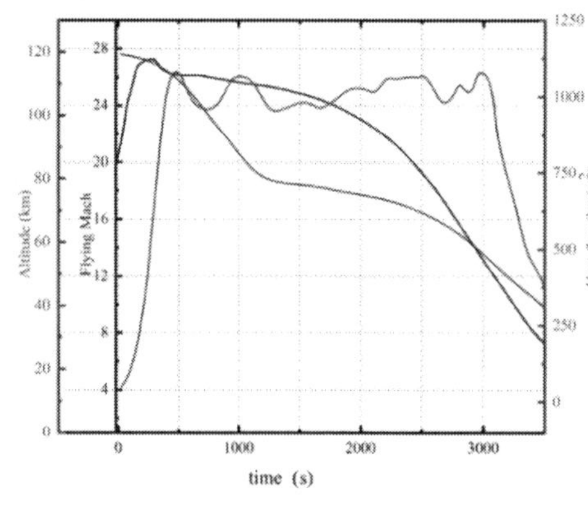

Figure 1. Time–altitude, time–velocity, and time–heat flux maps of an unmanned space vehicle.

This condition demands a new understanding of the aerodynamic performance of airfoil/vehicles. This can be done by applying the ABLE concept, as it is ideal for the near-space flow domain where the gas is no longer consistent with the continuous medium hypothesis. This paper reports a study that extends the application range of the ABLE concept. We studied the feasibility of the previous curved-channel ABLE concept applied to several classical supersonic/hypersonic symmetrical airfoils and verified the efficiency of the ABLE concept under flight conditions where gas rarefaction dominates the flow regime. If the ABLE concept can help reach the design point and increase the lift-to-drag ratio, a vehicle can exhibit a better control performance in terms of the flight time, flight speed, and vehicle range and even improved fuel efficiency. The heat transfer rate at the leading edge is a function of the flight speed, and therefore, by controlling the flight speed, we can limit the heat flux during re-entry.

This study focused primarily on reducing the wave drag force and increasing the lift force as a means of increasing the lift-to-drag ratio, aiming to establish the feasibility of the ABLE concept by quantifying the aerothermal and aerodynamic performance under a wide range of flow domain. This concept can be applied to any airfoil used on a supersonic/hypersonic vehicle, but for simplicity, we selected some basic symmetric airfoils. First, the curved-channel ABLE concept was applied to four classical supersonic/hypersonic symmetric airfoils in a low-altitude continuous flow, and the optimal airfoil geometry was determined by conducting Navier–Stokes simulations. The optimal airfoil was further evaluated in a high-altitude rarefied flow using the direct simulation Monte Carlo (DSMC) method. The Knudsen number and freestream Mach number were considered to characterize the aerodynamic performance of the airfoils employing the ABLE concept.

2. Simulation configuration and method

2.1. Model building

The concept of curved channel ABLE was first proposed by Gupta and Ruffin[3]. At the beginning, with the hypothesis of a negligible lateral flow, four groups of typical two-dimensional hypersonic or supersonic airfoils and corresponding ABLE airfoils derived from the previous baseline longitudinal symmetry blunted diamond airfoil[3] were considered. Relative to a non-channel airfoil, taking advantage of the ABLE concept is expected to help reduce the total drag. To avoid significantly increasing the wetted area, which will result in an increased skin-friction drag, the channel is made to slope downward near the leading edge of the airfoil. The curves for the channel walls need to follow those of the lower and upper channel walls to match the external (baseline geometry) slopes at their end points; the upper channel wall exit location is the design input. Figure 2 shows the geometric parameters of the airfoils. Here, C is the chord length, L is the channel length, t is the channel height, R is the leading-edge bluntness radius of the airfoil, and Rn is the leading-edge bluntness radius of the channel.

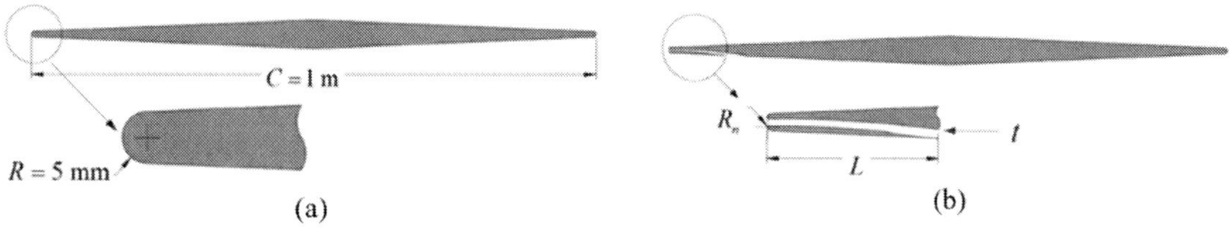

Figure 2. (a) Outline shape of no-channel baseline diamond airfoil; (b) that of curved channel ABLE baseline diamond airfoil.

Table 1. Two-dimensional airfoil shapes[a].

airfoil	Configuration of symmetrical airfoils	Curved channel ABLE
diamond		
biconvex		
quadrangle		
hexagon	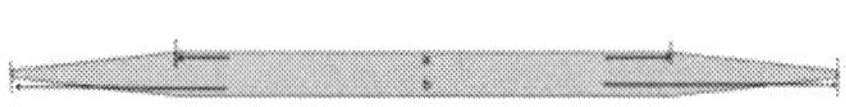	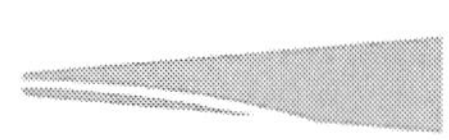

[a] The quadrilateral airfoil a/b is taken as 4/5; hexagonal airfoil a/b is 3/5.

It is convenient to nondimensionalize these parameters with L=10%C and t=4%L to ensure a sufficient expansion of the flow inside the channel and with Rn=0.8%L to reduce the leading-edge heat flux. For convenience, the following terminology is used to describe the airfoil. B-RN represents a symmetric curved-channel ABLE biconvex airfoil, which can be abbreviated as ABLE biconvex airfoil. NC, RN, D, B, Q, and H represent the no-channel baseline airfoil, round-nose curved-channel ABLE airfoil, diamond airfoil, biconvex airfoil, quadrangle airfoil, and hexagon airfoil, respectively.

Table 1 gives the configuration diagrams of the four typical supersonic symmetrical airfoils and the corresponding ABLE airfoils. The models are ensured to have the same main parameters as those of the baseline model, the only difference is that they have different two-dimensional plane shapes.

2.2. Simulation approach

The aerodynamic performance in the continuous flow was predicted using the commercial software Fluent. To this end, the k-ω turbulence model and density-based solver were used, and the compressible Navier–Stokes equations were solved using the finite volume method. The aerodynamic calculation is slightly affected by the grid. In this study, based on existing experience and for sufficient accuracy and efficiency, the near-wall and fore body grids were refined to some extent to ensure shock capturing capability.

A better alternative for practical applications is the open-source DSMC method called SPARTA, which has been well established by Sandia National Labs for rarefied and hypersonic conditions. This simulation approach is a particle method that does not rely on the continuous medium hypothesis. In the computational kernel, a large number of particles were simulated to represent real gas molecules, each having its velocity and internal energy. The grid generation has good flexibility because the DSMC method is computationally stable.

Based on previous experience, we take the grid dimension as $\Delta x \sim (1/3)\lambda$, which is called 'one third of the mean free path', and λ is the local mean free path of the gas molecules in the grid. Thus, a relatively ideal grid division can be obtained.

Currently, two main molecular models are used in the DSMC methodology: the variable hard sphere (VHS) model and the variable soft sphere (VSS) model. The only difference is that the deflection angle in the VHS model is given by [6]:

$$x = 2\cos^{-1}(b/d)^{1/\alpha} \tag{1}$$

In the other model, it is given by:

$$x = 2\cos^{-1}(b/d) \tag{2}$$

Here, α is a coefficient between 1 and 2.

The complete diffuse reflection model is an inelastic collision model that takes the temperature of the object surface as the equilibrium condition.

The distribution function of the diffuse reflection particles is given by:

$$f(C_r) = n_r\left(\frac{m}{2\pi kT}\right)^{1.5}\exp\left(-\frac{m}{2\pi kT}C_r^2\right) \tag{3}$$

Moreover, the Knudsen number (Kn) is defined as:

$$Kn = \frac{\lambda}{L} \tag{4}$$

The mean free path is a function of the pressure, viscosity, and most probable molecular velocity. Various models have been proposed to describe[6]. L is the characteristic length.

Besides Kn, the rarefaction parameter δ is another quantity that is also used to describe the flow regime and is defined in [7] as:

$$\delta = \frac{\sqrt{\pi}}{2}\frac{L}{\lambda} = \frac{\sqrt{\pi}}{2}\frac{1}{Kn} \tag{5}$$

IOP Conf. Series: Materials Science and Engineering **887** (2020) 012047 doi:10.1088/1757-899X/887/1/012047

3. Simulation validation

Table 2 gives the flow conditions used in the following test cases.

Table 2. Flow conditions.

Flow domain	V_x(m/s)	T_w(K)	$n_\infty(m^{-3})$	λ(mm)
supersonic continuous flow	708	300		
hypersonic rarefied flow	1504	290	1.05×10^{21}	1.6

3.1. Simulation validation

For validation in the first case, some comparisons of the force coefficients are made between the computational results of this study and the reference ones given in under the same flow condition. Figure 3 shows the corresponding plots. The drag and lift coefficients are in good agreement, with the maximum relative error of the lift–drag ratio reaching 11.16%. This discrepancy may be due to the systematic error in the simulation and the error in the numerical calculation when the lift-to-drag ratio was calculated using the lift and drag coefficients. Nevertheless, to some extent, the comparison results verify the validity of the method used for simulating the continuous flow.

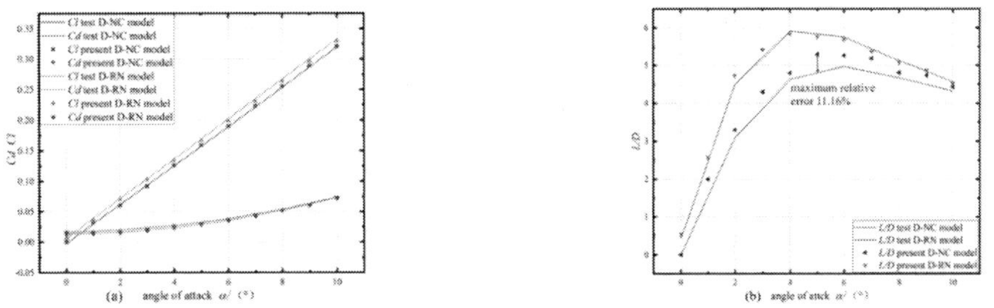

Figure 3. (a) Comparisons of *Cl* and *Cd* between the results obtained in this study and the experimental results obtained in[8]; (b) Comparison of *L/D*.

3.2. Validation in the case of hypersonic rarefied flow

For validation in the second case, a test model proposed in[8] was employed. The test model was defined by the following parameters: the length of the flat plate was 100 mm and the thickness was 5 mm. The VHS and specular diffusive reflection models were used. The computational grid used in the simulation is a 2D structural grid, and the simulation is performed in an orthogonal domain around the plate. The inflow boundary is defined on the left side of the domain, whereas the outflow boundary is defined on the right.

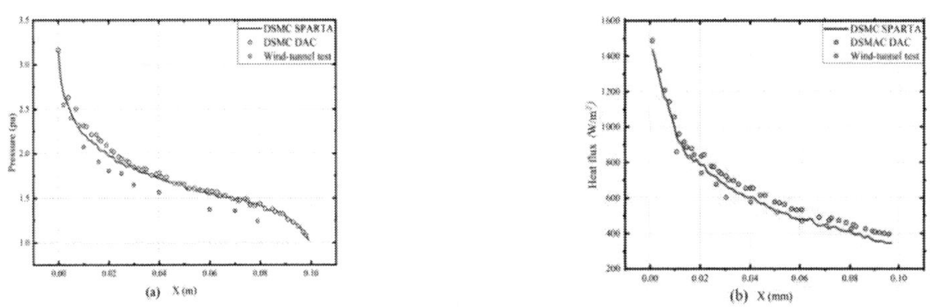

Figure 4. (a) Comparison of pressure between computational results of this study, the wind-tunnel experimental ones in [9], and DAC results in [10]; (b) Comparison of heat flux.

Referring to the low-density wind-tunnel experimental results in [9] and DAC (another DSMC method) results in [10], the pressure and heat flux distribution along the flat-plate upper surface are compared, as shown in figure 4. The results are in good agreement. Moreover, the results obtained using SPARTA are in better agreement with the wind tunnel experimental results; however, the heat flux readings are not predicted accurately enough by the DAC software. Previous studies[11,12] have verified SPARTA, and the reliability and validity of the computational kernel have been proven.

4. Results and discussion

4.1. Aerodynamic performance of airfoils in continuous flow

The influence of channel parameters on the aerodynamic performance in continuous flows has been preliminarily studied. The first set of results in continuous flow were obtained using Fluent with a freestream Ma number of 2.4 and an altitude of 12 km. These simulation tests can be viewed as modelling the low-altitude supersonic continuous flow around the airfoils. Figures 5(a)–6(c) show the computed results of the force coefficient and lift-to-drag ratio.

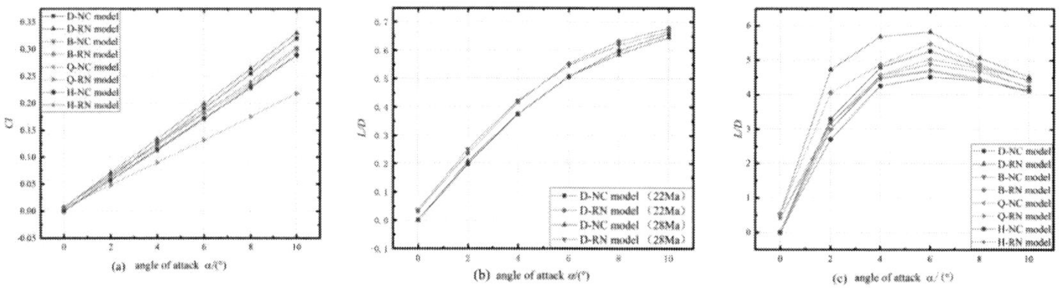

Figure 5. (a) *Cl*; (b) *Cd*; (c) *L/D* of the different airfoils and ABLE airfoils.

The lift coefficient of the airfoil increases linearly with the increase in the angle of attack, and the variation in the drag coefficient with the increase in the angle of attack first gradually increases and then rapidly increases. When the angle of attack reaches 4°, the drag coefficient increases more obviously, because with the increase in the angle of attack, the windward area increases, on which the pressure acting is greater than that on the leeward area. This also makes the lift-to-drag ratio to increase first and then decrease. Overall, among the four groups of investigated models, except in the cases of the ABLE quadrilateral airfoil and ABLE hexagonal airfoil, the curved-channel ABLE concept helps reduce the drag and increase the lift; moreover, the maximum lift-to-drag ratio increases by approximately 24%. This can be attributed to the sloping down of the curved channel near the leading edge, which reduces the high pressure acting on the stagnation point region; moreover, the jet force on the upper surface of the airfoil generates a positive normal force, resulting in an increased lift.

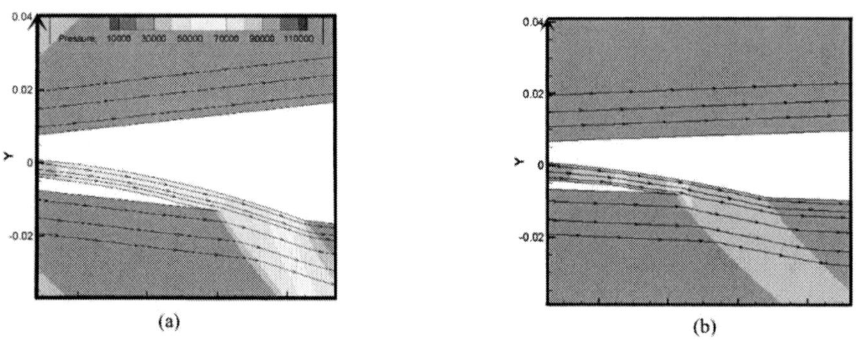

Figure 6. (a) Trace lines of the ABLE hexagonal airfoil; (b) that of the ABLE diamond airfoil.

The drag coefficient of the ABLE hexagonal airfoil is slightly greater than that of its baseline no-channel model. The lift coefficient and drag coefficient of the ABLE quadrilateral airfoil are lower than those of the baseline no-channel model; when the angle of attack reaches 4°, the lift coefficient of the ABLE quadrilateral airfoil is approximately 69.49% that of the no-channel one. To further analyse the possible reason for the abnormal characteristics, the trace lines near the outflow of the ABLE hexagonal and ABLE diamond airfoils are simulated, as shown in figure 6. Compared with the ABLE diamond airfoil, the leading edge of the ABLE hexagonal airfoil has a greater semi-apex angle and a thicker leading edge, resulting in a larger projected area in the x-direction at the outlet where the channel slopes down. The resulting jet force acting on this area induces a certain force in the x-direction and then converts to a drag part, leading to an increase in the total drag. However, the other ABLE airfoils generate a lower chordal force at the outlet, which is insufficient to increase the total drag.

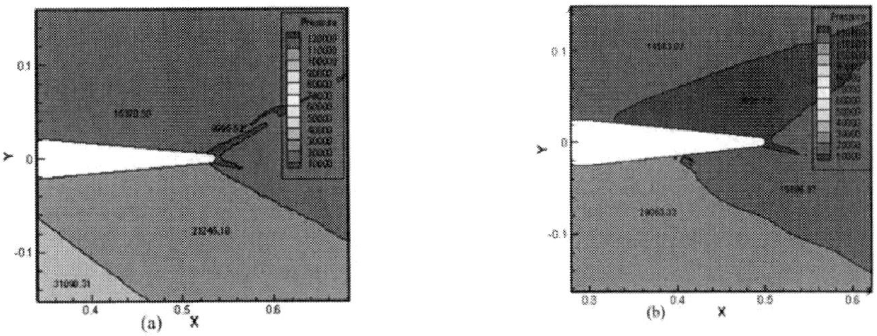

Figure 7. (a) Flow field contours of the ABLE quadrilateral airfoil near the tail edge; (b) Flow field contours of the baseline non-channel quadrilateral airfoil near the tail edge.

Figure 7 shows the results of the flow field contours near the tail edge of the ABLE quadrilateral and baseline no-channel airfoils. Because of the influence of jet-flow to flow field on the outlet of the channel, the expansion wave at the rear edge is weakened, resulting in a reduced low-pressure area, and the differential pressure between the upper and lower surfaces of the airfoil is reduced. This decreases the lift and drag coefficient of the ABLE quadrilateral airfoil.

The results confirm that the current configuration of the ABLE quadrilateral and ABLE hexagon airfoils may not reach the ideal design point. Therefore, we need to modify and optimise the shapes and parameters. Referring to figure 5(c), with such designs, the ABLE diamond airfoil exhibits an optimal aerodynamic performance in supersonic continuous flow with a maximum increase in L/D of approximately 25% at α=4°. Therefore, it is selected as the baseline airfoil for further analysis in hypersonic rarefied flow.

4.2. Aerodynamic performance of airfoils in rarefied flow

Taking the optimal case in the continuous flow as the diamond and ABLE diamond airfoils for further research, we obtained a series of results in rarefied flow using SPARTA under flow conditions based on an altitude of 70–90 km during re-entry. These simulations are intended for modelling the high-altitude hypersonic rarefied flow around the airfoils. With one of the parameters held constant, the effects of Mach and Knudsen Numbers, which represent different heights and air densities, on the aerodynamic performance are quantified.

First, the force coefficient and lift-to-drag ratio are computed with Ma=22 and 28, respectively, while maintaining an altitude of 80 km (with the Knudsen Number kept constant); the molecular density is 3.8317×10^{20} /m^3, and Kn is 0.044 with the slip-flow dominating the flow. The slip-flow starts to appear when Kn is greater than 0.01, as the gas becomes more and more rarefied; the flow is characterised to be in the transition and free molecular regimes when Kn reaches 0.1 and 10, respectively [13].

Figure 8 shows the aerodynamic parameters of the airfoils. The variations in Cl, Cd, and L/D with the angle of attack appear to be largely the same as those under the continuous flow condition, indicating there might be some similitude between the continuous and rarefied flows. With the increase in the angle of attack, the drag coefficient does not increase as much as it did under the continuous flow condition. This is because the flow regime is so rarefied that the freestream flows around the airfoil more smoothly, resulting in a lower differential pressure between the upper and lower surfaces of the airfoil relative to the one in the continuous flow.

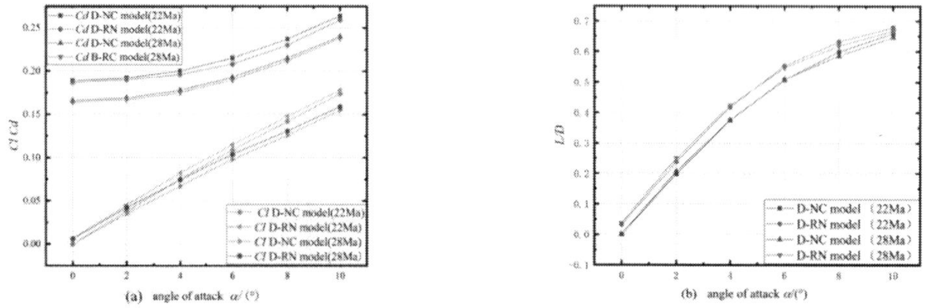

Figure 8. (a) Cl and Cd of baseline diamond airfoil and ABLE diamond airfoil; (b) *L/D* of airfoils.

At a certain Knudsen number, both *Cl* and *Cd* decrease with the increase in the Mach number. The ABLE concept can still yield reduced drag and increased lift and *L/D* ratios. Although the drag reduction ratio is only approximately 2.6%, the lift increment ratio reaches approximately 8–10%, thus increasing the *L/D* ratio. Table 3 gives the force coefficient values.

Under a such rarefied flow condition, for α=4°, the skin friction drag of the ABLE airfoil at 22 Ma reaches up to 75.12% of the total drag, with the wave drag ratio being less than 25%. With the curved-channel ABLE concept, the skin friction drag coefficient increases by only 3.35%, while the reduction ratio of the other drags (including the wave drag) increases by 20.08%. However, because of the much lower proportion of the wave drag, the drag reduction performance of the airfoil is much weaker when utilising the curved channel in the rarefied flow. Moreover, the drag reduction and lift increment rates hardly vary with Ma at any given angle of attack. In addition, because of the obvious rarefaction effect, the lift-to-drag ratio of the ABLE airfoil is much lower than that under the continuous flow condition. This is consistent with that reported previously [14, 15]: The maximum lift-drag ratio of a waverider reaches 7.5 when flying at an altitude of 30 km; however, at 90 km, the result calculated using the DSMC is only 0.3, which is much lower than that calculated under the continuous flow condition.

Table 3. Force coefficients with an angle of attack of 4° at 22 Ma.

	Skin friction drag	Other drags(including wave drag)	Total drag
Original airfoil	0.151	0.050	0.201
ABLE airfoil	0.156	0.040	0.196
Δ(•)	3.35%	−20.08%	−2.49%

Considering the airfoils with α=4°, parameters as listed in table 4, *Cd* does not change considerably for the ABLE airfoil; however, *Cl* and *L/D* increase to a great extent, and the lift-to-drag ratio increases by more than 12% while the maximum heat flux increases no more than 8% compared with the baseline airfoil. From this, we can conclude that the ABLE airfoil with a bluntness radius *Rn*=0.8%*L*, which is reasonable for continuous flows, meets the required design for increasing *L/D* while maintaining a relatively low heat flux increment in the rarefied flow.

Table 4. Heat flux and force coefficient with an angle of attack of 4° at 22 Ma.

	$Q_{max}(\times10^6)$	*Cd*	*Cl*	*L/D*
Original airfoil	1.001	0.201	0.075	0.373
ABLE airfoil	1.080	0.196	0.082	0.418
Δ(•)	7.89%	−2.49%	9.33%	12.06%

By analysing the pressure contour distributions of the diamond airfoil and the corresponding ABLE airfoil in figure 9, we can preliminarily reveal the aerodynamic characteristics of the ABLE airfoil in the rarefied flow. With the curved channel, most of the wall where the stagnation-region high pressure acts is removed, resulting in a suction high pressure at the throat of the channel, which is slightly greater than the stagnation pressure, and there is no obvious leading edge shock wave interaction like that in the continuous flow. Compared with the pressure distribution contour in the continuous flow given in [8], we find that the shock waves in the rarefied flow are very weak and relatively concentrated at the leading edge; the shock waves are very close to the airfoil surface, and the shock layer region is very narrow under the action of hypersonic airflow. Within the scope of the angle of attack studied, in the continuous flow, the lift coefficient is always greater than the drag coefficient; however, in the rarefied flow, the lift and drag coefficients decrease, with the drag coefficient being greater than the lift coefficient, resulting in a significant reduction in the lift-to-drag ratio.

For the second test, a shuttle is subjected to varying flow conditions at different flight heights across a typical trajectory. This requires understanding the effect of air density (*Kn*) on the aerodynamic performance of the airfoil. In fact, the lower the density, the greater the value of *Kn*.

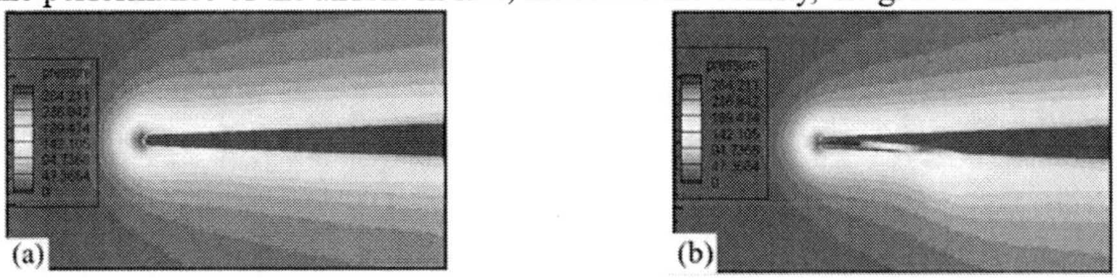

Figure 9. (a) Flow field of the diamond airfoil; (b) that of curved-channel ABLE airfoil.

Figure 10 shows the variation in the aerodynamic performance parameters with *Kn*. At the same *Ma*, with the increase in the Knudsen number, the lift and drag coefficients both first rapidly increase and then gradually increase, and the lift–drag ratio decreases gradually. The Knudsen number shows no obvious effect on the drag reduction and lift increment efficiency.

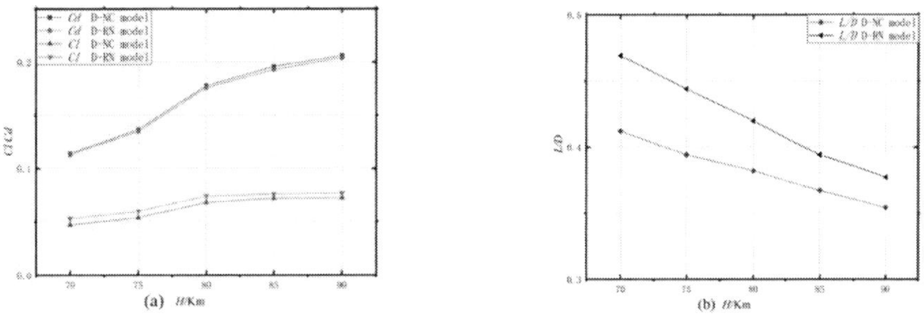

Figure 10. (a) *Cl* and *Cd* of airfoils with different *Kn*; (b) *L/D* of airfoils with different *Kn*.

However, compared with the influence of the Mach number, the effect of curved channel on the increment in the lift-to-drag ratio presents a decreasing trend with the increase in the Knudsen number.

5. Discussion and Conclusions

We investigated a set of two-dimensional supersonic/hypersonic symmetrical airfoils and corresponding curved-channel ABLE airfoils in this study. The continuous and rarefied flow fields around the ABLE airfoils during re-entry were simulated by solving the Navier–Stokes equations and using the DSMC method, respectively. The simulation results show that:

(1) In a supersonic continuous flow, creating a curved channel can help reduce the drag and increase the lift and effectively improve the lift-to-drag ratio. A curved-channel ABLE diamond airfoil provides an optimal aerodynamic performance relative to the other airfoils tested.

(2) In a hypersonic rarefied flow, because of the relatively low proportion of the wave drag in the total drag, the drag reduction performance of the channel is relatively insignificant. However, the lift

increment performance and the lift-to-drag ratio can still be effectively improved. Because of the rarefaction gas effect, the airfoil aerodynamic performance is seriously deteriorated compared with that in the continuous flow.

(3) Under the same Knudsen number, the lift and drag coefficient of the airfoil decrease with the increase in the Mach number, and the effect of the channel on the aerodynamic performance remains largely unaffected by the Mach number.

(4) Under the same Mach number, the lift and drag coefficient increase with the increase in the Knudsen number, while the Knudsen number has a relatively obvious influence on the channel effect: the increased lift-to-drag ratios decrease with the increase in the Knudsen number.

To summarize, we verified the feasibility of the previous curved-channel ABLE concept applied to several classical supersonic/hypersonic symmetrical airfoils. There might exist some differences and similitude between continuous and rarefied flows. Therefore, we verified the efficiency of the ABLE concept under flight conditions where the gas rarefaction dominates the flow regime. The paper provides insights into the requirements of the new design.

References

[1] Ruffin S M, Gupta A and Marshall D D 2000 Supersonic channel airfoils for reduced drag *J. AIAA* **38** 480–6

[2] Gupta A, Ruffin S, Newfield M and Yates L Aerothermodynamic performance enhancement and design of sphere-cones using the artificially blunted leading edge concept *37th Aerospace Sciences Meeting and Exhibit* 1999 Reno NV USA

[3] Gupta A and Ruffin S M Investigation of artificially blunted leading edge geometries with curved channels for high speed drag reduction *38th Aerospace Sciences Meeting and Exhibit* 2000 Reno NV USA

[4] Gupta A and Ruffin S M 1999 Optimal artificially blunted leading-edge airfoils for enhanced aerothermodynamic performance *J. Spacecraft Rockets* **36** 499–506

[5] Pezzella G, Filippone E and Serpico M Re-entry aerodynamics and aerothermodynamics analyses of the flying test bed USV-X in the framework of a high lift return *16th AIAA/DLR/DGLR Int. Space Planes and Hypersonic Systems and Technologies Conf.* 2009 Bremen Germany

[6] Kandlikar S G, Garimella S, Li D, Colin S and King M R 2005 *Heat transfer and fluid flow in minichannels and microchannels* (Oxford: Elsevier)

[7] Pitakarnnop J and Wongpithayadisai R 2014 Rarefied gas flow in pressure and vacuum measurements *Acta Imeko* **3** 60–3

[8] Geng Y F and Yan C 2010 Application of curved channel ABLE conception to supersonic airfoil drag reduction *Acta Aerodynamica Sinica* **28** 70–5

[9] Allegre J, Raffin M, Chpoun A and Gottesdiener L 1994 Rarefied hypersonic flow over a flat plate with truncated leading edge *Progress in Astronautics and Aeronautics* vol 160 ed A R Seebass (Washington: AIAA) pp 285–295

[10] Padilla J F 2010 Comparison of DAC and MONACO DSMC codes with flat plate simulation *NASA Technical Reports Server*

[11] Koppenwallner G 1993 Aerodynamics of rarefied gases and high-altitude flight *Rarefied Gas Dynamics: Space Sci. Eng.* ed B D Shizgal and D P Weaver (Washington: AIAA) pp 75–103

[12] Gallis M A, Koehler T P, Torczynski J R and Plimpton S J 2015 Direct simulation Monte Carlo investigation of the Richtmyer-Meshkov instability *Phys. Fluids* **27** 084105

[13] Pitakarnnop J, Wongpithayadisai R and Phanakulwijit S 2017 Alternative models and numerical simulations of rarefied gas flow in vacuum systems *Meas.* **126** 417–20

[14] Rault D F G 1992 Aerodynamic characteristics of a hypersonic viscous optimized waverider at high altitudes *J. Spacecraft Rockets* **31** 719–27

[15] Cui K, Zhao D and Yang G 2007 Waverider configurations derived from general conical flowfields *Acta Mech. Sinica* **23** 247–55

Error Analysis of High Precision Two-way Time Difference Measurement in Satellite-ground Links

Shuaihe GAO[1,a], Yuqi LIU[2], Lishu GUO[2,b], Leyuan SUN[3], Chen ZHOU[4] and Kun SONG[2]

[1]National Time Service Center, Chinese Academy of Science, China
[2]Science and Technology on Electronic Information Control Laboratory, China
[3]College of Intelligence Science and Technology, National University of Defense, China
[4]School of Electronic Information, Wuhan University, China

Corresponding author's e-mail: [a]gaoshuaihe@ntsc.ac.cn, [b]guolishu@aliyun.com

Abstract. High precision time difference measurement between satellite and earth is an important part of space-based time-frequency system application in the future. With the on-orbit operation of high-precision atomic clocks, the accuracy of satellite-ground time difference measurement link has become the bottleneck of space-based time-frequency distribution. In this paper, the basic principle of two-way time difference measurement is derived. The system error and equipment error involved are analyzed, and the error control strategy is given. The above work will provide engineering guidance for high-precision two-way time difference measurement in the design of satellite-ground links.

1. Introduction

The method of time difference measurement between satellite and ground station includes single frequency one-way measurement, dual frequency one-way measurement, dual frequency two-way time difference measurement etc.[1-4]. Taking the traditional GNSS time service as an example, the typical method is that the satellite continuous broadcasts navigation signal, and the ground user receives the single frequency signal to realize the time synchronization at 10 nanoseconds. In order to further improve the accuracy of time difference measurement, the dual frequency signal sent by GNSS is received to achieve time synchronization at nanoseconds, where the dual frequency signal can correct the ionospheric error introduced in the spatial transmission of navigation signal. However, for the needs of higher accuracy time difference measurement, such as the European Atomic Clock Ensemble in Space (ACES) plan, the measurement accuracy of picoseconds level is the upper limit, which is limited by the tropospheric dispersion effect in the high frequency satellite-ground link.

In this paper, the basic time difference measurement principle of microwave satellite-ground link is derived. The errors are analyzed and the corresponding control strategies are given, which will provide engineering guidance for high-precision two-way time difference measurement between satellite and ground, and technical means for application of space-based time-frequency system.

2. Two-way time difference measurement

The corresponding two-way ranging process[5] is shown in Figure 1, continuous receiving and sending ranging signal between satellite and ground station. When the two-way measurement is taken

Content from this work may be used under the terms of the Creative Commons Attribution 3.0 licence. Any further distribution of this work must maintain attribution to the author(s) and the title of the work, journal citation and DOI.
Published under licence by IOP Publishing Ltd

out, the clock difference solution can be completed. Assuming that the extracted downlink ranging signal is sent by satellite at t_S^{Snd} and received by the ground station at t_G^{Rev}, and the uplink ranging signal is sent by the ground station at t_G^{Snd} and received by the satellite at t_S^{Rev}.

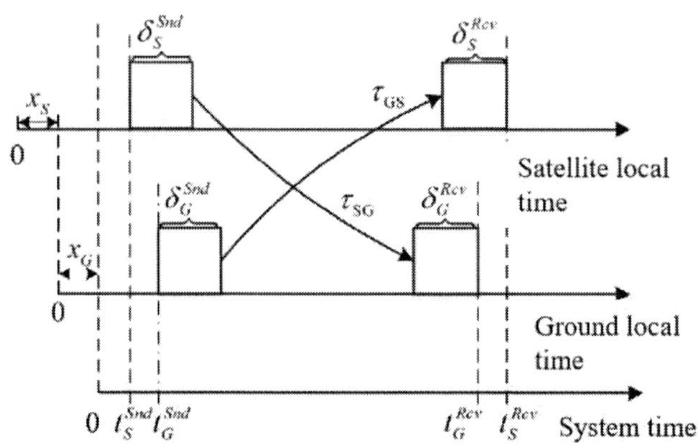

Figure 1. Process of two-way ranging between satellite and ground

The one-way distance measurement between satellite and ground station is calculated by the local time difference at the time of transmitting and receiving, including the clock difference information of satellite and ground station. The measurement between satellite and ground station is established as follows.

$$\begin{cases} \rho_{SG} = \left|R_G\left(t_G^{Rcv}\right) - R_S\left(t_S^{Snd}\right)\right| + \delta_{SG}^{sagnac} + cx_G\left(t_G^{Rcv}\right) - cx_S\left(t_S^{Snd}\right) + c\delta_G^{Rcv} + c\delta_S^{Snd} + \delta_{SG}^{ion}\left(R_S\left(t_S^{Snd}\right), R_G\left(t_G^{Rcv}\right)\right) \\ \quad + \delta_{KG}^{tro}\left(R_S\left(t_S^{Snd}\right), R_G\left(t_G^{Rcv}\right)\right) + \delta_{SG}^{rel\_period} + \delta_{SG}^{gra} + \delta_{SG}^{MP} - \delta_{SG}^{PCO} + \varepsilon_{SG} \\ \rho_{GS} = \left|R_S\left(t_S^{Rcv}\right) - R_G\left(t_G^{Snd}\right)\right| + \delta_{GS}^{sagnac} + cx_S\left(t_S^{Rcv}\right) - cx_G\left(t_G^{Snd}\right) + c\delta_S^{Rcv} + c\delta_G^{Snd} + \delta_{GS}^{ion}\left(R_G\left(t_G^{Snd}\right), R_S\left(t_S^{Rcv}\right)\right) \\ \quad + \delta_{GS}^{tro}\left(R_G\left(t_G^{Snd}\right), R_S\left(t_S^{Rcv}\right)\right) + \delta_{GS}^{rel\_period} + \delta_{GS}^{gra} + \delta_{GS}^{MP} - \delta_{GS}^{PCO} + \varepsilon_{GS} \end{cases} \tag{1}$$

Here, c is the speed of light, R_S and R_G are the position vectors of satellite and ground in the earth fixed system, x_S and x_G are the clock difference of satellite and ground station respectively, δ^{Snd} and δ^{Rcv} are the hardware transmission delay of signal transmission channel and receiving channel; δ^{rel_period} is the equivalent delay of relativistic periodic effect, δ^{gra} is the signal transmission delay caused by the gravitational field of the earth, δ^{sagnac} is the equivalent delay of Sagnac effect, δ^{MP} is the equivalent delay of multipath effect, δ^{PCO} is the equivalent delay of phase center offset, and ε is ranging noise.

In the time comparison between the satellite and the ground station, the time of the ground atomic clock is taken as the reference time, so the ground clock difference can be considered as $x_S = 0$. In the earth fixed system, the position coordinate of the ground station is constant, so the measurement equation can be simplified as:

$$\begin{cases} \rho_{SG} = \left|R_G - R_S\left(t_S^{Snd}\right)\right| + \delta_{SG}^{sagnac} - cx_S\left(t_S^{Snd}\right) + c\delta_G^{Rcv} + c\delta_S^{Snd} + \delta_{SG}^{ion}\left(R_S\left(t_S^{Snd}\right), R_G\right) \\ \quad + \delta_{SG}^{tro}\left(R_S\left(t_S^{Snd}\right), R_G\right) + \delta_{SG}^{rel\_period} + \delta_{SG}^{gra} + \delta_{SG}^{MP} + \delta_{SG}^{OF} + \varepsilon_{SG} \\ \rho_{GS} = \left|R_S\left(t_S^{Rcv}\right) - R_G\right| + \delta_{GS}^{sagnac} + cx_S\left(t_S^{Rcv}\right) + c\delta_S^{Rcv} + c\delta_G^{Snd} + \delta_{GS}^{ion}\left(R_G, R_S\left(t_S^{Rcv}\right)\right) \\ \quad + \delta_{GS}^{tro}\left(R_G, R_S\left(t_S^{Rcv}\right)\right) + \delta_{GS}^{rel\_period} + \delta_{GS}^{gra} + \delta_{GS}^{MP} + \delta_{GS}^{OF} + \varepsilon_{GS} \end{cases} \tag{2}$$

When the time of satellite receiving and transmitting ranging signal is inconsistent, the two unidirectional ranging signals contain the satellite clock difference at different times. However, it needs to be calculated to the same time.

$$x_S\left(t_S^{Rcv}\right)=x_S\left(t_S^{Snd}\right)+\Delta x_S \tag{3}$$

Here, Δx_S is the satellite clock difference correction quantity. The clock difference between the satellite and ground station can be calculated by making the difference between the two-way ranging equation.

$$\begin{aligned}x_S\left(t_S^{Snd}\right)=&\Big\{\left(\rho_{GS}-\rho_{SG}\right)+\left|R_G-R_S\left(t_S^{Snd}\right)\right|-\left|R_S\left(t_S^{Rcv}\right)-R_G\right|-c\Delta x_S+c\left(\delta_G^{Rcv}-\delta_G^{Snd}\right)-c\left(\delta_S^{Rcv}-\delta_S^{Snd}\right)\\&+\left[\delta_{SG}^{ion}\left(R_S\left(t_S^{Snd}\right),R_G\right)-\delta_{GS}^{ion}\left(R_G,R_S\left(t_S^{Rcv}\right)\right)\right]+\left[\delta_{SG}^{tro}\left(R_S\left(t_S^{Snd}\right),R_G\right)-\delta_{GS}^{tro}\left(R_G,R_S\left(t_S^{Rcv}\right)\right)\right]\\&+\delta_{SG}^{sagnac}-\delta_{GS}^{sagnac}+\delta_{SG}^{rel\_period}-\delta_{GS}^{rel\_period}+\delta_{SG}^{gra}-\delta_{GS}^{gra}+\delta_{SG}^{MP}-\delta_{GS}^{MP}+\delta_{SG}^{PCO}-\delta_{GS}^{PCO}+\varepsilon_{SG}-\varepsilon_{GS}\Big\}/2c\end{aligned} \tag{4}$$

In order to quantitatively analyze the accuracy level of two-way time difference measurement between satellite and ground station, the orbit related error, transmission path error and equipment characteristic error involved in the satellite orbit height of 400km and Ka band link are analyzed.

3. Error analysis related to orbit

Orbit related error items are related to the physical position of the satellite and the earth, including geometric delay correction error, clock error correction error, relativistic periodic term error, and gravitational delay.

3.1. *Geometric delay correction error*

In the geocentric earth fixed coordinate system, the space non-symmetry distance introduced by the relative motion of the satellite and the ground station can be equivalent to:

$$C_{orb}=\left(\left|R_G-R_S\left(t_S^{Snd}\right)\right|-\left|R_S\left(t_S^{Rcv}\right)-R_G\right|\right)/2c \tag{5}$$

According to the above formula, the space distance can be corrected by using the satellite precise orbit. If $d(t)=\left|R_G-R_S(t)\right|$ is the geometric distance between satellite and ground at time t, the above formula can be rewritten as follows:

$$C_{orb}=\left[\tilde{d}\left(t_K^{Snd}\right)-\tilde{d}\left(t_K^{Rcv}\right)+\delta d\left(t_K^{Snd}\right)-\delta d\left(t_K^{Rcv}\right)\right]/2c \tag{6}$$

Here, the superscript '~' represents the geometric distance calculated by the precise orbit, $\delta d(t)$ is the spatial distance error of the precise orbit at time t. Therefore, the space distance correction error is:

$$C_{orb}=\left[\delta d\left(t_K^{Snd}\right)-\delta d\left(t_K^{Rcv}\right)\right]/2c \tag{7}$$

The above formula shows that the space distance correction error is affected by the change rate of space distance error. $\sigma\left(C_{orb}\right)$ is better than 0.23ns when the precision orbit is centimeter scale and the error is random. In fact, the orbit error of the smoothed satellite is not completely shown as random error, that is, the space distance error $\delta d\left(t_K^{Snd}\right)$ and $\delta d\left(t_K^{Rcv}\right)$ at the two time points have correlation, which can be processed by two-way cancellation.

The equivalent time delay of earth rotation effect is the main geometric time delay in two-way time comparison.

$$C_{sagnac}=\frac{\omega_e}{c^2}\left[Y_G X_S\left(t_0\right)-X_G Y_S\left(t_0\right)\right] \tag{8}$$

The correction accuracy of Sagnac effect is affected by satellite orbit accuracy. Calculate the variance of the above formula.

$$D\left(C_{sagnac}\right)=\frac{\omega_e^2}{c^4}\left[Y_G^2 D\left(X_S\left(t_0\right)\right)+X_G^2 D\left(Y_S\left(t_0\right)\right)\right] \tag{9}$$

Taking Xi'an station as an example, if the orbit accuracy of the satellite is 10 cm(1 σ), it can be concluded that the two-way time synchronization error introduced by Sagnac effect is 4.3e-4ps.

3.2. Clock error correction

The uplink and downlink pseudo ranges contain the satellite clock difference at different receiving and transmitting times. In order to solve the clock difference, the uplink and downlink clock difference should be calculated as the clock difference at the same time, and the correction of clock difference is:

$$\Delta x_S = x_S\left(t_S^{Rcv}\right) - x_S\left(t_S^{Snd}\right) \tag{10}$$

When $t_G^{Rcv} = t_G^{Snd}$, the uplink and downlink pseudo distance includes the satellite clock difference at the same time of receiving and transmitting, where Δx_S is 0.

When $t_G^{Rcv} \neq t_G^{Snd}$, Δx_S depends on the frequency performance of satellite atomic clock and the time difference $\Delta t = t_S^{Rcv} - t_S^{Snd}$ between receiving and transmitting. For an atomic clock with a frequency accuracy of e-12, the correction value of the clock difference corresponding to the time difference of 10 ms is 0.01ps.

3.3. Relativistic periodic term correction error

The measurement of pseudo range and carrier phase between satellite and ground will be affected by general and special relativity effects. With the periodic motion of the satellite in the earth's gravitational field, the velocity and the gravitational potential change periodically, the equivalent time delay of the relativistic period term could be represented as:

$$\tau_{GS}^{rel\_period} = -\frac{2}{c^2}\sqrt{aGM_E} \cdot e\sin E \tag{11}$$

It needs to be corrected in the process of calculating the clock difference between satellite and ground. Its equivalent form is:

$$\tau_{GS}^{rel\_period} = -2R_S \cdot V_S / c^2 \tag{12}$$

R_S and V_S are the position and velocity vector of the satellite in the inertial system. The equivalent path delay of the relativistic effect can be calculated according to the orbit parameters of the satellite in the inertial reference system. For the special relativistic effect of the receiver and transmitter of the uplink and downlink pseudo range, the correction amount is as follows:

$$C_{rel\_period} = \left(\tau_{SG}^{rel\_period} - \tau_{GS}^{rel\_period}\right)/2 = 2R_S\left(t_0\right) \cdot V_S\left(t_0\right)/c^2 \tag{13}$$

When the three-axis position accuracy of satellite orbit is 10cm(1 σ) and the three-axis speed accuracy is 1mm/s(1 σ), the simulation calculation of the two-way time synchronization error introduced by the relativistic periodic effect is shown in Figure 2(a). The standard deviation of error is 0.15ps.

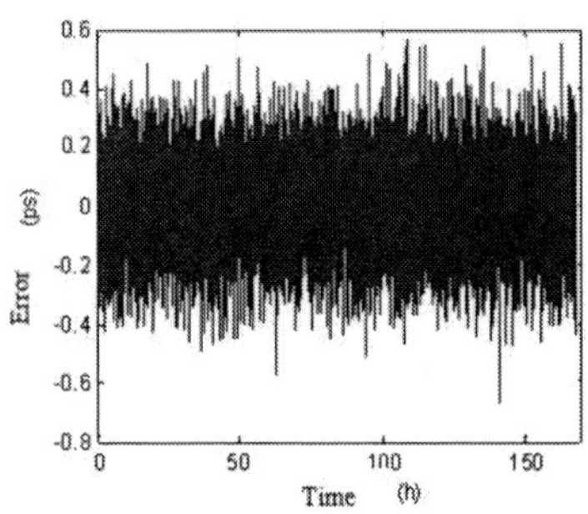

(a)speed accuracy is 1mm/s

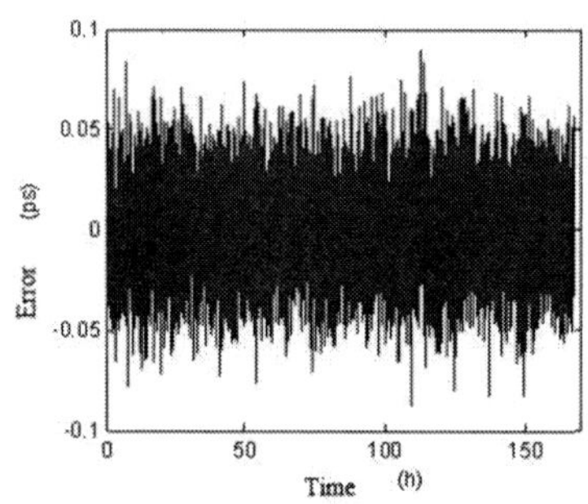

(b)speed accuracy is 0.1mm/s

Figure 2. Relativistic periodic effect correction error

When the three-axis position accuracy of satellite orbit is 10cm(1 σ) and the three-axis speed accuracy is 0.1mm/s(1 σ), the simulation calculation of the two-way time synchronization error introduced by the relativistic periodic effect is shown in Figure 2(b). The standard deviation of error is 0.02ps. The correction accuracy of relativistic effect is affected by the accuracy of satellite orbit.

3.4. Gravitational delay

The gravitational field will change the propagation speed of light or electromagnetic wave in space. The difference between signal propagation time in gravitational field and signal propagation time in vacuum (non gravitational field) is gravitational delay. The time delay of general relativity caused by the gravitational field of the earth is:

$$C_{gr} = \frac{2GM_E}{c^3} \ln \frac{r_S + r_G + d}{(r_S + r_G - d)} \tag{14}$$

Taking Xi'an station as an example, the gravitational delay between the satellite and the ground is shown in Figure 3. The maximum value can reach 300 ps, so the gravitational delay needs to be corrected. The gravitational delay correction is calculated by satellite and ground coordinates, so the gravitational delay correction error will be introduced along with the satellite orbit error. If the three-axis position accuracy of satellite orbit is 10cm(1 σ), the correction error of gravitational delay is shown in Figure 4. The correction error of gravitational delay introduced by satellite orbit is at the level of e-5ps.

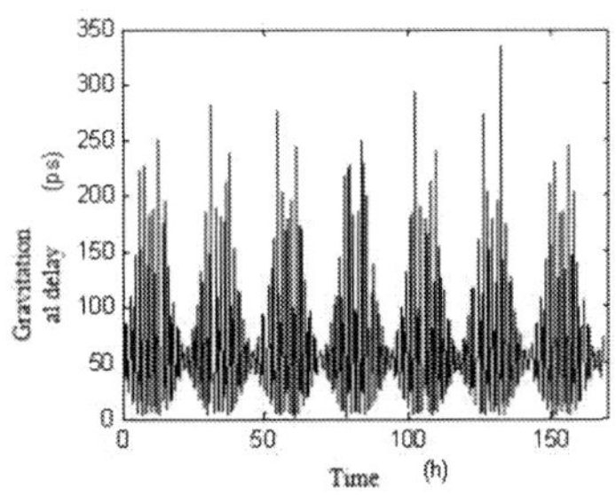

Figure 3. Gravitational delay

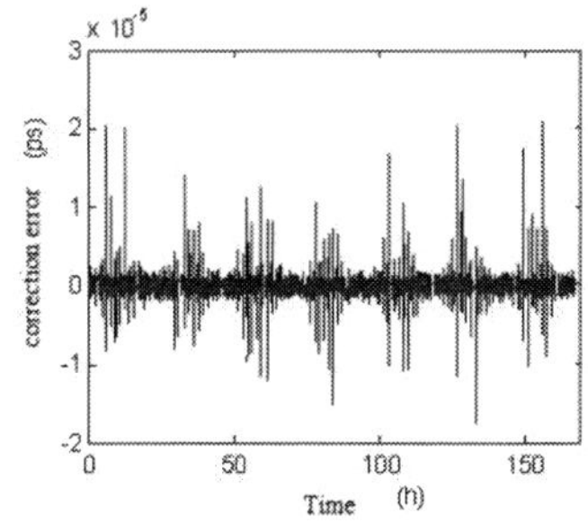

Figure 4. Gravitational delay correction error introduced by satellite orbit error

4. Transmission path error analysis

4.1. Tropospheric delay error analysis

Based on mesoscale meteorological model, the existing data is analyzed. From October 1 to 30, 2019, the nondispersive delay of the troposphere in Xi'an and Shanghai is in the range of 5-8ns as shown in Figure 5, which can be corrected by the model.

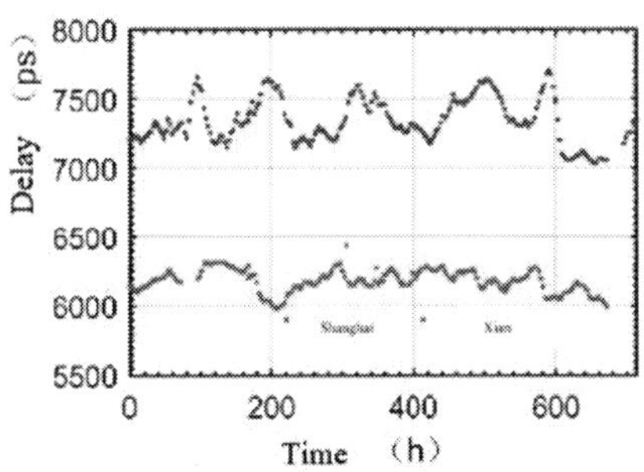

Figure 5. Tropospheric nondispersive delay

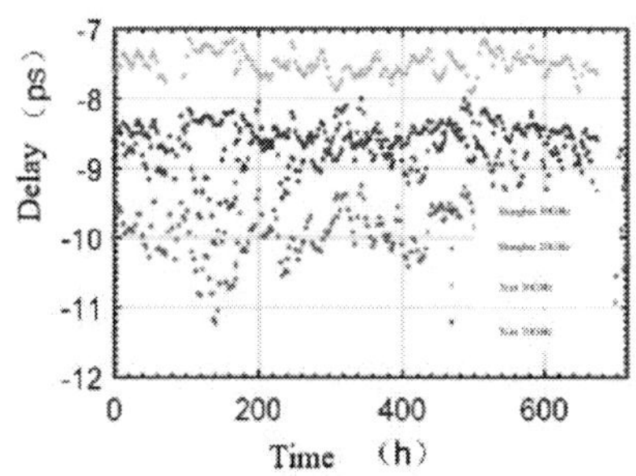

Figure 6. Tropospheric dispersion delay

Due to the continuous rotation of water vapor molecules, settling particles and oxygen molecules in the troposphere, the tropospheric dispersion delay is 11ps as shown in Figure 6, which could be corrected by atmospheric model or real-time temperature and humidity pressure data. The tropospheric dispersion should be comprehensively considered in high accuracy measurement.

4.2. Ionospheric delay error analysis

The ionospheric delay error is related to frequency. The first-order ionospheric error can be eliminated by multi frequency. The second-order effect of the ionosphere is determined by the earth's magnetic field and background electron density. In the year of great solar activity. The second-order dispersion effect of the ionosphere would be about 0.2ps. Considering the complexity of the system and the second-order effect, the ionospheric errors above the second-order can be processed by monitoring the solar activity.

5. Equipment characteristic error analysis

5.1. Receiver processing

Taking carrier measurement as an example, the phase measurement error sources of carrier tracking loop mainly include phase jitter and dynamic stress error. The error sources causing phase jitter include thermal noise σ_{tPLL}, oscillation frequency jitter σ_v and Allan mean square deviation $\sigma_A(\tau)$.

The estimation formula of σ_{tPLL} is as follows:

$$\sigma_{tPLL} = \frac{180^\circ}{\pi}\sqrt{\frac{B_L}{C/N_0}\left(1+\frac{1}{2T_{coh}C/N_0}\right)} \tag{15}$$

Here, B_L is the loop noise bandwidth, C/N_0 is the carrier noise ratio, T_{coh} is the coherent integration time. The above formula shows that reducing noise bandwidth and improving coherent integration time can reduce the tracking error.

The motion of the satellite and the mechanical vibration of the receiving device will cause the jitter of the reference oscillation frequency. In order to ensure the measurement accuracy, σ_v should be designed reasonably according to the carrier frequency.

The phase jitter noise will also be introduced into the oscillation frequency drift of Allan crystal along with the time accumulation. Mean square deviation σ_A is related to the Allan mean square deviation $\sigma_A(\tau)$ and coherent integration time T_{coh}. The formula is shown as follows.

$$\sigma_A = 360^\circ \frac{c}{\lambda} T_{coh}\sigma_A(\tau) \tag{16}$$

Here, c is the speed of light, λ is the carrier wavelength. Setting T_{coh} is 1ms, when $\sigma_A(\tau) < 10^{-12}$, $\sigma_A < 0.01ps$, it could be ignored.

In conclusion, the total mean square deviation of phase jitter caused by error sources is:

$$\sigma_i = \sqrt{\sigma_{tPLL}^2 + \sigma_v^2 + \sigma_A^2} \tag{17}$$

5.2. Multipath effect

Multipath effect is a technical problem that has not been solved effectively in the radio communication, and it is also an important factor that affects the high-precision measurement of the time difference between satellite and ground. Correlation output is affected by multipath, which will lead to zero crossing offset. For multipath effect, the most effective way is to control the environment around the equipment. It can also be achieved by improving the receiver's own suppression ability.

5.3. Phase center shift of Antennas

The measurement between the satellite and the ground station is corresponding to the distance between the phase centers of the two antennas. The propagation delay between the phase centers needs to be corrected to the center of mass. Generally, the antenna phase center is a function of antenna azimuth and pitch angle. The biggest change usually does not exceed the geometric size of the antenna, which is always close to the geometric center of the antenna.

5.4. Time delay error of channels

The hardware structure and working temperature will bring variation of channel delay. On the one hand, time delay is uncertain. On the other hand, the delay will change with the ambient temperature. Therefore, it is necessary to control the delay difference in the receiving and sending channels. The following measures should be taken for the control of time delay characteristics of the channels.

- In order to reduce the change of channel delay, the filter which is not sensitive to the change of component parameters could be selected.
- Control the change range of component parameters and minimize the change of parameters.
- A closed-loop self-tuning system should be designed, which could detect and compensate when the delay changes.
- To ensure small temperature change, which is an important error factor in the process of engineering design.

6. Conclusion

According to the above analysis, the following aspects need to be considered in the design of high-precision satellite-ground link.

- Obtaining high-precision satellite orbit data is the key content of satellite ground-link design.
- Transmission path error is an important factor to be considered in high-precision time difference measurement, especially the tropospheric dispersion effect of high-frequency signal.
- The key factors to be controlled in design phase include the equipment error, the satellite ground link margin, the surrounding environment, the stability of the equipment itself and so on.

Acknowledgments

This paper was supported by CAS "Light of West China" Program and Shaanxi Provincial Talents Plan.

References

[1] Shusen Tan 2006 *Engineering Science* **12** pp 74-78.
[2] Pavel Vyskocil and Jiri Sebesta 2009 *IWSSC* pp 230-234.
[3] M. P. Hess, J. Kehrer, M. Kufner and et al. 2011 *Proceedings of the IEEE International Frequency Control Symposium and Exposition* pp 1-8.
[4] M. Uhrich, P. Guillernot, P.Aubry and et al. 2000 *IEEE Transactions on Ultrasonics, Ferroelectrics, and Frequency Control* **47** pp 1134-1139.
[5] Li Liu and Chunhao Han 2004 *Progress in Astronomy* **3** pp 219-226.

Research Status and Expectation of Intelligent Deformable Wing Aircraft

Liangzheng Ma[1], Wei Huang[1], Yuntu Ao[1], Lin Yi[1,*] and Ying Wang[1]

[1] Air and missile defense college, Air Force Engineering University, Xi'an, China

[*]Corresponding author e-mail: 841420273@qq.com

Abstract. The performance requirements for military reconnaissance, military strikes, long-range transport and medical rescue are constantly increasing. The traditional wing has bottlenecks in terms of improving flight efficiency and mission adaptability, which restrict the performance of the aircraft. With the development of new intelligent materials and continuous pursue of aerodynamic performance, the concept of deformable wing emerges as the times require. This paper analyzes the advantages and disadvantages of deformable wing aircraft, summarizes the research results and current situation at home and abroad, and finally makes a prospect for the development of the technology.

1.Introduction

It is difficult to ensure that the guided ammunition has good maneuverability and control performance in a large range of flight speed envelope in the traditional flight mode with constant wing shape. Under this background, the concept of deformable wing came into being. The lift coefficient of airfoil is obviously affected by different deformation; the lift oscillation phenomenon can be reduced by the appropriate amount of deformation; the increase of deformation can restrain the vortex separation in the process of pitching, and thus delay the deep stall hysteresis effect. At present, domestic and international research on deformable wings is still mainly focused on the single direction of the wing large scale deformation and wing folding deformation. But for the smooth continuous deformation research and multi-dimensional deformation structure design and other aspects of the research is still limited, which need to further study. The existing deformable aircraft can meet the requirements of the changes of backward sweep angle and chord curvature under different conditions. When these deformable wings are deformed in the deformable direction, there is usually a beam, plate, or other structure or locking mechanism in the non-deformable direction to ensure the overall rigidity and load capacity of the wing. However, the single wing deformation type can hardly meet the requirements of wide speed range flight and multi-environmental adaptability of the aircraft. The true deformable wing concept is a seamless integration of new intelligent materials, actuators, exciters New design concepts. It is necessary to carry out research on deformable wings with certain structural stiffness and load-bearing capacity that can achieve multi-dimensional continuous deformation.

In this paper, based on the summary of scholars' research on the deformed wing, the research status of the deformed wing aircraft at home and abroad is summarized. Finally, the development of the deformed wing is prospected on the basis of the structural form and flight principle of the wing.

2. Development history at home and abroad

Content from this work may be used under the terms of the Creative Commons Attribution 3.0 licence. Any further distribution of this work must maintain attribution to the author(s) and the title of the work, journal citation and DOI.
Published under licence by IOP Publishing Ltd

In the early research, due to the lag of the development of material science, researchers can only control the relative motion of each rigid body of the wing to achieve the deformation of the wing, such as expansion, folding, torsion, bending, etc. So the structure designed is cumbersome and complex, which limits the improvement of the overall performance of the aircraft. With the development and application of smart materials and smart actuators, people have a strong interest in flexible wing, and have carried out extensive research. At the same time of ensuring the smoothness and continuity of the wing shape, the geometric parameters such as wing area, angle of attack, chord length and curvature are changed through flexible deformation, which effectively improves the aerodynamic performance of the aircraft. The airspace and speed domain of future vehicles are expanding, and a fixed aerodynamic shape may not be able to meet the aerodynamic and flight performance requirements of vehicles in different flight conditions. After entering the 1990s, the western developed countries have developed new concept wing design technologies such as active flexible wing technology (AFW) and adaptive wing technology (AWT).

In the 21st century, with the continuous emergence of new space missions, the requirements for inflatable wing aircraft are also increasing, especially in recent years, the most urgent demand for large-scale inflatable wing aircraft. The development of materials and electronic technology in the space age, as well as a deeper understanding of aerodynamics, prompted researchers to start to consider the role of deformable wings in high-performance aircraft.

3. Advantages and disadvantages of deformed wing[1-3]

Advantages of deformed wing:

- Expand flight envelope;
- Replace the traditional control surface to reduce the complexity of the system;
- Improve reliability;
- Adjust the flow field around the aircraft to reduce resistance;
- Reduce vibration and flutter;
- Reduce radar reflector and improve stealth performance;
- Reduce the empty weight of the aircraft and increase the payload;
- Improve the efficiency of fuel and increase the endurance of the aircraft.

Disadvantages of deformed wing:

- Complex design;
- Operational difficulty;
- High failure rate;
- Limited space for the actuators that deploy and retract the wings.
- Adverse effects of structural fatigue and kinematic hysteresis caused by repetitive motion of the wing;
- Adverse effects on the aircraft of the change in aerodynamic centre and the change in the centre of gravity due to deformation;
- New materials are needed to reduce aircraft weight;
- Complex variable wing mechanism limits the improvement of a series of performance of aircraft, such as load, shape, stealth, etc.

4. Research progress of intelligent deformable wing

4.1. Research progress abroad.

According to the existing foreign data, there are two main types of the layout of the deformable aircraft. One is the local deformation with "intelligent" materials; the other is the large-scale deformation with mechanical structure (or mechanical plus intelligent material structure). Local deformation belongs to the field of flow control, such as changing the local airfoil shape or bending degree of the wing to improve the aerodynamic efficiency of the wing. Large scale deformation mainly refers to the deformation in the form of changing the wing area and geometric parameters. Its deformation is fairly big, which can ensure that the aircraft can obtain excellent aerodynamic performance at different flight speeds.

The earliest concepts of deformable aircraft date back to the deformable vehicle MAK-10 , designed by Russian-born engineer Ivan Makhonine [4].

In 1951, American Bell built the X-5 variable swept-back wing aircraft, which could maintain a swept-back angle of 20°, 40° and 60° at three angles, with the wings changing from minimum to maximum swept-back angles in less than 30 s.

In 2016, Jenett et al. Of MIT designed and developed an active torsional deformation wing structure with discrete distribution and reversible assembly based on ultra light elastomer element[5] .

In 2010, Manzo and Garcia of Cornell University proposed a hyperelliptic arc span (HECs) wing structure with spanwise bending based on the angle of albatross wing deformation[6].

In 2011, Lesieutre et al. at Pennsylvania State University designed a flexible truss unit to replace the fixed support structure in a normal wing, allowing for wing span variation.

In 2011, Mestrinho et al. presented a variable-extension wing for small UAVs, made of composite materials and divided into a fixed part attached to the fuselage and a movable part that slides within the fixed part shell.

In 2012, Wang et al. from Nanyang Technological University proposed a single-degree-of-freedom mechanism based on sarrus connecting rods to achieve wing span variation.

In 2012, Woods and Werley of Swansea University, UK, designed a bending wing imitating fishbone structure according to the characteristics of small chord stiffness of fishbone structure[7].

In 2017, Santos et al. of the University of Beira, Portugal, designed wings that can change span length.

In 2018, Yu et al. proposed a design method for a rigid sliding skin. This rigid sliding skin is primarily used on six-degree-of-freedom motion stages.

In 2013, Heo et al. From Korea's Koryo university designed a passive deformable wing with flexible cell core based on deformable honeycomb structure. They respectively studied the deformation of airfoil with three honeycomb cores (chiral, regular and concave hexagon) under static load[8].The results show that the concave hexagon honeycomb core shows the highest flexibility , and the stress in the local honeycomb wall is lower than that of other parts under shear load, which means that the concave hexagon honeycomb core has the potential to be used in the deformation wing filling structure.

Cornerstone's research team studied a shape-memory polymer skin that could be used on variable chord length wings, using two air pistons filled with heated or cooled gas to control the stretching or shrinking of the shape-memory polymer.

Linkoping University of Sweden encouraged its students to carry out the research of the deformed aircraft, and put forward 12 plans. Through the review, three plans were determined for design, manufacture and demonstration flight, which are respectively: the expansion wing plan, the variable swept wing plan and the combination plan of the two wings. The final result is that the effect of expansion wing is good, the combination plan of the two wings is not obvious, and the test flight of the variable swept wing has not been carried out successfully.

Rediniotis et al. at Texas A&M University have designed a bionic hydrofoil that can be actively deformed by rotating six segments of a spine driven by a shape memory alloy, with metal skins mounted on the outside of the spine, which can interlock without interference during the deformation process.

The existing technologies of research results include: variable swept wing technology, tilt wing technology, X-Wing aircraft technology, etc.

4.2. Domestic research progress.

From the domestic data, the main research contents of the deformable wing focus on the internal deformable skeleton and the external deformable skin. In recent years, Harbin Institute of Technology,

Northwestern Polytechnical University, Nanjing University of Aeronautics and Astronautics and other units have carried out relevant research on the internal deformable structure and external deformable skin of the deformable wing. Bai Peng, Chen Qian, Guo Jianguo, Xu Guowu et al.[9-11] analyzed the aerodynamic and structural characteristics of the telescoping deformation aircraft and the variable swept back and variable swept forward aircraft. Zhang Jie et al [12] established the dynamic model of variable sweep and variable spread length combined deformation aircraft, and carried out the dynamic response analysis.

In 2016, Zhang Chengchun and others of Jilin University designed a bionic deformed wing of amphibious aircraft based on the characteristics of the wing posture during the diving process of Kingfisher. The wing folded and deformed through the contraction and extension of the expansion bar and its relative motion with the slide. By analyzing the structural form of deformable corrugated sandwich plate and the deformation principle of animal vertebrae, Lv Zhengyang[13] of Harbin Institute of Technology proposed the design idea of deformable wing structure: to realize the large deformation of the whole wing by accumulating small deformation in the deformable truss unit. Lili Gu et al[14]. of Nanjing University of Aeronautics and Astronautics took the pigeon as an object of study, abstracted a bionic structural model based on its flight attitude and skeletal characteristics, proposed a bionic deformation wing scheme with a two-stage layout, and simulated and analyzed pigeon's different flight states such as cruising and climbing. In addition, Chunlin Gong et al. of Northwestern Polytechnical University proposed a solution based on the Karhunen-Loève method and the Kriging method[15] for the optimization problem of distributed deformable vehicles, and verified the feasibility of the proposed methods.

5. Concluding Remarks

In this paper, the development process of deformable wing technology from its generation to now is summarized. At the same time, the technology of deformable wing is analyzed. It is understood that the existing technology is not enough to give full play to the advantages of deformed wing. At present, the application of deformable wing technology in large aircraft has been developed, but due to the continuous development of new technology and new materials, the technology still has great potential. If the deformable wing technology is popularized in the small self-adaptive aircraft, it will bring a new revolution in this field. At the same time, the role of UAV in civil and national defense will be greatly improved. In terms of development prospects, the technology of deformed wing can also bring new development to the traditional large-scale human controlled aircraft. From the current development point of view, the research of deformed wing UAV is still in the primary stage, and there are many aspects to be improved and in-depth. For example, in the aspect of mechanism design, the existing mechanism scheme can not fully achieve the ideal state of portability, accuracy and large transmission efficiency, and has great improvement space. Therefore, we should try to start from the aspect of new materials.

Further work :

- The dynamic characteristics of the deformable truss with external force constraints need to be further studied;
- A skin structure based on the deformable wing of tetrahedral truss needs to be designed, which can continuously adapt the deformation of truss;
- The method of multi-point control needs to be studied to realize the continuous deformation of plane truss structure, which provides a theoretical reference for the whole deformation of variant aircraft;
- At present, the research is limited to the topology optimization of the structure, without considering the driving mode, which belongs to the passive deformation mechanism. In the next research, the intelligent materials need to be considered, and the structure need to be given the ability of active deformation by controlling the deformation of the intelligent material;

- At present, isotropic materials are used instead of anisotropic materials. In the follow-up study, composite materials with better performance need to be considered to strengthen the mechanical properties of the structure;
- At present, the research of near-space vehicles is at an early stage. In the follow-up research, there is a need to strengthen the capability of the aircraft to achieve long-duration and long-range cruising in near space;
- Research the mesh reinforced membrane materials to enhance the bending load-bearing properties of inflatable beams and thus promote the development of UAV inflatable wings.

Reference

[1] Sofla A. Y. N., Meguid S. A., Tan K. T. and et al *Shape morphing of aircraft wing: Status and challenges*[J]. Materials and Design, 2010, 31(13): 1284~1292.

[2] Michael D. Skillen and William A. Crossley *Modeling and optimization for morphing wing concept generation*[R]. NASA CR-214860, 2007.

[3] Abhijit Hiraman Supekar *Design, analysis and development of a morphable wing structure for unmanned aerial vehicle performance augmentation*[D]. Arlington: The University of Texas, 2007.

[4] Ajaj R, Friswell M, Saavedra Flores E, et al. *Span morphing: a conceptual design study*[C]//53rd AIAA/ASME/ASCE/AHS/ASC Structures, Structural Dynamics and Materials Conference 20th AIAA/ASME/AHS Adaptive Structures Conference 14th AIAA. 2012: 1510.

[5] Jenett B, Calisch S, Cellucci D and et al *Digital Morphing Wing: Active Wing Shaping Concept Using Composite Lattice-Based Cellular Structures*[J]. Soft robotics, 2017(01): 33-48.

[6] Manzo J and Garcia E *Demonstration of an in situ morphing hyperelliptical cambered span wing mechanism*[J]. Smart Materials and Structures, 2010, 19(2): 025012.

[7] Woods B K S, Wereley N M and Friswell M *Preliminary investigation of a fishbone active camber concept*[C]. ASME conference on smart materials, adaptive structures and intelligent systems, Stone Mountain, Georgia, USA, 2012.

[8] Heo H, Ju J, Kim D M. *Compliant cellular structures: application to a passive morphing airfoil* [J]. Composite Structures, 2013, 106: 560-569.

[9] CHEN Qian,BAI Peng and LI Feng *Morphing aircraft wing variablesweep: two practical methods and their aerodynamic characteristics*[J]. Acta Aerodynamic Sinica,2012,30(5): 658 —663.(in Chinese)

[10] GUO Jianguo,CHEN Huijuan,ZHOU Jun and et al *Dynamics modeling and characteristic analysis for vehicle with asymmetric span morphing wing*[J]. Journal of Systems Engineering and Electronics,2016,38(8): 1951 — 1957. (in Chinese)

[11] XU Guowu,BAI Peng and CHEN Bingyan *Analysis on the liftdrag characteristics of new concept morphing aircraft*[J].Chinese Quarterly of Mechanics,2013,34(3): 444 — 450.(in Chinese)

[12] ZHANG Jie and WU Sentang *Dynamic modeling for a morphing aircraft and dynamic characteristics analysis*[P].Journal of Beijing University of Aeronautics and Astronautics,2015, 41(1): 58 —64 (in Chinese)

[13] Lv Zhengyang *Structural Design And Deformation Characteristics Analysis Of Morphing Wing Truss Structure*[D]. Harbin Institute of Technology, 2019.

[14] Gu Lili *Basic Research on Structure and Driving of Morphing Wing*[D]. Nan Jing: Nanjing University of Aeronautics and Astronautics, 2013.

[15] GONG chunlin, CHI fenghua, GU liangxian and et al *Optimal control method for distributed morphing aircraft based on Karhunen-Loève expansion*[J]. Acta Aeronautica et Astronautica Sinica, 2018, 39(2):121518. (in Chinese)

AUTHOR INDEX